Introduction to Optical Quantum Information Processing

Quantum information processing offers fundamental improvements over classical information processing, such as computing power, secure communication, and high-precision measurements. However, the best way to create practical devices is not yet known. This textbook describes the techniques that are likely to be used in implementing optical quantum information processors.

After developing the fundamental concepts in quantum optics and quantum information theory, this book shows how optical systems can be used to build quantum computers according to the most recent ideas. It discusses implementations based on single photons and linear optics, optically controlled atoms and solid-state systems, atomic ensembles, and optical continuous variables.

This book is ideal for graduate students beginning research in optical quantum information processing. It presents the most important techniques of the field using worked examples and over 120 exercises.

Pieter Kok is a Lecturer in Theoretical Physics in the Department of Physics and Astronomy, the University of Sheffield. He is a member of the Institute of Physics and the American Physical Society, and his Ph.D. thesis won the Institute of Physics Quantum Electronics and Photonics thesis award in 2001.

Brendon W. Lovett is a Royal Society University Research Fellow in the Department of Materials and a Fellow of St. Anne's College at the University of Oxford. He is a member of the Institute of Physics and the Materials Research Society. He has been a visiting Fellow at the University of Queensland, Australia, and is an Academic Visitor at the National University of Singapore.

Introduction to Optical Quantum Information Processing

Pieter Kok

University of Sheffield

Brendon W. Lovett

University of Oxford

CAMBRIDGE
UNIVERSITY PRESS

CAMBRIDGE
UNIVERSITY PRESS

Shaftesbury Road, Cambridge CB2 8EA, United Kingdom

One Liberty Plaza, 20th Floor, New York, NY 10006, USA

477 Williamstown Road, Port Melbourne, VIC 3207, Australia

314–321, 3rd Floor, Plot 3, Splendor Forum, Jasola District Centre, New Delhi – 110025, India

103 Penang Road, #05–06/07, Visioncrest Commercial, Singapore 238467

Cambridge University Press is part of Cambridge University Press & Assessment,
a department of the University of Cambridge.

We share the University's mission to contribute to society through the pursuit of
education, learning and research at the highest international levels of excellence.

www.cambridge.org
Information on this title: www.cambridge.org/9780521519144

First published 2010

A catalogue record for this publication is available from the British Library

ISBN   978-0-521-51914-4   Hardback

To Rose, Xander and Janet

Contents

Preface

The field of quantum information processing has reached a level of maturity, and spans such a wide variety of topics, that it merits further specialization. In this book, we consider quantum information processing with optical systems, including quantum communication, quantum computation, and quantum metrology. Optical systems are the obvious choice for quantum communication, since photons are excellent carriers of quantum information due to their relatively slow decoherence. Indeed, many aspects of quantum communication have been demonstrated to the extent that commercial products are now available. The importance of optical systems for quantum communication leads us to ask whether we can construct integrated systems for communication and computation in which all processing takes place in optical systems. Recent developments indicate that while full-scale quantum computing is still extremely challenging, optical systems are one of the most promising approaches to a fully functional quantum computer.

This book is aimed at beginning graduate students who are starting their research career in optical quantum information processing, and it can be used as a textbook for an advanced master's course. The reader is assumed to have a background knowledge in classical electrodynamics and quantum mechanics at the level of an undergraduate physics course. The nature of the topic requires familiarity with quantized fields, and since this is not always a core topic in undergraduate physics, we derive the quantum mechanical formulation of the free electromagnetic field from first principles. Similarly, we aim to present the topics in quantum information theory in a self-contained manner.

The book is organized as follows: in Part I, we develop the quantum theory of light, give an introduction to quantum communication and computation, and we present a number of advanced quantum mechanical techniques that are essential for the understanding of optical quantum information processing. In Part II, we consider quantum information processing using single photons and atoms. We first develop the theory of photodetection and explore what we mean by photon sources, followed by an exposition of quantum communication with single photons, quantum computation with single photons and linear optics, and quantum computing where the information carriers, the qubits, are encoded in atoms. In Part III, we explore quantum information processing in many-body systems. We revisit linear optical quantum communication and computation, but now in the context of quantum continuous variables, rather than qubits. We discuss how atomic ensembles can be used as quantum memories and repeaters, and we study in detail how to define robust qubits in solid-state systems such as quantum dots and crystal defects. The last chapter of the book deals with quantum metrology, where we explore how quantum states of light can be exploited to attain a measurement precision that outperforms classical metrology. As is inevitable in a book of this nature, a number of important topics have been omitted due to length restrictions. We have not included quantum information processing in ion traps, photonic band-gap materials, optical lattices, and Bose–Einstein condensates. We have also omitted the topic of quantum imaging.

We wish to thank a number of colleagues who have made valuable comments, and suggested many improvements: Charles Adams, Simon Benjamin, Samuel Braunstein, Earl Campbell, Jim Franson, Erik Gauger, Dominic Hosler, Nick Lambert, Peter van Loock, Janet Lovett, Ahsan Nazir, Todd Pittman, Nusrat Rafiq, Andrew Ramsay, Marshall Stoneham, Joachim Wabnig, David Whittaker, and Marcin Zwierz. We thank Joost Kok for suggesting the artist Victor Vasarely for the cover image. BWL thanks the Royal Society for financial support. Finally, we would like to thank Andrew Briggs and the Quantum Information Processing Interdisciplinary Research Collaboration (QIP IRC) for continued support.

PART I

QUANTUM OPTICS AND QUANTUM INFORMATION

1 The quantum theory of light

Classically, light is an electromagnetic phenomenon, described by Maxwell's equations. However, under certain conditions, such as low intensity or in the presence of certain nonlinear optical materials, light starts to behave differently, and we have to construct a 'quantum theory of light'. We can exploit this quantum behaviour of light for quantum information processing, which is the subject of this book. In this chapter, we develop the quantum theory of the free electromagnetic quantum field. This means that we do not yet consider the interaction between light and matter; we postpone that to Chapter 7. We start from first principles, using the canonical quantization procedure in the Coulomb gauge: we derive the field equations of motion from the classical Lagrangian density for the vector potential, and promote the field and its canonical momentum to operators and impose the canonical commutation relations. This will lead to the well-known creation and annihilation operators, and ultimately to the concept of the photon. After quantization of the free electromagnetic field we consider transformations of the mode functions of the field. We will demonstrate the intimate relation between these linear mode transformations and the effect of beam splitters, phase shifters, and polarization rotations, and show how they naturally give rise to the concept of squeezing. Finally, we introduce coherent and squeezed states.

The first two sections of this chapter are quite formal, and a number of subtleties arise when we quantize the electromagnetic field, such as the continuum of modes, the gauge freedom, and the definition of the creation and annihilation operators with respect to the classical modes. Readers who have not encountered field quantization procedures before may find these sections somewhat daunting, but most of the subtleties encountered here have very little bearing on the later chapters. We mainly include the full derivation from first principles to give the field of optical quantum information processing a proper physical foundation, and derive the annihilation and creation operators of the discrete optical modes from the continuum of modes that is the electromagnetic field.

1.1 The classical electromagnetic field

Classical electrodynamics is the theory of the behaviour of electric and magnetic fields in the presence of charge and current distributions. It was shown by James Clerk Maxwell (1831–1879) that the equations of motion for electric and magnetic fields, the Maxwell equations, allow for electromagnetic waves. In vacuum, these waves propagate with a velocity $c = 299\,792\,458$ ms^{-1}, and Maxwell therefore identified these waves in a certain frequency range with light. In this section, we define the electric and magnetic fields in terms of the scalar and vector potentials, and construct the field Lagrangian density in the presence of charge and current distributions. Variation of the Lagrangian density with respect to the potentials then leads to Maxwell's equations. Subsequently, we consider the Maxwell equations for the vacuum, and derive the wave equation and its plane-wave solutions. The source-free Lagrangian density is then used to define the canonical momenta to the potentials, which in turn allow us to give the Hamiltonian density for the free field. These are the ingredients we need for the canonical quantization procedure in Section 1.2.

The electric and magnetic fields $\mathbf{E}(\mathbf{r}, t)$ and $\mathbf{B}(\mathbf{r}, t)$ are related to a scalar and a vector potential $\Phi(\mathbf{r}, t)$ and $\mathbf{A}(\mathbf{r}, t)$:

$$\mathbf{E}(\mathbf{r}, t) = -\nabla\Phi(\mathbf{r}, t) - \frac{\partial\mathbf{A}(\mathbf{r}, t)}{\partial t} \quad \text{and} \quad \mathbf{B}(\mathbf{r}, t) = \nabla\times\mathbf{A}(\mathbf{r}, t). \tag{1.1}$$

The most elegant way to construct a classical field theory is via the Lagrangian density. We can use the potentials as the dynamical variables of our classical field theory, which means that we can write the Lagrangian density $\mathscr{L}$ as a function of the potentials and their time derivatives

$$\mathscr{L} = \mathscr{L}(\Phi, \dot{\Phi}; \mathbf{A}, \dot{\mathbf{A}}). \tag{1.2}$$

The equations of motion for the potentials Φ and $\mathbf{A}$ are then given by the Euler–Lagrange equations

$$\frac{d}{dt}\frac{\delta\mathscr{L}}{\delta\dot{\Phi}} - \frac{\delta\mathscr{L}}{\delta\Phi} = 0 \tag{1.3}$$

and

$$\frac{d}{dt}\frac{\delta\mathscr{L}}{\delta\dot{A}_j} - \frac{\delta\mathscr{L}}{\delta A_j} = 0. \tag{1.4}$$

Here δ denotes the functional derivative, since the potentials are themselves functions of space and time, and each component of $\mathbf{A}$, denoted by A_j, obeys a separate Euler–Lagrange equation.

In the presence of a charge density $\rho(\mathbf{r}, t)$ and a current density $\mathbf{J}(\mathbf{r}, t)$ the general Lagrangian density of classical electrodynamics can be written as

$$\mathscr{L} = \mathbf{J}(\mathbf{r}, t) \cdot \mathbf{A}(\mathbf{r}, t) - \rho(\mathbf{r}, t)\Phi(\mathbf{r}, t) + \frac{\varepsilon_0}{2}E^2(\mathbf{r}, t) - \frac{1}{2\mu_0}B^2(\mathbf{r}, t), \tag{1.5}$$

where $E^2 \equiv |\mathbf{E}|^2$ and $B^2 \equiv |\mathbf{B}|^2$ depend on Φ and $\mathbf{A}$ according to Eq. (1.1). When the Lagrangian density is varied with respect to Φ we obtain the Euler–Lagrange equation in Eq. (1.3), which can be written as Gauss' law

$$-\varepsilon_0\nabla \cdot \mathbf{E}(\mathbf{r}, t) + \rho(\mathbf{r}, t) = 0. \tag{1.6}$$

When we vary the Lagrangian density with respect to the components of $\mathbf{A}$, we find the Euler–Lagrange equations in Eq. (1.4). These can be reformulated as the Maxwell–Ampère law

$$\mathbf{J}(\mathbf{r}, t) + \varepsilon_0\frac{\partial\mathbf{E}}{\partial t}(\mathbf{r}, t) - \frac{1}{\mu_0}\nabla\times\mathbf{B}(\mathbf{r}, t) = 0. \tag{1.7}$$

The relations in Eq. (1.1) and Eqs. (1.6) and (1.7) are equivalent to Maxwell's equations, as can be seen by taking the curl of $\mathbf{E}$ in Eq. (1.1):

$$\nabla \times \mathbf{E} = -\frac{\partial \mathbf{B}}{\partial t} . \tag{1.8}$$

The last Maxwell equation, $\nabla \cdot \mathbf{B} = 0$, is implicit in $\mathbf{B} = \nabla \times \mathbf{A}$ since the divergence of any curl vanishes.

It is well known that we have a gauge freedom in defining the potentials Φ and $\mathbf{A}$ that constitute the fields $\mathbf{E}$ and $\mathbf{B}$. Since we are interested in radiation, it is convenient to adopt the Coulomb, or radiation, gauge

$$\nabla \cdot \mathbf{A} = 0 \quad \text{and} \quad \Phi = 0 . \tag{1.9}$$

In addition to the gauge choice, in this chapter we consider only the vacuum solutions of the electromagnetic fields:

$$\rho = 0 \quad \text{and} \quad \mathbf{J} = 0 . \tag{1.10}$$

When we now write Eq. (1.7) in terms of the potentials, we obtain the homogeneous wave equation for $\mathbf{A}$

$$\nabla^2 \mathbf{A} - \varepsilon_0 \mu_0 \frac{\partial^2 \mathbf{A}}{\partial t^2} = 0 . \tag{1.11}$$

The classical solutions to this equation can be written as

$$\mathbf{A}(\mathbf{r}, t) = \sum_\lambda \int \frac{d\mathbf{k}}{\sqrt{\varepsilon_0}} \frac{A_\lambda(\mathbf{k}) \boldsymbol{\epsilon}_\lambda(\mathbf{k}) e^{i\mathbf{k}\cdot\mathbf{r} - i\omega_\mathbf{k} t}}{\sqrt{(2\pi)^3 2\omega_\mathbf{k}}} + \text{c.c.,} \tag{1.12}$$

where $A_\lambda(\mathbf{k})$ denotes the amplitude of the mode with wave vector $\mathbf{k}$ and polarization λ, and c.c. denotes the complex conjugate. The vector $\boldsymbol{\epsilon}_\lambda$ gives the direction of the polarization, which we will discuss in Section 1.3. The dispersion relation for the free field is given by

$$|\mathbf{k}|^2 - \varepsilon_0 \mu_0 \, \omega_\mathbf{k}^2 \equiv k^2 - \frac{\omega_\mathbf{k}^2}{c^2} = 0 , \tag{1.13}$$

where c is the phase velocity of the wave with frequency $\omega_\mathbf{k}$. Any well-behaved potential $\mathbf{A}(\mathbf{r}, t)$ that can be expressed as a superposition of Fourier components is a solution to the wave equation. This is exemplified by the fact that we can see different shapes, colours, etc., rather than just uniform plane waves.

Finally, the Lagrangian density can be used to find the Hamiltonian density of the field. To this end, we define the canonical momenta of Φ and $\mathbf{A}$ as

$$\Pi_\Phi \equiv \frac{\delta \mathscr{L}}{\delta \dot{\Phi}} \quad \text{and} \quad \Pi_\mathbf{A} \equiv \frac{\delta \mathscr{L}}{\delta \dot{\mathbf{A}}} . \tag{1.14}$$

We can now take the Legendre transform of the Lagrangian density with respect to the dynamical variables $\dot{\Phi}$ and $\dot{\mathbf{A}}$ to obtain the Hamiltonian density of the free electromagnetic field

$$\mathcal{H}(\Phi, \mathbf{\Pi}_\Phi; \mathbf{A}, \mathbf{\Pi}_\mathbf{A}) = \mathbf{\Pi}_\Phi \dot{\Phi} + \mathbf{\Pi}_\mathbf{A} \cdot \dot{\mathbf{A}} - \mathcal{L}. \tag{1.15}$$

In the Coulomb gauge, the canonical momenta are

$$\mathbf{\Pi}_\Phi = 0 \quad \text{and} \quad \mathbf{\Pi}_\mathbf{A} = \varepsilon_0 \dot{\mathbf{A}}. \tag{1.16}$$

This leads to the Hamiltonian density for the free field

$$\mathcal{H} = \mathbf{\Pi}_\mathbf{A} \cdot \dot{\mathbf{A}} - \mathcal{L}|_{\rho=\mathbf{J}=0} = \frac{\varepsilon_0}{2} E^2 + \frac{1}{2\mu_0} B^2. \tag{1.17}$$

We now have all the necessary ingredients to proceed with the quantization of the electromagnetic field.

Exercise 1.1: Derive the homogeneous wave equation in Eq. (1.11) and show that the solutions are given by Eq. (1.12).

1.2 Quantization of the electromagnetic field

We are now ready to quantize the classical electromagnetic field. First, we have to decide which of the fields $\mathbf{A}$, $\mathbf{E}$ (or $\mathbf{B}$) we wish to quantize. In later chapters, we discuss the coupling between light and matter, and it is most convenient to express that coupling in terms of the vector potential $\mathbf{A}$. We therefore apply the quantization procedure to $\mathbf{A}$, rather than to $\mathbf{E}$. In the quantization procedure we have to ensure that the quantum fields obey Maxwell's equations in the classical limit, and this leads to the introduction of a modified Dirac delta function. After the formal quantization, we explore the properties of the mode functions and the mode operators that result from the quantization procedure, and establish a fundamental relationship between them. We then construct eigenstates of the Hamiltonian, and define the discrete, physical modes. This leads to the concept of the photon. The final part of this section is devoted to the construction of the quantum mechanical field observables associated with single modes.

1.2.1 Field quantization

We denote the difference between classical and quantum mechanical observables by writing the latter with a hat. In the quantum theory of light, $\mathbf{A}$ and $\mathbf{\Pi}_\mathbf{A}$ then become operators satisfying the equal-time commutation relations. In index notation these are written as

$$\left[\hat{A}^j(\mathbf{r}, t), \hat{A}^k(\mathbf{r}', t)\right] = \left[\hat{\Pi}_\mathbf{A}^j(\mathbf{r}, t), \hat{\Pi}_\mathbf{A}^k(\mathbf{r}', t)\right] = 0. \tag{1.18}$$

The field consists of four variables: three from $\mathbf{A}$ and one from Φ. We again work in the Coulomb gauge, where $\Phi = 0$ and $\nabla \cdot \mathbf{A} = 0$ ensures that we end up with only two dynamical variables.

Standard canonical quantization prescribes that, in addition to Eq. (1.18), we impose the following commutation relation:

$$\left[\hat{A}_i(\mathbf{r}, t), \hat{\Pi}_{\mathbf{A}}^j(\mathbf{r}', t) \right] = i\hbar \delta_{ij} \, \delta^3(\mathbf{r} - \mathbf{r}') \,, \tag{1.19}$$

where we must remember the difference between upper and lower indices, $A_j = -A^j$, because electrodynamics is, at heart, a relativistic theory. Unfortunately, given that in the Coulomb gauge $\Pi_{\mathbf{A}}^k \propto E^k$, this commutation relation is not compatible with Gauss' law in vacuum: $\nabla \cdot \mathbf{E} = 0$. If we take the divergence with respect to the variable $\mathbf{r}'$ on both sides of Eq. (1.19), the left-hand side will be zero, but the divergence of the delta function does not vanish. We therefore have to modify the delta function such that its divergence does vanish. For the ordinary Dirac delta function we use the following definition:

$$\delta_{ij}\delta^3(\mathbf{r} - \mathbf{r}') \equiv \int \frac{d\mathbf{k}}{(2\pi)^3} \, \delta_{ij} \, e^{i\mathbf{k} \cdot (\mathbf{r} - \mathbf{r}')} \,. \tag{1.20}$$

We have included the Kronecker delta δ_{ij}, because after the redefinition of the delta function the internal degree of freedom j and the external degree of freedom $\mathbf{r}$ may no longer be independent (in fact, they will not be). Taking the divergence of Eq. (1.20) with respect to $\mathbf{r}$ yields

$$\sum_i \partial_i \, \delta_{ij} \, \delta^3(\mathbf{r} - \mathbf{r}') = i \int \frac{d\mathbf{k}}{(2\pi)^3} \, k_j e^{i\mathbf{k} \cdot (\mathbf{r} - \mathbf{r}')} \,. \tag{1.21}$$

Therefore, we have to subtract something like this from the redefined delta function. We write

$$\Delta_{ij}(\mathbf{r} - \mathbf{r}') = \int \frac{d\mathbf{k}}{(2\pi)^3} \, \delta_{ij} \, e^{i\mathbf{k} \cdot (\mathbf{r} - \mathbf{r}')} - i \int \frac{d\mathbf{k}}{(2\pi)^3} \, \alpha_i k_j e^{i\mathbf{k} \cdot (\mathbf{r} - \mathbf{r}')} \,, \tag{1.22}$$

and we want to find α_i such that $\partial_i \Delta_{ij}(\mathbf{r} - \mathbf{r}') = 0$:

$$\sum_i \partial_i \Delta_{ij}(\mathbf{r} - \mathbf{r}') = \sum_i \partial_i \int \frac{d\mathbf{k}}{(2\pi)^3} \, e^{i\mathbf{k} \cdot (\mathbf{r} - \mathbf{r}')} \left(\delta_{ij} - ik_j \alpha_i \right)$$

$$= \sum_i \int \frac{d\mathbf{k}}{(2\pi)^3} \, e^{i\mathbf{k} \cdot (\mathbf{r} - \mathbf{r}')} \left[ik_i \delta_{ij} - (ik_j)(ik_i)\alpha_i \right]$$

$$= 0 \,. \tag{1.23}$$

We therefore have that

$$\sum_i \left(ik_i \delta_{ij} + k_i k_j \alpha_i \right) = 0 \,, \quad \text{or} \quad \alpha_i = -i \frac{k_i}{|\mathbf{k}|^2} \,, \tag{1.24}$$

and the 'transverse' delta function $\Delta_{ij}(\mathbf{r} - \mathbf{r}')$ becomes

$$\Delta_{ij}(\mathbf{r} - \mathbf{r}') = \int \frac{d\mathbf{k}}{(2\pi)^3}\, e^{i\mathbf{k}\cdot(\mathbf{r}-\mathbf{r}')} \left(\delta_{ij} - \frac{k_i k_j}{|\mathbf{k}|^2} \right). \tag{1.25}$$

Using this modified delta function, we can complete the quantization procedure by imposing the equal-time canonical commutation relation

$$\left[\hat{A}_i(\mathbf{r}, t), \hat{\Pi}_{\mathbf{A}}^j(\mathbf{r}', t) \right] = i\hbar\, \Delta_{ij}(\mathbf{r} - \mathbf{r}'). \tag{1.26}$$

That this leads to the correct covariant Hamiltonian and momentum is shown, for example, in Bjorken and Drell (1965). We can now write the three space components of the quantum field as

$$\hat{A}_j(\mathbf{r}, t) = \sum_{\lambda=1}^{2} \int d\mathbf{k} \sqrt{\frac{\hbar}{\varepsilon_0}} \left[\epsilon_{\lambda j}(\mathbf{k})\hat{a}_\lambda(\mathbf{k})u(\mathbf{k}; \mathbf{r}, t) + \epsilon_{\lambda j}^*(\mathbf{k})\hat{a}_\lambda^\dagger(\mathbf{k})u^*(\mathbf{k}; \mathbf{r}, t) \right], \tag{1.27}$$

where the $u(\mathbf{k}; \mathbf{r}, t)$ are mode functions that are themselves solutions to the wave equation in Eq. (1.11), and λ again indicates the polarization of the electromagnetic field. The classical amplitudes $A_\lambda(\mathbf{k})$ are replaced by the operators $\hat{a}_\lambda(\mathbf{k})$, and $\hat{\mathbf{A}}$ is now a *quantum* field. Note that $\hat{\mathbf{A}}$ is now an operator, and unlike its classical counterpart does not directly represent a particular vector potential. Specific quantum mechanical vector potentials are represented by quantum states.

The equal-time commutation relation in Eq. (1.26) determines the commutation relation for $\hat{a}_\lambda(\mathbf{k})$ and $\hat{a}_\lambda^\dagger(\mathbf{k})$, given the mode functions $u(\mathbf{k}; \mathbf{r}, t)$ and $u^*(\mathbf{k}, \mathbf{r}, t)$. From Eq. (1.12) we can read off the plane-wave solutions with continuum normalization:

$$u(\mathbf{k}; \mathbf{r}, t) = \frac{e^{i\mathbf{k}\cdot\mathbf{r} - i\omega_{\mathbf{k}}t}}{\sqrt{(2\pi)^3 2\omega_{\mathbf{k}}}}, \tag{1.28}$$

where $\mathbf{k}$ is the wave vector of a wave with frequency $\omega_{\mathbf{k}}$. Plane waves are of constant intensity throughout space and time, and are therefore unphysical. However, they are mathematically very convenient. When the mode functions are the plane waves defined in Eq. (1.28), we find explicitly that

$$\left[\hat{a}_\lambda(\mathbf{k}), \hat{a}_{\lambda'}^\dagger(\mathbf{k}') \right] = \delta_{\lambda\lambda'}\delta^3(\mathbf{k} - \mathbf{k}'), \tag{1.29}$$

and

$$\left[\hat{a}_\lambda(\mathbf{k}), \hat{a}_{\lambda'}(\mathbf{k}') \right] = \left[\hat{a}_\lambda^\dagger(\mathbf{k}), \hat{a}_{\lambda'}^\dagger(\mathbf{k}') \right] = 0. \tag{1.30}$$

The operator $\hat{a}_\lambda(\mathbf{k})$ and its Hermitian conjugate are the 'mode operators' of the quantized electromagnetic field. In the next section we will see that any operators that obey these commutation relations are good mode operators.

We are now done with the quantization of the classical electromagnetic field, and the remainder of this section is devoted to the exploration of the direct consequences of this procedure.

Exercise 1.2: Derive the commutation relations in Eqs. (1.29) and (1.30).

1.2.2 Mode functions and mode operators

We will now discuss some of the fundamental properties of the mode functions $u(\mathbf{k}; \mathbf{r}, t)$ and $u^*(\mathbf{k}; \mathbf{r}, t)$, and the mode operators $\hat{a}_\lambda(\mathbf{k})$ and $\hat{a}_\lambda^\dagger(\mathbf{k})$. In order to study the mode functions of the field (its 'shape', if you like), we must first define a scalar product that allows us to talk about orthogonal mode functions. This is given by the 'time-independent scalar product'

$$(\phi, \psi) \equiv i \int d\mathbf{r}\, \phi^* \overleftrightarrow{\partial}_t \psi = i \int d\mathbf{r} \left[\phi^*(\partial_t \psi) - (\partial_t \phi^*)\psi \right]. \tag{1.31}$$

From the general structure of the scalar product in Eq. (1.31) we see that

$$(\phi, \psi)^* = (\psi, \phi) \quad \text{and} \quad (\phi^*, \psi^*) = -(\psi, \phi). \tag{1.32}$$

This scalar product finds its origin in the continuity equation of the field, which determines the conserved currents (see Bjorken and Drell, 1965). It is therefore time-independent. The completeness relation of the mode functions $u(\mathbf{k}; \mathbf{r}, t)$ is then derived as follows: consider a function $f(\mathbf{r}, t)$ that is a superposition of different mode functions

$$f = \int d\mathbf{k} \left[\alpha(\mathbf{k}) u(\mathbf{k}) + \beta(\mathbf{k}) u^*(\mathbf{k}) \right], \tag{1.33}$$

where we have suppressed the dependence on $\mathbf{r}$ and t in $f(\mathbf{r}, t)$ and $u(\mathbf{k}; \mathbf{r}, t)$ for notational brevity. Using the orthogonality of the mode functions defined by the scalar product, we can write the coefficients $\alpha(\mathbf{k})$ and $\beta(\mathbf{k})$ as

$$\alpha(\mathbf{k}) = (u(\mathbf{k}), f) \quad \text{and} \quad \beta(\mathbf{k}) = -(u^*(\mathbf{k}), f). \tag{1.34}$$

This leads to an expression for f

$$f = \int d\mathbf{k} \left[(u(\mathbf{k}), f)\, u(\mathbf{k}) - (u^*(\mathbf{k}), f)\, u^*(\mathbf{k}) \right]. \tag{1.35}$$

For a second superposition of mode functions g the scalar product (g, f) can be written as

$$(g, f) = \int d\mathbf{k} \left[(g, u(\mathbf{k}))(u(\mathbf{k}), f) - (g, u^*(\mathbf{k}))(u^*(\mathbf{k}), f) \right]. \tag{1.36}$$

This constitutes the 'completeness relation' for the mode functions $u(\mathbf{k}; \mathbf{r}, t)$, and it holds only if the mode functions are, in fact, complete.

Using the definition of the time-independent scalar product, we can show that plane-wave solutions are orthonormal:

$$
\begin{aligned}
(u_\mathbf{k}, u_{\mathbf{k}'}) &\equiv i \int d\mathbf{r}\, \frac{e^{-i\mathbf{k}\cdot\mathbf{r} + i\omega_\mathbf{k} t}}{\sqrt{(2\pi)^3 2\omega_\mathbf{k}}} \overleftrightarrow{\partial}_t \frac{e^{i\mathbf{k}'\cdot\mathbf{r} - i\omega_{\mathbf{k}'} t}}{\sqrt{(2\pi)^3 2\omega_{\mathbf{k}'}}} \\
&= \int \frac{d\mathbf{r}}{(2\pi)^3} \frac{(\omega_\mathbf{k} + \omega_{\mathbf{k}'}) e^{-i(\mathbf{k}-\mathbf{k}')\cdot\mathbf{r} + i(\omega_\mathbf{k} - \omega_{\mathbf{k}'})t}}{2\sqrt{\omega_\mathbf{k} \omega_{\mathbf{k}'}}} \\
&= \delta^3(\mathbf{k} - \mathbf{k}').
\end{aligned} \tag{1.37}
$$

We also find by direct evaluation that $(u_{\mathbf{k}}, u^*_{\mathbf{k}'}) = 0$. This can be understood physically as the orthogonality of waves moving forward in time, and waves moving backwards in time and in opposite directions. The plane waves therefore form a complete orthonormal set of mode functions.

We can define a new set of mode functions $v(\boldsymbol{\kappa}; \mathbf{r}, t)$, which are a linear combination of plane waves

$$v(\boldsymbol{\kappa}; \mathbf{r}, t) = \int d\mathbf{k}\, V(\boldsymbol{\kappa}, \mathbf{k}) u(\mathbf{k}; \mathbf{r}, t) = \int d\mathbf{k}\, V(\boldsymbol{\kappa}, \mathbf{k}) \frac{e^{i\mathbf{k}\cdot\mathbf{r} - i\omega_{\mathbf{k}} t}}{\sqrt{(2\pi)^3 2\omega_{\mathbf{k}}}}. \tag{1.38}$$

Two symbols are needed for wave vectors here, namely $\mathbf{k}$ and $\boldsymbol{\kappa}$. We emphasize that we will normally reserve $\mathbf{k}$ for describing wave vectors. When $V(\boldsymbol{\kappa}, \mathbf{k})$ is unitary, the new mode functions $v(\boldsymbol{\kappa}; \mathbf{r}, t)$ are also orthonormal. When we express $\hat{\mathbf{A}}$ in terms of the new mode functions, we should also change the operators $\hat{a}_\lambda(\mathbf{k})$ to $\hat{b}_{\lambda'}(\boldsymbol{\kappa})$, since the mode operators depend on $\mathbf{k}$ and will generally change due to the transformation $V(\boldsymbol{\kappa}, \mathbf{k})$. The field operator then becomes

$$\hat{A}_j(\mathbf{r}, t) = \sum_{\lambda'=1}^{2} \int d\boldsymbol{\kappa}\, \sqrt{\frac{\hbar}{\varepsilon_0}} \left[\epsilon_{\lambda' j}(\boldsymbol{\kappa})\, \hat{b}_{\lambda'}(\boldsymbol{\kappa}) v(\boldsymbol{\kappa}; \mathbf{r}, t) + \epsilon^*_{\lambda' j}(\boldsymbol{\kappa})\, \hat{b}^\dagger_{\lambda'}(\boldsymbol{\kappa}) v^*(\boldsymbol{\kappa}; \mathbf{r}, t) \right]. \tag{1.39}$$

Note that here we have also included a possible change in the polarizarion degree of freedom λ', which can be incorporated straightforwardly in the time-independent scalar product.

Exercise 1.3: Prove the orthonormality of $v(\boldsymbol{\kappa}; \mathbf{r}, t)$ if V is unitary.

We next explore the precise relationship between mode functions and mode operators. The mode operators $\hat{a}_\lambda(\mathbf{k})$ and $\hat{a}^\dagger_\lambda(\mathbf{k})$ are related to the mode functions $u(\mathbf{k}; \mathbf{r}, t)$ and $u^*(\mathbf{k}; \mathbf{r}, t)$ via the time-independent scalar product

$$\hat{a}_{\lambda'}(\mathbf{k}) \equiv \sqrt{\frac{\varepsilon_0}{\hbar}} \left(u(\mathbf{k}) \boldsymbol{\epsilon}_{\lambda'}, \hat{\mathbf{A}} \right)$$

$$= i\sqrt{\frac{\varepsilon_0}{\hbar}} \int d\mathbf{r}\, u^*(\mathbf{k}; \mathbf{r}, t) \overleftrightarrow{\partial}_t \boldsymbol{\epsilon}^*_{\lambda'}(\mathbf{k}) \cdot \hat{\mathbf{A}}(\mathbf{r}, t). \tag{1.40}$$

We can then extract the operator $\hat{b}_{\lambda'}(\boldsymbol{\kappa})$, associated with the mode function $v(\boldsymbol{\kappa}; \mathbf{r}, t)$, using the procedure

$$\hat{b}_{\lambda'}(\boldsymbol{\kappa}) \equiv \sqrt{\frac{\varepsilon_0}{\hbar}} \left(v(\boldsymbol{\kappa}) \boldsymbol{\epsilon}_{\lambda'}, \hat{\mathbf{A}} \right) = i\sqrt{\frac{\varepsilon_0}{\hbar}} \int d\mathbf{r}\, v^*(\boldsymbol{\kappa}; \mathbf{r}, t) \overleftrightarrow{\partial}_t \boldsymbol{\epsilon}^*_{\lambda'}(\boldsymbol{\kappa}) \cdot \hat{\mathbf{A}}(\mathbf{r}, t). \tag{1.41}$$

This is a *definition* of the operator $\hat{b}_{\lambda'}(\boldsymbol{\kappa})$, and is completely determined by the mode function $v(\boldsymbol{\kappa})$ and polarization vector $\boldsymbol{\epsilon}_{\lambda'}(\boldsymbol{\kappa})$. Now suppose that we have an expression for $\hat{\mathbf{A}}(\mathbf{r}, t)$ in terms of mode functions $u(\mathbf{k})$, mode operators $\hat{a}_\lambda(\mathbf{k})$, and polarization vectors $\boldsymbol{\epsilon}_\lambda(\mathbf{k})$ given in Eq. (1.27). The mode operator $\hat{b}_{\lambda'}(\boldsymbol{\kappa})$ then becomes

$$\hat{b}_{\lambda'}(\boldsymbol{\kappa}) = \sum_{\lambda} \int d\mathbf{k} \left[\boldsymbol{\epsilon}^*_{\lambda'}(\boldsymbol{\kappa}) \cdot \boldsymbol{\epsilon}_\lambda(\mathbf{k})\, (v, u)\, \hat{a}_\lambda(\mathbf{k}) + \boldsymbol{\epsilon}^*_{\lambda'}(\boldsymbol{\kappa}) \cdot \boldsymbol{\epsilon}^*_\lambda(\mathbf{k})\, (v, u^*)\, \hat{a}^\dagger_\lambda(\mathbf{k}) \right], \tag{1.42}$$

where we express the mode operator $\hat{b}$ in terms of mode operators $\hat{a}$ and $\hat{a}^{\dagger}$. The spatial integration in the scalar products (v, u) and (v, u^{*}) must be evaluated before the integration over $\mathbf{k}$ in order to make the scalar product of the two polarization vectors $\boldsymbol{\epsilon}_{\lambda'}(\boldsymbol{\kappa}) \cdot \boldsymbol{\epsilon}_{\lambda}(\mathbf{k})$ definite. This demonstrates that the mode *operators* have a notion of orthogonality that is directly inherited from the orthogonality of the mode *functions*. Up to addition of a complex constant, Eq. (1.42) is the most general linear transformation of the mode operators, and is called the 'Bogoliubov transformation'. In principle it can mix the mode operators with their adjoints when the scalar product (v, u^{*}) is non-zero.

Exercise 1.4: Using Eq. (1.42), show that

$$\left[\hat{b}_{\lambda}(\boldsymbol{\kappa}), \hat{b}_{\lambda'}^{\dagger}(\boldsymbol{\kappa}')\right] = \delta_{\lambda\lambda'}\, \delta^{3}(\boldsymbol{\kappa} - \boldsymbol{\kappa}')\,, \tag{1.43}$$

and

$$\left[\hat{b}_{\lambda}(\boldsymbol{\kappa}), \hat{b}_{\lambda'}(\boldsymbol{\kappa}')\right] = \left[\hat{b}_{\lambda}^{\dagger}(\boldsymbol{\kappa}), \hat{b}_{\lambda'}^{\dagger}(\boldsymbol{\kappa}')\right] = 0 \tag{1.44}$$

These are the expected commutation relations for the mode operators.

1.2.3 Photons as excitations of the electromagnetic field

The revolutionary aspect of the quantum mechanical description of the electromagnetic field is the notion that the field can deliver its energy only in discrete amounts. This leads to the concept of the 'photon'. In order to derive this from the quantum theory, we first consider the Hamiltonian and momentum operators for the quantum field. We then construct energy eigenstates, and regularize them to obtain well-behaved physical states of the electromagnetic field.

From the quantum mechanical version of Eq. (1.17) we can formally derive the Hamiltonian operator $\mathcal{H}$ of the free field as

$$\mathcal{H} = \sum_{\lambda} \int d\mathbf{k}\, \frac{\hbar\omega_{\mathbf{k}}}{2} \left[\hat{a}_{\lambda}^{\dagger}(\mathbf{k})\hat{a}_{\lambda}(\mathbf{k}) + \hat{a}_{\lambda}(\mathbf{k})\hat{a}_{\lambda}^{\dagger}(\mathbf{k})\right]$$

$$\equiv \sum_{\lambda} \int d\mathbf{k}\, \mathcal{H}_{\lambda}(\mathbf{k})\,, \tag{1.45}$$

where $\mathcal{H}_{\lambda}(\mathbf{k})$ will be called the 'single-mode Hamiltonian operator'. Similarly, the 'field momentum operator' is

$$\hat{\mathbf{P}} = \sum_{\lambda} \int d\mathbf{k}\, \frac{\hbar\mathbf{k}}{2} \left[\hat{a}_{\lambda}^{\dagger}(\mathbf{k})\hat{a}_{\lambda}(\mathbf{k}) + \hat{a}_{\lambda}(\mathbf{k})\hat{a}_{\lambda}^{\dagger}(\mathbf{k})\right]. \tag{1.46}$$

This operator is similar to $\mathcal{H}$, but $\omega_{\mathbf{k}}$ is replaced by $\mathbf{k}$. Therefore, the properties we derive for the Hamiltonian can easily be translated into properties for the momentum. The field momentum can be formally derived from Maxwell's stress tensor, but this is beyond the scope of this book.

The Hamiltonian and the operators $\hat{a}_\lambda(\mathbf{k})$ and $\hat{a}_\lambda^\dagger(\mathbf{k})$ obey the following commutation relations:

$$\left[\mathcal{H}, \hat{a}_\lambda(\mathbf{k})\right] = -\hbar\omega_\mathbf{k}\,\hat{a}_\lambda(\mathbf{k}) \quad \text{and} \quad \left[\mathcal{H}, \hat{a}_\lambda^\dagger(\mathbf{k})\right] = \hbar\omega_\mathbf{k}\,\hat{a}_\lambda^\dagger(\mathbf{k})\,. \tag{1.47}$$

We can define $|\psi_n\rangle$ as an eigenstate of the Hamiltonian $\mathcal{H}$ such that

$$\mathcal{H}\,|\psi_n\rangle = E_n\,|\psi_n\rangle\,, \tag{1.48}$$

with E_n the corresponding energy eigenvalue (which depends on $\omega_\mathbf{k}$). From the commutation relations in Eq. (1.47) we find

$$\mathcal{H}\hat{a}_\lambda^\dagger(\mathbf{k})\,|\psi_n\rangle = \hat{a}_\lambda^\dagger(\mathbf{k})\,(\mathcal{H} + \hbar\omega_\mathbf{k})\,|\psi_n\rangle = (E_n + \hbar\omega_\mathbf{k})\,\hat{a}_\lambda^\dagger(\mathbf{k})\,|\psi_n\rangle\,. \tag{1.49}$$

This means that $\hat{a}_\lambda^\dagger(\mathbf{k})\,|\psi_n\rangle$ is again an eigenstate of the Hamiltonian $\mathcal{H}$ with energy $E_n + \hbar\omega_\mathbf{k}$. Similarly, $\hat{a}_\lambda(\mathbf{k})\,|\psi_n\rangle$ is again an eigenstate of the Hamiltonian $\mathcal{H}$ with energy $E_n - \hbar\omega_\mathbf{k}$. The eigenvalues of $\mathcal{H}$ must be bounded from below, and there is therefore a ground state $|\psi_0\rangle$ for which $\hat{a}_\lambda(\mathbf{k})\,|\psi_0\rangle = 0$ for any $\mathbf{k}$ and λ. We call this state the 'vacuum state' for that mode. This leads to the vacuum energy

$$\mathcal{H}\,|\psi_0\rangle = \frac{\hbar}{2}\sum_\lambda \int d\mathbf{k}\,\omega_\mathbf{k}\hat{a}_\lambda(\mathbf{k})\hat{a}_\lambda^\dagger(\mathbf{k})\,|\psi_0\rangle\,. \tag{1.50}$$

Using the commutation relation in Eq. (1.29), we find that this expression diverges. The vacuum energy is therefore infinite, and needs to be subtracted in all practical calculations of energy expectation values. However, we cannot completely ignore the vacuum energy, since it leads to so-called vacuum fluctuations. It has observable consequences, such as the Lamb shift and vacuum noise. This will become important in certain applications of optical quantum information processing, e.g., in Chapters 8 and 9.

For every mode denoted by $\mathbf{k}$ and λ, we can construct a set of basis states that are eigenstates of the Hamiltonian by repeatedly applying the operator $\hat{a}_\lambda^\dagger(\mathbf{k})$ to the ground state $|\psi_0\rangle$. After removing the infinite vacuum energy, we can find the relative energy of each mode:

$$E_n(\mathbf{k}) = n\,\hbar\omega_\mathbf{k}\,, \tag{1.51}$$

and we can write the energy eigenstate of mode $\{\mathbf{k}, \lambda\}$ as $|n\rangle_{\mathbf{k},\lambda}$ or $|n_{\mathbf{k},\lambda}\rangle$. These states are commonly known as 'Fock states'.

We can similarly define the operator

$$\hat{n} = \sum_\lambda \int d\mathbf{k}\,\hat{a}_\lambda^\dagger(\mathbf{k})\hat{a}_\lambda(\mathbf{k})\,. \tag{1.52}$$

The Fock states are clearly also eigenstates of $\hat{n}$:

$$\hat{n}|n\rangle_{\mathbf{k}',\lambda} = n|n\rangle_{\mathbf{k}',\lambda}\,, \tag{1.53}$$

and for obvious reasons $\hat{n}$ is called the number operator. We have seen that $a_\lambda^\dagger(\mathbf{k})$ and $a_\lambda(\mathbf{k})$ move us up and down the ladder of Fock states, and are therefore called the 'creation' and 'annihilation' operators, respectively.

However, the state created by applying the creation operator $\hat{a}_\lambda^\dagger(\mathbf{k})$, associated with a plane wave, to the vacuum is not normalizable, and therefore unphysical. For example, the scalar product of two single-excitation Fock states is

$$\langle 1_{\mathbf{k}',\lambda'} | 1_{\mathbf{k},\lambda} \rangle = \delta_{\lambda,\lambda'} \, \delta(\mathbf{k} - \mathbf{k}') \,. \tag{1.54}$$

In order to define physical states, we must construct well-behaved mode functions. Earlier, we have seen that we can superpose plane waves

$$f(\mathbf{r}, t) = \int d\mathbf{k} \left[\alpha^*(\mathbf{k}) \, u(\mathbf{k}; \mathbf{r}, t) + \beta^*(\mathbf{k}) \, u^*(\mathbf{k}; \mathbf{r}, t) \right], \tag{1.55}$$

which can be normalized according to

$$(f, f) = 1 \quad \Rightarrow \quad \int d\mathbf{k} \left[|\alpha(\mathbf{k})|^2 - |\beta(\mathbf{k})|^2 \right] = 1 \,. \tag{1.56}$$

In general, f can be any well-behaved function, but in practice we often assume that f is sharply peaked around a central wave vector $\mathbf{k}_0$. This can then be considered a normalizable frequency mode, even though it is strictly an approximation.

Using the definition of the mode operators in Eq. (1.40), we can then write the mode operator for the mode f with polarization λ as

$$\hat{b}_{f\lambda} = \sqrt{\frac{\varepsilon}{\hbar}} \left(\boldsymbol{\epsilon}_\lambda f, \hat{\mathbf{A}} \right) = \int d\mathbf{k} \left[\alpha(\mathbf{k}) \, \hat{a}_\lambda(\mathbf{k}) + \beta(\mathbf{k}) \, \hat{a}_\lambda^\dagger(\mathbf{k}) \right]. \tag{1.57}$$

By definition, the mode operators $\hat{b}_{f\lambda}$ and $\hat{b}_{f\lambda}^\dagger$ obey the commutation relation

$$\left[\hat{b}_{f\lambda}, \hat{b}_{f\lambda'}^\dagger \right] = \delta_{\lambda\lambda'} \,, \tag{1.58}$$

and if f is part of an orthonormal set of mode functions $\{ f_1, f_2, \ldots \}$, the commutation relations become

$$\left[\hat{b}_{j\lambda}, \hat{b}_{k\lambda'}^\dagger \right] = \delta_{\lambda\lambda'} \, \delta_{jk} \quad \text{and} \quad \left[\hat{b}_{j\lambda}, \hat{b}_{k\lambda'} \right] = \left[\hat{b}_{j\lambda}^\dagger, \hat{b}_{k\lambda'}^\dagger \right] = 0 \,, \tag{1.59}$$

where the subscripts j and k indicate the mode functions f_j and f_k, respectively. In the remainder of this book, we will mostly consider these well-behaved, discretized mode functions with their associated mode operators.

Assuming that $\beta(\mathbf{k}) = 0$, a single excitation in mode f_j with polarization λ can be written as

$$|1_{j\lambda}\rangle = \hat{b}_{j\lambda}^\dagger |0\rangle = \int d\mathbf{k} \, \alpha_j^*(\mathbf{k}) \, \hat{a}_\lambda^\dagger(\mathbf{k}) |0\rangle \,, \tag{1.60}$$

where $\alpha_j(\mathbf{k})$ is associated with f_j, c.f., Eq. (1.55). This is the state of a 'wave packet', albeit of infinite duration. It follows that

$$_{j\lambda}\langle 1 | 1 \rangle_{j\lambda'} = \delta_{\lambda,\lambda'} \int d\mathbf{k}\, d\mathbf{k}'\, \alpha_j^*(\mathbf{k}) \alpha_j(\mathbf{k}') \hat{a}_\lambda(\mathbf{k}) \hat{a}_{\lambda'}^\dagger(\mathbf{k}')$$

$$= \delta_{\lambda,\lambda'} \int d\mathbf{k}\, |\alpha_j(\mathbf{k})|^2 \equiv \delta_{\lambda,\lambda'}, \qquad (1.61)$$

where we used Eq. (1.56). Wave packets with higher numbers of excitations are easily generated by applying the discrete creation operator to the vacuum several times

$$|n\rangle_{j\lambda} = \frac{(\hat{b}_{j\lambda}^\dagger)^n}{\sqrt{n!}} |0\rangle. \qquad (1.62)$$

The factor $\sqrt{n!}$ is found by normalizing the states and using the commutation relations. We can now be precise about the action of the discrete operators as follows

$$\hat{b}_{j\lambda} |n\rangle_{j\lambda} = \sqrt{n} |n-1\rangle_{f\lambda}$$

$$\hat{b}_{j\lambda}^\dagger |n\rangle_{j\lambda} = \sqrt{n+1} |n+1\rangle_{j\lambda}. \qquad (1.63)$$

These relations strongly suggest an interpretation of $|n\rangle_{j\lambda}$ as a state of 'n particles'. The particle is created by the action of $\hat{b}_{j\lambda}^\dagger$ and annihilated by the action of $\hat{b}_{j\lambda}$, and is called the 'photon'. Photons in sharply peaked frequency modes have well-defined energy and momentum.

The interpretation of $|1\rangle_{j\lambda}$ as a physical particle is not without problems. The photon is massless, and is therefore the ultimate relativistic particle, travelling at the speed of light. As a consequence, there is no position operator for the photon, unlike for traditional particles. Another subtlety is that the superposition f can in principle be chosen with non-zero $\beta(\mathbf{k})$, which means that $\hat{b}$ is a superposition of annihilation and creation operators $\hat{a}(\mathbf{k})$ and $\hat{a}^\dagger(\mathbf{k})$. The concept of a photon therefore depends on the mode functions and the coordinate system. We will encounter this mixing of creation and annihilation operators later in this chapter, when we describe optical squeezing.

1.2.4 Quadrature operators

We now return to the continuous mode operators $\hat{a}(\mathbf{k})$ and $\hat{a}^\dagger(\mathbf{k})$, and momentarily suppress the polarization degree of freedom λ for notational brevity. The creation and annihilation operators are not Hermitian, and are therefore not associated directly with physical observables. However, they can be used to construct Hermitian operators. For example, for any operator $\hat{F}$ we can construct a Hermitian operator $\hat{F}^\dagger \hat{F}$, and with $\hat{F} = \hat{a}$ this led to the number operator $\hat{a}^\dagger \hat{a}$. We can also construct the Hermitian operator $\hat{F} + \hat{F}^\dagger$. Including an extra phase freedom, this leads us to define the 'quadrature operator' $\hat{x}_\zeta$ of the general form

$$\hat{x}_\zeta(\mathbf{k}) = \frac{e^{-i\zeta} \hat{a}(\mathbf{k}) + e^{i\zeta} \hat{a}^\dagger(\mathbf{k})}{\sqrt{2}}, \qquad (1.64)$$

for a particular mode $\mathbf{k}$. The natural observables associated with $\hat{x}_\zeta(\mathbf{k})$ are not dimensionless, and in order to find the constant of proportionality, we use the well-known fact that the free field single-mode Hamiltonian operator is formally identical to the Hamiltonian of the simple harmonic oscillator. It can therefore be written as

$$\mathcal{H}_\lambda(\mathbf{k}) = \omega_{\mathbf{k}}^2 \, \hat{q}_\lambda^2(\omega_{\mathbf{k}}) + \hat{p}_\lambda^2(\omega_{\mathbf{k}}) \,. \tag{1.65}$$

This must be of the same form as the definition in Eq. (1.45). It is not difficult to show that

$$\hat{q}_\lambda(\mathbf{k}) = \sqrt{\frac{\hbar}{2\omega_{\mathbf{k}}}} \left[\hat{a}_\lambda(\mathbf{k}) + \hat{a}_\lambda^\dagger(\mathbf{k}) \right] \tag{1.66}$$

and

$$\hat{p}_\lambda(\mathbf{k}) = -i \sqrt{\frac{\hbar\omega_{\mathbf{k}}}{2}} \left[\hat{a}_\lambda(\mathbf{k}) - \hat{a}_\lambda^\dagger(\mathbf{k}) \right] \,. \tag{1.67}$$

These operators obey the commutation relation

$$\left[\hat{q}_\lambda(\mathbf{k}), \hat{p}_{\lambda'}(\mathbf{k}') \right] = i\hbar \, \delta_{\lambda\lambda'} \, \delta^3(\mathbf{k} - \mathbf{k}') \,, \tag{1.68}$$

and the operators $\hat{q}$ and $\hat{p}$ are therefore often called the 'position' and 'momentum' quadratures. They are not really associated with the position and momentum of a particle; they merely have the same commutation relations.

The quadratures are Hermitian by construction, and this therefore raises the question as to how these observables should be interpreted physically. To answer this, consider a single mode of the quantum field $\hat{A}$ given in Eq. (1.27):

$$\hat{A}_{\mathbf{k}}(\mathbf{r}, t) = \sqrt{\frac{\hbar}{2(2\pi)^3 \varepsilon_0 \omega_{\mathbf{k}}}} \left[\hat{a}(\mathbf{k}) \, e^{i\mathbf{k}\cdot\mathbf{r} - i\omega_{\mathbf{k}}t} + \hat{a}^\dagger(\mathbf{k}) \, e^{-i\mathbf{k}\cdot\mathbf{r} + i\omega_{\mathbf{k}}t} \right] \,, \tag{1.69}$$

where we have suppressed the polarization degree of freedom λ by considering just one linear polarization direction. At a given time t and position $\mathbf{r}$, we can write $\zeta = \omega_{\mathbf{k}}t - \mathbf{k}\cdot\mathbf{r}$, and the field becomes

$$\hat{A}_{\mathbf{k}} = \sqrt{\frac{\hbar}{(2\pi)^3 \varepsilon_0 \omega_{\mathbf{k}}}} \left[\frac{\hat{a}(\mathbf{k}) \, e^{-i\zeta} + \hat{a}^\dagger(\mathbf{k}) \, e^{i\zeta}}{\sqrt{2}} \right] = \sqrt{\frac{\hbar}{(2\pi)^3 \varepsilon_0 \omega_{\mathbf{k}}}} \, \hat{x}_\zeta(\mathbf{k}) \,. \tag{1.70}$$

Similarly, the canonically conjugate momentum to $\hat{A}_{\mathbf{k}}$ is $\Pi_{\hat{A}} = \varepsilon_0 \partial_t \hat{A}_{\mathbf{k}}$, which can be written as

$$\hat{\Pi}_{\hat{A}_{\mathbf{k}}} = -i \sqrt{\frac{\hbar\varepsilon_0\omega_{\mathbf{k}}}{(2\pi)^3}} \left[\frac{\hat{a}(\mathbf{k}) \, e^{-i\zeta} - \hat{a}^\dagger(\mathbf{k}) \, e^{i\zeta}}{\sqrt{2}} \right] = \sqrt{\frac{\hbar\varepsilon_0\omega_{\mathbf{k}}}{(2\pi)^3}} \, \hat{x}_{\zeta+\pi/2}(\mathbf{k}) \,. \tag{1.71}$$

So the quadratures correspond to the physical observables that are the single-mode field amplitudes of $\mathbf{A}$ and $\mathbf{E} \propto \mathbf{\Pi}_{\mathbf{A}}$.

Exercise 1.5: Show that the commutation relations of the quadrature operators are

$$\left[\hat{x}_\zeta(\mathbf{k}), \hat{x}_{\zeta+\pi/2}(\mathbf{k}')\right] = i\delta^3(\mathbf{k} - \mathbf{k}') \quad \text{and} \quad \left[\hat{x}_\zeta(\mathbf{k}), \hat{x}_\zeta(\mathbf{k}')\right] = 0. \quad (1.72)$$

For any discrete mode, the expectation values of the quadrature operators $\hat{x}_\zeta(\mathbf{k})$ are dimensionless, whereas the expectation values of position and momentum quadratures $\hat{q}$ and $\hat{p}$ have dimensions $\sqrt{ML}$ and $\sqrt{ML}T^{-1}$, respectively.

1.3 Mode functions and polarization

In this section we study in more detail some aspects of the mode functions and the polarization. This is essentially a classical study, since the mode functions and the polarization are inherited directly from classical electrodynamics. We first describe the structure of the polarization of the field, and subsequently derive the orthonormal transverse mode functions in terms of Hermite–Gaussian and Laguerre–Gaussian mode shapes.

1.3.1 Polarization

One of the most popular degrees of freedom for quantum information processing with light is polarization. Usually, the polarization is defined in terms of the vector behaviour of the electric field $\mathbf{E}$. Here, however, we treat the vector potential $\hat{\mathbf{A}}$ as the fundamental quantized field. Since the electric field is proportional to the canonical momentum of the vector potential, and the time derivative of $\hat{\mathbf{A}}$ does not change the vector behaviour, we can define the polarization based on the vector potential without problems.

We already established that polarization is closely related to the vector character of $\hat{\mathbf{A}}(\mathbf{r}, t)$, carried by the vector $\boldsymbol{\epsilon}_\lambda$, and we will now explore this in more detail. Using the plane-wave expansion and the Coulomb gauge condition that $\nabla \cdot \hat{\mathbf{A}} = 0$, it follows that

$$\boldsymbol{\epsilon}_\lambda(\mathbf{k}) \cdot \mathbf{k} = 0 \qquad \text{for each } \lambda. \quad (1.73)$$

This means that the vectors $\boldsymbol{\epsilon}_\lambda$ point transversely to the direction of propagation $\mathbf{k}$. Furthermore, since the two dynamical variables indicated by λ are independent, we can choose

$$\boldsymbol{\epsilon}_\lambda(\mathbf{k}) \cdot \boldsymbol{\epsilon}_{\lambda'}^*(\mathbf{k}) = \delta_{\lambda\lambda'}. \quad (1.74)$$

The vectors $\boldsymbol{\epsilon}_\lambda$ have unit length, and are now identified with the polarization of the field. They may in general have complex components, since the vector potential will still be Hermitian if the term with the creation operator contains the complex conjugate of $\boldsymbol{\epsilon}_\lambda$.

Suppose, without loss of generality, that the direction of propagation of the light is in the z direction: $\mathbf{k} = k\hat{\mathbf{z}}$. We can then immediately construct two orthogonal vectors $\boldsymbol{\epsilon}_1$ and $\boldsymbol{\epsilon}_2$

according to

$$\epsilon_1 = \begin{pmatrix} 1 \\ 0 \\ 0 \end{pmatrix} \quad \text{and} \quad \epsilon_2 = \begin{pmatrix} 0 \\ 1 \\ 0 \end{pmatrix}. \tag{1.75}$$

Furthermore, we can rotate these vectors around the z axis over an angle θ to find a new set of orthonormal vectors. In addition, the third component of the vector is always zero, and we therefore omit it from the description. The two independent polarization vectors then become

$$\epsilon_1 = \begin{pmatrix} \cos\theta \\ \sin\theta \end{pmatrix} \quad \text{and} \quad \epsilon_2 = \begin{pmatrix} -\sin\theta \\ \cos\theta \end{pmatrix}. \tag{1.76}$$

These are real vectors, and they are associated with 'linear polarization' since they point in an unambiguous spatial direction.

What happens when the vectors ϵ_λ have complex components? Two possible orthonormal polarization vectors are

$$\epsilon_L = \frac{1}{\sqrt{2}} \begin{pmatrix} 1 \\ i \end{pmatrix} \quad \text{and} \quad \epsilon_R = \frac{1}{\sqrt{2}} \begin{pmatrix} 1 \\ -i \end{pmatrix}. \tag{1.77}$$

Due to the imaginary entries $\pm i$, it is not straightforward to interpret the spatial direction of these polarization vectors. It turns out that the imaginary entries cause a rotation of the polarization in time. The polarization vectors in Eq. (1.77) therefore describe 'circular' polarization, and the two orthonormal vectors correspond to left-handed and right-handed rotations. When the magnitudes of the real and imaginary components become unequal, we speak of 'elliptical polarization'. We will discuss the polarization of single photons in detail in Chapter 5.

1.3.2 Transverse mode functions

The mode functions of the electromagnetic field, or in this case the vector potential, are intrinsically a continuum, which has associated difficulties of normalization and precision in addressing. For quantum information processing purposes it is essential that we have good control over all possible degrees of freedom. We will derive an expression for properly confined modes in the transverse direction, instead of infinitely extending plane waves, and we then construct complete orthonormal sets of transverse mode functions. Finally, we use these sets to construct the quantum mechanical creation and annihilation operators via the procedure in Eq. (1.40).

First, we return to plane waves. The exact translational symmetry in the transverse plane of the plane wave means that the propagation of the plane wave in the z direction is exact. In reality, however, beams of light do not stretch out to infinity in the transverse directions. As a result, translational invariance is lost, and the direction of propagation is no longer strictly in the z direction. One way to understand this intuitively is to note that the localization of the intensity at the position $x = x_0$ and $y = y_0$ requires some uncertainty in the transverse momentum k_x and k_y. We next introduce the paraxial wave approximation, and then find the minimum uncertainty localized transverse mode. These modes can, in turn, be used as a generating function when we construct the Hermite–Gaussian and Laguerre–Gaussian transverse modes.

When the transverse momentum is small we can treat the beam in the 'paraxial approximation'. We start with the Helmholtz equation for the jth spatial component of the classical vector potential in the frequency domain $A_j(\mathbf{r}, \omega)$, given by the Fourier transform

$$A_j(\mathbf{r}, t) = \int_0^\infty d\omega \, A_j(\mathbf{r}, \omega) e^{-i\omega t}. \tag{1.78}$$

Using $\omega = ck$, this leads to the wave equation for the Fourier components

$$\left(\nabla^2 + k^2\right) A_j(\mathbf{r}, \omega) = 0. \tag{1.79}$$

This is the Helmholtz equation. For a wave propagating in the z direction, we use the following 'Ansatz' for A_j:

$$A_j(\mathbf{r}, \omega) = \Psi_j(\mathbf{r}, \omega) \, e^{ikz}. \tag{1.80}$$

The differential operator acting on A_j can then be written as

$$\nabla^2 A_j = \partial_x^2 \Psi_j \, e^{ikz} + \partial_y^2 \Psi_j \, e^{ikz} + \partial_z \left[\left(\partial_z \Psi_j + ik \Psi_j \right) e^{ikz} \right]. \tag{1.81}$$

Substituting into Eq. (1.79), this leads to

$$\partial_x^2 \Psi_j + \partial_y^2 \Psi_j + 2ik \partial_z \Psi_j + \partial_z^2 \Psi_j = 0. \tag{1.82}$$

Our discussion has been exact up to this point. We can now make the paraxial approximation when $2ik\partial_z\Psi_j \gg \partial_z^2\Psi_j$, that is, when the variation of the field in the direction of propagation z is much smaller than the wavelength of the light. In addition, we require $\partial_z^2\Psi_j \ll \partial_x^2\Psi_j, \partial_y^2\Psi_j$. The second derivative with respect to z can be dropped, leading to the 'paraxial wave equation':

$$\frac{\partial^2}{\partial x^2} \Psi_j + \frac{\partial^2}{\partial y^2} \Psi_j = -2ik \frac{\partial}{\partial z} \Psi_j. \tag{1.83}$$

We will almost exclusively use the paraxial approximation in this book.

We note something interesting about the paraxial wave equation: when we treat z as a time variable, Eq. (1.83) is formally identical to the Schrödinger equation. Consequently, we can use well-known results from quantum mechanics in our discussion of the transverse mode functions. In particular, it will allow us to derive the orbital angular momentum properties of paraxial light beams.

Gaussian modes

Rather than constructing solutions to the paraxial wave equation directly, we will instead derive a complete set of orthonormal mode functions for the classical case, and subsequently show that they are solutions to Eq. (1.83). We first derive the mode function of the lowest-order Gaussian mode, and only then make the paraxial approximation.

In the scalar field approximation[1] (where the polarization degree of freedom has been suppressed, and A has become a scalar) we can write the transverse components for an arbitrary solution of the wave equation with frequency ω in Cartesian coordinates

$$
\begin{aligned}
A(\mathbf{r}, \omega) &= \int d\mathbf{k} \sqrt{\frac{\hbar}{\varepsilon_0}} \, \delta\left(\frac{\omega^2}{c^2} - \mathbf{k}^2\right) \alpha(\mathbf{k}) \, e^{i\mathbf{k}\cdot\mathbf{r}} + \text{c.c.} \\
&= \int d\mathbf{k} \sqrt{\frac{\hbar}{\varepsilon_0}} \, \frac{\delta\left(k_z - \sqrt{k^2 - |\mathbf{k}_\perp|^2}\right)}{|2\sqrt{k^2 - |\mathbf{k}_\perp|^2}|} \alpha(\mathbf{k}) \, e^{i\mathbf{k}\cdot\mathbf{r}} + \text{c.c.} \\
&= \int d\mathbf{k}_\perp \sqrt{\frac{\hbar}{\varepsilon_0}} f(\mathbf{k}_\perp) \, e^{ik_x x + ik_y y + i\sqrt{k^2 - k_x^2 - k_y^2}\, z} + \text{c.c..}
\end{aligned} \tag{1.84}
$$

In the first line the delta function is introduced to constrain the $\mathbf{k}$ vector to the 'dispersion shell' $k^2 = \omega^2/c^2$. In the second line we have simplified the delta function according to

$$
\delta\left(h(x)\right) = \frac{\delta(x_0)}{|h'(x_0)|}, \tag{1.85}
$$

where h' is the derivative of h with respect to x, and x_0 is defined by $h(x_0) = 0$. There are two solutions x_0 in the above case, and we have selected the term that corresponds to propagation in the positive z direction. In the last line of Eq. (1.84) we evaluated the integral over k_z and absorbed the factor $|2\sqrt{k^2 - |\mathbf{k}_\perp|^2}|^{-1}$ into the amplitude function f, yielding

$$
f(\mathbf{k}_\perp) \equiv \frac{\alpha\left(\mathbf{k}_\perp, \sqrt{k^2 - |\mathbf{k}_\perp|^2}\right)}{2\left|\sqrt{k^2 - |\mathbf{k}_\perp|^2}\right|}, \tag{1.86}
$$

where $\mathbf{k}_\perp = (k_x, k_y)$ and $d\mathbf{k}_\perp = dk_x \, dk_y$. For convenience we define the mode function u as

$$
u(\mathbf{k}, \mathbf{r}) = f(\mathbf{k}_\perp) \exp\left(i\mathbf{k}_\perp \cdot \mathbf{r}_\perp + i\sqrt{k^2 - k_\perp^2}\, z\right). \tag{1.87}
$$

Even though this expression is still completely general, it will be most useful when we consider propagation in the z direction, such that $f(\mathbf{k}_\perp)$ becomes small very quickly when $|\mathbf{k}_\perp|$ becomes large.

The beam defined by the mode functions $f(\mathbf{k}_\perp)$ will have a certain divergence, or 'spread' of the amplitudes $f(\mathbf{k}_\perp)$ in k-space. This means that the transverse mode area of the beam will become larger (or smaller) with increasing z. The divergence is given by the variance in the transverse wave vector

$$
(\Delta \mathbf{k}_\perp)^2 = \int \frac{d\mathbf{k}_\perp}{(2\pi)^2} \left(k_x^2 + k_y^2\right) |f(\mathbf{k}_\perp)|^2, \tag{1.88}
$$

---

[1] We will often call A a field, rather than a potential, because we do not want to call A the scalar potential (which strictly means Φ), and in the scalar approximation it is inappropriate to call A a *vector* potential.

where the domain of integration is now taken as the entire transverse wave vector space $-\infty < k_x, k_y < \infty$. At the same time, the transverse extension in real space can be written as

$$(\Delta \mathbf{r}_\perp)^2 = \frac{\varepsilon_0}{\hbar} \int dx\, dy\, \left(x^2 + y^2\right) |A(\mathbf{r})|^2 = \int \frac{d\mathbf{k}_\perp}{(2\pi)^2} \left(\left|\frac{\partial f}{\partial k_x}\right|^2 + \left|\frac{\partial f}{\partial k_y}\right|^2\right), \quad (1.89)$$

where the last equality follows from Eq. (1.84) and integrating over x and y. We use the Cauchy–Schwarz inequality for square-integrable complex functions g and h:

$$\int |g(\mathbf{a})|^2\, d\mathbf{a} \cdot \int |h(\mathbf{a})|^2\, d\mathbf{a} \geq \left|\int d\mathbf{a}\, g(\mathbf{a}) \cdot h^*(\mathbf{a})\right|^2 \quad (1.90)$$

to obtain the classical 'uncertainty relations'

$$\Delta k_x\, \Delta x \geq \frac{\|f\|^2}{4\pi} \quad \text{and} \quad \Delta k_y\, \Delta y \geq \frac{\|f\|^2}{4\pi}, \quad (1.91)$$

where $\|f\|^2$ is the norm of the mode function:

$$\|f\|^2 = \int d\mathbf{k}_\perp\, |f(\mathbf{k}_\perp)|^2. \quad (1.92)$$

For the minimum uncertainty in the beam (maximum transverse localization and minimum divergence) the equality in Eq. (1.91) holds. This can only happen when $k_i f$ and $\partial f / \partial k_i$ differ by a constant factor. It is straightforward to verify that in this case f must be a Gaussian function of k_i. The (exact) minimum uncertainty mode function $u_0^{(ex)}(\mathbf{k}, \mathbf{r})$ therefore becomes

$$u_0^{(ex)}(\mathbf{k}, \mathbf{r}) = f(\mathbf{k}_\perp)\, e^{i\mathbf{k}\cdot\mathbf{r}} = \exp\left(i\mathbf{k}_\perp \cdot \mathbf{r}_\perp + ik_z z - \frac{w_0^2}{4} |\mathbf{k}_\perp|^2\right), \quad (1.93)$$

for some real number w_0. Indeed, it is clear that $f(\mathbf{k}_\perp)$ drops to zero exponentially fast as $|\mathbf{k}_\perp|$ becomes large. We now make the paraxial approximation that $|\mathbf{k}_\perp|$ is small compared to k, and that led to the paraxial wave equation in Eq. (1.83):

$$k_z = \sqrt{k^2 - |\mathbf{k}_\perp|^2} \simeq k - \frac{|\mathbf{k}_\perp|^2}{2k}. \quad (1.94)$$

This leads to the following expression for the approximate mode function $u_0(\mathbf{k}, \mathbf{r})$:

$$u_0(\mathbf{k}, \mathbf{r}) = \exp\left[i\mathbf{k}_\perp \cdot \mathbf{r}_\perp + ikz - \left(\frac{w_0^2}{4} + \frac{iz}{2k}\right)|\mathbf{k}_\perp|^2\right]$$

$$= \exp\left[i\mathbf{k}_\perp \cdot \mathbf{r}_\perp + ikz - \frac{s^2(z)}{4}|\mathbf{k}_\perp|^2\right]. \quad (1.95)$$

The parameter w_0 is the beam waist and we have defined the complex parameter s:

$$s^2(z) \equiv w_0^2 + \frac{2iz}{k}, \quad (1.96)$$

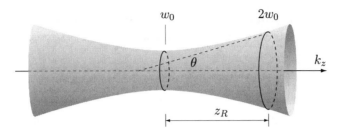

Fig. 1.1. The mode shape of a beam of light localized in the transverse direction. The beam waist is w_0 and the Rayleigh length is z_R. The angle of divergence is θ, which must be small for the paraxial approximation to hold.

which can be interpreted as the curvature of the wave front. Since the divergence of the beam is non-zero, the surfaces of constant phase are no longer planes of constant z extending in the xy direction; they are curved surfaces. The surfaces of constant phase can be extracted by writing the complex mode function $u_0(\mathbf{k}, \mathbf{r})$ in polar notation and setting the phase equal to a constant. The mode has a characteristic length in the direction of propagation, called the 'Rayleigh length':

$$z_R = \frac{k w_0^2}{2} = \frac{\pi w_0^2}{\lambda}, \qquad (1.97)$$

which is the distance along the beam in which the waist becomes twice as large (in linear dimensions), as shown in Fig. 1.1. Here, $\lambda = 2\pi/k$ is the wavelength of the light. The (scalar) field $A_0(\mathbf{r}, \omega)$ in the fundamental Gaussian transverse mode can now be written as ($\omega = ck$)

$$A_0(\mathbf{r}, \omega) = \int \frac{d\mathbf{k}_\perp}{\sqrt{\varepsilon_0}} u_0(\mathbf{k}, \mathbf{r}) + \text{c.c.}$$

$$= \frac{4\pi}{\sqrt{\varepsilon_0}\, s^2(z)} \exp\left(ikz - \frac{x^2 + y^2}{s^2(z)} \right) + \text{c.c.}. \qquad (1.98)$$

Exercise 1.6: Show that Eq. (1.98) is a solution to the paraxial wave equation.

In the quantum theory, the single-field mode in Eq. (1.98) is associated with the annihilation operator $\hat{a}(k)$. The Gaussian function then describes the appropriate amplitude function in the transverse to the beam direction. A physical state of a photon would require a further regularization of the continuum of modes in the direction of propagation. For a mode function strongly peaked around wave vector k_0 we can then assign the photon a frequency $\omega_0 = ck_0$.

Hermite–Gaussian modes

The Gaussian beam is only one possible transverse mode shape. In particular, we expect that we can construct basis functions that can be used to describe *any* transverse amplitude distribution. For example, such a basis can be given in terms of Hermite polynomials.

We can write a general transverse mode shape of a field mode with (central) frequency ω as a power expansion in k_x and k_y:

$$u(\mathbf{k}, \mathbf{r}) = \sum_{n,m=0}^{\infty} c_{nm} k_x^n k_y^m \, u_0(\mathbf{k}, \mathbf{r}). \tag{1.99}$$

Suppose that a particular $c_{nm} = 1$, and all the other coefficients are zero. The higher-order mode A_{nm} of the field can then be written as

$$\begin{aligned} A_{nm}(\mathbf{r}, \omega) &= \int \frac{d\mathbf{k}_\perp}{\sqrt{\varepsilon_0}} (ik_x)^n (ik_y)^m u_0(\mathbf{k}, \mathbf{r}) + \text{c.c.} \\ &= \int \frac{d\mathbf{k}_\perp}{\sqrt{\varepsilon_0}} \frac{\partial^{n+m}}{\partial x^n \partial y^m} u_0(\mathbf{k}, \mathbf{r}) + \text{c.c.} \\ &= \frac{\partial^{n+m}}{\partial x^n \partial y^m} A_0(\mathbf{r}, \omega). \end{aligned} \tag{1.100}$$

Given Eq. (1.98), it is clear that the field can be written in terms of Hermite polynomials:

$$H_q(x) = (-1)^q \, e^{x^2} \frac{d^q}{dx^q} e^{-x^2}. \tag{1.101}$$

This leads to a frequency component of the field in mode n, m

$$A_{nm}(\mathbf{r}, \omega) = \frac{4\pi(-1)^{n+m}}{\sqrt{\varepsilon_0}\, s^{n+m+2}(z)} H_n\left(\frac{x}{s}\right) H_m\left(\frac{y}{s}\right) \exp\left[ikz - \frac{x^2 + y^2}{s^2(z)}\right] + \text{c.c.}. \tag{1.102}$$

If we take the Fourier transform in the frequency domain again, we find that the real scalar field becomes

$$A_{nm}(\mathbf{r}, t) = \int dk \, u_{nm}^{\text{HG}}(k; \mathbf{r}, t) + \text{c.c.}, \tag{1.103}$$

with

$$u_{nm}^{\text{HG}}(k; \mathbf{r}, t) = \frac{4\pi(-1)^{n+m}}{s^{n+m+2}(z)} H_n\left(\frac{x}{s}\right) H_m\left(\frac{y}{s}\right) \exp\left[ikz - i\omega_k t - \frac{x^2 + y^2}{s^2(z)}\right], \tag{1.104}$$

the Hermite–Gaussian mode functions.

We can now construct the quantum theory by associating an annihilation operator $\hat{a}_{nm}(k)$ with the functions $u_{nm}^{\text{HG}}(k; \mathbf{r}, t)$ using Eq. (1.40)

$$\hat{A}(\mathbf{r}, t) = \sqrt{\frac{\hbar}{\varepsilon_0}} \sum_{n,m=0}^{\infty} \int dk \left[\hat{a}_{nm}(k) \, u_{nm}^{\text{HG}}(k; \mathbf{r}, t) + \hat{a}_{nm}^\dagger(k) \, u_{nm}^{\text{HG}*}(k; \mathbf{r}, t)\right]. \tag{1.105}$$

The modes $u_{nm}^{\text{HG}}(k; \mathbf{r}, t)$ are complete and orthonormal (see Hochstadt, 1971), and depend intrinsically on the waist dimension w_0^2 (or, equivalently, the wave-front curvature $s^2(z)$). Some of the lower-order modes are shown in Fig. 1.2.

The argument of the Hermite polynomials in $u_{nm}^{\text{HG}}(k;\mathbf{r},t)$ is complex because $s^2(z)$ is complex. These are the so-called elegant Hermite–Gaussian modes. There are also 'standard' Hermite–Gaussian modes, in which the argument of the Hermite polynomials is real. The elegant modes can be written in terms of the standard modes, and the imaginary part is then captured by a phase factor, called the 'Gouy phase'. Most lasers create fields that are near standard Hermite–Gaussian modes. For the translation from the elegant to the standard modes, we refer the reader to Enderlein and Pampaloni (2004).

Laguerre–Gaussian modes

The Hermite–Gaussian functions were found using a power expansion of the wave vectors k_x and k_y of the transverse mode function in Cartesian coordinates. We can also choose a rotated coordinate system x' and y' and expand the transverse mode functions in terms of $k_{x'}$ and $k_{y'}$, which leads simply to rotated Hermite–Gaussian modes. However, instead of Cartesian coordinates in the transverse plane, we can also use 'polar' coordinates (r,ϕ):

$$x = r\cos\phi \quad \text{and} \quad y = r\sin\phi \quad \text{or}$$

$$x + iy = re^{i\phi} \equiv \sigma \quad \text{and} \quad x - iy = re^{-i\phi} \equiv \bar{\sigma}. \tag{1.106}$$

The transverse mode function can then be expanded as

$$u(\mathbf{k},\mathbf{r}) \propto \sum_{l,m=0}^{\infty} d_{lm} k_+^l k_-^{l+m} u_0(\mathbf{k},\mathbf{r}), \tag{1.107}$$

with $k_\pm = k_x \pm ik_y$. We have chosen the powers l and $l+m$ such that our equations will simplify later on. In polar coordinates the fundamental Gaussian mode becomes

$$A_0(\mathbf{r},\omega) = \frac{1}{s^2(z)} \exp\left(ikz - \frac{\sigma\bar{\sigma}}{s^2(z)}\right). \tag{1.108}$$

We can then write for the field with a particular l and m:

$$\begin{aligned}
A_{lm}(\mathbf{r},\omega) &\propto \int d\mathbf{k}_\perp \, (ik_+)^l (ik_-)^{l+m} u_0(\mathbf{k},\mathbf{r}) \\
&= \int d\mathbf{k}_\perp \, \frac{\partial^{2l+m}}{\partial\sigma^l \partial\bar{\sigma}^{l+m}} u_0(\mathbf{k},\mathbf{r}) \\
&= \frac{\partial^{2l+m}}{\partial\sigma^l \partial\bar{\sigma}^{l+m}} A_0(\mathbf{r},\omega).
\end{aligned} \tag{1.109}$$

Using the definition of the generalized Laguerre polynomials

$$L_l^m(\xi) = \frac{e^\xi \xi^{-m}}{l!} \frac{d^l}{d\xi^l}\left(e^\xi \, \xi^{l+m}\right) \tag{1.110}$$

this leads, after some algebra, to the *elegant* Laguerre–Gaussian mode functions

$$A_{lm}(\mathbf{r}, \omega) = \frac{4\pi(-1)^{l+m}l!}{\sqrt{\varepsilon_0}\, s^{2(l+m+1)}(z)} r^{|m|} e^{im\phi} L_l^m\left(\frac{r^2}{s^2(z)}\right) \exp\left(ikz - \frac{r^2}{s^2(z)}\right) + \text{c.c..} \quad (1.111)$$

Taking the Fourier transform over the frequency domain, this leads to the classical field

$$A_{lm}(\mathbf{r}, t) = \int dk\, u_{lm}^{\text{LG}}(k; \mathbf{r}, t) + \text{c.c.,} \quad (1.112)$$

with

$$u_{lm}^{\text{LG}}(k; \mathbf{r}, t) = \frac{4\pi(-1)^{l+m}l!}{s^{2(l+m+1)}(z)} r^{|m|} e^{im\phi} L_l^m\left(\frac{r^2}{s^2(z)}\right) \exp\left(ikz - i\omega_k t - \frac{r^2}{s^2(z)}\right). \quad (1.113)$$

Analogous to the Hermite–Gaussian case, we can now construct the quantum field for the elegant Laguerre–Gaussian modes:

$$\hat{A}(\mathbf{r}, t) = \sqrt{\frac{\hbar}{\varepsilon_0}} \sum_{l,m=0}^{\infty} \int dk \left[\hat{a}_{lm}(k)\, u_{lm}^{\text{LG}}(k; \mathbf{r}, t) + \hat{a}_{lm}^{\dagger}(k)\, u_{lm}^{\text{LG}*}(k; \mathbf{r}, t) \right]. \quad (1.114)$$

The Laguerre–Gaussian mode functions are orthogonal and complete (see, for example, Hochstadt, 1971), and the lowest-order modes are shown in Fig. 1.2.

As noted before, the paraxial wave equation has a similar form to the Schrödinger equation, where the time parameter is replaced with z. This allows us to use key results from quantum mechanics. Specifically, we can construct an angular momentum operator $\hat{L}_z$ using the transverse position and momentum operators:

$$\hat{L}_z = \hat{x}\hat{p}_y - \hat{y}\hat{p}_x = \frac{\hbar}{i}\frac{\partial}{\partial\phi}. \quad (1.115)$$

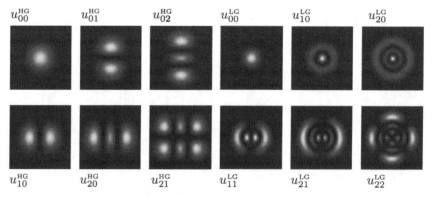

$$u_{00}^{\text{HG}} \qquad u_{01}^{\text{HG}} \qquad u_{02}^{\text{HG}} \qquad u_{00}^{\text{LG}} \qquad u_{10}^{\text{LG}} \qquad u_{20}^{\text{LG}}$$

$$u_{10}^{\text{HG}} \qquad u_{20}^{\text{HG}} \qquad u_{21}^{\text{HG}} \qquad u_{11}^{\text{LG}} \qquad u_{21}^{\text{LG}} \qquad u_{22}^{\text{LG}}$$

Fig. 1.2.   The lowest-order transverse mode functions.

It is immediately clear from Eq. (1.113) that the Laguerre–Gaussian mode functions u_{lm}^{LG} are eigenfunctions of $\hat{L}_z$:

$$\hat{L}_z\, u_{lm}^{\text{LG}}(\mathbf{k}, \mathbf{r}) = m\hbar\, u_{lm}^{\text{LG}}(\mathbf{k}, \mathbf{r}) \,. \qquad (1.116)$$

The Laguerre–Gaussian modes therefore carry orbital angular momentum in that the beam exerts a torque on the medium it travels through. These are sometimes also called 'helical' modes. The angular momentum is quantized (with quantum numbers l and m) such that each photon carries an orbital angular momentum of $m\hbar$. This effect is different from the spin angular momentum, or polarization of the light. In particular, the orbital angular momentum is not restricted to $\pm\hbar$. In some circumstances it is difficult to separate spin and orbital angular momentum, since both are determined by the direction and phase of the electric and magnetic fields in the beam.

1.4 Evolution of the field operators

So far, we have concentrated mostly on the spatial aspects of the electromagnetic quantum field. However, it is important for all physical applications, and for quantum information processing in particular, to find the time evolution of the quantum fields. There are two equivalent ways of achieving this. We can either take the time dependence into account when we give a description of the quantum state (the Schrödinger picture), or we can describe the time dependence in terms of the operators (the Heisenberg picture). Later in the book we will adopt a hybrid approach, where the free evolution of the systems is described in the state, and the deviation from this is described by the interaction Hamiltonian (the interaction picture). For now, we will use the Heisenberg picture.

1.4.1 The Heisenberg equations of motion

In the Heisenberg picture an arbitrary Hermitian operator A evolves in time under the influence of a unitary operator $U(t)$ such that

$$A(t) = U(t)AU^{\dagger}(t) \quad \text{with} \quad U(t) = \exp\left(-\frac{i}{\hbar}\mathcal{H}t\right), \qquad (1.117)$$

where t is the time and $\mathcal{H}$ the Hamiltonian. In the exponential form of $U(t)$, the Hamiltonian is often called the 'generator' of U. For convenience, we define $\mu = -it/\hbar$, such that we have

$$A(\mu) = \exp(\mu\mathcal{H})A\exp(-\mu\mathcal{H}) \,. \qquad (1.118)$$

A general Taylor expansion of $A(\mu)$ around $\mu = 0$ can be written as

$$A(\mu) = A(0) + \mu\left.\frac{dA}{d\mu}\right|_{\mu=0} + \frac{\mu^2}{2!}\left.\frac{d^2A}{d\mu^2}\right|_{\mu=0} + \cdots \qquad (1.119)$$

Next, we evaluate the derivatives of A to μ in Eq. (1.119) using Eq. (1.118) and

$$i\hbar \frac{dU(t)}{dt} = \mathcal{H} U(t) \,. \tag{1.120}$$

In terms of the commutators with $\mathcal{H}$ the derivatives of A become:

$$\frac{dA}{d\mu} = [\mathcal{H}, A] \quad \text{and} \quad \frac{d^2 A}{d\mu^2} = [\mathcal{H}, [\mathcal{H}, A]] \,, \tag{1.121}$$

and so forth. This leads to the Baker–Campbell–Hausdorff relation

$$A(\mu) = A(0) + \mu[\mathcal{H}, A(0)] + \frac{\mu^2}{2!} [\mathcal{H}, [\mathcal{H}, A(0)]] + \cdots \tag{1.122}$$

This equation is true for all Hermitian operators $\mathcal{H}$ and complex parameters μ. Hermiticity of A is not required. In compact notation it reads

$$e^{\mu B} A e^{-\mu B} = A + \mu[B, A] + \frac{\mu^2}{2!} [B, [B, A]] + \cdots \,, \tag{1.123}$$

where B is Hermitian. Taking the time derivative of Eq. (1.117) we find the Heisenberg equation of motion for the operator $A(t)$:

$$\frac{dA(t)}{dt} = \frac{i}{\hbar} [\mathcal{H}, A(t)] + \frac{\partial A(t)}{\partial t} \,. \tag{1.124}$$

The Heisenberg equations of motion for the creation and annihilation operators are then

$$\frac{d\hat{a}_\lambda(\mathbf{k})}{dt} = \frac{i}{\hbar} \left[\mathcal{H}, \hat{a}_\lambda(\mathbf{k}) \right] \quad \text{and} \quad \frac{d\hat{a}_\lambda^\dagger(\mathbf{k})}{dt} = \frac{i}{\hbar} \left[\mathcal{H}, \hat{a}_\lambda^\dagger(\mathbf{k}) \right] \,, \tag{1.125}$$

where we have exploited the fact that the annihilation and creation operators do not have an explicit time dependence. Using the expression in Eq. (1.47) we can solve the differential equations for the creation and annihilation operators to get

$$\hat{a}_\lambda(\mathbf{k}, t) = \hat{a}_\lambda(\mathbf{k}) \, e^{-i\omega_\mathbf{k} t} \quad \text{and} \quad \hat{a}_\lambda^\dagger(\mathbf{k}, t) = \hat{a}_\lambda^\dagger(\mathbf{k}) \, e^{i\omega_\mathbf{k} t} \,. \tag{1.126}$$

The time-dependent annihilation operator is obtained by integrating over all wave vectors $\mathbf{k}$:

$$\hat{a}_\lambda(t) = \int d\mathbf{k} \, \hat{a}_\lambda(\mathbf{k}) \, e^{-i\omega_\mathbf{k} t} \,, \tag{1.127}$$

and the time-dependent number operator can be written as

$$\hat{n}_\lambda(t) = \hat{a}_\lambda^\dagger(t) \hat{a}_\lambda(t) = \int d\mathbf{k} \int d\mathbf{k}' \, \hat{a}_\lambda^\dagger(\mathbf{k}) \, \hat{a}_\lambda(\mathbf{k}') \, e^{-i(\omega_{\mathbf{k}'} - \omega_\mathbf{k}) t} \,. \tag{1.128}$$

Exercise 1.7: Derive the Heisenberg equations of motion in Eq. (1.124).

As an example, suppose we have a wave packet containing exactly one photon with polarization λ, defined in Eq. (1.60)

$$|1_{j,\lambda}\rangle = \int d\mathbf{k}\,\alpha_j(\mathbf{k})\hat{a}_\lambda^\dagger(\mathbf{k})|0\rangle\,, \tag{1.129}$$

where, as in Section 1.2 the state is normalized according to a mode function $\alpha_j(\mathbf{k})$ and $|0\rangle$ is the global vacuum state (i.e., no excitations in any mode). The number density $\langle \hat{n} \rangle$ at time t is readily calculated to be

$$\langle 1_{j,\lambda}|\hat{n}_\lambda(t)|1_{j,\lambda}\rangle = \left| \int d\mathbf{k}\,\alpha_j(\mathbf{k})\,e^{-i\omega_\mathbf{k}t} \right|^2. \tag{1.130}$$

This is a function in t that depends on the amplitudes $\alpha_j(\mathbf{k})$. We can construct temporally localized, single-photon wave packets by choosing a suitable $\alpha_j(\mathbf{k})$ in Eq. (1.129). For example, a Lorentzian wave packet is defined by

$$\alpha_L(\mathbf{k}) = \frac{1}{\sqrt{\pi}}\frac{\sqrt{\gamma}}{\gamma + i(\omega_\mathbf{k} - \omega_0)} \tag{1.131}$$

with median ω_0, and γ parameterizes the width of the distribution. Alternatively, Gaussian wave packets can be written as

$$\alpha_G(\mathbf{k}) = \frac{1}{\sqrt[4]{2\pi\sigma^2}}\exp\left[-\frac{(\omega_\mathbf{k} - \omega_0)^2}{4\sigma^2} \right], \tag{1.132}$$

which amounts to a wave packet with central frequency ω_0 and spectral width 2σ. These modes must be broadband, and when we speak of 'the frequency' of the wave packet, we mean the central (mean) frequency, or median in the case of the Lorentzian.

The evolution of the quadrature operators is similarly determined by the Heisenberg equation of motion

$$\frac{d\hat{x}_\zeta(\mathbf{k})}{dt} = \frac{i}{\hbar}\left[\mathcal{H}, \hat{x}_\zeta(\mathbf{k}) \right], \tag{1.133}$$

the solution of which is given by

$$\hat{x}_\zeta(\mathbf{k},t) = \frac{e^{-i(\zeta+\omega_\mathbf{k}t)}\,\hat{a}(\mathbf{k}) + e^{i(\zeta+\omega_\mathbf{k}t)}\,\hat{a}^\dagger(\mathbf{k})}{\sqrt{2}}. \tag{1.134}$$

Integrating over all wave vectors then yields the time-dependent quadrature operator:

$$\hat{x}_\zeta(t) = \frac{e^{-i\zeta}\,\hat{a}(t) + e^{i\zeta}\,\hat{a}^\dagger(t)}{\sqrt{2}}. \tag{1.135}$$

The position and momentum quadratures are then given by

$$\hat{q}(t) = \sqrt{\frac{\hbar}{2\omega_\mathbf{k}}}\left[\hat{a}(t) + \hat{a}^\dagger(t) \right] \quad \text{and} \quad \hat{p}(t) = -i\sqrt{\frac{\hbar\omega_\mathbf{k}}{2}}\left[\hat{a}(t) - \hat{a}^\dagger(t) \right]. \tag{1.136}$$

1.4.2 Time-bin mode operators

We have shown how to construct temporally localized mode functions, with the Lorentzian and Gaussian wave packets as examples. Here, we will construct mode operators that create and annihilate photons in finite time intervals, or 'time bins', which are important in quantum communication. It is most convenient to construct the time-bin operators from discrete, one-dimensional plane-wave operators $\hat{a}_n$ in a cavity of length L. The index n relates to the discrete frequency $\omega_n = n\pi c/L$, and the corresponding wave number is $k_n = n\pi/L$. The field operator in terms of discrete modes is then given by

$$\hat{A} = \frac{1}{\sqrt{N+1}} \sqrt{\frac{\hbar}{\varepsilon_0}} \sum_{n=0}^{N} \left[\hat{a}_n u_n(\mathbf{r}, t) + \hat{a}_n^\dagger u_n^*(\mathbf{r}, t) \right], \tag{1.137}$$

where we have suppressed the polarization for notational simplicity, and we cut off the frequency at some sufficiently high value.

The finite time interval τ is determined by the cut-off frequency ω_N, above which the modes are unoccupied. This allows us to truncate the plane-wave expansion, and set

$$\tau = \frac{2L}{c(N+1)} \quad \text{and} \quad \omega_n \tau = \frac{2\pi n}{N+1}. \tag{1.138}$$

We define the time-bin annihilation operator as

$$\hat{b}_\mu = \frac{1}{\sqrt{N+1}} \sum_{n=0}^{N} e^{-i\mu\omega_n\tau} \hat{a}_n, \tag{1.139}$$

which has the inverse

$$\hat{a}_n = \frac{1}{\sqrt{N+1}} \sum_{\mu=0}^{N} e^{i\mu\omega_n\tau} \hat{b}_\mu. \tag{1.140}$$

The integer μ indicates the time bin. To see that this can really be interpreted as a time-bin operator, we relate $\hat{b}_\mu$ to the annihilation operator $\hat{a}(t)$ in the time domain

$$\hat{a}(t) = \sum_{n=0}^{\infty} e^{-i\omega_n t} \hat{a}_n = \frac{1}{\sqrt{N+1}} \sum_{n=0}^{\infty} \sum_{\mu=0}^{N} e^{-i\omega_n(t-\mu\tau)} \hat{b}_\mu$$

$$= \sum_{\mu=0}^{N} \left(\frac{1}{\sqrt{N+1}} \sum_{n=0}^{\infty} e^{-i\omega_n(t-\mu\tau)} \right) \hat{b}_\mu$$

$$= \sum_{\mu=0}^{N} \alpha_\mu(t) \hat{b}_\mu, \tag{1.141}$$

where, using $\omega = \pi c/L$

$$\alpha_\mu(t) = \frac{1}{\sqrt{N+1}} \left[1 - e^{-i\omega(t-\mu\tau)} \right]^{-1} \tag{1.142}$$

are the time-bin expansion functions. They depend only on $t - \mu\tau$ and are discrete translations of $\alpha_\mu(0)$ with period τ: this is the requirement for $\hat{b}_\mu$ to be a time-bin operator. Since we constructed the time-bin operators from (a discrete set of) infinite plane waves, the $\alpha_\mu(t)$ are highly singular when $t \to \mu\tau$. Discrete mode functions, which have localized transverse and longitudinal profiles and a regularized frequency distribution, can be used to remove the singularity. The time-bin expansion function that is most natural for a given application depends on the mode functions that describe the photon source.

We still need to verify if the operator $\hat{b}_\mu$ and its Hermitian adjoint $\hat{b}_\mu^\dagger$ are valid mode operators. In other words, we need to verify that the commutation relations hold. Assuming that $[\hat{a}_n, \hat{a}_m^\dagger] = \delta_{nm}$, we can write

$$\left[\hat{b}_\mu, \hat{b}_\nu^\dagger\right] = \frac{1}{N+1} \sum_{n,m=0}^{N} e^{-i\mu\omega_n\tau + i\nu\omega_m\tau} \left[\hat{a}_n, \hat{a}_m^\dagger\right]$$

$$= \frac{1}{N+1} \sum_{n=0}^{N} e^{-i(\mu-\nu)\omega_n\tau}$$

$$= \frac{1}{N+1} \sum_{n=0}^{N} e^{-2\pi i(\mu-\nu)n/(N+1)}$$

$$= \delta_{\mu\nu}, \tag{1.143}$$

which is indeed the correct commutation relation for photons. The last equation can be regarded as a definition of the Kronecker delta symbol. The remaining commutation relations are also satisfied by the time-bin operators.

1.4.3 Mode transformations and optical elements

We have seen that the mode operators are defined with respect to mode functions, and that linear mode transformations lead to Bogoliubov transformations of the mode operators. The question is then how we generate Bogoliubov transformations. Since they are unitary, they must be generated by Hermitian Hamiltonians. This will lead to a physical interpretation of these Hamiltonians in terms of ideal beam splitters and phase shifters. In the remainder of this chapter we will use discrete physical modes.

Let us assume that the mode functions are sharply peaked such that the discrete modes are labelled with j, and can be associated with central wave vectors $\mathbf{k}_j$, and label the operators $\hat{a}_{j\lambda}$ and $\hat{a}_{j\lambda}^\dagger$. The time evolution of a discrete single-mode operator depends only on the discrete free-field Hamiltonian $\mathcal{H}_{j\lambda} = \hbar\omega_{\mathbf{k}}\hat{a}_{j\lambda}^\dagger\hat{a}_{j\lambda}$, and this generates the natural 'phase shift' $\exp(-i\omega_j t)$ accumulated by the free evolution. This gives us a clue how to describe lossless phase shifts in optical modes that are induced by materials with a refractive index n_r. When a dielectric of length ℓ is inserted in the optical beam, the phase shift is $\phi = n_r k\ell$, where $k = |\mathbf{k}|$ is the wave number of the optical mode. We can construct a dimensionless

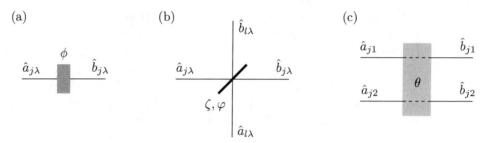

Fig. 1.3. Graphical representations of optical elements: (a) the phase shifter; (b) the beam splitter; and (c) the polarization rotation.

single-mode 'interaction' Hamiltonian[2] $\mathcal{H}_{j\lambda}$

$$\mathcal{H}_{j\lambda}(\phi) \equiv \hbar\phi\, \hat{a}_{j\lambda}^\dagger \hat{a}_{j\lambda} \quad \text{and} \quad U(\phi) = \exp\left[-\frac{i}{\hbar}\mathcal{H}_{j\lambda}(\phi)\right]. \tag{1.144}$$

To stress that this is not ordinary time evolution, we have removed the parameter t from the unitary operator and replaced it with ϕ. This will induce the transformation

$$\hat{b}_{j\lambda} = \exp\left[-\frac{i}{\hbar}\mathcal{H}_{j\lambda}(\phi)\right] \hat{a}_{j\lambda} \exp\left[\frac{i}{\hbar}\mathcal{H}_{j\lambda}(\phi)\right] = \hat{a}_{j\lambda}\, e^{-i\phi}. \tag{1.145}$$

The interaction Hamiltonian in Eq. (1.144) can be written in terms of the number operator for mode $\mathbf{k}_j$, and the number operator is therefore the generator of phase shifts. See Fig. 1.3a for a graphical representation of the phase shift.

Similarly, we can construct an interaction Hamiltonian for two modes $\mathbf{k}_j$ and $\mathbf{k}_l$ and fixed polarization λ

$$\mathcal{H}_{jl}(\zeta,\varphi) = \hbar\zeta e^{i\varphi}\, \hat{a}_{j\lambda}^\dagger \hat{a}_{l\lambda} + \hbar\zeta e^{-i\varphi}\, \hat{a}_{j\lambda} \hat{a}_{l\lambda}^\dagger. \tag{1.146}$$

Both terms must be present to ensure that $\mathcal{H}_{jl}(\zeta,\varphi)$ is Hermitian. If we interpret $\mathcal{H}_{jl}(\zeta,\varphi)$ physically, it describes the creation of a photon in mode $\{\mathbf{k}_j, \lambda\}$ and the annihilation of a photon in mode $\{\mathbf{k}_l, \lambda\}$, and vice versa. We therefore expect that $\mathcal{H}_{jl}(\zeta,\varphi)$ is the interaction Hamiltonian for a beam splitter mixing modes $\mathbf{k}_j$ and $\mathbf{k}_l$. Indeed, when we calculate the transformations of $\hat{a}_{j\lambda}$ and $\hat{a}_{l\lambda}$ using the Baker–Campbell–Hausdorff relation, we find that

$$\hat{b}_{j\lambda} = e^{\frac{i}{\hbar}\mathcal{H}_{jl}}\hat{a}_{j\lambda}e^{-\frac{i}{\hbar}\mathcal{H}_{jl}} = \cos\zeta\, \hat{a}_{j\lambda} - ie^{i\varphi}\sin\zeta\, \hat{a}_{l\lambda},$$

$$\hat{b}_{l\lambda} = e^{\frac{i}{\hbar}\mathcal{H}_{jl}}\hat{a}_{l\lambda}e^{-\frac{i}{\hbar}\mathcal{H}_{jl}} = -ie^{-i\varphi}\sin\zeta\, \hat{a}_{j\lambda} + \cos\zeta\, \hat{a}_{l\lambda}. \tag{1.147}$$

---

[2] It is not an interaction Hamiltonian in the standard sense of interactions, since we are considering free fields. There is a microscopic interaction of the field with the dielectric, but the details are averaged out in this Hamiltonian.

This can be expressed in terms of a unitary matrix:

$$\begin{pmatrix} \hat{b}_{j\lambda} \\ \hat{b}_{l\lambda} \end{pmatrix} = \begin{pmatrix} \cos\zeta & -ie^{i\varphi}\sin\zeta \\ -ie^{-i\varphi}\sin\zeta & \cos\zeta \end{pmatrix} \begin{pmatrix} \hat{a}_{j\lambda} \\ \hat{a}_{l\lambda} \end{pmatrix}. \tag{1.148}$$

A physical beam splitter is typically described with $\varphi = 0$ or $\varphi = \pi/2$. When $\zeta = \pi/4$, we have a 50:50 beam splitter. The two-mode transformation given here is therefore a generalized beam-splitter transformation. The beam-splitter and phase-shifter elements are shown in Fig. 1.3b.

In the above example of a beam splitter we chose two modes with identical polarization λ, but different wave vectors $\mathbf{k}_j$ and $\mathbf{k}_l$. However, we can also vary the polarization ($\lambda \in \{1, 2\}$) and keep the wave vector constant:

$$\mathcal{H}_j(\theta) = \hbar\theta e^{i\varphi}\hat{a}_{j1}^\dagger \hat{a}_{j2} + \hbar\theta e^{-i\varphi}\hat{a}_{j1}\hat{a}_{j2}^\dagger. \tag{1.149}$$

This will lead to a polarization rotation

$$\begin{pmatrix} \hat{b}_1 \\ \hat{b}_2 \end{pmatrix} = \begin{pmatrix} \cos\theta & -ie^{i\varphi}\sin\theta \\ -ie^{-i\varphi}\sin\theta & \cos\theta \end{pmatrix} \begin{pmatrix} \hat{a}_1 \\ \hat{a}_2 \end{pmatrix}, \tag{1.150}$$

where we have suppressed the dependence on $\mathbf{k}_j$ for notational brevity. The graphical representation of a polarization rotation is given in Fig. 1.3c.

If we also include a relative phase shift ϕ in the two modes a_1 and a_2, the corresponding matrix

$$U(\phi, \theta, \varphi) = \begin{pmatrix} e^{i\phi/2}\cos\theta & -ie^{i\varphi}\sin\theta \\ -ie^{-i\varphi}\sin\theta & e^{-i\phi/2}\cos\theta \end{pmatrix} \tag{1.151}$$

is the most general two-mode unitary transformation. In other words, any 2×2 unitary matrix can be written as a particular trio of angles (ϕ, θ, φ).

The beam splitter and the polarization rotation described here are lossless, and are therefore idealizations of real beam splitters and polarization rotations, as implemented by e.g., half-silvered mirrors and quarter wave plates. Nevertheless, we will see later that we can account for photon loss quite easily in this formalism, and considering that the losses must be low for quantum information processing anyway, this is an excellent approximation.

As a practical example of linear mode transformations, consider the Mach–Zehnder interferometer (Fig. 1.4). It consists of two input and output modes, two 50:50 beam splitters BS_1 and BS_2, and a relative phase shift ϕ between the internal arms in the interferometer. Suppose the input modes are $\hat{a}_1$ and $\hat{a}_2$, and the output modes are $\hat{b}_1$ and $\hat{b}_2$. The mode transformations are

$$\begin{pmatrix} \hat{b}_1 \\ \hat{b}_2 \end{pmatrix} = \frac{1}{2} \begin{pmatrix} 1 & 1 \\ -1 & 1 \end{pmatrix} \begin{pmatrix} 1 & 0 \\ 0 & e^{i\phi} \end{pmatrix} \begin{pmatrix} 1 & -1 \\ 1 & 1 \end{pmatrix} \begin{pmatrix} \hat{a}_1 \\ \hat{a}_2 \end{pmatrix},$$

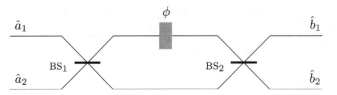

The Mach–Zehnder interferometer.

where the first matrix represents BS_2, the second is the phase shift, and the third represents BS_1. The intensities (proportional to the number operators of the modes) in the output can be related to the intensities in the input:

$$\hat{b}_1^\dagger \hat{b}_1 - \hat{b}_2^\dagger \hat{b}_2 = \cos\phi \left(\hat{a}_1^\dagger \hat{a}_1 + \hat{a}_2^\dagger \hat{a}_2 \right) - i\sin\phi \left(\hat{a}_1^\dagger \hat{a}_2 + \hat{a}_2^\dagger \hat{a}_1 \right),$$

$$\hat{b}_1^\dagger \hat{b}_1 + \hat{b}_2^\dagger \hat{b}_2 = \hat{a}_1^\dagger \hat{a}_1 + \hat{a}_2^\dagger \hat{a}_2 . \tag{1.152}$$

Taking the expectation values on the right-hand side with respect to the input state gives the expectation value of the intensity sum and difference in the output modes. The second equality expresses photon number conservation in the Mach–Zehnder interferometer. Measuring the intensity difference of the output modes gives information about the (unknown) phase shift ϕ. We will return to this topic in Chapter 13.

1.4.4 Normal modes

A beam splitter is an example of the mixing of two optical modes. Often, we can think of the modes as two Gaussian beams that have near-perfect overlap on the beam splitter. It is then a good approximation to treat the system as an 'interferometer' of two discrete modes, described by a unitary 2×2 matrix. This can be generalized to many modes. For N discrete modes, the transformation is an $N \times N$ unitary matrix. The physical system is sometimes called an 'N-port interferometer' (Fig. 1.5). The question is now whether we can create any conceivable N-port interferometer out of regular beam splitters and phase shifters, or whether we need more arbitrary M-point beam splitters ($2 < M \leq N$) that are not reducible to beam splitters and phase shifters.

Reck *et al.* (1994) proved that every N-mode unitary transformation can be constructed from at most $N(N-1)/2$ beam splitters and phase shifters. A sketch of their proof runs as follows: let's denote the unitary transformation that describes the N-port interferometer by $U(N)$. In addition, we define the matrix T_{nm} as the N-dimensional identity operator $I(N)$ with the elements I_{nn}, I_{nm}, I_{mn}, and I_{mm} replaced by the four elements of a 2×2 unitary matrix. We can then use the T_{nm} to reduce the size of $U(N)$:

$$U(N) \cdot T_{N,N-1} \cdot T_{N,N-2} \cdots T_{N,1} = U(N-1) \oplus e^{i\phi} . \tag{1.153}$$

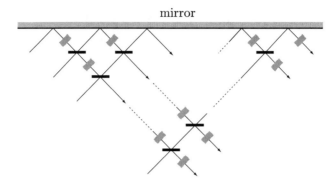

mirror

The *N*-port interferometer in terms of generalized beam splitters.

In block schematics this is

$$
\begin{pmatrix} \\ & U(N) & \\ \\ \end{pmatrix} \cdot T_{N,N-1} \cdots T_{N,1} = \left(\begin{array}{c|c} U(N-1) & 0 \\ \hline 0 & e^{i\phi_N} \end{array} \right). \tag{1.154}
$$

We therefore need $N-1$ matrices T to reduce the unitary $U(N)$ to $U(N-1)$. We can repeat this procedure to reduce $U(N-1)$ to $U(N-2)$, and so on. In total, we have to use at most $N(N-1)/2$ matrices T to reduce $U(N)$ to a diagonal matrix with only phases as non-zero elements:

$$
U(N) \cdot T_{N,N-1} \cdot T_{N,N-2} \cdots T_{2,1} = \begin{pmatrix} e^{i\phi_1} & & 0 \\ & \ddots & \\ 0 & & e^{i\phi_N} \end{pmatrix}. \tag{1.155}
$$

Since the T matrices are unitary (and therefore invertible), we can write $U(N)$ as a series of $N(N-1)/2$ matrices acting on the diagonal matrix in Eq. (1.155). Because every T_{nm} matrix is associated with a beam splitter and phase shift on modes n and m, this means that any N-port interferometer can be constructed with only beam splitters and phase shifters.

Since $U(N)$ can be deconstructed into a series of beam splitters and phase shifters on discrete modes a_n, we can write its generator in terms of a Hamiltonian

$$
\mathcal{H}_{1\ldots N}(B) = \hbar \sum_{n,m=1}^{N} \hat{a}_n^\dagger B_{nm} \hat{a}_m, \tag{1.156}
$$

where B is a Hermitian matrix. In the following argument, we will suppress the subscripts and argument of $\mathcal{H}$ for brevity. Under the influence of this Hamiltonian a mode operator $\hat{a}_n$ becomes

$$
\hat{a}_n \to \hat{b}_n = e^{\frac{i}{\hbar}\mathcal{H}} \hat{a}_n e^{-\frac{i}{\hbar}\mathcal{H}} = \sum_{m=1}^{N} U_{nm} \hat{a}_m. \tag{1.157}
$$

The U_{nm} are elements of the unitary matrix $U(N)$. We can diagonalize $\mathcal{H}$ by diagonalizing B using a unitary matrix V:

$$VHV^\dagger = \hbar \sum_{j,n,m=1}^{N} V_{jn}\hat{a}_n^\dagger V_{nj}^* \, V_{jn}B_{nm}V_{mj}^* \, V_{jm}\hat{a}_m V_{mj}^*$$

$$= \hbar \sum_{j=1}^{N} \hat{b}_j^\dagger D_{jj} \hat{b}_j , \qquad \qquad (1.158)$$

where D is a diagonal matrix. This gives us the 'normal modes' $\hat{b}_n$ given in Eq. (1.157) with $U_{nm} = V_{nm}V_{mn}^*$. By construction, these modes diagonalize the Hamiltonian, which in turn describes a set of non-interacting modes. Since beam splitters can be transformed away using linear Bogoliubov transformations, their operation does not constitute a real interaction between modes (and by implication between photons). This will have profound implications for optical quantum information processing.

1.4.5 Non-photon-number-preserving transformations

So far, the 'interaction' Hamiltonians have generated transformations of annihilation operators into other annihilation operators (and creation operators into other creation operators). However, we have seen in Eq. (1.42) that the most general mode transformations allow for the transformation of annihilation operators into creation operators, and vice versa. The question is therefore which Hamiltonian can generate such Bogoliubov transformations.

There is one type of quadratic Hamiltonian that we have not yet considered, namely the Hamiltonians that are proportional to $\hat{a}^2$ and $\hat{a}^{\dagger 2}$. First, we consider the Hamiltonian

$$\mathcal{H}_j(\xi,\varphi) = \hbar\xi e^{i\varphi} \hat{a}_{j\lambda}^2 + \hbar\xi e^{-i\varphi} \hat{a}_{j\lambda}^{\dagger 2} . \qquad (1.159)$$

We use the Baker–Campbell–Hausdorff relation from Eq. (1.123) to calculate the transformed mode operators

$$\hat{b}_{j\lambda} = \cosh(2\xi)\,\hat{a}_{j\lambda} - ie^{-i\varphi}\sinh(2\xi)\,\hat{a}_{j\lambda}^\dagger$$

$$\hat{b}_{j\lambda}^\dagger = ie^{i\varphi}\sinh(2\xi)\,\hat{a}_{j\lambda} + \cosh(2\xi)\,\hat{a}_{j\lambda}^\dagger . \qquad (1.160)$$

In matrix form this is

$$\begin{pmatrix} \hat{b}_{j\lambda} \\ \hat{b}_{j\lambda}^\dagger \end{pmatrix} = \begin{pmatrix} \cosh(2\xi) & -ie^{-i\varphi}\sinh(2\xi) \\ ie^{i\varphi}\sinh(2\xi) & \cosh(2\xi) \end{pmatrix} \begin{pmatrix} \hat{a}_{j\lambda} \\ \hat{a}_{j\lambda}^\dagger \end{pmatrix} . \qquad (1.161)$$

We now observe that the quadratic Hamiltonian in Eq. (1.159) mixes the creation and annihilation operators of the mode $a_{j\lambda}$. This is called 'single-mode squeezing'.

Alternatively, we can define 'two-mode squeezing' as the mode transformation that is induced by the Hamiltonian

$$\mathcal{H}_{jl}(\xi, \varphi) = \hbar\xi e^{i\varphi}\, \hat{a}_{j\lambda}\hat{a}_{l\lambda'} + \hbar\xi e^{-i\varphi}\, \hat{a}_{j\lambda}^{\dagger}\hat{a}_{l\lambda'}^{\dagger}. \tag{1.162}$$

The mode transformation then becomes

$$\hat{b}_{j\lambda} = \cosh\xi\, \hat{a}_{j\lambda} - ie^{i\varphi}\sinh\xi\, \hat{a}_{l\lambda'}^{\dagger}$$
$$\hat{b}_{l\lambda'} = -ie^{-i\varphi}\sinh\xi\, \hat{a}_{j\lambda}^{\dagger} + \cosh\xi\, \hat{a}_{l\lambda'}. \tag{1.163}$$

In terms of the matrix representation this transformation can be written as

$$\begin{pmatrix} \hat{b}_{j\lambda} \\ \hat{b}_{j\lambda}^{\dagger} \\ \hat{b}_{l\lambda'} \\ \hat{b}_{l\lambda'}^{\dagger} \end{pmatrix} = \begin{pmatrix} \cosh\xi & 0 & 0 & -ie^{i\varphi}\sinh\xi \\ 0 & \cosh\xi & ie^{-i\varphi}\sinh\xi & 0 \\ 0 & ie^{i\varphi}\sinh\xi & \cosh\xi & 0 \\ -ie^{-i\varphi}\sinh\xi & 0 & 0 & \cosh\xi \end{pmatrix} \begin{pmatrix} \hat{a}_{j\lambda} \\ \hat{a}_{j\lambda}^{\dagger} \\ \hat{a}_{l\lambda'} \\ \hat{a}_{l\lambda'}^{\dagger} \end{pmatrix}.$$

We now have all the ingredients for a general theory of discrete linear mode transformations: a general Bogoliubov transformation is given by

$$\hat{b}_i = \sum_j A_{ij}\, \hat{a}_j + B_{ij}\, \hat{a}_j^{\dagger}$$
$$\hat{b}_i^{\dagger} = \sum_j B_{ij}^*\, \hat{a}_j + A_{ij}^*\, \hat{a}_j^{\dagger}, \tag{1.164}$$

where the labels i and j include the polarization degree of freedom. The operators $\hat{b}_i$ and $\hat{b}_i^{\dagger}$ must again be proper annihilation and creation operators, which means they have to obey the commutation relations

$$\left[\hat{b}_i, \hat{b}_j^{\dagger}\right] = \delta_{ij} \quad \text{and} \quad \left[\hat{b}_i, \hat{b}_j\right] = \left[\hat{b}_i^{\dagger}, \hat{b}_j^{\dagger}\right] = 0. \tag{1.165}$$

This leads to the following restrictions on A and B:

$$AB^T = \left(AB^T\right)^T \quad \text{and} \quad AA^{\dagger} = BB^{\dagger} + \hat{\mathbb{1}}. \tag{1.166}$$

The second equation indicates that $AA^{\dagger}$ and $BB^{\dagger}$ are simultaneously diagonalized, for example by a matrix U. The matrices A and B are then diagonalized according to the singular value decomposition theorem:

$$A = UA_D V^{\dagger} \quad \text{and} \quad B = UB_D W^{\dagger}, \tag{1.167}$$

where A_D and B_D are diagonal matrices. To determine the relation between V and W, we consider the inverse transformation:

$$\hat{a}_i = \sum_j A_{ij}^*\, \hat{b}_j - B_{ij}^*\, \hat{b}_j^{\dagger}. \tag{1.168}$$

Following the same argumentation as before, we find the restriction

$$A^\dagger B = \left(A^\dagger B\right)^T \quad \text{and} \quad A^\dagger A = \left(B^\dagger B\right)^T + \hat{\mathbb{1}}. \tag{1.169}$$

This leads to the relation $V^* = W$, and

$$A = U A_D V^\dagger \quad \text{and} \quad B = U B_D V^T. \tag{1.170}$$

When we write the Bogoliubov transformation in matrix notation with $\vec{a} = (\hat{a}_1, \dots, \hat{a}_N)$ and $\vec{b} = (\hat{b}_1, \dots, \hat{b}_N)$, we obtain

$$\begin{pmatrix} \vec{b} \\ \vec{b}^\dagger \end{pmatrix} = \begin{pmatrix} U & 0 \\ 0 & U^* \end{pmatrix} \begin{pmatrix} A_D & B_D \\ B_D^* & A_D^* \end{pmatrix} \begin{pmatrix} V^\dagger & 0 \\ 0 & V^T \end{pmatrix} \begin{pmatrix} \vec{a} \\ \vec{a}^\dagger \end{pmatrix}$$

$$= \begin{pmatrix} A & B \\ B^* & A^* \end{pmatrix} \begin{pmatrix} \vec{c} \\ \vec{c}^\dagger \end{pmatrix}, \tag{1.171}$$

which agrees with Eq. (1.170). The three matrices in the top line are three successive components in a nonlinear optical interferometer. The two block-diagonal matrices in the first line do not mix creation and annihilation operators, and therefore describe linear optical interferometers. The central matrix *does* mix creation and annihilation operators. However, the components A_D, B_D, A_D^*, and B_D^* are themselves diagonal matrices, and the entire matrix therefore corresponds to a set of single-mode squeezers. This procedure is called the 'Bloch–Messiah reduction', and is shown in Fig. 1.6.

The generator that is responsible for the Bogoliubov transformation in Eq. (1.164) is, in terms of the normal modes,

$$\mathcal{H} = \frac{\hbar}{2} \sum_i \hat{b}_i^\dagger \hat{b}_i. \tag{1.172}$$

(a)

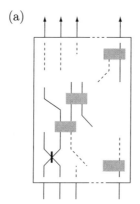

(b)

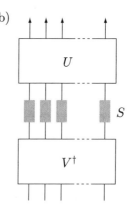

Fig. 1.6. The Bloch–Messiah reduction: (a) a general multi-port interferometer including multi-mode squeezing; (b) the reduction of the interferometer in two linear optical interferometers U and $V^\dagger$, and a set of single-mode squeezers S.

We can now substitute the mode transformation in Eq. (1.164) into $\mathcal{H}$ in order to obtain

$$\mathcal{H} = \frac{\hbar}{2} \sum_{ijk} \left[\left(A_{ij}\, \hat{a}_j + B_{ij}\, \hat{a}_j^\dagger \right) \left(B_{ik}^*\, \hat{a}_k + A_{ik}^*\, \hat{a}_k^\dagger \right) \right]$$

$$= \frac{\hbar}{2} \sum_{ijk} \left[\hat{a}_j A_{ij} B_{ik}^* \hat{a}_k + \hat{a}_j^\dagger B_{ij} A_{ik}^* \hat{a}_k^\dagger + \hat{a}_j A_{ij} A_{ik}^* \hat{a}_k^\dagger + \hat{a}_j^\dagger B_{ij} B_{ik}^* \hat{a}_k \right]$$

$$= \frac{\hbar}{2} \sum_{jk} \left[\hat{a}_j F_{jk} \hat{a}_k + 2\hat{a}_j^\dagger G_{jk} \hat{a}_k + \hat{a}_j^\dagger F_{jk}^* \hat{a}_k^\dagger \right], \qquad (1.173)$$

where

$$F_{jk} = \sum_i A_{ij} B_{ik}^* \qquad (1.174)$$

and

$$G_{jk} = \frac{1}{2} \sum_i \left(A_{ij} A_{ik}^* + B_{ij}^* B_{ik} \right) + \frac{\delta_{jk}}{2} = \sum_i A_{ij} A_{ik}^*. \qquad (1.175)$$

The last equation follows from Eq. (1.166). The Hamiltonian in Eq. (1.173) is the most general quadratic Hamiltonian. It governs the dynamics of free fields, where the mixing of modes is given by the Bogoliubov transformations. These transformations can be implemented with beam splitters and phase shifters, as well as single-mode squeezers.

At this point we should mention a possible source of confusion in the nomenclature. The Bogoliubov transformation in Eq. (1.42) is *linear*, since it transforms a mode operator into a linear combination of other mode operators. As we have seen, these linear transformations are induced by generators (Hamiltonians) that are *quadratic* in the mode operators. In particular, both (generalized) beam-splitter and squeezing transformations are generated by quadratic Hamiltonians. On the other hand, when people mention linear optics, they often refer specifically to optical elements that are described by generalized beam-splitters, and not squeezers. The reason is that beam-splitters and phase-shifters are implemented with linear dielectric media, whereas squeezing requires a *nonlinear* dielectric medium. When using the term *linear*, we must therefore always give its context, namely either the mode transformations of optical elements or the physical implementation.

1.5 Quantum states of the electromagnetic field

We have seen how the creation and annihilation operators produce and destroy photons in their respective optical modes. In this section we study how linear and quadratic functions of the mode operators can be used to define two important classes of states of the electromagnetic field. Linear Hamiltonians give rise to 'coherent states', while quadratic Hamiltonians produce 'squeezed states'.

1.5.1 Coherent states

We can define states according to the operators that produce them out of the vacuum. For example, as in Eq. (1.62), the n-photon Fock state in a normalized mode a is produced by the operator

$$|n\rangle = \frac{[\hat{a}^\dagger]^n}{\sqrt{n!}} |0\rangle .$$

(1.176)

In general, a mode can be in a superposition of Fock states, which means that a single-mode state can be written in the form

$$|\psi\rangle = \sum_{n=0}^{\infty} c_n |n\rangle ,$$

(1.177)

where the c_n are complex amplitudes[3] and $\sum_n |c_n|^2 = 1$. A particular class of states that we are interested in is created with an interaction Hamiltonian that is linear in the creation and annihilation operators, that is, it is generated by the operator

$$D(\alpha) \equiv \exp\left[\alpha\hat{a}^\dagger - \alpha^*\hat{a}\right],$$

(1.178)

where α is a complex number. We need both the creation and annihilation operator in the exponent for D to be unitary. This is called the 'Glauber displacement operator'. Furthermore, we have:

$$D^\dagger(\alpha) = D(-\alpha) = D^{-1}(\alpha) .$$

(1.179)

We make use of the relation

$$\exp(A)\exp(B) = \exp\left(A + B + \frac{1}{2}[A,B]\right) ,$$

(1.180)

which holds when $[A,B]$ commutes with both A and B. Using this relation we find that

$$D(\alpha) = \exp\left[\alpha\hat{a}^\dagger\right]\exp\left[-\alpha^*\hat{a}\right]\exp\left(-\frac{|\alpha|^2}{2}\right) .$$

(1.181)

Letting this act on the vacuum state $|0\rangle$ gives

$$D(\alpha)|0\rangle = e^{-|\alpha|^2/2}\exp\left[\alpha\hat{a}^\dagger\right]|0\rangle = e^{-|\alpha|^2/2}\sum_{n=0}^{\infty}\frac{\alpha^n[\hat{a}^\dagger]^n}{n!}|0\rangle$$

$$= e^{-|\alpha|^2/2}\sum_{n=0}^{\infty}\frac{\alpha^n}{\sqrt{n!}}|n\rangle \equiv |\alpha\rangle .$$

(1.182)

The states $|\alpha\rangle$ with $\alpha \in \mathbb{C}$ are called 'coherent states'. It is straightforward to verify that $|\alpha\rangle$ is an eigenstate of the annihilation operator $\hat{a}$:

$$\hat{a}|\alpha\rangle = \alpha|\alpha\rangle ,$$

(1.183)

---

[3] This is the most general *pure* state for a single mode. We will consider mixed states in Chapter 3.

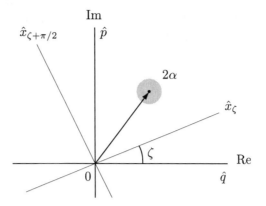

Fig. 1.7. The complex state space of coherent states. The state $|\alpha\rangle$ is represented by a point 2α. It is surrounded by a circle, which indicates the region of uncertainty. The displacement operator moves the point 2α around the plane.

and that the time evolution of the state is given by

$$U(t)|\alpha\rangle = \exp\left(-\frac{i}{\hbar}\mathcal{H}t\right)|\alpha\rangle = e^{-i\omega t\,\hat{n}}|\alpha\rangle = |\alpha\,e^{-i\omega t}\rangle\,. \tag{1.184}$$

where $\mathcal{H}$ is the free-field Hamiltonian, n the number operator ($\equiv \hat{a}^\dagger\hat{a}$) and successive Fock states are separated by an energy $\hbar\omega$. We label the state $|\alpha\rangle$ because it depends only on the complex number α. The state can therefore be represented by a point α in the complex plane (see Fig. 1.7), and the time evolution is described by a counter-clockwise circular movement of the point around the origin. In addition, the origin $|0\rangle$ is identified with the vacuum state $\alpha = 0$. The operator $D(\alpha)$ therefore *displaces* the state from zero to α. A true displacement operator would also induce the displacement on the coherent state $|\beta\rangle$:

$$D(\alpha)|\beta\rangle = |\psi\rangle = |\beta + \alpha\rangle\,. \tag{1.185}$$

Indeed, this is how the displacement operator acts on arbitrary coherent states $|\beta\rangle$. We can prove this by noting that

$$D(-\alpha)\,\hat{a}\,D(\alpha) = \hat{a} + \alpha\,, \tag{1.186}$$

and write

$$D(\alpha)\hat{a}D(-\alpha)\,|\psi\rangle = D(\alpha)\hat{a}D(-\alpha)D(\alpha)\,|\beta\rangle = \beta D(\alpha)\,|\beta\rangle = \beta\,|\psi\rangle\,, \tag{1.187}$$

and

$$D(\alpha)\hat{a}D(-\alpha)\,|\psi\rangle = \left(\hat{a} - \alpha\right)|\psi\rangle\,. \tag{1.188}$$

Equating the two expressions yields the eigenvalue equation

$$\hat{a}\,|\psi\rangle = (\beta + \alpha)\,|\psi\rangle\,, \tag{1.189}$$

which is satisfied for $|\psi\rangle = |\beta + \alpha\rangle$, as required.

We can rewrite $D(\alpha)$ as a function of the canonical position and momentum operators $\hat{q}$ and $\hat{p}$ using the relations

$$\hat{a} = \sqrt{\frac{\omega}{2\hbar}} \, \hat{q} + \frac{i}{\sqrt{2\hbar\omega}} \, \hat{p} \quad \text{and} \quad \hat{a}^{\dagger} = \sqrt{\frac{\omega}{2\hbar}} \, \hat{q} - \frac{i}{\sqrt{2\hbar\omega}} \, \hat{p}, \tag{1.190}$$

which yields

$$\begin{aligned} D(\alpha) &= \exp\left[\alpha\hat{a}^{\dagger} - \alpha^{*}\hat{a}\right] \\ &= \exp\left[i\mathrm{Im}(\alpha)\sqrt{\frac{2\omega}{\hbar}} \, \hat{q} - i\mathrm{Re}(\alpha)\sqrt{\frac{2}{\hbar\omega}} \, \hat{p}\right]. \end{aligned} \tag{1.191}$$

Along $\hat{q}$ we have the displacement of the real part of $\sqrt{2\omega/\hbar}\,\alpha$ (generated by the operator $\hat{p}$), and along $\hat{p}$ the displacement is over the imaginary part of $\sqrt{2/\hbar\omega}\,\alpha$, generated by the operator $-\hat{q}$ (see Fig. 1.7). We can therefore choose $\hat{q}$ and $\hat{p}$ as two orthogonal axes in the complex state space of the coherent states. When we rescale the position and momentum quadratures by $\sqrt{\omega}$ and $\hbar$, the dimensionless quadratures $\hat{x}_{\zeta}$ are then represented by straight lines through the origin:

$$\hat{x}_{\zeta} = \cos\zeta \, \hat{x}_{0} + i\sin\zeta \, \hat{x}_{\pi/2} \,. \tag{1.192}$$

We will explore a variation of the state space depicted in Fig. 1.7 in more detail in Chapter 8.

The expectation value of the number operator $\hat{n}$ is

$$\langle\alpha|\,\hat{n}\,|\alpha\rangle = |\alpha|^{2} \,, \tag{1.193}$$

while the variance of the number operator with respect to the coherent state $|\alpha\rangle$ is

$$(\Delta\hat{n})^{2} = \langle\hat{n}^{2}\rangle - \langle\hat{n}\rangle^{2} = |\alpha|^{4} + |\alpha|^{2} - |\alpha|^{4} = |\alpha|^{2}. \tag{1.194}$$

Similarly, the expectation value of an arbitrary quadrature is

$$\langle\alpha|\,\hat{x}_{\zeta}\,|\alpha\rangle = \frac{1}{\sqrt{2}}\left(\alpha e^{-i\zeta} + \alpha^{*}e^{i\zeta}\right), \tag{1.195}$$

and the variance is $\Delta\hat{x}_{\zeta}^{2} = 1/2$. For two orthogonal quadratures $\hat{x}_{\zeta}$ and $\hat{x}_{\zeta+\pi/2}$, the commutation relation $[\hat{x}_{\zeta},\hat{x}_{\zeta+\pi/2}] = i$ leads to the general Heisenberg uncertainty relation for the quadratures of the electromagnetic field

$$\Delta\hat{x}_{\zeta}\,\Delta\hat{x}_{\zeta+\pi/2} \geq \frac{1}{2}\left|\langle\left[\hat{x}_{\zeta},\hat{x}_{\zeta+\pi/2}\right]\rangle\right| = \frac{1}{2}\,. \tag{1.196}$$

The uncertainty relation for the position and momentum quadratures is then

$$\Delta\hat{q}\,\Delta\hat{p} \geq \frac{1}{2}\left|\langle\left[\hat{q},\hat{p}\right]\rangle\right| = \frac{\hbar}{2}\,. \tag{1.197}$$

The coherent states satisfy the equality, and are therefore minimum uncertainty states.

Finally, we note that the scalar product of two coherent states is given by

$$\langle \alpha | \beta \rangle = \exp\left[-\frac{1}{2}\left(|\alpha|^2 + |\beta|^2 - 2\alpha^* \beta \right) \right], \qquad (1.198)$$

and the completeness relation for a single discrete mode is

$$\int_{\mathbb{C}} \frac{d^2\alpha}{\pi} \, |\alpha\rangle \langle \alpha| = \hat{\mathbb{I}}. \qquad (1.199)$$

We can see immediately from Eq. (1.198) that the scalar product between two coherent states is never exactly zero; it approaches zero only when α and β become very different from each other (that is, when at least one of them becomes large). The states $|\alpha\rangle$ therefore do *not* form an orthonormal set. Moreover, the factor π^{-1} in the completeness relation indicates that the set of all coherent states is 'overcomplete'. For more details about coherent states, see Perelomov (1986).

The coherent states are a very good approximation of the state produced by lasers operating high above threshold. Due to the photon statistics in the coherent state, it is also often regarded as a 'classical' state of light, even though here we have given a quantum description of the state.

1.5.2 Squeezed states

The next class of states we consider in this section is the 'squeezed' states, which are created by quadratic Hamiltonians. As we have seen, there are two different ways to create a single-parameter squeezing operator. There is single-mode squeezing:

$$S(\xi) = \exp\left[-\xi \frac{\hat{a}^{\dagger 2}}{2} + \xi^* \frac{\hat{a}^2}{2} \right], \qquad (1.200)$$

and there is two-mode squeezing:

$$S(\xi) = \exp\left[-\xi \, \hat{a}^\dagger \hat{b}^\dagger + \xi^* \, \hat{a}\hat{b} \right]. \qquad (1.201)$$

We have used the two operators $\hat{a}$ and $\hat{b}$ to distinguish the two modes. In both cases, the squeezing operator is unitary and obeys the equations

$$S^\dagger(\xi) = S(-\xi) = S^{-1}(\xi), \qquad (1.202)$$

where the parameter ξ is a complex number. There are, however, qualitative differences between the two types of squeezing.

Single-mode squeezing

In order to find out the effect of the single-mode squeezing operator $S(\xi)$ on the vacuum state $|0\rangle$ we have to use a special operator-ordering theorem (a Baker–Campbell–Hausdorff relation). The theorem we need requires operators that obey the commutation relations $[K_-, K_+] = 2K_0$ and $[K_0, K_\pm] = \pm K_\pm$, and also satisfy $K_+^\dagger = K_-$, and $K_0^\dagger = K_0$. We can then write the ordering formula

$$e^{\kappa K_+ - \kappa^* K_-} = e^{\tau K_+} e^{-2\nu K_0} e^{-\tau^* K_-}, \tag{1.203}$$

where

$$\tau = \frac{\kappa}{|\kappa|} \tanh|\kappa| \quad \text{and} \quad \nu = \ln(\cosh|\kappa|). \tag{1.204}$$

We prove this relation in Appendix 1. It is readily verified that with the definitions

$$K_+ = \frac{\hat{a}^{\dagger 2}}{2} \quad \text{and} \quad K_- \equiv \frac{\hat{a}^2}{2}, \tag{1.205}$$

the above conditions are satisfied. In particular, for K_0 we find

$$2K_0 = [K_-, K_+] = \hat{a}^\dagger \hat{a} + \frac{1}{2}. \tag{1.206}$$

We therefore have

$$S(\xi) = \exp\left[-\tau \frac{\hat{a}^{\dagger 2}}{2}\right] \exp\left[-\nu\left(\hat{a}^\dagger \hat{a} + \frac{1}{2}\right)\right] \exp\left[\tau^* \frac{\hat{a}^2}{2}\right], \tag{1.207}$$

with $\tau = \xi/|\xi| \tanh|\xi|$ and $\nu = \ln(\cosh|\xi|)$. When we act with the squeezing operator on the vacuum, we obtain the squeezed state

$$S(\xi)|0\rangle \equiv |\xi\rangle = e^{-\nu/2} \exp\left[-\tau \frac{\hat{a}^{\dagger 2}}{2}\right] |0\rangle$$

$$= \frac{1}{\sqrt{\cosh|\xi|}} \sum_{n=0}^{\infty} \frac{\sqrt{(2n)!}}{n!} \left(\frac{-\xi}{2|\xi|}\right)^n \tanh^n|\xi| |2n\rangle. \tag{1.208}$$

The squeezed vacuum state $|\xi\rangle$ has only even photon number states in the superposition.

Exercise 1.8: Prove that $\frac{1}{2}\hat{a}^2$ and $\frac{1}{2}\hat{a}^{\dagger 2}$ obey the algebra defined by the commutation relations for $K_\pm$ and K_0.

Since $S(\xi)$ is a unitary evolution, we can ask how it evolves the creation and annihilation operators. We have seen in the previous section that

$$S(-\xi)\hat{a}S(\xi) = \hat{a}\cosh|\xi| - \hat{a}^\dagger \frac{\xi}{|\xi|}\sinh|\xi|,$$

$$S(-\xi)\hat{a}^\dagger S(\xi) = \hat{a}^\dagger \cosh|\xi| - \hat{a}\frac{\xi^*}{|\xi|}\sinh|\xi|. \tag{1.209}$$

This leads to the eigenvalue equation (with eigenvalue zero)

$$\left[\hat{a} \cosh |\xi| + \hat{a}^{\dagger} \frac{\xi}{|\xi|} \sinh |\xi| \right] |\xi\rangle = 0 . \tag{1.210}$$

Exercise 1.9: Prove Eq. (1.210).

In terms of the canonical position and momentum operators $\hat{q}$ and $\hat{p}$ we can write the squeezing operator as

$$S(\xi) = \exp\left[-\frac{i \operatorname{Im}(\xi)}{2\hbar\omega} \left(\omega^2 \hat{q}^2 - \hat{p}^2 \right) + \frac{i \operatorname{Re}(\xi)}{2\hbar} \left(\hat{q}\hat{p} + \hat{p}\hat{q} \right) \right] . \tag{1.211}$$

This expression simplifies considerably when ξ is either purely real or imaginary:

$$S_r(r) = \exp\left[\frac{ir}{2\hbar} \left(\hat{q}\hat{p} + \hat{p}\hat{q} \right) \right] \qquad \text{for } \xi = r \text{ real,} \tag{1.212}$$

and

$$S_i(r) = \exp\left[-\frac{ir}{2\hbar\omega} \left(\omega^2 \hat{q}^2 - \hat{p}^2 \right) \right] \qquad \text{for } \xi = r \text{ imaginary.} \tag{1.213}$$

The next question we ask is how the squeezing operator acts on coherent states. It is easy to show that the displacement operator and the squeezing operator do not commute, so we have to be careful how we define a squeezed coherent state. We follow the convention in the literature and apply the squeezing operator first:

$$|\alpha, \xi\rangle \equiv D(\alpha) S(\xi) |0\rangle . \tag{1.214}$$

Using Eq. (1.210) we can deduce that for the reverse order of the displacement and squeezing operator we have

$$|\alpha, \xi\rangle = S(\xi) D \left(\alpha \cosh |\xi| + \alpha^* \frac{\xi}{|\xi|} \sinh |\xi| \right) |0\rangle . \tag{1.215}$$

Using the definition of the squeezed coherent state Eq. (1.214), we can evaluate the expectation values of various important operators. First, the expectation value of the annihilation and creation operators are

$$\langle \alpha, \xi | \hat{a} | \alpha, \xi \rangle = \alpha \quad \text{and} \quad \langle \alpha, \xi | \hat{a}^{\dagger} | \alpha, \xi \rangle = \alpha^* . \tag{1.216}$$

The average photon number in a squeezed coherent state is

$$\langle \alpha, \xi | \hat{n} | \alpha, \xi \rangle = \sinh^2 |\xi| + |\alpha|^2 . \tag{1.217}$$

The expectation values of the mode operators also immediately allow us to calculate the expectation value of an arbitrary field quadrature operator:

$$\langle \alpha, \xi | \hat{x}_\zeta | \alpha, \xi \rangle = \frac{1}{\sqrt{2}} \left(\alpha e^{-i\zeta} + \alpha^* e^{i\zeta} \right) \equiv x_\zeta(\alpha). \tag{1.218}$$

Comparing this to Eq. (1.195), we see that squeezing has no effect on the expectation value of the quadrature operator. However, it *does* have an effect on the variance of the quadrature operator. To see this, we first calculate

$$\langle \alpha, \xi | \hat{x}_\zeta^2 | \alpha, \xi \rangle = \frac{1}{2} \left[e^{2|\xi|} \sin^2 \left(\zeta - \frac{\varphi}{2} \right) + e^{-2|\xi|} \cos^2 \left(\zeta - \frac{\varphi}{2} \right) \right]$$
$$+ \frac{1}{2} \left(\alpha e^{-i\zeta} + \alpha^* e^{i\zeta} \right)^2, \tag{1.219}$$

where we have used the definition $\exp(i\varphi) = \xi/|\xi|$. This leads to the variance

$$\Delta \hat{x}_\zeta^2 = \frac{1}{2} \left[e^{2|\xi|} \sin^2 \left(\zeta - \frac{\varphi}{2} \right) + e^{-2|\xi|} \cos^2 \left(\zeta - \frac{\varphi}{2} \right) \right]. \tag{1.220}$$

If we line up the quadrature and the direction of squeezing such that $\zeta = \varphi/2$, the variance can be made smaller than $1/2$, the variance for a coherent state. This can lead to an improved precision in determining the value of the quadrature. We will explore this further in Chapter 13.

Finally, we consider the overlap of the single-mode squeezed coherent state $|\alpha, \xi\rangle$ and the quadrature eigenstate $|x_\zeta\rangle$. These states have an eigenvalue equation

$$\hat{x}_\zeta |x_\zeta\rangle = x_\zeta |x_\zeta\rangle. \tag{1.221}$$

We define the wave function

$$\psi_{\alpha,\xi}(x_\zeta) = \langle x_\zeta | \alpha, \xi \rangle, \tag{1.222}$$

and we want to derive a closed functional form for it in terms of x_ζ, α, and ξ. This is not a trivial task, but we can solve this problem by converting the eigenvalue equation in Eq. (1.210) into a differential equation. To this end we first write

$$\langle x_\zeta | \left[(\hat{a} - \alpha) \cosh |\xi| + \left(\hat{a}^\dagger - \alpha^* \right) \frac{\xi}{|\xi|} \sinh |\xi| \right] | \alpha, \xi \rangle = 0, \tag{1.223}$$

and make the substitution

$$\hat{a} = \frac{e^{i\zeta}}{\sqrt{2}} \left(\hat{x}_\zeta + i\hat{x}_{\zeta+\pi/2} \right) \quad \text{and} \quad \hat{a}^\dagger = \frac{e^{-i\zeta}}{\sqrt{2}} \left(\hat{x}_\zeta - i\hat{x}_{\zeta+\pi/2} \right). \tag{1.224}$$

The next step is to recognize that the commutation relation $[\hat{x}_\zeta, \hat{x}_{\zeta+\pi/2}] = i$ implies that

$$\langle x_\zeta | \hat{x}_{\zeta+\pi/2} | \psi \rangle = -i \frac{d}{dx_\zeta} \psi(x_\zeta), \tag{1.225}$$

for any quantum state $|\psi\rangle$. Combining all of this leads to the differential equation

$$\left(A\frac{d}{dx_\zeta} + Bx_\zeta + C\right)\psi_{\alpha,\xi}(x_\zeta) = 0, \tag{1.226}$$

with

$$A = \frac{e^{i\zeta}}{\sqrt{2}}\cosh|\xi| - \frac{\xi}{|\xi|}\frac{e^{-i\zeta}}{\sqrt{2}}\sinh|\xi|,$$

$$B = \frac{e^{i\zeta}}{\sqrt{2}}\cosh|\xi| + \frac{\xi}{|\xi|}\frac{e^{-i\zeta}}{\sqrt{2}}\sinh|\xi|,$$

$$C = -\alpha\cosh|\xi| - \alpha^*\frac{\xi}{|\xi|}\sinh|\xi|. \tag{1.227}$$

Solving this differential equation gives

$$\psi_{\alpha,\xi}(x_\zeta) = \frac{e^{ip_\zeta x_\zeta}}{\sqrt[4]{2\pi\Delta\hat{x}_\zeta^2}}\exp\left(-\frac{[x_\zeta - x_\zeta(\alpha)]^2}{4\Delta\hat{x}_\zeta^2}[1 - i\sin(2\zeta - \varphi)\sinh 2|\xi|]\right), \tag{1.228}$$

up to a phase, and where we have defined p_ζ as the expectation value $\langle\hat{x}_{\zeta+\pi/2}\rangle$. In the limit when $|\xi| \to \infty$ we find that

$$\langle x_\zeta \,|\, \alpha,\xi\rangle \to \delta[x_\zeta - x_\zeta(\alpha)]. \tag{1.229}$$

Therefore, in this limit the squeezed coherent state becomes an eigenstate of the quadrature operator. However, this can never be attained physically, since it requires an infinite amount of energy:

$$E = \hbar\omega\langle\hat{n}\rangle = \hbar\omega\left(|\alpha|^2 + \sinh|\xi|\right) \to \infty. \tag{1.230}$$

The squeezed coherent states are useful approximations to quadrature eigenstates, and will be used in quantum information processing with continuous variables in Chapters 8 and 9.

Two-mode squeezing

The operator that creates two-mode squeezing is

$$S(\xi) = \exp\left[-\xi\,\hat{a}^\dagger\hat{b}^\dagger + \xi^*\,\hat{a}\hat{b}\right]. \tag{1.231}$$

When we define $K_+ = \hat{a}^\dagger\hat{b}^\dagger$ and $K_- = K_+^\dagger$, then the same commutation relations hold as for the case of single-mode squeezing:

$$[K_-, K_+] = 2K_0 \quad\text{and}\quad [K_0, K_\pm] = \pm K_\pm \quad\text{and}\quad K_0^\dagger = K_0, \tag{1.232}$$

where $2K_0 = \hat{a}^\dagger\hat{a} + \hat{b}^\dagger\hat{b} + 1$. This means that we can use the same operator-ordering formula to find the two-mode squeezed vacuum state:

$$S(\xi) = e^{-\tau K_+}\,e^{-2\nu K_0}\,e^{\tau^* K_-}, \tag{1.233}$$

with

$$\tau = \frac{\xi}{|\xi|} \tanh |\xi| \quad \text{and} \quad \nu = \ln(\cosh |\xi|) . \tag{1.234}$$

This leads to

$$S(\xi) |0\rangle = e^{-\nu} \sum_{n=0}^{\infty} (-\tau)^n \frac{K_+^n}{n!} |0\rangle$$

$$= \frac{1}{\cosh |\xi|} \sum_{n=0}^{\infty} \frac{(-\xi)^n}{|\xi|^n} \tanh^n |\xi| |n_a, n_b\rangle . \tag{1.235}$$

In other words, the two-mode squeezing operation on the vacuum creates photon *pairs*. The annihilation and creation operators transform under the two-mode squeezing operator according to

$$S(-\xi)\hat{a}S(\xi) = \hat{a} \cosh |\xi| - \hat{b}^\dagger \frac{\xi}{|\xi|} \sinh |\xi| ,$$

$$S(-\xi)\hat{a}^\dagger S(\xi) = \hat{a}^\dagger \cosh |\xi| - \hat{b} \frac{\xi}{|\xi|} \sinh |\xi| . \tag{1.236}$$

Finally, we can define two-mode squeezing on multiple modes. For example, consider

$$K_+ = \frac{1}{2} \sum_{j,k} U_{jk} \hat{a}_j^\dagger \hat{a}_k^\dagger , \tag{1.237}$$

where the creation operator for mode i is denoted $\hat{a}_i^\dagger$. We can require K_+ and $K_- = K_+^\dagger$ to obey the commutation relations in Eq. (1.232). This places a restriction on U.

Exercise 1.10: Show that the commutation relations in Eq. (1.232) are satisfied when U is unitary in Eq. (1.237).

1.6 References and further reading

In this chapter we have quantized the electromagnetic field in a non-relativistic setting in the Coulomb gauge, and constructed normalizable discrete modes that can be populated with photons. The earlier parts of the chapter relied heavily on results from classical and quantum field theory. For a detailed introduction of field Lagrangians, see for example Chapter 12 of Goldstein (1980). Canonical quantization of the vector potential is discussed in detail in Bjorken and Drell (1965). There are many books on quantum optics, and a few recent examples are Gerry and Knight (2005), Garrison and Chiao (2008), Fox (2006) and the second edition of Walls and Milburn (2008). For a clear exposition of more advanced methods in quantum optics, see Barnett and Radmore (1997). Orbital angular momentum of light is a fairly large research area in its own right, and many of the important papers in the field are collected in a volume edited by Allen *et al.* (2003). For the specific aspects of Hermite– and Laguerre–Gaussian modes we have discussed here, we refer to Enderlein and Pampaloni (2004). The theory of photon-number preserving networks was developed by Reck *et al.* (1994), and the Bloch–Messiah reduction for general linear networks is given in Braunstein (2005).

2 Quantum information processing

In general, all forms of information processing can be considered in a quantum mechanical context, including for example communication, channel capacities, and the quantum limits to information extraction. In this chapter we introduce the basic features of quantum information processing. We first present qubits as abstractions of two-level addressable quantum systems, before generalizing this concept to higher-dimensional qudit systems, and continuous variables. We introduce the stabilizer formalism towards the beginning of our discussion as a convenient way to track quantum information. In Section 2.2 we derive the no-cloning theorem, which leads us to cryptography, teleportation, and quantum repeaters. Section 2.3 describes quantum computing with qubits. We start with a brief description of the circuit model, and use it to define cluster states and the one-way model. We study the properties of cluster states and discuss error correction. In the last section we introduce quantum computing over continuous variables.

2.1 Quantum information

Classically, information is carried by 'bits', which are physical systems that have two (macroscopic) states denoted by 0 and 1.[1] These bits are described classically, which immediately raises the question of how we should extend this information-theoretic description to the quantum mechanical case. This will lead to the quantum unit of information, the quantum bit or 'qubit'.

2.1.1 Classical and quantum bits

We can define the qubit as a two-dimensional quantum system, which means that the state space of the system is a two-dimensional complex Hilbert space $\mathcal{H}$. It is customary to define a standard basis $\{|0\rangle, |1\rangle\}$, called the 'computational basis', which is used for the preparation and read-out of the quantum information. In addition, the qubits must have some level of addressability: if we want to manipulate quantum information, we need a high level of control over the qubits.

---

[1] In this book we denote the logical values of bits and qubits by a dedicated typeface in order to distinguish them from photon number states.

In matrix notation, the computational basis can be written as

$$|0\rangle = \begin{pmatrix} 1 \\ 0 \end{pmatrix} \quad \text{and} \quad |1\rangle = \begin{pmatrix} 0 \\ 1 \end{pmatrix}. \tag{2.1}$$

The basic operations on a qubit are a bit flip (X), a phase flip (Z), and a combination of a bit flip and phase flip (Y):

$$X = \begin{pmatrix} 0 & 1 \\ 1 & 0 \end{pmatrix}, \quad Y = \begin{pmatrix} 0 & -i \\ i & 0 \end{pmatrix} \quad \text{and} \quad Z = \begin{pmatrix} 1 & 0 \\ 0 & -1 \end{pmatrix}. \tag{2.2}$$

These are the 'Pauli matrices', which are traceless and have a determinant of minus one. The square of the Pauli matrices is the 2×2 identity matrix $\hat{\mathbb{I}}$. Another important property of the Pauli matrices is that they anti-commute with each other:

$$XZ = -ZX, \quad YX = -XY \quad \text{and} \quad ZY = -YZ. \tag{2.3}$$

It is worth memorizing these relations, as they will be used again and again in the description of the one-way model of quantum computation.

The eigenstates of the Pauli X matrix are

$$|+\rangle = \frac{|0\rangle + |1\rangle}{\sqrt{2}} \quad \text{and} \quad |-\rangle = \frac{|0\rangle - |1\rangle}{\sqrt{2}}, \tag{2.4}$$

such that

$$X|\pm\rangle = \pm|\pm\rangle. \tag{2.5}$$

The eigenvalues are $+1$ and -1. Similarly, the eigenstates of the Pauli Y matrix are

$$|\circlearrowleft\rangle = \frac{|0\rangle + i|1\rangle}{\sqrt{2}} \quad \text{and} \quad |\circlearrowright\rangle = \frac{|0\rangle - i|1\rangle}{\sqrt{2}}, \tag{2.6}$$

and the eigenvalue equations are

$$Y|\circlearrowleft\rangle = |\circlearrowleft\rangle \quad \text{and} \quad Y|\circlearrowright\rangle = -|\circlearrowright\rangle. \tag{2.7}$$

Finally, the eigenvalue equations for the Pauli Z matrix are

$$Z|0\rangle = |0\rangle \quad \text{and} \quad Z|1\rangle = -|1\rangle. \tag{2.8}$$

When the Pauli matrices are exponentiated, they generate a continuous family of single-qubit rotations:

$$U_X(\theta) = \exp(-i\theta X) = \cos\theta\,\hat{\mathbb{I}} - i\sin\theta\,X = \begin{pmatrix} \cos\theta & -i\sin\theta \\ -i\sin\theta & \cos\theta \end{pmatrix},$$

$$U_Y(\phi) = \exp(-i\phi Y) = \cos\phi\,\hat{\mathbb{I}} - i\sin\phi\,Y = \begin{pmatrix} \cos\phi & -\sin\phi \\ \sin\phi & \cos\phi \end{pmatrix},$$

$$U_Z(\varphi) = \exp(-i\varphi Z) = \cos\varphi\,\hat{\mathbb{I}} - i\sin\varphi\,Z = \begin{pmatrix} e^{-i\varphi} & 0 \\ 0 & e^{i\varphi} \end{pmatrix}. \tag{2.9}$$

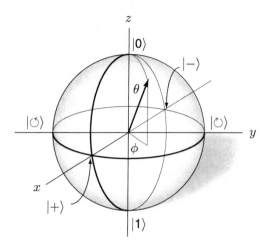

Fig. 2.1. The Bloch sphere. The x axis intersects the sphere at the eigenstates of the Pauli X matrix: $|\pm\rangle$, the y axis intersects the sphere at the eigenstates of the Y matrix: $|\circlearrowleft\rangle$, $|\circlearrowright\rangle$, and the z axis intersects the Bloch sphere at the eigenstates of the Z matrix: the computational basis $|0\rangle$ and $|1\rangle$.

When we apply arbitrary U_X, U_Y, and U_Z rotations to a qubit state, for example $|0\rangle$, we generate a qubit-state space that can be represented by a three-dimensional sphere, called the 'Poincaré' or 'Bloch sphere', shown in Fig. 2.1. Every point on the sphere corresponds to a qubit state that is unique (up to a global phase). The two eigenstates of any Pauli operator can be identified with antipodal points on the Bloch sphere, and the axis through these points is the axis of rotation associated with that Pauli operator. The three Pauli operators define three orthogonal rotation axes. For example, a rotation $U_X(\theta)$ rotates the Bloch sphere around the axis defined by $|\pm\rangle$ over an angle 2θ (because orthonormal states in the Hilbert space are antipodal in the Bloch sphere). It applies only an unobservable global phase to the states $|\pm\rangle$, and they are therefore invariant under the rotation.

Note that the construction of the Pauli matrices does not depend on the physical representation of the two-level quantum system. This is a very important property, because it allows us to consider a large variety of physical qubit systems, and treat each implementation formally as a two-level quantum system.

There are a number of important single-qubit operations that have special names: the 'Hadamard gate', the 'phase gate', and the '$\pi/8$ gate'. The Hadamard gate H is given by

$$H = \frac{1}{\sqrt{2}} \begin{pmatrix} 1 & 1 \\ 1 & -1 \end{pmatrix}, \tag{2.10}$$

with $H = H^\dagger$ and $H^2 = \hat{\mathbb{1}}$. An important relation between the Pauli matrices and the Hadamard operator is

$$HX = ZH \quad \text{or} \quad HXH = Z \quad \text{and} \quad HZH = X. \tag{2.11}$$

In addition, we have $HYH = -Y$. The Hadamard operation is the Fourier transform in two dimensions. It acts on the computational basis states as follows:

$$H\,|0\rangle = |+\rangle \quad \text{and} \quad H\,|1\rangle = |-\rangle \,. \tag{2.12}$$

The phase gate Φ is given by

$$\Phi = \begin{pmatrix} 1 & 0 \\ 0 & i \end{pmatrix}. \tag{2.13}$$

Acting on the Pauli matrices, it behaves as follows:

$$\Phi X \Phi^\dagger = Y \quad \text{and} \quad \Phi Y \Phi^\dagger = -X \quad \text{and} \quad \Phi Z \Phi^\dagger = Z\,. \tag{2.14}$$

Finally, the $\pi/8$ gate T is given by

$$T = \exp(i\pi/8) \begin{pmatrix} e^{-i\pi/8} & 0 \\ 0 & e^{i\pi/8} \end{pmatrix} = \begin{pmatrix} 1 & 0 \\ 0 & e^{i\pi/4} \end{pmatrix}, \tag{2.15}$$

which acts on the Pauli operators according to

$$TXT^\dagger = \frac{X+Y}{\sqrt{2}} \quad \text{and} \quad TYT^\dagger = \frac{Y-X}{\sqrt{2}} \quad \text{and} \quad TZT^\dagger = Z\,. \tag{2.16}$$

The $\pi/8$ gate is therefore different from the Hadamard gate and the phase gate, in that it does not generally transform Pauli operators into other Pauli operators. This will become important later, when we discuss the stabilizer formalism. Both the phase gate and the $\pi/8$ gate are special cases of rotations around the z axis in the Bloch sphere.

We can generalize the results above from two-level systems to d-level systems. To do this, observe first that we can construct the qubit Pauli operator Y from matrix multiplication of X and Z, up to a constant. To determine this constant, we require that Y is Hermitian, so that exponentiation of Y yields a unitary transformation (that is, a rotation). This means that $Y = \alpha XZ$, and $Y^\dagger = Y$. Solving for α gives $\alpha = \pm i$. The sign determines the chirality of the three axes in the Bloch sphere. We traditionally choose $\alpha = -i$, so that for cyclic permutation of the Pauli operators, the commutator $[X/2, Y/2] = iZ/2$ always has a factor $+i$. We can generalize this construction for d-level quantum systems, or 'qudits'. First, we define the X_d and Z_d operators on the computational basis states $|j\rangle$ with $j = 0, \ldots, d-1$ as

$$X_d\,|j\rangle = |(j+1) \bmod d\rangle \quad \text{and} \quad Z_d\,|j\rangle = e^{2\pi i j/d}\,|j\rangle \,. \tag{2.17}$$

Raising these operators to the power d gives the identity: $X_d^d = Z_d^d = \hat{\mathbb{1}}$. We can construct Hermitian operators $\alpha_{a,b}^{(d)} X_d^a Z_d^b$ with phase factors $\alpha_{a,b}^{(d)}$ determined up to a minus sign. It is convenient to represent all operators $X_d^a Z_d^b$ on a grid, shown in Fig. 2.2b. Fig. 2.2a depicts the special case of the qubit.

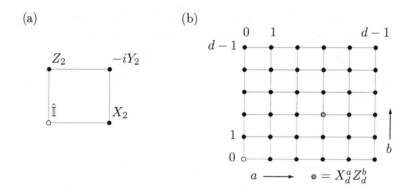

Fig. 2.2.
(a) The Pauli matrices X_2 and Z_2 can be combined via matrix multiplication to form $X_2 Z_2 = -iY_2$. The subscript '2' indicates that they act on qubits. (b) The discrete Heisenberg–Weyl matrices in d dimensions are created via repeated matrix multiplication $X_d^a Z_d^b$. The prefactor of $X_d^a Z_d^b$ is determined by Hermiticity (up to a minus sign). The open dot is the identity $\hat{\mathbb{1}}$. There are $d^2 - 1$ nontrivial operators.

The operators $X_d^a Z_d^b$ form the so-called discrete Heisenberg–Weyl matrices. There are d^2 of them, including the identity. Each operator $X_d^a Z_d^b$ generates qudit rotations around an orthogonal axis labeled by (a, b), and the qudit-state space is $(d^2 - 1)$-dimensional. For $d = 2$ we recover the familiar Bloch sphere. In the last section of this chapter we define the generalized Pauli operators $X(x)$ and $Z(p)$, which are continuous Heisenberg–Weyl operators for systems called 'qunats'.

Exercise 2.1: Show that $X_d Z_d = e^{2\pi i/d} Z_d X_d$, and calculate $[X_d^a Z_d^b, X_d^{a'} Z_d^{b'}]$. Prove that each point in Fig. 2.2b corresponds to a different operator, and that there are $d^2 - 1$ different operators.

2.1.2 Two-qubit logical operations and gates

Single-qubit operations are not sufficient to unlock all the computational power that is contained in a collection of N qubits prepared in a computational basis state $|0\rangle$ or $|1\rangle$. In other words, if we apply *only* single-qubit operations to the qubits, whatever computation we perform can be simulated efficiently[2] on a classical computer. We therefore have to include many-qubit operations (or 'gates') in our arsenal of primitive computational operations. It turns out that it is sufficient to add a *two-qubit* gate to the single-qubit gates, in order to probe the region that is no longer efficiently simulable on a classical computer. In fact, *almost any* two-qubit gate is sufficient to enter the regime of quantum computing, when added to the single-qubit gates. A set of single- and two-qubit gates that can simulate any other multi-qubit gate efficiently is called a 'universal' set of gates.

---

[2] Here, 'efficient' means that the classical computer uses resources that scale as a polynomial in N: $O(N^k)$ with k a finite number (as opposed to exponential scaling $O(e^N)$). These resources include everything, from the physical systems used, to the space and time requirements.

The Hilbert space $\mathcal{H}_2$ of two qubits is given by the tensor product of the single-qubit Hilbert space $\mathcal{H}$ with itself: $\mathcal{H}_2 = \mathcal{H} \otimes \mathcal{H}$. The computational basis for $\mathcal{H}_2$ is given by the states

$$|00\rangle, \quad |01\rangle, \quad |10\rangle, \quad \text{and} \quad |11\rangle. \tag{2.18}$$

We assume that this is the basis we use to prepare and read the state of the two qubits. However, since $\mathcal{H}_2$ is a four-dimensional complex vector space, we can construct an infinite number of bases, all connected by 4×4 unitary matrices U. One particular basis stands out in that it is a set of four 'maximally entangled' states:

$$\left|\Psi^{\pm}\right\rangle = \frac{1}{\sqrt{2}}(|01\rangle \pm |10\rangle) \quad \text{and} \quad \left|\Phi^{\pm}\right\rangle = \frac{1}{\sqrt{2}}(|00\rangle \pm |11\rangle). \tag{2.19}$$

These states are called 'Bell states'. They form the complete orthonormal 'Bell basis'.

It is impossible to obtain the Bell states from the computational basis states with (local) single-qubit operations alone. To achieve this, we need two-qubit operations. These can be given in matrix form, where the matrix is represented in the computational basis $\{|00\rangle, |01\rangle, |10\rangle, |11\rangle\}$ in that order. Here, we give two examples of two-qubit gates, namely the controlled-not gate (denoted CX here, and often written as CNOT) and the controlled-Z gate (denoted CZ, and often written as CPHASE). The CX gate applies a bit-flip X to the target qubit j when the control qubit i is in the state $|1\rangle$:

$$\text{CX}_{ij} = \begin{pmatrix} 1 & 0 & 0 & 0 \\ 0 & 1 & 0 & 0 \\ 0 & 0 & 0 & 1 \\ 0 & 0 & 1 & 0 \end{pmatrix}, \tag{2.20}$$

while the CZ gate applies a phase-flip Z to the target qubit j when the control qubit i is in state $|1\rangle$:

$$\text{CZ}_{ij} = \begin{pmatrix} 1 & 0 & 0 & 0 \\ 0 & 1 & 0 & 0 \\ 0 & 0 & 1 & 0 \\ 0 & 0 & 0 & -1 \end{pmatrix}. \tag{2.21}$$

This gate has the important properties that it is symmetric in the control and target qubits, and that it is diagonal in the computational basis. The final two-qubit gate we discuss here is the root-swap ($\sqrt{\text{SWAP}}$) gate:

$$\sqrt{\text{SWAP}}_{ij} = \frac{1}{\sqrt{2}} \begin{pmatrix} \sqrt{2} & 0 & 0 & 0 \\ 0 & 1 & i & 0 \\ 0 & i & 1 & 0 \\ 0 & 0 & 0 & \sqrt{2} \end{pmatrix}. \tag{2.22}$$

When this gate is applied twice in a row on the same qubits, it executes a swap of the two qubit states: $(\sqrt{\text{SWAP}}_{ij})^2 |\psi, \phi\rangle_{ij} = |\phi, \psi\rangle_{ij}$.

2.1.3 The stabilizer formalism

An important part of quantum information processing is the transformation of input states into output states in some prescribed way. The stabilizer formalism is a way to describe quantum states in terms of a collection of operators, called the 'stabilizer'. Relations between the operators can then be used to give a very compact description of the states and how they evolve. The stabilizer formalism was developed by Daniel Gottesman, and is traditionally used to describe quantum error correction. It has, however, many more uses. In particular, we will use it to describe cluster states later in this chapter.

Suppose that $|\psi\rangle$ is the quantum state of a set of qubits. A 'stabilizer $\mathfrak{S}$' is a set of operators S_i that leave the state invariant:

$$S_i |\psi\rangle = |\psi\rangle \ . \tag{2.23}$$

When two operators S_i and S_j are members of a stabilizer, then their products $S_i S_j$ and $S_j S_i$ are again members of the stabilizer, and $[S_i, S_j] = 0$. The identity operator is a (trivial) member of the stabilizer, and so is the inverse S_i^{-1} of every member S_i:

$$S_i^{-1} |\psi\rangle = S_i^{-1} (S_i |\psi\rangle) = \left(S_i^{-1} S_i \right) |\psi\rangle = |\psi\rangle \ . \tag{2.24}$$

The stabilizer $\mathfrak{S}$ therefore forms a finite 'abelian group', in which the group elements all commute with each other. The minimal subset of the stabilizer that can be used to retrieve the complete group is called the set of stabilizer 'generators'. This subset is generally not unique.

Let's make this construction more concrete with a few examples: consider a single qubit in the state $|0\rangle$. One element of the stabilizer is the Pauli Z operator, since $Z |0\rangle = |0\rangle$. Z is also the generator of the stabilizer, since $Z^2 = \hat{\mathbb{1}}$. The stabilizer for this state is therefore the group $\mathfrak{S} = \{\hat{\mathbb{1}}, Z\}$. Similarly, the stabilizer generator for the qubit state $|-\rangle$ is $-X$, since $(-X) |-\rangle = |-\rangle$. Note that the minus sign is important, since the state stabilized by $+X$ is $|+\rangle$.

For a less trivial example, consider the singlet state

$$\left|\Psi^-\right\rangle = \frac{|01\rangle - |10\rangle}{\sqrt{2}} \ . \tag{2.25}$$

What is the stabilizer of this state? First, we can apply a bit flip on both qubits and obtain the singlet state up to an overall minus sign. One generator is therefore $-X_1 X_2 \equiv -X_1 \otimes X_2$. Second, a phase flip can be applied to both qubits and incurs again an overall minus sign. The other generator is therefore $-Z_1 Z_2$. The entire stabilizer for the singlet state is given by $\mathfrak{S} = \{\hat{\mathbb{1}}, -X_1 X_2, -Y_1 Y_2, -Z_1 Z_2\}$.

The minus signs are important here; unlike in a state vector description they do not correspond to overall phase factors that can be ignored. To illustrate this further, let us construct the state corresponding to the generators $S_1 = X_1 X_2$ and $S_2 = Z_1 Z_2$. Starting with the requirement that $X_1 X_2 |\psi\rangle = |\psi\rangle$, with the most general expression for $|\psi\rangle$:

$$|\psi\rangle = c_{00} |00\rangle + c_{01} |01\rangle + c_{10} |10\rangle + c_{11} |11\rangle \ , \tag{2.26}$$

we immediately deduce that

$$c_{00} = c_{11} \equiv \frac{c_{\text{even}}}{\sqrt{2}} \quad \text{and} \quad c_{01} = c_{10} \equiv \frac{c_{\text{odd}}}{\sqrt{2}} . \tag{2.27}$$

In other words, $|\psi\rangle$ can be written as

$$|\psi\rangle = c_{\text{even}} |\Phi^+\rangle + c_{\text{odd}} |\Psi^+\rangle . \tag{2.28}$$

Next, we calculate

$$S_2 |\psi\rangle = c_{\text{even}} Z_1 Z_2 |\Phi^+\rangle + c_{\text{odd}} Z_1 Z_2 |\Psi^+\rangle = c_{\text{even}} |\Phi^+\rangle - c_{\text{odd}} |\Psi^+\rangle . \tag{2.29}$$

It is a requirement that $Z_1 Z_2 |\psi\rangle = |\psi\rangle$ and so $c_{\text{odd}} = -c_{\text{odd}} = 0$. The state stabilized by $X_1 X_2$ and $Z_1 Z_2$ is therefore $|\Phi^+\rangle$, and not the singlet state $|\Psi^-\rangle$.

Exercise 2.2: Show that $X_1 X_2$ and $Z_1 Z_2$ commute.

This example also illustrates another interesting property of the stabilizer generators. When there are not enough generators to uniquely determine the state, a d-dimensional subspace (with $d > 1$) of the Hilbert space is carved out. In Eq. (2.28) the subspace is two-dimensional, and spanned by the Bell states $|\Phi^+\rangle$ and $|\Psi^+\rangle$. With every generator S_i we can associate a subspace $\mathcal{V}_i$ of the Hilbert space of the total system. A multi-qubit state therefore always lies in the intersection of the subspaces associated with its stabilizer generators. In the example above the subspaces of S_1 and S_2 are $\mathcal{V}_1$ and $\mathcal{V}_2$. We already determined the subspace $\mathcal{V}_1$ in Eq. (2.28), and $\mathcal{V}_2$ is spanned by $|\Phi^+\rangle$ and $|\Phi^-\rangle$. The intersection $\mathcal{V}_1 \cap \mathcal{V}_2$ is the one-dimensional subspace associated with the state $|\Phi^+\rangle$. To define an N-qubit state uniquely, we need N stabilizer generators on N linearly independent (but not orthogonal) subspaces.

States are completely determined by the generators of their stabilizer group. All the other group members of $\mathfrak{G}$ can be constructed from the generators (by definition), and therefore do not add any additional restrictions the state must meet. In other words, the subspaces associated with the stabilizer generators form a set of linearly independent subspaces that is complete in the subspace spanned by $\mathfrak{G}$ (but not the total Hilbert space $\mathcal{H}$). This is a very desirable property of the stabilizer, because the number of elements in $\mathfrak{G}$, denoted by $|\mathfrak{G}|$, can be quite large. The number of generators, however, scales with $\log_2 |\mathfrak{G}|$ or less. To see this, imagine an operator $T \notin \mathfrak{G}$ and $S_i \in \mathfrak{G}$. This implies that $S_i T \notin \mathfrak{G}$ $\forall S_i$, for if it was, then we would have $T = S_i^{-1} S_i T \in \mathfrak{G}$, which is false by assumption. However, when we add T to the set of generators of $\mathfrak{G}$, then $S_i T \in \mathfrak{G}$ for all S_i. We therefore at least double the size of the group whenever we add a new generator. This means that the number of (independent) generators of a group can be at most $\log_2 |\mathfrak{G}|$. Describing a state in terms of stabilizer generators can therefore be very efficient.

So far we considered examples where the S_i are tensor products of Pauli operators. This means that the stabilizer $\mathfrak{G}$ was a subgroup of the 'Pauli group' for one and two qubits. The Pauli group for a single qubit $\mathfrak{P}_1$ is given by the Cartesian product of the numbers ± 1, $\pm i$, and the Pauli operators (including the identity):

$$\mathfrak{P}_1 = \{\pm 1, \pm i\} \times \{\hat{\mathbb{I}}, X, Y, Z\}$$

$$= \{\pm\hat{\mathbb{I}}, \pm i\,\hat{\mathbb{I}}, \pm X, \pm iX, \pm Y, \pm iY, \pm Z, \pm iZ\}. \tag{2.30}$$

Sometimes the Pauli group is called the Heisenberg–Weyl group. We can also define the Pauli group for N qubits $\mathfrak{P}_N = \mathfrak{P}_1^{\otimes N}$. In the above example of the stabilizer of the singlet state, the generators $-X_1 X_2$ and $-Z_1 Z_2$ are elements of $\mathfrak{P}_2$. Moreover, $\mathfrak{S} \subset \mathfrak{P}_2$ for the singlet state. No stabilizer ever includes $-\hat{\mathbb{I}}$ or $\pm i\,\hat{\mathbb{I}}$.

How can we describe other states that are not so obviously stabilized by tensor products of Pauli operators? The equivalent question is how to construct the stabilizer of general states that are unitary transformations of states with a stabilizer $\mathfrak{S} \subset \mathfrak{P}_N$. Let $|\psi\rangle$ be stabilized by $\mathfrak{S} \subset \mathfrak{P}_N$ with generators S_i. A state $U |\psi\rangle$ then has stabilizer generators $\widetilde{S}_i = U S_i U^\dagger$:

$$U |\psi\rangle = U S_i |\psi\rangle = U S_i U^\dagger U |\psi\rangle \equiv \widetilde{S}_i U |\psi\rangle. \tag{2.31}$$

The group $\widetilde{\mathfrak{S}} = \{\widetilde{S}_i\}$ is the stabilizer of the state $U |\psi\rangle$.

An important question is: which set of transformations U yields new generators $\widetilde{S}_i$ such that $\widetilde{\mathfrak{S}}$ is a subgroup of the Pauli group, $\widetilde{\mathfrak{S}} \subset \mathfrak{P}_N$? By construction, these unitary operations transform $|\psi\rangle$ into another state $|\psi'\rangle$ that is stabilized by (possibly different) Pauli operators. We answer this by considering another, more general question: let P_i, $P_j \in \mathfrak{P}_N$ be elements of the Pauli group. What is the structure of the set of unitary transformations $\{U_k\}$ that satisfy the relation

$$U_k P_i U_k^\dagger = P_j, \tag{2.32}$$

for all $P_i \in \mathfrak{P}_N$? The answer is that the U_k form again a group under multiplication. To see this, first, suppose that U_1 and U_2 both satisfy Eq. (2.32). Then we can find a P_j and a P_k such that

$$(U_2 U_1) P_i \left(U_1^\dagger U_2^\dagger \right) = U_2 \left(U_1 P_i U_1^\dagger \right) U_2^\dagger = U_2 P_j U_2^\dagger = P_k. \tag{2.33}$$

In other words, the product $U_2 U_1$ is also an element of the group. A similar argument shows that $U_1 U_2$ is an element of the group, but it is generally *not* the same as $U_2 U_1$, and the group is therefore not abelian. Second, the inverse of U_k, denoted by $U_k^\dagger$, is an element of the group:

$$P_i = (U_k^\dagger U_k) P_i (U_k^\dagger U_k) = U_k^\dagger \left(U_k P_i U_k^\dagger \right) U_k = U_k^\dagger P_j U_k. \tag{2.34}$$

Finally, the unit element is in the group: $\hat{\mathbb{I}} P_i \hat{\mathbb{I}} = P_i$ for all $P_i \in \mathfrak{P}_N$. The name of this group is the 'Clifford group', denoted by the symbol $\mathfrak{C}_N$. It is again a finite group, with a particularly compact set of generators, consisting of the Hadamard operator H_i and the phase gate Φ_i on qubit i, and the cz gate on two arbitrary qubits i and j. Any $U \in \mathfrak{C}_N$ on N qubits can be constructed from Hadamards, phase gates, and cz gates in at most $O(N^2)$ steps. The proof of this is not too difficult, but it is a bit lengthy. It is given as an exercise in Nielsen and Chuang (2000). Note that the $\pi/8$ gate given in Eq. (2.15) is not part of the Clifford group.

2.2 Quantum communication

In this section we explore various aspects of quantum communication. We first present the no-cloning theorem and the famous BB84 and EKERT91 protocols for quantum cryptography, and then we give the procedure for quantum teleportation. We conclude this section with a discussion about quantum repeaters.

2.2.1 The no-cloning theorem and cryptography

An important operation in classical information processing is copying. In quantum mechanics, on the other hand, it is not possible to copy an arbitrary qubit state $|\psi\rangle = \alpha |0\rangle + \beta |1\rangle$. The proof of this is simple. Imagine first that a copying operation does exist and is given by the unitary operation U. It acts on the state $|\psi\rangle_1$ and on a second qubit in some initial state $|i\rangle_2$. The output of the copying procedure should be $|\psi\rangle_1 |\psi\rangle_2$. Since $|0\rangle$ and $|1\rangle$ are legitimate qubit states, the unitary operation must yield

$$U |0\rangle_1 |i\rangle_2 = |0\rangle_1 |0\rangle_2 \quad \text{and} \quad U |1\rangle_1 |i\rangle_2 = |1\rangle_1 |1\rangle_2 . \tag{2.35}$$

However, due to the linearity of quantum mechanics this implies

$$U |\psi\rangle_1 |i\rangle_2 = \alpha \, U |0\rangle_1 |i\rangle_2 + \beta \, U |1\rangle_1 |i\rangle_2 = \alpha |0\rangle_1 |0\rangle_2 + \beta |1\rangle_1 |1\rangle_2$$
$$\neq |\psi\rangle_1 |\psi\rangle_2 . \tag{2.36}$$

We therefore cannot make a perfect copy of an unknown qubit. Any copying procedure *must* introduce some form of noise in the qubits. This result has become known as the 'no-cloning theorem', and it leads to perfectly secure key distribution for cryptography.

The objective of most quantum cryptography protocols is to establish a random bit string between two parties, traditionally known as Alice and Bob. The random bit string is called a 'key', and Alice and Bob can use it for secure communication by adding the key to the message in binary (modulo 2). If the key is at least as long as the message, the encrypted message is now completely randomized. However, decoding is straightforwardly performed by performing the same modulo 2 addition to the key and the encrypted message. Therefore, if Alice and Bob share a random key that is unknown to anybody else, they can engage in secure communication. The earliest quantum protocol for key distribution was invented in 1984 by Charles Bennett and Gilles Brassard, and has become known as BB84.

In the BB84 protocol, Alice sends a sequence of qubits to Bob. Alice prepares each qubit randomly in one of four states; $|0\rangle$, $|1\rangle$, $|+\rangle$, or $|-\rangle$. Bob measures the qubits in the Z basis $\{|0\rangle, |1\rangle\}$ (the computational basis), or in the X basis $\{|+\rangle, |-\rangle\}$. He varies his measurement basis randomly. After Alice has sent a sufficiently long string of qubits, she publicly announces the *basis* for each qubit that she sent (but not the state itself). This is a classical bit string. Bob responds by publicly disclosing which of the qubits he measured in *the same basis* as Alice. The (undisclosed) outcomes of Bob's measurements of these qubits must be correlated perfectly with Alice's input state, and these form a random bit string shared by Alice and Bob. In turn, they can use this string as a key for cryptography.

In principle, there may be an eavesdropper on the line (called Eve), who can intercept the qubits sent from Alice to Bob. She can do whatever is permitted by the laws of quantum mechanics in order to gain (partial) information about the random bit string, and this could render the protocol insecure. However, due to the no-cloning theorem, she cannot make a perfect copy of the qubits. In fact, whenever she gains *any* information about the qubits, she necessarily disrupts the state of the qubits sent on to Bob, and as a consequence the correlation between the bit string held by Alice and Bob will deteriorate. Alice and Bob therefore publicly compare part of the random bit string to verify that the correlation is perfect. If they detect an eavesdropper, they abandon this communication channel and try a different one. An outline of the security proof of BB84 is given in Section 5.4. Note that this protocol does not protect against 'denial of service' attacks, nor does it offer any guarantee that Alice is talking to Bob, as opposed to someone posing as Bob.

Exercise 2.3: Eve measures randomly in the X or Z basis and sends the measured qubits on to Bob. Show that the error rate in the verification procedure is 25%.

Another protocol for random key distribution is due to Ekert (1991), and uses entanglement to protect against eavesdroppers. Alice prepares sets of two qubits in a maximally entangled state, for example the singlet state

$$|\Psi^-\rangle = \frac{|0\rangle_1 |1\rangle_2 - |1\rangle_1 |0\rangle_2}{\sqrt{2}} = \frac{|+\rangle_1 |-\rangle_2 - |-\rangle_1 |+\rangle_2}{\sqrt{2}}, \tag{2.37}$$

and sends the second qubit in each singlet state to Bob. Next, they both measure their qubits randomly in the X or the Z basis and publicly compare bases. From Eq. (2.37) it is clear that they should find a perfect anti-correlation when they measure in the same basis. Eavesdropper detection proceeds in very much the same way as in BB84: any information extraction that Eve attempts will disrupt the state $|\Psi^-\rangle$ and will show up in the subsequent verification of the random key.

2.2.2 Teleportation

One of the most striking protocols in quantum information processing is quantum teleportation, in which a quantum state is transferred from one system to another. Teleportation was invented by Bennett *et al.* in 1993, and it was demonstrated experimentally with light in 1997 and 1998, and with massive particles (ions in traps) in 2004. Here, we describe the protocol for qubits. Suppose that one communicating party, Alice, has a qubit in the state $|\psi\rangle = \alpha |0\rangle + \beta |1\rangle$ with $|\alpha|^2 + |\beta|^2 = 1$. This state may be unknown to her, meaning that she has no information about α and β. In addition to the qubit she wishes to teleport, she is given one part of an entangled system of two qubits. The other part of the entanglement is held by Bob. The shared entangled state is the singlet state[3]

$$\left|\Psi^-\right\rangle = \frac{|0\rangle_A |1\rangle_B - |1\rangle_A |0\rangle_B}{\sqrt{2}}, \tag{2.38}$$

where the subscripts A and B indicate the parts of the system held by Alice and Bob. The total three-qubit system can then be written as

$$\begin{aligned}
|\psi\rangle_A \left|\Psi^-\right\rangle_{AB} &= (\alpha |0\rangle_A + \beta |1\rangle_A) \otimes \frac{|0\rangle_A |1\rangle_B - |1\rangle_A |0\rangle_B}{\sqrt{2}} \\
&= \frac{\alpha |00\rangle_A |1\rangle_B - \alpha |01\rangle_A |0\rangle_B + \beta |10\rangle_A |1\rangle_B - \beta |11\rangle_A |0\rangle_B}{\sqrt{2}}. \tag{2.39}
\end{aligned}$$

Alice next makes a measurement in the Bell basis. In other words, she makes a specific *joint measurement* on the two qubits that she holds. To understand the effect of this, we first write the computational basis states as a superposition of Bell states:

$$|00\rangle = \frac{\left|\Phi^+\right\rangle + \left|\Phi^-\right\rangle}{\sqrt{2}} \quad \text{and} \quad |11\rangle = \frac{\left|\Phi^+\right\rangle - \left|\Phi^-\right\rangle}{\sqrt{2}},$$

$$|01\rangle = \frac{\left|\Psi^+\right\rangle + \left|\Psi^-\right\rangle}{\sqrt{2}} \quad \text{and} \quad |10\rangle = \frac{\left|\Psi^+\right\rangle - \left|\Psi^-\right\rangle}{\sqrt{2}}. \tag{2.40}$$

Now substitute these equations into Eq. (2.39) and collect the terms with the same Bell states:

$$\begin{aligned}
|\psi\rangle_A \left|\Psi^-\right\rangle_{AB} &= \frac{\left|\Phi^+\right\rangle_A}{2} (\alpha |1\rangle - \beta |0\rangle) + \frac{\left|\Phi^-\right\rangle_A}{2} (\alpha |1\rangle + \beta |0\rangle) \\
&\quad - \frac{\left|\Psi^+\right\rangle_A}{2} (\alpha |0\rangle - \beta |1\rangle) - \frac{\left|\Psi^-\right\rangle_A}{2} (\alpha |0\rangle + \beta |1\rangle). \tag{2.41}
\end{aligned}$$

Alice's measurement can give all four Bell states with equal probability, regardless of the values of α and β. It therefore reveals no information about the original state $|\psi\rangle$. Moreover, when Alice obtains her measurement result, she can tell Bob exactly how to retrieve the original qubit state using only the communication of two classical bits: when the outcome is $\left|\Phi^+\right\rangle$, Alice needs to tell Bob to perform both a Z operation and an X operation; when the outcome is $\left|\Phi^-\right\rangle$, Bob needs to perform only an X operation; for $\left|\Psi^+\right\rangle$ the corrective operation is the Z operation, and finally, when Alice's measurement outcome indicates $\left|\Psi^-\right\rangle$, she needs to tell Bob to do nothing. After the corrective operation, Bob's qubit is in the state $|\psi\rangle$.

This is a remarkable protocol. We have shown that it is possible in quantum mechanics to transfer the state of a quantum system to another quantum system perfectly, without disturbing it and without gaining any information about the state. Neither the entanglement channel, nor the classical communication of two classical bits from Alice to Bob, tells us anything about the state $|\psi\rangle$. One may object that this violates the no-signalling principle

---

[3] The singlet state is not essential; any maximally entangled Bell state will do.

(no information can travel faster than light), and the no-cloning theorem (we cannot make a perfect identical copy of an unknown quantum state). However, on closer inspection we realize that neither principle is violated: first, the no-cloning theorem is satisfied, because Alice destroys her copy of $|\psi\rangle$ by performing the Bell measurement. Second, the no-signalling principle is satisfied because Bob needs to know the corrective operations in order to make use of the quantum state $|\psi\rangle$. The operations occur with equal probability (they are related to Alice's measurement outcome), and applying these operators with equal probability completely randomizes the state $|\psi\rangle$. In other words, without the corrective operations, Bob does not know how to interpret the results of any measurements that he may perform on his qubit. He has to wait for Alice's classical message, which cannot travel faster than the speed of light.

When Alice and Bob are located at approximately the same position we can simplify the teleportation protocol considerably. Suppose again that Alice holds the unknown quantum state $|\psi\rangle = \alpha |0\rangle + \beta |1\rangle$, and Bob holds a second qubit prepared in the state $|0\rangle$. Since Alice and Bob are no longer spatially separated, we can apply a two-qubit gate such as the cx gate. This will give the following state:

$$|\psi\rangle_A |0\rangle_B \; \rightarrow_{cx} \; \alpha |00\rangle_{AB} + \beta |11\rangle_{AB} \;. \tag{2.42}$$

Next, Alice applies a Hadamard operation to her qubit such that the total state becomes

$$\alpha |+, 0\rangle + \beta |-, 1\rangle = |0\rangle_A (\alpha |0\rangle + \beta |1\rangle)_B + |1\rangle_A (\alpha |0\rangle - \beta |1\rangle)_B \;. \tag{2.43}$$

A measurement of Alice's qubit in the computational basis then gives the state $\alpha |0\rangle \pm \beta |1\rangle$ for Bob's qubit, where the sign is determined by the measurement outcome $m = 0, 1$. Bob can retrieve the original state by applying a corrective Z operation depending on the measurement result. We can write this operation as Z^m.

Note that in this case we use fewer qubits (two, rather than three), and Alice has to communicate only one classical bit to Bob, rather than two in the original teleportation protocol. Later in this chapter we will see how this two-qubit 'local' teleportation protocol can be used to translate between the standard circuit model of quantum computation and the one-way model using cluster states.

2.2.3 Quantum repeaters

When we want to implement any quantum communication protocol over a considerable distance, we have to pay attention to the fact that the physical system that carries the quantum information may be degraded or even lost in the process. For example, when a single photon carrying a qubit is sent through an optical fibre, the small but finite absorption coefficient of the fibre will cause a considerable photon loss at long distances. When the probability of photon transmission over a fibre of length l_0 is $1/e$, the probability that a photon travels through a fibre of length l without being absorbed is $\exp(-l/l_0)$. This exponential attenuation means that a different approach is required to allow quantum information to travel over distances much longer than l_0.

Classically, this problem is solved by adding repeater stations to a communication line. First, the signal is encoded such that it can be faithfully reconstructed after a certain amount of noise and loss. Before the signal becomes too degraded for noiseless decoding, a repeater reads the information in the signal and re-encodes it in a new, strong signal. This is then sent to the next repeater station. In quantum mechanics, however, the no-cloning theorem forbids this type of signal amplification. A quantum repeater can nevertheless be constructed, using 'entanglement swapping' and 'entanglement purification'.

We do not yet have the language to describe the degradation of the state of a qubit, so we will describe the quantum repeater strictly in terms of qubit loss. This simplified version will allow us to understand the essence of the mechanism. The repeater protocol begins with entanglement swapping, which is a variation on quantum teleportation. However, it differs from teleportation in that the unknown input qubit is itself part of an entangled qubit pair. Suppose parties A, B share a singlet state, and B and C share another singlet state. The total state is then

$$
\begin{aligned}
|\Psi^-\rangle_{AB} |\Psi^-\rangle_{B'C} = \frac{1}{2} & \left(|0101\rangle - |0110\rangle - |1001\rangle + |1010\rangle \right)_{ABB'C} \\
= \frac{1}{2} & \left(|\Psi^+\rangle_{BB'} |\Psi^+\rangle_{AC} - |\Psi^-\rangle_{BB'} |\Psi^-\rangle_{AC} \right. \\
& \left. - |\Phi^+\rangle_{BB'} |\Phi^+\rangle_{AC} + |\Phi^-\rangle_{BB'} |\Phi^-\rangle_{AC} \right).
\end{aligned}
\tag{2.44}
$$

A Bell measurement on systems B and B' will leave parties A and C with an entangled state (provided the measurement outcome is communicated to A and C), even though their systems never interacted.

Assume that A and B are a distance l apart, and B and C are also a distance l apart. To gain the full benefit of the repeater, we take A and C to be a distance $2l$ apart. The probability for the successful transmission of a qubit from A to B and from C to B is given by $p = \exp(-l/l_0)$. That means that on average at least p^{-1} entangled qubit pairs must be shared between A and B, and the same number of pairs must be shared between B and C. Party B next determines which qubits have survived the transmission *without* collapsing the qubit state. In other words, B checks whether the system carrying the qubit is there, without measuring the qubit value. Once B has identified one qubit sent by A and one qubit sent by C, he performs the Bell measurement and tells A and C both the measurement outcome and the labels of the qubits he measured. The parties A and C now share an entangled state.

The physical cost of this procedure is $2p^{-1}$ singlet states. By contrast, if A had sent entangled qubits directly to C, the transmission probability would have been $\exp(-2l/l_0) = p^2$, and she would have had to send at least p^{-2} qubits. This procedure can be repeated N times to stretch a distance Nl. The cost in entangled qubit pairs with this procedure is Np^{-1}, whereas the cost of direct transmission is p^N. This is an exponential speed-up. A note of warning: when N is very large, the probability of losing all qubits in one leg of the protocol becomes appreciable. The number of attempts to share entangled qubits must therefore also grow. If the number of shared qubits in one leg is n, and the total failure probability of the protocol is smaller than ϵ, we require

$$
N(1-p)^n < \epsilon.
\tag{2.45}
$$

For fixed $p = \exp(-l/l_0)$, ϵ, and N, the number of pairs n must satisfy:

$$n > \frac{\log \epsilon - \log N}{\log(1 - p)}. \tag{2.46}$$

Therefore n grows only logarithmically in N, and the exponential gain of the quantum repeater is preserved: $n \simeq N \log N p^{-1}$ as opposed to $n \simeq p^{-N}$.

When the qubits are degraded during the transmission over distance l, the multiple entangled pairs are purified before they engage in the entanglement swapping protocol. This increases the resource requirement, but it does not alter the overall scaling of the resource requirements. We will revisit the quantum repeater and entanglement purification in Chapter 10.

2.3 Quantum computation with qubits

Quantum computation is a generic name for a class of operations on quantum systems that start in a known input state, and evolve into the output state while implementing a certain computational algorithm. Measuring the output state then gives the correct outcome of the computation with a probability that is larger than some fixed bound (this probabilistic aspect makes quantum computing useful only when the answer can be verified efficiently on a classical computer). For some algorithms, such as factoring, quantum computing offers an exponential speed-up over the best known classical algorithms. Specific architectures for quantum computers differ in the way they evolve the input state into the output state. For our purposes we distinguish between a unitary evolution, where the output state is a unitary transformation of the input state, and a measurement-based evolution, in which adaptive measurements are used to drive the computation along. In practice, the distinction between them is somewhat blurred, since measurements are an essential component of quantum error correction, which is needed in both models. Examples of the unitary evolution type are the circuit model and adiabatic quantum computing, while examples of measurement-based quantum computing are the one-way model and the Gottesman–Chuang teleportation model. The quantum systems carrying the computation can be (among others) qubits, qudits, or continuous variables. Here, we mainly consider quantum computation using qubits and continuous variables in the circuit model and in the one-way model of quantum computing. We first briefly review the circuit model for qubits, and subsequently translate this into the one-way model. This leads directly to the Gottesman–Knill theorem and error correction. In the next section, we introduce quantum computing using continuous variables.

2.3.1 The circuit model

As stated above, the circuit model essentially describes a unitary transformation to a collection of input systems, here taken to be qubits. Any unitary transformation on N qubits can be decomposed into single-qubit operations and a number of well-chosen, identical, two-qubit gates. Often the two-qubit gate is taken to be the CX, but the CZ or the $\sqrt{\text{SWAP}}$ is equally suitable. To keep track of (large) quantum computations, it is convenient to write them as a collection of their constituent single-qubit and two-qubit gates. Every qubit will undergo a subset of these operations, and their input state will be transformed into the output state. The qubits themselves are the carriers of the information, and typically they are taken to be *distinguishable* objects. This means that we can associate a 'world line' (i.e. a trajectory in spacetime) with every qubit, drawn from left to right, with the operations, or gates, as symbols on the line. For example, here is a single-qubit gate, transforming $|\psi_{\text{in}}\rangle$ to $|\psi_{\text{out}}\rangle = U |\psi_{\text{in}}\rangle$:

$$|\psi_{\text{in}}\rangle \quad \boxed{U} \quad |\psi_{\text{out}}\rangle$$

U can be any single-qubit gate, for example a Hadamard H, one of the Pauli operators X, Y, or Z, or any other rotation in the Bloch sphere.

Apart from single-qubit operations, a quantum computation needs two-qubit gates. We have already encountered the CX, the CZ, and the $\sqrt{\text{SWAP}}$. In circuit language, these gates are represented as follows:

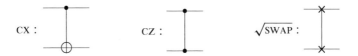

The CX and CZ gates are related via Hadamard gates:

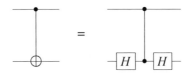

This is a very important identity that becomes indispensable when we construct cluster states in the next section.

There are also three-qubit gates, the most famous being the Fredkin gate and the Toffoli gate.

As an example of how quantum circuits work, consider the quantum teleportation protocols that we studied in the previous section. The top line in the circuit below is the qubit in the unknown state $|\psi\rangle$, held by Alice. The lower two qubits are prepared in state $|00\rangle$, after which a Hadamard and a CX transform the state into the maximally entangled Bell state $|\Psi^+\rangle$. This combination of a Hadamard and a CX can be used to transform between the computational basis and the Bell basis. We use this transformation again on Alice's two qubits to implement the Bell measurement.

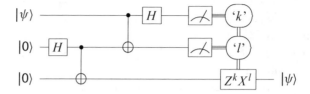

Depending on the measurement outcome (k, l), Bob performs a single-qubit operation $Z^k X^l$ on his qubit, after which he retrieves the original qubit state $|\psi\rangle$. The ancilla state $|00\rangle$ can be chosen differently (in particular $|11\rangle$ is a common choice), which would change the corrective single-qubit gate for Bob.

Exercise 2.4: Give the circuit diagram of the teleportation protocol in which the two measurements are replaced by one- and two-qubit unitary operators. What are the three output states?

Similarly, we can give the circuit for the two-qubit local teleportation protocol:

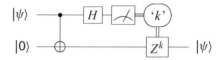

Next, we use this two-qubit protocol to construct the one-way model of quantum computation, which turns out to be particularly useful for optical implementations of quantum computers.

2.3.2 Cluster states and the one-way model

The one-way model was introduced by Raussendorf and Briegel in 2001, and forms an alternative approach to quantum computing. The heart of this architecture is to create a large entangled resource state,[4] called a 'cluster state'. The computation then proceeds as a series of single-qubit measurements. In this section, we first derive the one-way model from the circuit model following Nielsen (2006), and in the process we will find a constructive definition of the cluster state.

Consider the following circuit diagram of local single-qubit teleportation:

$$|\psi\rangle \quad \bullet \quad \boxed{H} \quad \measuredangle \quad \text{`}m\text{'}$$
$$|0\rangle \quad \oplus \quad\quad\quad\quad\quad Z^m |\psi\rangle$$

The measurement is in the computational basis, and the outcome $m = 0, 1$ determines whether we apply a Pauli Z operation to the teleported qubit. The next step is to translate

---

[4] The philosophy behind this is somewhat similar to the Ekert protocol for cryptography, where entanglement is created and shared beforehand, and subsequently used as a resource.

the CX gate into the CZ gate. This procedure incurs two Hadamard gates, which are absorbed into the state of the ancilla ($|0\rangle \rightarrow |+\rangle$) and the teleported qubit:

$$
\begin{array}{l}
|\psi\rangle \quad \text{———} H \text{———} \measuredangle \text{——— 'm'} \\
H|0\rangle = |+\rangle \text{————} HZ^m |\psi\rangle = X^m H |\psi\rangle
\end{array}
$$

A single-qubit rotation around the z axis over an angle α is written as $U_Z(\alpha) = \exp(i\alpha Z/2)$. We can apply this rotation to the input qubit in state $|\psi\rangle$, which means that after the local teleportation protocol the output is in the state $U_Z(\alpha)|\psi\rangle$. In a circuit diagram this becomes

$$
\begin{array}{l}
|\psi\rangle \text{—} \boxed{U_Z(\alpha)} \text{—} \boxed{H} \text{—} \measuredangle \text{— 'm'} \\
|+\rangle \text{————} X^m H U_Z(\alpha)|\psi\rangle
\end{array}
$$

Since the single-qubit gate commutes with the CZ gate, we can write the circuit as

$$
\begin{array}{l}
|\psi\rangle \text{—} \boxed{H U_Z(\alpha)} \text{—} \measuredangle \text{— 'm'} \\
|+\rangle \text{————} X^m H U_Z(\alpha)|\psi\rangle
\end{array}
$$

We can reinterpret this diagram as an entangled state $|\psi\rangle = U_{CZ}|\psi, +\rangle$, followed by a single-qubit measurement on the first qubit that performs the single-qubit gate $H U_Z(\alpha)$ on the state $|\psi\rangle$. The measured observable is $M(\alpha) = U_Z(\alpha) X U_Z(-\alpha)$. In other words, we can apply a single-qubit gate $H U_Z(\alpha)$ to a qubit in state $|\psi\rangle$ by entangling the qubit with an ancilla qubit in state $|+\rangle$ via a CZ gate, and then performing a measurement of the observable $M(\alpha)$ on the initial qubit.

Exercise 2.5: Verify $M(\alpha)$ and show that it lies in the equatorial plane of the Bloch sphere, depicted in Fig. 2.1.

So far, we have shown that we can apply rotations around the z axis (followed by a Hadamard) via this measurement technique, but what about arbitrary single-qubit operations? Single-qubit operations are equivalent to arbitrary rotations in the Bloch sphere. Any such rotation $U(\boldsymbol{\theta})$ over an angle θ around the axis $\hat{\boldsymbol{\theta}}$ can be decomposed into three Euler angles α, β, and γ:

$$U(\boldsymbol{\theta}) = U_Z(\gamma) U_X(\beta) U_Z(\alpha). \tag{2.47}$$

Since $X = HZH$, we can rewrite

$$
\begin{aligned}
U_X(\beta) = \exp\left(-i\frac{\beta}{2}X\right) &= \sum_{n=0}^{\infty} \left(-i\frac{\beta}{2}\right)^n \frac{X^n}{n!} \\
&= \sum_{n=0}^{\infty} \left(-i\frac{\beta}{2}\right)^n \frac{(HZH)^n}{n!} = \sum_{n=0}^{\infty} H\left(-i\frac{\beta}{2}\right)^n \frac{Z^n}{n!} H \\
&= H U_Z(\beta) H.
\end{aligned} \tag{2.48}
$$

Therefore

$$U(\boldsymbol{\theta}) = H\, HU_Z(\gamma)\, HU_Z(\beta)\, HU_Z(\alpha)\,. \tag{2.49}$$

In circuit language, this becomes

$$\boxed{HU_Z(\alpha)} - \boxed{HU_Z(\beta)} - \boxed{HU_Z(\gamma)} - \boxed{H}$$

In order to implement an arbitrary rotation $U(\boldsymbol{\theta})$, the teleportation procedure presented above must be concatenated three times

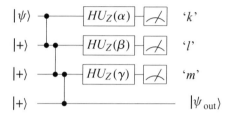

with $|\psi_{\text{out}}\rangle = (X^m HU_Z(\gamma))\, (X^l HU_Z(\beta))\, (X^k HU_Z(\alpha))\, |\psi\rangle$. However, the operators X^k, X^l, and X^m depend on the measurement outcomes, and we should try to get rid of them by commuting them through the Pauli gates and Hadamards. This will give rise to an adjustment of the measurement bases depending on the previous measurement outcomes. This is often called 'feed-forward'.

Exercise 2.6: Show that the effective single-qubit operation due to the three measurements above corresponds to

$$\boxed{HU_Z(\alpha)} - \boxed{HU_Z((-1)^k \beta)} - \boxed{HU_Z((-1)^l \gamma)} - \boxed{X^k} - \boxed{Z^l} - \boxed{X^m}$$

The rotation angle of U_Z depends on the measurement outcome of the previous qubit, and determines a temporal order in which to perform the measurements. This is why this model of computation is called the 'one-way' model.

The four-qubit state in the above example can be regarded as an entangled *resource* for the rotation $U(\boldsymbol{\theta})$. It is convenient to introduce a graphical representation of this entangled state: every qubit in the state $|+\rangle$ is represented by a circle (the 'vertex'), while a CZ gate is represented by a line (the 'edge') connecting the qubits it operates on. Since any two CZ gates (on possibly overlapping qubit pairs) always commute, we do not have to take the order into account. Any state that can be represented in this graphical way is a 'cluster state'. For example, in the four-qubit case above the initial entangled state is a cluster state if $|\psi\rangle = |+\rangle$. In graphical language, this becomes

After the consecutive measurements $M(\alpha)$, $M(\pm\beta)$, and $M(\pm\gamma)$, and the corrective Pauli operations, the final qubit on the right is in the state $U(\boldsymbol{\theta})\,|+\rangle$.

Exercise 2.7: Write out the state of a three- and four-qubit linear cluster in the computational basis.

A distinction is often made between cluster states and 'graph states', where the former is any d-dimensional regular lattice of qubits with nearest-neighbour CZ operations, while graph states are more general, and include any topology of vertices.

We can now determine how an arbitrary N-qubit computation defined by a quantum circuit can be translated into a computation in the one-way model. First, all multi-qubit gates in the circuit must be translated into single-qubit gates and CZ gates. This is always possible. Each 'logical' qubit in the circuit model is represented by a horizontal line. In the one-way model the logical qubit becomes a string of 'physical qubits' in the state $|+\rangle$, which are linearly connected (entangled) via CZ gates:

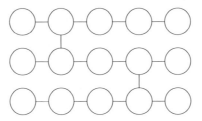

Each logical qubit in the circuit model is associated with such a one-dimensional cluster state. Just as in the circuit model, in the one-way model the logical qubits are arranged in a vertical stacking, such that the CZ gates between two logical qubits become lines connecting physical qubits on different horizontal levels:

This reasoning shows by construction that two-dimensional cluster states can be a resource for quantum computing. A natural question is if there are cluster states that are a *universal* resource for quantum computing, meaning that the cluster allows the implementation of *any* quantum computation up to a certain size (related to the size of the cluster). Before this question can be answered, a framework is needed to describe the manipulation of cluster states.

2.3.3 Manipulating cluster states

The manipulation of cluster states is most efficiently described using the stabilizer formalism. The question is therefore: what are the stabilizer generators of a cluster state of N qubits? In the case of a cluster state without any edges (that is, without CZ gates), this question is easily answered: each qubit is disconnected from every other qubit, and is therefore in the $|+\rangle$ state. The stabilizer generator for $|+\rangle_j$ is $S_j = X_j$, so the stabilizer generator for the maximally disconnected cluster is $\{X_j\}_{j=1}^N$. To find the stabilizer generators for less trivial clusters we investigate how S_j (and S_k) change when a CZ_{jk} gate is applied to qubits j and k.

First, the gate CZ_{jk} can be written in terms of 2×2 matrices $P^{(0)}$, $P^{(1)}$, Z, and $\hat{\mathbb{I}}$, where

$$P^{(0)} = \begin{pmatrix} 1 & 0 \\ 0 & 0 \end{pmatrix} \quad \text{and} \quad P^{(1)} = \begin{pmatrix} 0 & 0 \\ 0 & 1 \end{pmatrix}. \tag{2.50}$$

These are 'projectors' onto the states $|0\rangle$ and $|1\rangle$, respectively. The CZ_{jk} gate then becomes

$$\mathrm{CZ}_{jk} = P_j^{(0)} \otimes \hat{\mathbb{I}}_k + P_j^{(1)} \otimes Z_k. \tag{2.51}$$

This means that when qubit j is projected onto $|0\rangle$, the operation on k is the identity, and when qubit j is projected onto $|1\rangle$, the operation on k is the Pauli Z gate. By definition this is the CZ gate, even though its symmetry is no longer manifest. The projectors $P^{(0)}$ and $P^{(1)}$ obey the following relations:

$$P^{(0)} X = X P^{(1)} \quad \text{and} \quad P^{(1)} X = X P^{(0)}. \tag{2.52}$$

Exercise 2.8: Prove Eqs. (2.51) and (2.52).

The effect of the CZ gate on the stabilizer generator S_j is:

$$\begin{aligned}
\mathrm{CZ}_{jk} S_j \mathrm{CZ}_{jk}^\dagger &= \mathrm{CZ}_{jk} S_j \left(P_j^{(0)} \otimes \hat{\mathbb{I}}_k + P_j^{(1)} \otimes Z_k \right) \\
&= \mathrm{CZ}_{jk} \left(P_j^{(1)} \otimes \hat{\mathbb{I}}_k + P_j^{(0)} \otimes Z_k \right) S_j \\
&= \left(P_j^{(0)} \otimes \hat{\mathbb{I}}_k + P_j^{(1)} \otimes Z_k \right) \left(P_j^{(1)} \otimes \hat{\mathbb{I}}_k + P_j^{(0)} \otimes Z_k \right) S_j \\
&= \left(P_j^{(1)} \otimes Z_k + P_j^{(0)} \otimes Z_k \right) S_j \\
&= S_j Z_k,
\end{aligned} \tag{2.53}$$

where in the second line we have used $S_j \propto X_j$, and in the last line we have used $P^{(0)} + P^{(1)} = \hat{\mathbb{I}}$. Since the CZ gate is symmetric, we can immediately find the transformed stabilizer generator for S_k:

$$S_k \rightarrow \mathrm{CZ}_{jk} S_k \mathrm{CZ}_{jk}^\dagger = S_k Z_j. \tag{2.54}$$

The CZ gate therefore multiplies the stabilizer generator S_j with the Pauli Z_k and vice versa. If the S_j already contained a Z_k, the effect of the second Z_k is to remove the first one from S_j. This agrees with the observation that applying the CZ gate twice is the same as the identity operator $\mathrm{CZ}_{jk}^2 = \hat{\mathbb{I}}_{jk}$. From this we can construct the stabilizer for any N-qubit cluster state: Each qubit j in a cluster state carries a stabilizer generator S_j that is given by

$$S_j = X_j \prod_{k \in n(j)} Z_k, \tag{2.55}$$

where $n(j)$ is the 'neighbourhood' of qubit j, i.e., the set of qubits that are directly connected to j via a single CZ gate. There are N stabilizer generators of this form, which completely determines the cluster. Given the stabilizer generator $\{S_j\}_{j=1}^N$, we can easily find the graphical representation, and vice versa. This makes the stabilizer formalism a tremendously powerful tool for tracking changes in cluster states.

A cluster state is defined as the state $|C\rangle$ for which $S_j |C\rangle = |C\rangle$ for all j, where S_j is defined as in Eq. (2.55). However, there are also states for which $S_j |C'\rangle = -|C'\rangle$ for one or more j. These states are orthogonal to $|C\rangle$:

$$\langle C|C'\rangle = \langle C| S_j^\dagger S_j |C'\rangle = \langle C| \left(-|C'\rangle\right) = -\langle C|C'\rangle, \qquad (2.56)$$

and this inner product must therefore be zero. An N-qubit cluster state will have N stabilizer generators S_j (with eigenvalue 1), and by choosing eigenvalues -1 for subsets of S_j we can construct 2^N orthonormal states. These stabilizer states thus form a basis for the N-qubit Hilbert space. Only the states with all eigenvalues of $+1$ for the stabilizer generators are strictly called a cluster state. However, any of the stabilizer states $|C'\rangle$ are often called cluster states with a number of negative stabilizer eigenvalues.

As a simple example, we can calculate the effect of a Pauli operator on a cluster state. From Eq. (2.31) we find that $\widetilde{S}_j U_l |\psi\rangle = U_l |\psi\rangle$ with $\widetilde{S}_j = U_l S_j U_l^\dagger$. If U_l is a Pauli operator $P_l^{(k)} = X, Y, Z$ on qubit l, and $S_j \propto P_l^{(m)}$ is a stabilizer generator of the cluster state $|\psi\rangle$, then by Eq. (2.55) we have

$$P_l^{(k)} P_l^{(m)} P_l^{(k)} = \begin{cases} P_l^{(m)} & \text{if } m = k, \\ -P_l^{(m)} & \text{if } m \neq k. \end{cases} \qquad (2.57)$$

A single Pauli operation on a cluster state will therefore take some stabilizer generators S_j to $-S_j$, and the cluster state $|C\rangle$ transforms into a cluster state $|C'\rangle$ with a few negative stabilizer eigenvalues.

We now also see the reason for introducing the Clifford operations: these gates (Hadamard, phase, cz, and cx gates) transform Pauli operators into (other) Pauli operators. When we act on a cluster state with a Clifford operation, the resulting state is again a stabilizer state, i.e., the stabilizer generators are products of Pauli operators. When a stabilizer state can be transformed into a cluster state using a Clifford operation made of tensor products of single-qubit operations, the stabilizer state is called 'LC-equivalent' to the cluster state. Here, LC stands for 'local Clifford'.

Exercise 2.9: Show that the GHZ state $|000\rangle + |111\rangle$ is LC-equivalent to a three-qubit cluster state. Construct the stabilizer of the GHZ state and transform it into the stabilizer of the cluster state using local Clifford gates.

An equally important question is how cluster states transform when some of their qubits are measured. We will restrict the discussion to single-qubit measurements. We have already seen that arbitrary single-qubit measurements in the equatorial plane of the Bloch sphere can be used to implement single-qubit rotations. We can also infer from Exercise 2.6 that in the special case where $U_j(\alpha) = U_j(-\alpha)$, with $j = X, Y, Z$, there is no need to wait for the measurement outcome k and l. These special unitary transformations correspond to the respective Pauli operators. However, we have not yet determined the effect of measurements in the eigenbasis of Pauli operators, and these turn out to be very important tools for shaping cluster states. For pedagogical reasons we will first investigate measuring the Pauli Z operator, and then the X operator. Measurement of the Y operator is left as an exercise for the reader.

Consider again the linear cluster state

where we measure the third qubit in the computational basis (indicated by Z_3). The stabilizer of the cluster before measurement is generated by

$$S_1 = X_1 Z_2$$
$$S_2 = Z_1 X_2 Z_3$$
$$S_3 = \quad Z_2 X_3 Z_4$$
$$S_4 = \qquad Z_3 X_4 Z_5$$
$$S_5 = \qquad\quad Z_4 X_5 \,. \tag{2.58}$$

After the (projective) measurement of Z_3, the stabilizer generator of qubit 3 is $\pm Z_3$. Assume for definiteness that the measurement outcome was $|0\rangle$, so the stabilizer generator is $+Z_3$ (the measurement outcome $|1\rangle$ will only introduce overall minus signs on some generators, which we can keep track of efficiently). To find out how the cluster has changed we need to construct from Eq. (2.58) four more stabilizers that commute with Z_3. We immediately see that the generators S_1, S_2, S_4, and S_5 meet this requirement, and in principle we have solved the problem. However, the stabilizer generator is not in the standard form of Eq. (2.55) because in S_2 we have a factor $X_2 Z_3$, but there is no matching stabilizer that contains the factor $Z_2 X_3$. We therefore cannot yet make the identification with the graphical representation. To proceed, we use the group structure of stabilizers: the outcome of multiplying any two stabilizer generators is another valid stabilizer. In particular, we can multiply both S_2 and S_4 with Z_3 to remove the factors Z_3 (since $Z^2 = \hat{\mathbb{I}}$). Furthermore, we can apply a Hadamard gate to qubit 3 to change the stabilizer into X_3. The stabilizer generator then becomes

$$S_1' = X_1 Z_2$$
$$S_2' = Z_1 X_2$$
$$S_3' = \qquad X_3$$
$$S_4' = \qquad\quad X_4 Z_5$$
$$S_5' = \qquad\qquad Z_4 X_5 \,, \tag{2.59}$$

and the cluster state can be drawn as

Generally, a Z measurement removes the qubit from a cluster and breaks all the bonds with its neighbours. A cluster like this, where there exist pairs of qubits that are not connected via a continuous path of edges, is called a 'disconnected cluster'.

Let us now consider again the linear five-qubit cluster with the stabilizer given by Eq. (2.58), but this time we wish to determine the effect of a Pauli X operator measurement on qubit 3. We assume that the measurement result is '+', such that the stabilizer of qubit 3 is generated by X_3. In order to determine what the new cluster state looks like, we have to find the new stabilizer generator, and put it into the form of Eq. (2.55) using LC operations. First, note that again S_1 and S_5 commute with X_3, so these are valid generators. So is S_3 – but we multiply it by X_3 to obtain $S_3'' = Z_2 Z_4$. Second, we construct the remaining stabilizer by multiplying S_2 and S_4 to obtain $S_2'' = Z_1 X_2 X_4 Z_5$. This removes the troublesome Z_3 factor. Third, we multiply S_3'' with either S_1 or S_5 (we choose S_1) to find $S_1' = X_1 Z_4$. Finally, to cast the stabilizer generator in the form of Eq. (2.55) we apply a Hadamard gate to qubit 2. This results in the stabilizer generator

$$
\begin{aligned}
S_1' &= X_1 && Z_4 \\
S_2' &= & X_2 & Z_4 \\
S_3' &= & X_3 \\
S_4' &= Z_1 Z_2 & X_4 Z_5 \\
S_5' &= & & Z_4 X_5 \, .
\end{aligned}
\tag{2.60}
$$

Graphically, this is represented by the cluster state

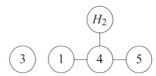

A general rule for finding the results of X measurements will be given below.

Exercise 2.10: Show that two neighbouring X measurements in a linear cluster shorten the cluster by two qubits and connect the neighbour of the first measured qubit with the neighbour of the second.

Exercise 2.11: Show that the measurement of the Pauli Y operator on qubit 3 with outcome "↻" results in the cluster

where Φ_j is the phase gate applied to qubit j.

We can give the rules for how clusters change under Pauli measurements directly in terms of operations on the graph. Define 'inversion' of a graph as the addition modulo 2 of edges between every node in the graph. This has the effect of removing the existing edges, and creating new edges where there were none previously. In graph language we then have

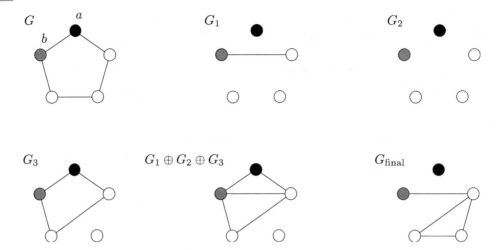

Fig. 2.3. An *X* measurement of qubit *a*, using a designated neighbour *b*.

Z measurement: Remove all edges between the measured qubit and its neighbours. If the measurement outcome is '1', a *Z* operation is applied to all former neighbours of the measured qubit, otherwise no corrective unitary is needed.

Y measurement: Invert the subgraph of the measured qubit and its neighbours. This automatically removes the measured qubit from the graph. For the measurement outcome '↻' or '↺', the corrective operation on each neighbour is the phase gate Φ or $\Phi^\dagger$, respectively (up to a global phase).

X measurement: Choose one of the neighbouring qubits (call it *b*) of the measured qubit *a* in graph state *G* (see Fig. 2.3). First, construct a subgraph G_1 by drawing edges between *b* and the other neighbours of *a*. Second, construct the subgraph G_2 by drawing edges between all qubits that are neighbours of both *a* and *b* (with respect to *G*). Third, construct G_3 by connecting each member of *a*'s neighbourhood with each member of *b*'s neighbourhood (again, with respect to *G*). Add G_1, G_2, and G_3 to the original graph modulo 2 to obtain the final graph G_{final} after measurement. If the measurement outcome is '+', apply a Hadamard gate to qubit *b*, and a *Z* gate to the neighbours of *a* that are not neighbours of *b* (or *b* itself). Similarly, if the measurement outcome is '−', apply a Hadamard gate to qubit *b*, and a *Z* operation to the neighbours of *b* that are not neighbours of *a* (or *a* itself).

General single-qubit measurements will change the stabilizer such that the generators can typically no longer be expressed as tensor products of Pauli operators.

2.3.4 The Gottesman–Knill theorem

Using the stabilizer formalism and the knowledge about cluster states we just developed, it is straightforward to prove a very profound and general theorem about quantum computing, namely the 'Gottesman–Knill theorem':

> Gottesman–Knill theorem: Suppose a computation can be cast into a form with the following properties:
>   (i) The input state is a computational basis state;
> (ii) only gates from the Clifford group are applied;
> (iii) measurements of Pauli operators are allowed; and
> (iv) the measurement results may be used in classical control of subsequent Clifford gates,
> then such a computation can be simulated efficiently on a classical computer.

The proof of this theorem is almost immediate, given what we already know: an input state in the computational basis is efficiently described by a stabilizer ($S_j = \pm Z_j$ for each qubit j). Subsequent Clifford gates (i.e., unitary gates that are members of the Clifford group) and Pauli measurements transform stabilizer states into other stabilizer states, so at any point in the calculation the quantum state is determined by a stabilizer. Any classical control of Clifford gates based on outcomes of the Pauli measurements is turned into a classical conditional transformation of the stabilizer.

A corollary of the Gottesman–Knill theorem is that the use of entanglement and non-commuting observables is in itself not sufficient for a computational speed-up over classical computing. In Chapter 9, we will derive the continuous-variable version of the Gottesman–Knill theorem, which has important consequences for quantum computing with linear optical elements.

2.3.5 Properties of cluster states

Now we consider a number of practical issues related to measurement-based quantum computing. First, we need to know what cluster or graph state we need to prepare for the quantum computation. Second, from the perspective of optical quantum computing, we want to know whether measurement-based quantum computing can solve problems that are difficult in the circuit model.

Universality of different classes of clusters

For any given computation in the circuit model, we can immediately find the corresponding cluster state via the translation presented on page 64. The question therefore becomes more general: what types of cluster state are *universal* for quantum computations up to a certain size? In Fig. 2.4 we depict five examples of cluster state classes. The linear cluster is not universal, because any quantum computation that can be performed on this cluster is efficiently simulable on a classical computer. The same is true for tree clusters, which are defined as graph states without 'loops', that is, there is no path on the graph such that by starting from a vertex and following the edges (traversing each edge at most once) we return to that vertex after a finite number of steps.

Two-dimensional and higher-dimensional regular clusters are universal resources for quantum computing. To see this, note that we can write every quantum computation in terms of single-qubit operations and cz gates on nearest neighbours. We can then 'punch out' qubits from the regular cluster by means of Z measurements, such that we retrieve the cluster state needed for a particular computation. By virtue of the translation procedure we used to introduce clusters, we know that we can always do this. The universal cluster is analogous to a large piece of marble: the sculptor removes the excess marble to carve out a statue of his design. Three-dimensional clusters (and higher) offer new possibilities over two-dimensional clusters in that they allow for topological quantum computing, which has particularly desirable error-correcting properties. For more information, see Raussendorf *et al.* (2006).

When a two-dimensional cluster is large in one direction (say the horizontal direction in Fig. 2.4), and short in the other (vertical) direction, we expect that for some ratio the cluster becomes too much like the linear cluster and ceases to be a universal resource for quantum computation. It turns out that when the vertical depth is a logarithmic function of the horizontal width, the cluster can support only classically simulable quantum computations. The intuition behind this is as follows: suppose the vertical depth is the number n of logical qubits, and the horizontal width N is the length of the computation. When $n \propto \log N$, the

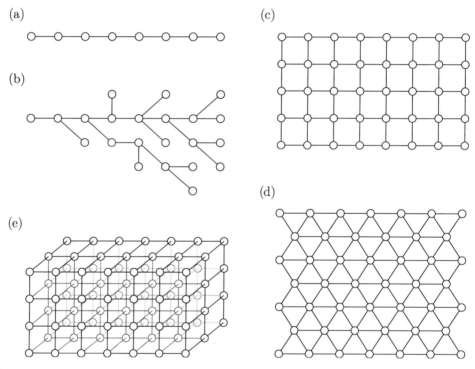

Fig. 2.4.   Different classes of cluster state: (a) linear clusters are not universal; (b) tree clusters are not universal; (c) two-dimensional clusters such as the square and (d) the hexagonal clusters are universal; and (e) three- (and higher-) dimensional clusters and higher are also universal.

length of the computation is exponential in the number of qubits, which scales in the same way as a classical simulation.

Practical considerations

There are many practical reasons why the one-way model of quantum computing might be preferable to the circuit model. The circuit model relies critically on the ability to perform a universal set of quantum gates, both single- and two-qubit gates, with errors well below the fault-tolerant threshold. The errors include noise and gate failure. In short, the circuit model requires the quantum gates to be near perfect. Cluster states, on the other hand, can be constructed before the computation takes place, which gives a considerable relaxation of gate requirements: a near-perfect, deterministic CZ gate is not needed to create the cluster; any entangling operation will do, as long as we end up with the cluster state. In particular, the entangling gate may have a large failure rate. Cluster state growth can then proceed via a statistical procedure, which merely adds an extra overhead cost to the state construction.

As a simple example, consider a CZ gate with success probability p. Assume that it is always known whether the gate has failed or succeeded. We want to add a qubit to a linear cluster of length N. In many implementations of the CZ gate, failure scrambles the qubits it acted on. As a result, we have to remove the qubit from the linear cluster, yielding a shorter cluster of length $N - 1$. When $p < 1/2$, the CZ gate fails more often than it succeeds, and it becomes exponentially unlikely that we end up with a long cluster. However, we can still grow long clusters efficiently. Rather than trying to add a single qubit, we prepare a linear cluster of length m at some cost of p^{-m} gate operations. This is a large but constant overhead per added qubit. Given the success probability p, we want to know the size m of the chains we have to add to the main cluster. This can be found by considering the average length of the new cluster, and the growth requirement becomes

$$p(N + m) + (1 - p)(N - 1) > N \quad \Longrightarrow \quad m > \frac{1 - p}{p} \,. \tag{2.61}$$

Different entangling procedures for making cluster states can have different growth requirements, and there is a variety of different strategies for creating universal cluster states (see for example Gross *et al.* 2006).

Another practical advantage of the one-way model over the circuit model is that the qubits may be completely isolated from each other. If we are allowed to use probabilistic entangling operations to create the cluster, we can create gates that do not rely on a direct interaction between qubits. This may significantly reduce noise, and the lifetime of the qubits can be extended by storing them in individually optimized environments. We will return to this approach in Chapter 7.

2.3.6 Error correction and fault tolerance

In any practical implementation of a quantum computer we will encounter errors. The qubits will have some interaction with the environment that, however small, can cause bit flips, phase shifts, or other unwanted changes in the computation. When those errors occur with probability p and accumulate in a computation of length N, the success probability of finding the correct outcome of the computation generally deteriorates as $\sim p^N$. Thus, any exponential speed-up over classical computation is lost, and so some form of error correction becomes essential. Many error-correction codes have been developed over the years, most notably the nine-qubit Shor code, Calderbank–Shor–Steane (css) codes, and the minimal five-qubit code that corrects any single qubit error. These are examples of so-called 'stabilizer' codes. However, not all error-correcting codes are stabilizer codes. Another class is that of the so-called non-additive error-correcting codes, which we will not discuss in this book. In this section, we give a short introduction to stabilizer codes, followed by a discussion of how to make quantum computers fault-tolerant.

Stabilizer codes

Error correction requires three stages: first, the information is encoded in some (larger) system. Second, after a noisy operation the 'error syndrome' is determined; it is a coded message detailing which kind of error (if any) has occurred. Finally, the syndrome is used to apply a correction to the system that eliminates the error. The remarkable feat of *quantum* error correction is that these steps can be performed without affecting any of the quantum information contained in the logical qubits to the environment.

The stabilizer formalism was used in Section 2.1.3 to uniquely define quantum states in terms of a small set of operators, and more generally to define subspaces of Hilbert space. Recall that the stabilizer $\mathfrak{S}$ is the group generated by the operators S_j, with j ranging from 1 to n. We found that $[S_i, S_j] = 0$, and

$$S_j |\psi\rangle = |\psi\rangle \quad \text{for all } j. \tag{2.62}$$

This means that the states $|\psi\rangle$ span a subspace $\mathcal{V}$. When dim $\mathcal{V} = 1$, the stabilizer determines a unique quantum state, as was the case when we constructed cluster states. In quantum error correction, on the other hand, the dimensionality of $\mathcal{V}$ is typically larger than one: to encode a qudit, the dimension of $\mathcal{V}$ must be d, and we need d orthonormal basis states $|\psi_0\rangle, \ldots, |\psi_d\rangle$, such that

$$S_j |\psi_k\rangle = |\psi_k\rangle \quad \text{for all } j, k. \tag{2.63}$$

Our single-qudit operations on $\{|\psi_0\rangle, \ldots, |\psi_d\rangle\}$ must commute with all S_j, but must not themselves be members of the stabilizer. We often specify the generalized Pauli X and Z operators on the qudit space, from which any single-qudit operation can be constructed. The stabilizer codes are now constructed as follows: the quantum information is encoded in states $|\psi\rangle \in \mathcal{V}$, and the syndrome is extracted by measuring the stabilizer generators. The measurement outcomes should tell us whether or not an error occurred, but should still protect the quantum information.

For a particular state $|\psi\rangle$ that lives in the subspace $\mathcal{V}$ of Hilbert space, a measurement of *any* stabilizer generator S_j must by construction give the outcome $+1$. Moreover, since all S_j commute, we can measure all generators simultaneously. Suppose an error E happens (with some probability). This will transform the state $|\psi\rangle$ into $E|\psi\rangle$. In the case of a qubit, E may be a bit-flip X, a phase-flip Z, or any other unitary error. We want the stabilizer to pick up on this error, so for at least one S_j we want the measurement outcome to be different from 1. Also, $E|\psi\rangle$ must be another eigenstate of S_j, otherwise the error detection is not deterministic. Without loss of generality we assume that the eigenvalue of S_j for $E|\psi\rangle$ is -1. This leads to

$$S_j E|\psi\rangle = -E|\psi\rangle = -E S_j |\psi\rangle \ . \tag{2.64}$$

Therefore, S_j and E must *anti-commute* for some j: $\{S_j, E\} = 0$. When the error does not affect a particular S_k, the generator and the error commute: $[S_k, E] = 0$. This is the core of the error-correction mechanism of stabilizer codes. When we measure a predetermined set of members of the stabilizer, we obtain a series of $+1$ and -1 that tells us exactly which error occurred. The anti-commutation relation between the error and the stabilizer generator suggests that we should write the error operator E as a superposition of Pauli errors, and the stabilizer as a set of Pauli operators from the Pauli group $\mathfrak{P}_N$.

We give two simple examples of stabilizer codes, namely the three-qubit majority code, and the minimal five-qubit code. First, in the majority code the logical qubits $|\bar{0}\rangle$ and $|\bar{1}\rangle$ are encoded in three physical qubits:

$$|\bar{0}\rangle = |000\rangle \quad \text{and} \quad |\bar{1}\rangle = |111\rangle \ . \tag{2.65}$$

The stabilizer of the code space spanned by $|\bar{0}\rangle$ and $|\bar{1}\rangle$ is given by

$$\mathfrak{S} = \left\{ \hat{\mathbb{1}}_1 \otimes \hat{\mathbb{1}}_2 \otimes \hat{\mathbb{1}}_3, Z_1 \otimes Z_2 \otimes \hat{\mathbb{1}}_3, Z_1 \otimes \hat{\mathbb{1}}_2 \otimes Z_3, \hat{\mathbb{1}}_1 \otimes Z_2 \otimes Z_3 \right\} , \tag{2.66}$$

and the logical X and Z operations (denoted by $\bar{X}$ and $\bar{Z}$) are given by

$$\bar{X} = X_1 X_2 X_3 \quad \text{and} \quad \bar{Z} = Z_1 \ . \tag{2.67}$$

Notice that there is a freedom in the choice of $\bar{Z}$. The dimension of the Hilbert space of the three qubits is $2^3 = 8$, and $\mathfrak{S}$ is generated by two operators, which we can choose from the last three entries in the set. Each generator halves the size of the Hilbert space (see Section 2.1.3), so the resulting stabilizer space $\mathcal{V}$ is two-dimensional. It therefore supports a qubit, as required.

Now suppose that with some (small) probability a bit-flip X_i occurs on one of the three qubits. We do not know whether the error occurred, nor do we know which qubit is affected. By symmetry, we can assume that the error occurred on qubit 1: $E_1 = X_1 \otimes \hat{\mathbb{1}}_2 \otimes \hat{\mathbb{1}}_3$. An unknown quantum state $\alpha |\bar{0}\rangle + \beta |\bar{1}\rangle$ then becomes $\alpha |100\rangle + \beta |011\rangle$, where the first qubit is flipped. The eigenvalue equations for the stabilizer in this state are:

$$
\begin{aligned}
Z_1 Z_2 \hat{\mathbb{1}}_3 |100\rangle &= -|100\rangle & Z_1 Z_2 \hat{\mathbb{1}}_3 |011\rangle &= -|011\rangle \\
Z_1 \hat{\mathbb{1}}_2 Z_3 |100\rangle &= -|100\rangle \quad \text{and} \quad & Z_1 \hat{\mathbb{1}}_2 Z_3 |011\rangle &= -|011\rangle \\
\hat{\mathbb{1}}_1 Z_2 Z_3 |100\rangle &= |100\rangle & \hat{\mathbb{1}}_1 Z_2 Z_3 |011\rangle &= |011\rangle \ .
\end{aligned}
\tag{2.68}
$$

The eigenvalues of the stabilizer are the same for the states $|100\rangle$ and $|011\rangle$. This is not a coincidence. It needs to be in order for the stabilizer measurement not to reveal information about α and β. The outcomes allow us to determine that we need to apply a corrective bit flip on the first qubit. Note also that these are *joint* measurements, and not just three independent measurements in the computational basis!

The three-qubit majority code is simple and small, but it cannot correct a phase flip. The smallest code that can correct for any single-qubit error consists of five qubits. The stabilizer generators are

$$S_1 = X_1 Z_2 Z_3 X_4$$
$$S_2 = X_2 Z_3 Z_4 X_5$$
$$S_3 = X_1 X_3 Z_4 Z_5$$
$$S_4 = Z_1 X_2 X_4 Z_5 \tag{2.69}$$

and the qubit operations $\bar{X}$ and $\bar{Z}$ are

$$\bar{X} = X_1 X_2 X_3 X_4 X_5 \quad \text{and} \quad \bar{Z} = Z_1 Z_2 Z_3 Z_4 Z_5 \ . \tag{2.70}$$

Exercise 2.12: Find the logical qubit states of the five-qubit code $|\bar{0}\rangle$ and $|\bar{1}\rangle$, and verify that $\bar{X}$ and $\bar{Z}$ are the Pauli operators in the logical qubit space that commute with the stabilizer.

There are many more quantum error-correcting codes, most notably the seven-qubit Steane code and the class of CSS codes. For further details on quantum error-correction codes, we refer to Chapter 10 of Nielsen and Chuang (2000).

You may object that errors are generally not unitary. Indeed, in the next chapter we will see that general errors are so-called completely positive maps. However, these maps can be written in terms of products of Pauli errors (in the case of qubits) and the codes outlined above still work. This is rather remarkable, and in Chapter 6 we will discover a particularly counterintuitive aspect of stabilizer codes.

Fault tolerance

We have seen how the general principle of stabilizer codes can protect a logical qubit from a single error. However, the codes themselves consist of quite a few physical qubits, all of which are prone to errors. What guarantees that quantum error correction does more harm than good? The answer is 'fault tolerance', and we will sketch the general idea in the remainder of this section.

Suppose we wish to implement a certain operation in a quantum computation (a 'computational element'). This may be a state-preparation procedure, simple propagation in time, a quantum gate, or a measurement. There will be a failure probability p associated with the unencoded element. After encoding, syndrome extraction, and the recovery operation, the total failure probability of the encoded computational element will be q. By virtue of the error-correction code, q will be proportional to p^2: one error occurs with probability p, but can be corrected, so the encoded procedure fails if there is a second error in the code, also with probability p. The constant of proportionality α in $q = \alpha p^2$ is simply the total number of distinct places an error can occur in the code (and p must be small enough such that $q < 1$). The larger the code, the larger α will be. In general, quantum error correction is helpful only if for a specific code $q < p$, or

$$p < \frac{1}{\alpha} \quad \text{and} \quad p_{\text{th}} = \frac{1}{\alpha}. \tag{2.71}$$

The probability p_{th} is the 'threshold' of the error-correction code. For the Steane code α is approximately 10^4, and the corresponding error p must be very small to begin with. Raussendorf and Harrington (2007) showed that two-dimensional cluster states have a relatively high error threshold of $p_{\text{th}} = 0.75\%$. In many ways achieving fault tolerance is the hardest part of implementing quantum computing with real systems.

The above analysis shows that error correction can be used to lower the failure rate of a computational element, but it does not make it zero. Accumulated errors will still cause an exponential decay in the success probability of the computation. To circumvent this, we use the concept of 'concatenation': each logical qubit is encoded in a set of qubits, and in turn these qubits are encoded again. The codes can be nested many times. If m is the maximum number of operations needed to implement a computational element in one level of encoding, k-fold concatenation means that the number of operations per computational element becomes m^k. This grows quite rapidly, so concatenation works only if the error decreases even quicker. Indeed, the failure probability q_k becomes

$$q_k = \frac{(\alpha p^2)^{2^k}}{\alpha}. \tag{2.72}$$

If a particular computation contains $c(N)$ computational elements, where N indicates the size of the problem, and we want the quantum computation to succeed with probability $1 - \epsilon$, we require that

$$\frac{(\alpha p^2)^{2^k}}{\alpha} \leq \frac{\epsilon}{c(N)}. \tag{2.73}$$

Note that $c(N)$ is typically a polynomial function of N. The number of operations per computational element in the fully encoded implementation is

$$m^k = \left(\frac{\log(\alpha \epsilon) - \log[c(N)]}{\log(\alpha p)} \right)^{\log m} = O\left(\text{Poly}\left[\log\left(\frac{c(N)}{\epsilon} \right) \right] \right). \tag{2.74}$$

When the error probability p is right at the threshold ($p = 1/\alpha$), the factor $\log(\alpha p) = 0$ causes the number of operations to diverge towards infinity. Indeed, p parameterizes a critical phenomenon, with a phase transition at $p_{\text{th}} = \alpha$. When $p < p_{\text{th}}$, the above construction is called 'fault tolerant'.

A subtlety of fault tolerance is that error propagation and multiplication must be kept in check. Consider for example a CX gate U, and a Pauli X error on the control qubit just before the gate. We have

$$UX_1(\alpha\,|00\rangle + \beta\,|10\rangle) = U(\alpha\,|10\rangle + \beta\,|00\rangle) = X_1 X_2(\alpha\,|00\rangle + \beta\,|11\rangle). \tag{2.75}$$

Suddenly, one error is turned into two errors, even though the cx worked perfectly! In this case we are saved because the second error happens in a different qubit, inhabiting a different code space. Nevertheless, we need to be careful that the gates we wish to implement do not multiply errors in the code space of a qubit. It turns out that some gates can be constructed in such a way that this error multiplication is kept under control. As well as the Clifford operations H, Φ, and cz, the $\pi/8$ gate can be constructed this way. Given a suitable error-correction code and sufficiently small errors, and the fact that the Clifford operations plus the $\pi/8$ gate can be used to efficiently construct any n-qubit unitary operator with some precision δ, fault-tolerant 'universal' scalable quantum computation is indeed possible.

2.4 Quantum computation with continuous variables

As in classical computing, most research in quantum computing is based on the discrete qubit model. However, it is also possible to do useful computations with 'continuous variables'. Classically, these computers are known as 'analogue computers', and their origin can be traced back to the ancient Greeks. The continuous-variable variant of quantum computing is due to Lloyd and Braunstein. In this section we present the basics of this approach.

Analogous to the discrete qubit states $|0\rangle$ and $|1\rangle$ we can construct the states $|x\rangle_x$ with $x \in \mathbb{R}$. These are eigenstates of an operator $\hat{x}$ with a continuous spectrum, such as the position operator, and are called 'qunats'. A quantum computation over continuous variables is then given by an input string $|x_1, \ldots, x_N\rangle_x$, an evolution due to an interaction Hamiltonian $\mathcal{H}$: $\exp(-i\mathcal{H}t/\hbar)$, and the output state $|f_1(x_1, \ldots, x_N), \ldots, f_N(x_1, \ldots, x_N)\rangle_x$. The result of the computation is read out by measuring the observables $\hat{x}_i$.

As an example, consider the addition of x_1 and x_2. We assume that we can create a unitary transformation $U_\oplus = \exp(-i\mathcal{H}_\oplus t/\hbar)$, such that

$$|x_1, x_2\rangle_x \rightarrow U_\oplus |x_1, x_2\rangle_x = |x_1, x_2 + x_1\rangle_x . \tag{2.76}$$

The first slot in the output must be there to keep the computation unitary. The outcome of the computation (addition) of the two input states is written in the output state of the second system. We now have to construct the Hamiltonians that correspond to the computation we wish to perform.

Before we consider more general Hamiltonians, we define the canonically conjugate operator $\hat{p}$ to the operator $\hat{x}$

$$[\hat{x}, \hat{p}] = i\hbar . \tag{2.77}$$

When $\hat{x}$ is a position operator, $\hat{p}$ is the canonical momentum to $\hat{x}$. It also has a continuous unbounded spectrum. We know from quantum mechanics that the eigenstates of $\hat{x}$ and $\hat{p}$ are related by a Fourier transform:

$$|p\rangle_p = \int \frac{dx}{\sqrt{2\pi\hbar}} \exp\left(\frac{i}{\hbar}px\right) |x\rangle_x ; \qquad |x\rangle_x = \int \frac{dp}{\sqrt{2\pi\hbar}} \exp\left(-\frac{i}{\hbar}px\right) |p\rangle_p . \tag{2.78}$$

Next, we will construct Hamiltonians from $\hat{x}$ and $\hat{p}$.

2.4.1 Linear and quadratic interaction Hamiltonians

Now we are ready to consider some important operators on the states $|x\rangle$. First, we need the ability to create an arbitrary input state $|x\rangle$ from some initial state $|0\rangle$, which can be taken as the eigenstate of the eigenvalue $x = 0$. In other words, we need some 'displacement operator' that changes the numerical value from 0 to x. Such an operator is given by

$$X(x) = \exp\left(-\frac{i}{\hbar}x\hat{p}\right). \tag{2.79}$$

We distinguish operators from numbers by a hat. To prove that this is a displacement operator, we calculate the effect of $X(x)$ on an eigenstate $|y\rangle_x$ of $\hat{x}$:

$$
\begin{aligned}
X(x)|y\rangle_x &= \int \frac{dp}{\sqrt{2\pi\hbar}}\,\exp\left(-\frac{i}{\hbar}py\right)\exp\left(-\frac{i}{\hbar}x\hat{p}\right)|p\rangle_p \\
&= \int \frac{dp}{\sqrt{2\pi\hbar}}\,\exp\left[-\frac{i}{\hbar}p(x+y)\right]|p\rangle_p \\
&= |x+y\rangle_x .
\end{aligned} \tag{2.80}
$$

This is the continuous-variable generalization of the Pauli X operator, in that it adds a number to the logical value of the computational basis state: the discrete Pauli X adds the number 1 to the logical qubit value, modulo 2:

$$X|0\rangle = |0 \oplus 1\rangle = |1\rangle \quad \text{and} \quad X|1\rangle = |1 \oplus 1\rangle = |0\rangle , \tag{2.81}$$

where $\oplus$ denotes addition modulo 2.

The construction of the $X(x)$ operator immediately suggests another construction, namely of the $Z(p)$ operator:

$$Z(p) = \exp\left(\frac{i}{\hbar}p\hat{x}\right), \tag{2.82}$$

which acts on a position eigenstate $|x\rangle_x$ as follows:

$$Z(p)|x\rangle_x = \exp\left(\frac{i}{\hbar}p\hat{x}\right)|x\rangle_x = \exp\left(\frac{i}{\hbar}px\right)|x\rangle_x . \tag{2.83}$$

It is clear that this is the generalization of the Pauli Z operator for continuous variables: the operator adds a state-dependent phase to the computational basis states. The generalized Pauli operators for continuous variables are elements of the Heisenberg–Weyl group, and to avoid confusion we will refer to them as Heisenberg–Weyl operators.

It is now a straightforward exercise to show that the Heisenberg–Weyl operators act on momentum eigenstates $|r\rangle_p$ according to

$$X(x)|r\rangle_p = \exp\left(-\frac{i}{\hbar}xr\right)|r\rangle_p \quad \text{and} \quad Z(p)|r\rangle_p = |r+p\rangle_p . \tag{2.84}$$

It is tempting to calculate the commutator of $X(x)$ and $Z(p)$ and call it $Y(s)$, but here the analogy with the qubit case breaks down.

Exercise 2.13: Show that $X(x)Z(p) = \exp(-ixp/\hbar)Z(p)X(x)$.

The $X(x)$ and $Z(p)$ operators give us all the linear interaction Hamiltonians, but no combination of X and Z can ever give us higher-order polynomials in $\hat{x}$ and $\hat{p}$ in the Hamiltonian. To see this, we use the Baker–Campbell–Hausdorff relation

$$e^A B e^{-A} = B + \frac{1}{1!}[A,B] + \frac{1}{2!}[A,[A,B]] + \ldots \tag{2.85}$$

If A is linear in $\hat{x}$ and $\hat{p}$, and B is linear in $\hat{x}$ and $\hat{p}$, we can never obtain higher orders in $\hat{x}$ and $\hat{p}$ because $[A,B]$ is at most a constant. If we start in an eigenstate of $\hat{x}$, all we can do is displace the system to another eigenstate of $\hat{x}$, or add an overall state-dependent phase. Clearly, this is not enough for computation. We need to add interactions with quadratic Hamiltonians.

Consider the unitary transformation $\mathcal{F}$:

$$\mathcal{F} = \exp\left[\frac{i}{\hbar}\frac{\pi}{4}\left(\hat{x}^2 + \hat{p}^2\right)\right]. \tag{2.86}$$

We want to find the effect of this operator on a position eigenstate $|x\rangle$. This is not at all straightforward to calculate directly, so we first compute the effect of $\mathcal{F}$ on the Heisenberg–Weyl operators:

$$\begin{aligned}
\mathcal{F}X(x)\mathcal{F}^\dagger &= \exp\left[\frac{i}{\hbar}\frac{\pi}{4}\left(\hat{x}^2 + \hat{p}^2\right)\right]\exp\left(-\frac{i}{\hbar}x\hat{p}\right)\exp\left[-\frac{i}{\hbar}\frac{\pi}{4}\left(\hat{x}^2 + \hat{p}^2\right)\right] \\
&= \sum_{n=0}^\infty \frac{1}{n!}\left(-\frac{ix}{\hbar}\right)^n \exp\left[\frac{i}{\hbar}\frac{\pi}{4}\left(\hat{x}^2 + \hat{p}^2\right)\right]\hat{p}^n \exp\left[-\frac{i}{\hbar}\frac{\pi}{4}\left(\hat{x}^2 + \hat{p}^2\right)\right] \\
&= \sum_{n=0}^\infty \frac{1}{n!}\left(-\frac{ix}{\hbar}\right)^n\left\{\exp\left[\frac{i}{\hbar}\frac{\pi}{4}\left(\hat{x}^2 + \hat{p}^2\right)\right]\hat{p}\exp\left[-\frac{i}{\hbar}\frac{\pi}{4}\left(\hat{x}^2 + \hat{p}^2\right)\right]\right\}^n \\
&= \sum_{n=0}^\infty \frac{1}{n!}\left(-\frac{ix}{\hbar}\right)^n(-\hat{x})^n = \exp\left(\frac{i}{\hbar}x\hat{x}\right) \\
&= Z(x). \tag{2.87}
\end{aligned}$$

In the third line we have used $UA^nU^\dagger = UAU^\dagger UA \ldots AU^\dagger = (UAU^\dagger)^n$, and we evaluated

$$\begin{aligned}
\mathcal{F}\hat{p}\mathcal{F}^\dagger &= \exp\left[\frac{i}{\hbar}\frac{\pi}{4}\left(\hat{x}^2 + \hat{p}^2\right)\right]\hat{p}\exp\left[-\frac{i}{\hbar}\frac{\pi}{4}\left(\hat{x}^2 + \hat{p}^2\right)\right] \\
&= \hat{p} + \frac{i\pi}{4\hbar}\left[\hat{x}^2 + \hat{p}^2, \hat{p}\right] + \ldots \\
&= \cos(\pi/2)\hat{p} - \sin(\pi/2)\hat{x} = -\hat{x} \tag{2.88}
\end{aligned}$$

Similarly, we find that

$$\mathcal{F}Z(p)\mathcal{F}^\dagger = X(-p) = X^{-1}(p). \tag{2.89}$$

Comparing this to the discrete case, we are reminded of the Hadamard gate:

$$HZH = X \quad \text{and} \quad HXH = Z. \tag{2.90}$$

We therefore suspect that $\mathcal{F}$ is the Fourier transform. We can prove this using the stabilizer formalism: an eigenstate $|x\rangle_x$ of $\hat{x}$ is stabilized by the operator

$$\exp\left(-\frac{i}{\hbar}px\right) Z(p) |x\rangle_x = |x\rangle_x, \tag{2.91}$$

while an eigenstate $|p\rangle$ of $\hat{p}$ is stabilized by

$$\exp\left(\frac{i}{\hbar}px\right) X(x) |p\rangle_p = |p\rangle_p. \tag{2.92}$$

Applying the transformation $\mathcal{F}$ to the state $|x\rangle$ then gives

$$\begin{aligned}
\mathcal{F}|x\rangle_x &= \mathcal{F}\exp\left(-\frac{i}{\hbar}px\right) Z(p) |x\rangle_x \\
&= \mathcal{F}\exp\left(-\frac{i}{\hbar}px\right) Z(p)\mathcal{F}^\dagger \mathcal{F}|x\rangle_x \\
&= \exp\left(-\frac{i}{\hbar}px\right) \mathcal{F}Z(p)\mathcal{F}^\dagger \mathcal{F}|x\rangle_x \\
&= \exp\left(-\frac{i}{\hbar}px\right) X(-p)\mathcal{F}|x\rangle_x.
\end{aligned} \tag{2.93}$$

So the stabilizer S_x of $\mathcal{F}|x\rangle_x$ is

$$S_x = \exp\left(-\frac{i}{\hbar}px\right) X(-p). \tag{2.94}$$

Comparing this to Eq. (2.92), we see that $\mathcal{F}|x\rangle_x$ is the 'momentum' eigenstate $|x\rangle_p$. Similarly, the transformation $\mathcal{F}$ applied to the momentum eigenstate $|p\rangle_p$ yields a position eigenstate $|p\rangle_x$. Therefore, $\mathcal{F}$ is the Fourier transform.

Another unitary transformation that is generated by a quadratic Hamiltonian for a single system is the phase gate $\Phi(\theta)$:

$$\Phi(\theta) = \exp\left(\frac{i}{\hbar}\frac{\theta\hat{x}^2}{2}\right), \tag{2.95}$$

where we have included the factor $1/2$ for later convenience. Its effect on the Heisenberg–Weyl operators is determined in the following exercise:

Exercise 2.14: Since $[\Phi(\theta), Z(p)] = 0$, $Z(p)$ remains unaffected by the phase gate. Show that the effect of $\Phi(\theta)$ on $X(x)$ is given by

$$\Phi(\theta)X(x)\Phi^{-1}(\theta) = X(x)Z\left(\frac{x\theta}{\hbar}\right) \exp\left(\frac{i\theta x^2}{2\hbar^2}\right). \tag{2.96}$$

Next, we move up in the world, and consider quadratic Hamiltonians that act on two qunats: the continuous-variable version of the CX is given by

$$\mathrm{CX}_{ij} = \exp\left(-\frac{i}{\hbar}\hat{x}_i \otimes \hat{p}_j\right). \tag{2.97}$$

Its action on two eigenstates of the position operator is

$$
\begin{aligned}
\mathrm{CX}\,|x_1,x_2\rangle_x &= \exp\left(-\frac{i}{\hbar}\hat{x}_1 \otimes \hat{p}_2\right)|x_1,x_2\rangle_x \\
&= \exp\left(-\frac{i}{\hbar}x_1\hat{p}_2\right)|x_1,x_2\rangle_x = X_2(x_1)\,|x_1,x_2\rangle_x \\
&= |x_1, x_1 + x_2\rangle_x.
\end{aligned} \tag{2.98}
$$

This is the operator used in the example in Eq. (2.76), and it allows us to do simple arithmetic. In general, the operator CX carries two indices that determine which two systems it operates on. How does CX act on tensor products of generalized Pauli operators? It is straightforward to verify that

$$
\begin{aligned}
\mathrm{CX}\left[X_1(x) \otimes \hat{\mathbb{I}}_2\right]\mathrm{CX}^\dagger &= X_1(x) \otimes X_2(x), \\
\mathrm{CX}\left[Z_1(p) \otimes \hat{\mathbb{I}}_2\right]\mathrm{CX}^\dagger &= Z_1(p) \otimes \hat{\mathbb{I}}_2, \\
\mathrm{CX}\left[\hat{\mathbb{I}}_1 \otimes X_2(x)\right]\mathrm{CX}^\dagger &= \hat{\mathbb{I}} \otimes X_2(x), \\
\mathrm{CX}\left[\hat{\mathbb{I}} \otimes Z_2(p)\right]\mathrm{CX}^\dagger &= Z_1(p)^{-1} \otimes Z_2(p).
\end{aligned} \tag{2.99}
$$

Finally, the continuous-variable analog of the CZ gate is

$$\mathrm{CZ}_{ij} = \exp\left(\frac{i}{\hbar}\hat{x}_i \otimes \hat{x}_j\right). \tag{2.100}$$

The CZ gate is symmetric and commutes with the $Z(p)$ operator. The nontrivial transformation of the Heisenberg–Weyl operators then becomes

$$\mathrm{CZ}\left[X_1(x) \otimes \hat{\mathbb{I}}_2\right]\mathrm{CZ}^\dagger = X_1(x) \otimes Z_2(x). \tag{2.101}$$

This may remind you of the discussion leading up to Eq. (2.55), and indeed we may regard Eqs. (2.99) and (2.101) as transformations of stabilizer generators.

We can construct a complete orthonormal set of EPR states, which are the generalization of the Bell states for continuous variables. In addition, we can obtain all possible unitary evolutions corresponding to quadratic interaction Hamiltonians over arbitrarily many systems with the Heisenberg–Weyl operators $X_i(x)$ and $Z_i(p)$, the Fourier transform $\mathcal{F}_i$, the phase gate $\Phi_i(\theta)$, and the operators CX and CZ.

2.4.2 Universal quantum computation

The stabilizer generators of continuous-variable quantum computing are tensor products of $X(x)$ and $Z(p)$. We have seen that the quadratic operators $\mathcal{F}$, $\Phi(\theta)$, and CX transform tensor products of $X(x)$ and $Z(p)$ into other tensor products of $X(x)$ and $Z(p)$, and can therefore be regarded as the continuous-variable equivalent of the Clifford operations for qubit systems. The mathematical structure of these transformations is identical to that presented in Section 1.4 from page 34 onwards. Quadratic interaction Hamiltonians cause linear transformations of the canonical position and momentum. This suggests that we can derive the Gottesman–Knill theorem for continuous-variable quantum computing. Indeed, the generalized Gottesman–Knill theorem states that a computation that proceeds by means of linear and quadratic operators, and employs classical feed-forward, can be simulated efficiently on a classical computer. We will return to this topic in Chapter 9.

Quantum computing with continuous variables requires Hamiltonians that are higher-order polynomials in the canonical positions and momenta. In the remainder of this section we will show that *any* such higher-order Hamiltonian suffices to generate a universal set of gates for continuous variables efficiently. Mathematically, the reason why quadratic Hamiltonians produce linear transformations of the canonical position and momentum operators is that the commutator $[\hat{x}, \hat{p}]$ is a constant, and $[\mathcal{H}(\hat{x}, \hat{p}), \hat{x}]$ can create polynomials in $\hat{x}$ and $\hat{p}$ of at most order one if $\mathcal{H}$ is quadratic in $\hat{x}$ and $\hat{p}$. The Baker–Campbell–Hausdorff relation used to determine the operator transformations relies critically on these commutators.

To create higher-order functions of $\hat{x}$ and $\hat{p}$ we need Hamiltonians that are at least cubic in $\hat{x}$ and $\hat{p}$. Here, we consider the Hamiltonian $\mathcal{H} = (\hat{x}^2 + \hat{p}^2)^2$, which can immediately be seen to increase the order of polynomials in $\hat{x}$ and $\hat{p}$:

$$[\mathcal{H}, \hat{x}] = \frac{i}{2}\left(\hat{x}^2\hat{p} + \hat{p}\hat{x}^2 + 2\hat{p}^3\right) \quad \text{and} \quad [\mathcal{H}, \hat{p}] = -\frac{i}{2}\left(\hat{p}^2\hat{x} + \hat{x}\hat{p}^2 + 2\hat{x}^3\right). \quad (2.102)$$

Repeated commutators generate higher and higher order polynomials in $\hat{x}$ and $\hat{p}$, as required. A polynomial of order n and precision ϵ in $\hat{p}$ and $\hat{x}$ can be created by invoking $\mathcal{H}$ a number of times. This number scales as a small polynomial in n and ϵ.

It is now not too difficult to see that we can create a quantum computation of the form

$$|x_1, \ldots, x_N\rangle_x \xrightarrow[U_F]{} \sum_k c_k \, |x_1, \ldots, x_{N-1}, f_k(x_1, \ldots, x_N)\rangle_x \,, \quad (2.103)$$

where one of the f_k gives the outcome of the computation with probability $|c_k|^2$. The evolution U_F is generated by a (large) polynomial F over multiple continuous variables:

$$U_F = \exp\left[-\frac{i}{\hbar}F(\hat{x}_1, \ldots, \hat{x}_N, \hat{p}_1, \ldots, \hat{p}_N)\right]. \quad (2.104)$$

This unitary transformation can be decomposed into a sequence of linear, quadratic, and higher-order interactions, the length of which is polynomial in the number N of continuous-variable systems (qunats). The difficulty is to find polynomials F that can implement interesting quantum computations. However, this problem is beyond the scope of this book.

2.4.3 Cluster states for continuous variables

The continuous-variable model of quantum computation introduced in the previous section is essentially a circuit model, where we apply gates to qunats. However, by introducing the CZ gate and its action on the Heisenberg–Weyl operators $X(x)$ and $Z(p)$ in Eq. (2.101) we are strongly reminded of the stabilizer formalism that led to cluster states for qubits. A natural question is therefore whether we can construct a one-way model for continuous-variable quantum computing. We will now show that this is the case.

The most straightforward way to construct the one-way model is to exploit once more the local teleportation protocol for qubits:

$$|\psi\rangle \quad \boxed{U_Z(\alpha)}\ \boxed{H}\ \boxed{\diagup}\quad \text{`}m\text{'}$$
$$|+\rangle \quad\text{—————————}\quad X^m H U_Z(\alpha)\,|\psi\rangle$$

We translate this into a protocol for continuous variables as follows: first, the input state $|\psi\rangle$ can be any superposition of position eigenstates

$$|\psi\rangle = \int dx\,\psi(x)\,|x\rangle_x \quad \text{with} \quad \int dx\,|\psi(x)|^2 = 1. \tag{2.105}$$

Second, the ancilla system is in a zero-momentum eigenstate $|0\rangle_p$

$$|0\rangle_p = \int dx\,|x\rangle_x , \tag{2.106}$$

which is the continuous-variable analog of the $|+\rangle$ state for qubits. Third, the two systems are entangled using a CZ gate, after which we apply the single-qunat operation. This gate must commute with the CZ gate, and we therefore have

$$U_Z(\alpha) \;\rightarrow\; U_f = \exp\left[\frac{i}{\hbar}f(\hat{x})\right], \tag{2.107}$$

where f is some polynomial in $\hat{x}$. Fourth, the Hadamard gate for qubits becomes the Fourier transform $\mathcal{F}^\dagger$, and finally the top qunat is measured in the position basis. The circuit for local qunat teleportation then becomes

$$|\psi\rangle \quad \boxed{U_f}\ \boxed{\mathcal{F}^\dagger}\ \boxed{\diagup}\quad \text{`}q\text{'}$$
$$|0\rangle_p \quad\text{—————————}\quad X(x)\mathcal{F}U_f\,|\psi\rangle$$

The output state of the circuit is the input state with a gate $\mathcal{F}U_f$ applied to it, and a corrective Heisenberg–Weyl operator $X(x)$ must be applied.

Exercise 2.15: Show that the above circuit gives the output state $X(x)\mathcal{F}U_f\,|\psi\rangle$.

The cluster-state model is obtained by preparing a set of qunats in the state $|0\rangle_p$ and applying the CZ gates to the appropriate pairs of qunats. It is a straightforward exercise to give the stabilizers for the continuous-variable cluster states, and show that quadratic polynomials in $\hat{x}$ transform the cluster into another stabilizer state. Just as in the circuit model for continuous variables, we need single-qunat transformations, the generators of which are at least cubic in $\hat{x}$ in order to achieve universality in our quantum computer.

The measurement bases of the qunats in the one-way model are given by

$$M_f = U_f^\dagger \mathcal{F} \hat{x} \, \mathcal{F}^\dagger U_f = U_f^\dagger \hat{p} \, U_f \,. \tag{2.108}$$

When f is linear, U_f is a simple displacement of $\hat{p}$, and when f is quadratic, U_f is a simple rotation in phase space.

2.4.4 Approximate position eigenstates and error correction

There is a problem with continuous-variable quantum computing that we have so far completely ignored: the position and momentum eigenstates $|x\rangle_x$ and $|p\rangle_p$ are unphysical. There are many ways to define approximate position and momentum eigenstates, but the most straightforward way is to construct a normalized Gaussian wave function that is sharply peaked around a central position value x:

$$|\psi(x)\rangle_x = \int_{-\infty}^{\infty} \frac{dy}{\sqrt[4]{\pi \Delta^2}} \exp\left[-\frac{(y-x)^2}{2\Delta^2}\right] |y\rangle_x \,, \tag{2.109}$$

and similarly for a momentum value p:

$$|\psi(p)\rangle_p = \int_{-\infty}^{\infty} \frac{ds}{\sqrt[4]{\pi/\Delta^2}} \exp\left[-\frac{(s-p)^2\Delta^2}{2\hbar^2}\right] |s\rangle_p \,. \tag{2.110}$$

We will see in Chapter 9 that these are very similar to the wave functions that are used in optical quantum information processing with continuous variables. For our purposes here, it will be convenient to define the Gaussian state centered at the origin:

$$\begin{aligned}|\psi(0)\rangle &= \int_{-\infty}^{\infty} \frac{dx}{\sqrt[4]{\pi\Delta^2}} \exp\left[-\frac{x^2}{2\Delta^2}\right] |x\rangle_x \\ &= \int_{-\infty}^{\infty} \frac{dp}{\sqrt[4]{\pi/\Delta^2}} \exp\left[-\frac{p^2\Delta^2}{2\hbar^2}\right] |p\rangle_p \,. \end{aligned} \tag{2.111}$$

A translation operator $T(x)$ acting on $|\psi(0)\rangle$ is defined as

$$T(x) |\psi(0)\rangle = |\psi(x)\rangle_x \,. \tag{2.112}$$

Since the continuous-variable states now have finite precision, a translation x must be larger than the width of the Gaussian $\sqrt{2}\Delta$ if we are to distinguish between the two approximate position eigenvalues. A quantum computation of the form f_k in Eq. (2.103) is also restricted by this precision. This situation is reminiscent of classical analogue computing, in which classical continuous variables have an intrinsic precision.

We can still do robust quantum computing if we encode a finite-dimensional quantum system in the continuous variable. In particular, we will now show how to encode a qubit in a continuous variable. This encoding was introduced by Gottesman *et al.* (2001). An error in a continuous variable can be written as $x \rightarrow x+\delta$ for any $\delta \in \mathbb{R}$. Small errors mean that δ is small on some suitable metric. We may not be able to protect the value x for finite errors, but we can encode a qubit in the continuous-variable system. The code will be reminiscent of the three-qubit majority code:

$$|\bar{0}\rangle = \sum_{s=-\infty}^{+\infty} |2s\alpha\rangle \quad \text{and} \quad |\bar{1}\rangle = \sum_{s=-\infty}^{+\infty} |(2s+1)\alpha\rangle . \tag{2.113}$$

The state $|\bar{0}\rangle$ forms a comb of delta functions along the q axis, with a spacing of 2α between the delta functions. The state $|\bar{1}\rangle$ has the same structure, but the delta functions are shifted by a distance α along the x axis. A Fourier transform turns these states into combs of delta functions in p space, and the separation between the delta functions becomes $2\pi/\alpha$. A measurement along the x axis can yield either an even or an odd multiple of α, which indicates $|\bar{0}\rangle$ or $|\bar{1}\rangle$, respectively. When the translation error δ is smaller than $\alpha/2$, we can infer the qubit state from the measurement outcome.

With approximate position and momentum eigenstates, the delta functions become sharply peaked Gaussians with variance Δ, and the amplitude of the terms in the superposition decreases according to a Gaussian profile centered around $x = 0$, and with width κ. The encoded state therefore becomes

$$|\bar{0}\rangle \propto \sum_{s=-\infty}^{+\infty} \exp(-2\kappa^2 s^2 \alpha^2) \, T(2s\alpha) \, |\psi(0)\rangle$$

$$|\bar{1}\rangle \propto \sum_{s=-\infty}^{+\infty} \exp\left[-\frac{\kappa^2 (2s+1)^2 \alpha^2}{2} \right] T[(2s+1)\alpha] \, |\psi(0)\rangle , \tag{2.114}$$

up to normalization constants. The amplitudes of $|\bar{0}\rangle$ and $|\bar{1}\rangle$ are shown in Fig. 2.5. For this code to work, Δ must be small compared to α, and κ is small compared to π/α. In

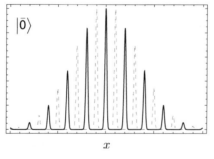
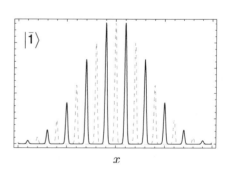

Fig. 2.5. The Gottesman–Kitaev–Preskill code for $|\bar{0}\rangle$ and $|\bar{1}\rangle$, respectively. The dashed line in the plot of $|\bar{0}\rangle$ is $|\bar{1}\rangle$ and vice versa. A small displacement error in the position x will not cause a Z error in the code space.

momentum space, the Hadamard transformed qubits $|\bar{+}\rangle = H\,|\bar{0}\rangle$ and $|\bar{-}\rangle = H\,|\bar{1}\rangle$ are again a superposition of regularly spaced Gaussians:

$$|\bar{+}\rangle \propto \sum_{m=-\infty}^{+\infty} \int_{-\infty}^{+\infty} dp \, \exp\left[-\frac{p^2\Delta^2}{2} - \frac{1}{2\kappa^2}\left(p - \frac{2\pi m}{\alpha}\right)^2\right] |p\rangle_p \,. \tag{2.115}$$

Exercise 2.16: Find the expression for $|\bar{-}\rangle$ in terms of momentum eigenstates.

2.5 References and further reading

In this chapter we introduced some of the basic ideas in quantum information processing. Our treatment here is far from complete, and for a more in-depth treatment of the various topics we refer to the general literature. Currently, the standard work on quantum information processing is the book by Nielsen and Chuang (2000). For a more recent text see Bruß and Leuchs (2007), and for a recent textbook focused mainly on the computer science aspect of quantum computing, see Mermin (2007). Cluster-state quantum computing is not covered in these books. For more details on the one-way model we refer to Raussendorf and Briegel (2001), Raussendorf *et al.* (2003), and Hein *et al.* (2004). The stabilizer formalism was developed by Gottesman (1996 and 1997).

The no-cloning theorem was first published by Wootters and Zurek (1982), and by Dieks (1982). It led to the quantum key distribution protocol by Bennett and Brassard (1984), and later to the entanglement-assisted key distribution protocol by Ekert (1991). Quantum teleportation was discovered in 1993 by Bennett *et al.*, and the first experimental demonstrations were performed by the end of the decade by Bouwmeester *et al.* (1997), Boschi *et al.* (1998), and Furusawa *et al.* (1998). See also Chapters 5 and 8 for more details on the optical implementation of quantum teleportation. Quantum repeaters were introduced by Briegel *et al.* (1998).

Continuous-variable quantum computing was developed by Lloyd and Braunstein (1999). Two edited books on continuous-variable quantum information have been published, by Braunstein and Pati (2003), and by Cerf *et al.* (2007). For a comprehensive review of continuous-variable quantum information, see Braunstein and van Loock (2005). Cluster-state quantum computing with continuous variables was developed by Menicucci *et al.* (2006), and error-correcting codes for continuous variables were constructed by Gottesman *et al.* (2001).

3

Figures of merit

Before we begin our detailed discussion of optical quantum information processing, we need to introduce certain figures of merit that quantify how well our information processor is performing. For a quantum computer, the time and resources it takes to complete a task are good measures, but we are faced with the problem that we do not know what the final design for a quantum computer will be. We therefore need additional figures of merit that are applicable in a wide range of situations. We first introduce the concept of the 'density operator', which will be used to describe quantum states about which we have incomplete knowledge. We then define the 'fidelity', which is used for assessing how close we are to a particular desired quantum state, and we discuss different measures of entanglement. The later part of the chapter will focus on figures of merit that are particularly relevant for assessing optical states, namely the first-order correlation functions, and the visibility of interference phenomena.

3.1 Density operators and superoperators

Classical physics often confronts us with situations where we can say only a limited amount about the state of a system: we can measure certain 'bulk' variables such as the temperature and pressure, but we do not know all of the details of the microscopic make-up of the state. For example, we typically have very little knowledge of the positions and velocities of all the atoms that constitute the system. Nonetheless, we can still make definite statements and predictions about the behaviour of such a system, using the methods of thermodynamics and statistical mechanics. Similarly, situations commonly arise in quantum information processing where we do not have complete knowledge of the quantum state of a system. Nevertheless, we can still apply the laws of quantum mechanics to make definite predictions about the system's behaviour. In order to do this, we define a formal way in which statistical uncertainty is represented in our description of a quantum state. We will define pure states that contain all possible information about a system, and mixed states that contain only partial information.

We will first construct an alternative representation of a pure state, $|\psi\rangle$, that can be extended to deal with classical uncertainty. For this, we need the concept of the 'trace of an operator' $\hat{A}$, which can be defined as

$$\text{Tr}\left(\hat{A}\right) = \sum_j \langle\phi_j| \hat{A} |\phi_j\rangle, \tag{3.1}$$

for *any* complete orthonormal basis $\{|\phi_j\rangle\}$: the trace is independent of the choice of basis. In matrix form, the trace is the sum of the elements on the diagonal. The trace of an operator is therefore the sum of its eigenvalues. Another important property of the trace is that it is 'cyclic':

$$\text{Tr}(\hat{A}\hat{B}\hat{C}) = \text{Tr}(\hat{B}\hat{C}\hat{A}) = \text{Tr}(\hat{C}\hat{A}\hat{B}), \tag{3.2}$$

regardless of the commutation relations between $\hat{A}$, $\hat{B}$, and $\hat{C}$. Finally, the trace operation is linear

$$\text{Tr}(a\hat{A}) = a\text{Tr}(\hat{A}) \quad \text{and} \quad \text{Tr}(\hat{A} + \hat{B}) = \text{Tr}(\hat{A}) + \text{Tr}(\hat{B}), \tag{3.3}$$

for any $a \in \mathbb{C}$.

Exercise 3.1: Prove that the trace is independent of the basis $\{|\phi_j\rangle\}$ in Eq. (3.1), and prove the cyclic property and linearity.

Consider now the expectation value of $\hat{A}$ with respect to the state $|\psi\rangle$:

$$\langle \hat{A} \rangle = \langle \psi | \hat{A} | \psi \rangle . \tag{3.4}$$

By expressing the state in an orthonormal basis $\{|\chi_n\rangle\}$ such that $|\psi\rangle = \sum_n c_n |\chi_n\rangle$ with $c_n = \langle \chi_n | \psi \rangle$, we find

$$\langle \hat{A} \rangle = \sum_{n,m} c_n^* c_m \langle \chi_n | \hat{A} | \chi_m \rangle , \tag{3.5}$$

which can be written as

$$\langle \hat{A} \rangle = \sum_{n,m} \langle \chi_m | \psi \rangle \langle \psi | \chi_n \rangle \langle \chi_n | \hat{A} | \chi_m \rangle$$

$$= \sum_m \langle \chi_m | \psi \rangle \langle \psi | \hat{A} | \chi_m \rangle . \tag{3.6}$$

In the second line we have used the completeness relation $\sum_n |\chi_n\rangle \langle \chi_n| = \hat{\mathbb{I}}$. We can now define an operator ρ such that

$$\rho = |\psi\rangle \langle \psi| , \tag{3.7}$$

and this leads to the following form of the expectation value of $\hat{A}$:

$$\langle \hat{A} \rangle = \text{Tr}\left(\rho \hat{A}\right) . \tag{3.8}$$

It is clear that ρ then gives us a very natural way of expressing the expectation value of operators. However, we have considered only pure states so far. A straightforward extension of the definition in Eq. (3.8) allows us to describe a statistical mixture of states. Consider one such mixture that is composed of states $|\psi_i\rangle$, not necessarily orthogonal, and each with a certain probability p_i. The expectation value $\langle \hat{A} \rangle$ of an operator $\hat{A}$ should also be a weighted average of the expectation values $\langle \hat{A} \rangle_i$ with respect to each state $|\psi_i\rangle$. We can write this expectation value as

$$\langle \hat{A} \rangle = \sum_i p_i \langle \hat{A} \rangle_i = \sum_i p_i \text{Tr}\left(\rho_i \hat{A}\right) = \text{Tr}\left(\sum_i p_i \rho_i \hat{A}\right) \equiv \text{Tr}\left(\rho \hat{A}\right), \tag{3.9}$$

where $\rho_i = |\psi_i\rangle\langle\psi_i|.$[1] We have generalized our definition of ρ as

$$\rho = \sum_i p_i \rho_i = \sum_i p_i |\psi_i\rangle\langle\psi_i|, \qquad (3.10)$$

where the $|\psi_i\rangle$ do not have to be orthonormal. Since an expectation value is essentially an averaging procedure, the operator ρ can be interpreted as a weight function, or a 'density'. We therefore call ρ the density operator. By construction, ρ represents the state of a *single* system that is prepared in one of the states $|\psi_i\rangle$ with probability p_i. This is sometimes called a 'proper mixture'. We will consider improper mixtures shortly.

We now consider three important properties of density operators. Since ρ is a well-behaved operator, it can be diagonalized. In other words, the matrix representation of the density operator (called the 'density matrix') can be written as a diagonal matrix. This is called the 'spectral decomposition' of the density operator

$$\rho = \sum_j \lambda_j |\lambda_j\rangle\langle\lambda_j|. \qquad (3.11)$$

As we now show, ρ must be Hermitian, which means that the λ_j are real. First, consider a pure state in the form of a projector $|\psi_i\rangle\langle\psi_i|$. Taking the Hermitian adjoint of this operator means exchanging the arguments of the bra and the ket. Clearly, the projector is invariant under the Hermitian adjoint, so it must be a Hermitian operator. Second, the sum of two Hermitian operators is also a Hermitian operator. And finally, multiplying a Hermitian operator by a real number again yields a Hermitian operator. Using these composition rules, it is clear that ρ must be Hermitian, and its eigenvalues are therefore real.

Second, we prove that $\mathrm{Tr}(\rho) = 1$. First, we calculate the trace of a pure state $|\psi_i\rangle\langle\psi_i|$. To this end, we choose an orthonormal basis $\{|\phi_j\rangle\}$ of which $|\psi_i\rangle$ is a member. This yields

$$\mathrm{Tr}(|\psi_i\rangle\langle\psi_i|) = \sum_j \langle\phi_j|\psi_i\rangle\langle\psi_i|\phi_j\rangle = \sum_j \delta_{ij} = 1. \qquad (3.12)$$

Next, we use the linearity of the trace to show that

$$\mathrm{Tr}(\rho) = \mathrm{Tr}\left(\sum_i p_i |\psi_i\rangle\langle\psi_i|\right) = \sum_i p_i \mathrm{Tr}(|\psi_i\rangle\langle\psi_i|) = \sum_i p_i = 1, \qquad (3.13)$$

where the last equality follows from the fact that probabilities must sum to unity. Therefore, the trace of any density matrix must always be one.

Third, the density operator is positive, or $\langle\psi|\rho|\psi\rangle \geq 0$ for any state $|\psi\rangle$. We can show this by noting that the eigenvalues of ρ can be interpreted as probabilities, and are therefore non-negative. If the eigenbasis of ρ is $\{|\phi_j\rangle\}$, we can write $|\psi\rangle = \sum_j c_j |\phi_j\rangle$ in this basis

---

[1] The operator ρ_i is 'idempotent', which means that $\rho_i^2 = \rho_i$. Since ρ_i is also an operator in a vector space, we call this a 'projection operator'.

and evaluate

$$\langle\psi|\,\rho\,|\psi\rangle = \sum_{jk} c_j^* c_k \langle\phi_j|\,\rho\,|\phi_k\rangle$$

$$= \sum_{jk} c_j^* c_k \langle\phi_j|\left(\sum_l p_l\,|\phi_l\rangle\,\langle\phi_l|\right)|\phi_k\rangle$$

$$= \sum_{jkl} p_l c_j^* c_k \langle\phi_j|\phi_l\rangle\langle\phi_l|\phi_k\rangle$$

$$= \sum_{jkl} p_l c_j^* c_k \delta_{jl}\delta_{kl}$$

$$= \sum_l p_l |c_l|^2$$

$$\geq 0. \tag{3.14}$$

To summarize:

  (i) the density operator has real eigenvalues, and is therefore Hermitian;
 (ii) the trace of ρ is one: $\mathrm{Tr}(\rho) = 1$;
(iii) the density operator is positive, i.e., for any $|\psi\rangle$, $\langle\psi|\,\rho\,|\psi\rangle \geq 0$.

Conversely, any operator that has these three properties can be considered a valid density operator.

Exercise 3.2: Show that the convex sum of two density operators ρ_1 and ρ_2 is again a valid density operator

$$\rho = w_1\rho_1 + w_2\rho_2, \tag{3.15}$$

where w_1 and w_2 are real, and $w_1 + w_2 = 1$.

3.1.1 The qubit density operator on the Bloch sphere

We saw earlier that any qubit state $|\psi\rangle$ can be represented by a point on the surface of the Bloch sphere, and similarly the pure-state density operator $|\psi\rangle\langle\psi|$ must correspond to the same point on the sphere. Equivalently, we can think of a pure-state representation as a vector that connects the centre of the sphere to the point of representation on the sphere's surface. It is then straightforward to represent a mixed density operator. We again begin at the centre of the sphere and then add the vectors corresponding to each state of a particular decomposition of the density operator, and scale the length of each vector by its corresponding probability in the decomposition. The point at the end of the final vector is another point in the state space. Since a density operator is a convex combination of pure states, these points must lie inside the sphere. A 'maximally' mixed state $\frac{1}{2}|0\rangle\langle0| + \frac{1}{2}|1\rangle\langle1|$ lies at the centre of the sphere.

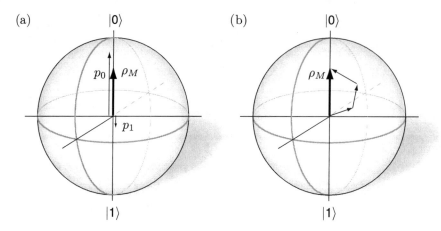

Fig. 3.1. A mixed state is represented by points inside the Bloch sphere. (a) Representation of the state $\rho_M = p_0\, |0\rangle\, \langle 0| + p_1\, |1\rangle\, \langle 1|$, in terms of the states $|0\rangle$ and $|1\rangle$. (b) An alternative decomposition of the same density matrix, in terms of three component states.

In Fig. 3.1a, we show what the state $\rho_M = p_0\, |0\rangle\, \langle 0| + p_1\, |1\rangle\, \langle 1|$ looks like in this picture. The Bloch picture also shows us that the decomposition of a given density operator into different states with certain probabilities is not unique. In Fig. 3.1b we show another possible representation of ρ_M, this time composed of three different states. The only constraint on any particular representation of a density matrix ρ is that the vectors that compose it, when added together, finish at the point that represents ρ, and, that their total length is unity.[2] Using the construction just outlined, it is obvious that there are an infinite number of possible decompositions of the density operator. However, apart from subtleties regarding degeneracies in the eigenvalues of ρ, the spectral decomposition into orthogonal states is unique.

It is tempting to ascribe a notion of physical reality to the spectral decomposition, in the sense that the state may be regarded as a real statistical distribution of the basis states of the spectral decomposition. However, our physical construction of ρ in terms of potentially non-orthogonal states $|\psi_i\rangle$ in Eq. (3.10) shows that this is inappropriate. Deriving physical consequences from elevating one decomposition of a density operator over another is called the 'partition ensemble fallacy', or PEF for short. In this sense, a quantum mechanical mixture described by a density operator is fundamentally different from a classical statistical distribution. A classical statistical distribution over states represents *our* lack of knowledge, but the system really is in one unique state. A quantum mechanical mixture also represents a lack of knowledge, but we can no longer assert that the system 'really' is in one pure state.

We have introduced the density operator as a way to describe states about which we have incomplete knowledge. For example, the state ρ may have been created by an evasive experimentalist, who prepares a quantum system in a particular pure state, but divulges only a set of several possible states with a probability distribution. She would be hiding some information, and the best we could do is to construct a statistical model of the behaviour of

---

[2] This latter constraint arises simply because the probabilities of the states in the decomposition must sum to unity.

that state. As a consequence of avoiding the PEF, there is no operational procedure based on measuring the system with arbitrary precision that will reveal which state the experimenter chose to prepare. This remains true even if the experimenter sends many states that have been prepared according to her probability distribution. We will see in Chapter 5 that this will have profound consequences for quantum communication.

The density operator can also be interpreted in a much more fundamental way. Our evasive experimenter may also be described as a (rather complex) quantum system that has interacted with the system we are studying. Since we really do not have enough information to describe the system and the experimentalist as a whole, we need a procedure that allows us to ignore the experimentalist altogether. As we shall see, it turns out that this procedure will lead directly to mixing in the quantum state. Density operators that arise in this way are sometimes called 'improper mixtures'.

In order to see how to do this, we give an abstract representation of the experimentalist, or any other large system that we cannot describe in detail. We will call this the 'environment'. Consider the case of a qubit interacting with its environment. Assume that the qubit and environment are initially in a highly correlated entangled state $|\Psi\rangle$, given by

$$|\Psi\rangle = c_{00} |0\rangle_q \otimes |\psi_0\rangle_e + c_{11} |1\rangle_q \otimes |\psi_1\rangle_e \,, \tag{3.16}$$

where the subscripts q and e denote the qubit and the environment, respectively. We will assume that through measurements we can directly observe the behaviour of the qubit alone. Next, suppose that we wish to make a measurement on our qubit of the particular observable $\hat{A}$. Since we are measuring a property of the qubit and not the environment, the measurement operator takes the form $\hat{A}_{qe} = \hat{A}_q \otimes \hat{\mathbb{I}}_e$. The expectation value is

$$\langle\Psi| \hat{A}_{qe} |\Psi\rangle = |c_{00}|^2 \langle 0|_q \hat{A}_q |0\rangle_q + |c_{11}|^2 \langle 1|_q \hat{A}_q |1\rangle_q \,. \tag{3.17}$$

This is exactly what we would have calculated if we had used a density matrix for q, with the probability $|c_{00}|^2$ for state $|0\rangle$ and $|c_{11}|^2$ for state $|1\rangle$. This result is completely general, and there is a well-defined operation that takes us from a pure density operator $|\Psi\rangle \langle\Psi|$ for a larger system to the density operator of a smaller part of that system. It is called the 'partial trace', denoted by a subscript indicating the system that is traced over:

$$
\begin{aligned}
\rho_q &= \text{Tr}_e(\rho_{qe}) \\
&= \text{Tr}_e \left(\sum_{ijkl} r_{ij} |\phi_i\rangle_q \otimes |\psi_j\rangle_e \langle\phi_k|_q \otimes \langle\psi_l|_e r_{kl}^* \right) \\
&= \text{Tr}_e \left(\sum_{ijkl} r_{ij} r_{kl}^* |\phi_i\rangle_q \langle\phi_k| \otimes |\psi_j\rangle_e \langle\psi_l| \right) \\
&= \sum_{ijklm} r_{ij} r_{kl}^* |\phi_i\rangle_q \langle\phi_k| \langle\chi_m|\psi_j\rangle \langle\psi_l|\chi_m\rangle \\
&= \sum_{ijkl} r_{ij} r_{kl}^* \langle\psi_l|\psi_j\rangle \, |\phi_i\rangle_q \langle\phi_k| \,.
\end{aligned}
\tag{3.18}
$$

It is commonly the case that the density matrix is expressed in an orthonormal basis, in which case $\langle \psi_l | \psi_j \rangle = \delta_{lj}$, and the above expression simplifies to

$$\rho_q = \sum_{ik} s_{ik} |\phi_i\rangle_q \langle\phi_k| , \qquad (3.19)$$

with $s_{ik} = \sum_l r_{kl}^* r_{il}$. This is generally a mixed state. Note that we started with a pure state for the qubit and its environment, and the partial trace over the environment produces a mixed state of the qubit. The partial trace over the environment is equivalent to discarding any information about the environment. We often have no choice but to take the partial trace when we do not *have* any information about the environment. When the qubit and the environment are in a separable state, the partial trace over the environment does not mix the state of the qubit. For the qubit state to become mixed, it must be entangled with the environment.

This interpretation of the density operator, as the state of a system that has interacted and become entangled with other systems, is of fundamental importance in quantum information processing. For example, a single electron spin that is embedded in a solid-state environment is not isolated. It interacts with a whole range of different things in the environment such as phonons, other electron spins, or nuclear spins. Indeed, it is never possible to completely isolate the behaviour of a quantum system from its surroundings. In quantum information processing, this often manifests itself as errors that must be corrected or protected against.

3.1.2 Superoperators and completely positive maps

Next, we consider the mathematical properties of general operators acting on density operators. Suppose that we have a physical system described by a set of accessible states $\{|\phi_j\rangle\}$. The superposition principle and the linearity of quantum mechanics imply that this set spans a Hilbert space $\mathcal{H}$ of dimension d. This is a complex vector space with an orthonormal basis

$$\begin{pmatrix} 1 \\ 0 \\ \vdots \\ 0 \end{pmatrix}, \begin{pmatrix} 0 \\ 1 \\ \vdots \\ 0 \end{pmatrix}, \ldots, \begin{pmatrix} 0 \\ 0 \\ \vdots \\ 1 \end{pmatrix}.$$

We can define a set of linear operators $\{\hat{A}_k\}$ on $\mathcal{H}$ that transform states into states:

$$\hat{A}_k : |\phi_j\rangle \longrightarrow |\phi_j'\rangle , \qquad (3.20)$$

where $|\phi'\rangle$ is again a state in $\mathcal{H}$. A choice of a basis in $\mathcal{H}$ puts Hermitian non-negative operators into a one-to-one correspondence with Hermitian non-negative matrices. The set of all linear operators $\{\hat{A}_k\}$ in turn define a Hilbert space $\mathcal{H} \otimes \mathcal{H}$ of dimension d^2, one orthonormal basis of which can be written as

$$
\begin{pmatrix} 1 & 0 & \cdots & 0 \\ 0 & \ddots & & \\ \vdots & & \vdots & \\ 0 & 0 & \cdots & 0 \end{pmatrix} , \begin{pmatrix} 0 & 1 & \cdots & 0 \\ 0 & \ddots & & \\ \vdots & & \vdots & \\ 0 & 0 & \cdots & 0 \end{pmatrix} , \ldots , \begin{pmatrix} 0 & 0 & \cdots & 0 \\ 0 & \ddots & & \\ \vdots & & \vdots & \\ 0 & 0 & \cdots & 1 \end{pmatrix} . \tag{3.21}
$$

We can now define an even higher set of objects called 'maps', denoted by $\{\mathcal{L}\}$, the elements of which linearly transform the set of linear operators into itself:

$$
\mathcal{L} : \hat{A} \longrightarrow \hat{A}' = \mathcal{L}(\hat{A}) . \tag{3.22}
$$

As an example, consider the Schrödinger equation

$$
i\hbar \frac{d}{dt} |\psi\rangle = \mathcal{H} |\psi\rangle , \tag{3.23}
$$

which holds for any isolated system in a pure state $|\psi\rangle$ (in the absence of measurements). Using the definition of the density operator in Eq. (3.10), we can write

$$
i\hbar \frac{d}{dt} \rho = [\mathcal{H}, \rho] \equiv \mathcal{L}(\rho). \tag{3.24}
$$

This is the von Neumann equation, and defines the Liouvillian $\mathcal{L}$.

Exercise 3.3: Using Eq. (3.10) show that the equation of motion for the density matrix is indeed given by the von Neumann equation.

The linear maps $\mathcal{L}$ are sometimes called 'superoperators'. They are operators on the Hilbert space $\mathscr{H} \otimes \mathscr{H}$, but we give them a different name to avoid confusion. A simple example of a map corresponds to a unitary transformation $U : \hat{A} \rightarrow \hat{A}' = U^\dagger \hat{A} U$. The corresponding map $\mathcal{L}_U$ may be written as $\mathcal{L}_U = U^\dagger \otimes U$. The set of all maps thus constitutes a Hilbert space $\mathscr{H}^{\otimes 4}$ of dimension d^4. A map $\mathcal{L}$ is called *positive* if for every non-negative operator $\hat{A}$ the operator $\hat{A}' = \mathcal{L}(\hat{A})$ is again a non-negative operator. When $\mathrm{Tr}(\hat{A}') = \mathrm{Tr}(\hat{A})$ for every $\hat{A}$, the map $\mathcal{L}$ is 'trace-preserving'. Trace-preserving positive maps are important in quantum mechanics, since they transform the set of density operators to itself. This property may lead one to expect that these maps correspond to physical processes or symmetries. Interestingly, not all positive maps can be associated with physical processes in quantum mechanics. This subtle fact becomes important when we consider composite systems.

Suppose we have two systems 1 and 2 with respective accessible states $\{|\phi_j^{(1)}\rangle\}$ and $\{|\psi_k^{(2)}\rangle\}$. These states span two Hilbert spaces $\mathscr{H}^{(1)}$ and $\mathscr{H}^{(2)}$ with dimensions d_1 and d_2, respectively. The accessible states of the composite system can be written on the basis of the tensor product of the states $|\phi_j^{(1)}\rangle \otimes |\psi_k^{(2)}\rangle$, generating a Hilbert space $\mathscr{H}^{(1)} \otimes \mathscr{H}^{(2)}$ of dimension $d_1 \times d_2$. Similarly, the set of linear operators $\{\hat{A}_j\}$ on $\mathscr{H}^{(1)} \otimes \mathscr{H}^{(2)}$ generates a Hilbert space of dimension $(d_1 \times d_2)^2$, and the set of maps generates a Hilbert space of dimension $(d_1 \times d_2)^4$.

Consider a map $\mathcal{L}_1$, defined for subsystem 1. When this map is positive (and trace-preserving) it transforms density operators of the subsystem to density operators. When system 1 is part of a composite system $1 + 2$, we want to know when a positive map of system 1 (leaving system 2 unchanged) would transform a density operator defined on the composite system again into a density operator. In other words, we ask when the 'extended' map $\mathcal{L}_{12} = \mathcal{L}_1 \otimes \hat{\mathbb{I}}_2$, with $\hat{\mathbb{I}}_2$ the identity map of system 2, is again positive. Maps with this property have a special name: a map $\mathcal{L}_1$ is called 'completely positive' if all its extensions are positive.

There exist maps that are positive, but not completely positive. One such map is the transpose. Take, for example, the singlet state of a two-qubit system written in the computational basis

$$|\Psi\rangle = \frac{1}{\sqrt{2}} (|01\rangle - |10\rangle) \ . \tag{3.25}$$

The density operator of this state can be written as

$$\rho = |\Psi\rangle \langle\Psi| = \frac{1}{2} (|01\rangle \langle 01| - |01\rangle \langle 10| - |10\rangle \langle 01| + |10\rangle \langle 10|) \ . \tag{3.26}$$

The transpose of a general density operator for a single system in this notation is given by

$$\mathcal{T} : a\,|0\rangle \langle 0| + b\,|0\rangle \langle 1| + c\,|1\rangle \langle 0| + d\,|1\rangle \langle 1|$$
$$\longrightarrow \quad a\,|0\rangle \langle 0| + b\,|1\rangle \langle 0| + c\,|0\rangle \langle 1| + d\,|1\rangle \langle 1| \ , \tag{3.27}$$

that is, we *exchange* the entries of the bras and kets. This is a positive map. The extended transpose (or 'partial transpose') on a compound system $\mathcal{T}_\wp = \mathcal{T}_1 \otimes \hat{\mathbb{I}}_2$, however, is *not* positive. To see this, apply the extended transpose to the density operator given in Eq. (3.26), and we obtain

$$\mathcal{T}_\wp : \rho \longrightarrow \rho' = \frac{1}{2} (|01\rangle \langle 01| - |11\rangle \langle 00| - |00\rangle \langle 11| + |10\rangle \langle 10|) \ . \tag{3.28}$$

If the eigenvalues of ρ' are non-negative, ρ' is again a density operator. In order to find the eigenvalues of this operator we write ρ' in matrix representation on the computational basis:

$$\rho' = \frac{1}{2} \begin{pmatrix} 0 & 0 & 0 & -1 \\ 0 & 1 & 0 & 0 \\ 0 & 0 & 1 & 0 \\ -1 & 0 & 0 & 0 \end{pmatrix} \ . \tag{3.29}$$

It is easily found that this matrix has eigenvalues $1/2$ (with multiplicity 3) and $-1/2$. Therefore, ρ' is *not* a density operator, and $\mathcal{T}$, although positive, is not a completely positive map. The fact that positive but not completely positive maps on a subsystem do not necessarily transform density operators on the composite system to density operators can be exploited to detect (or witness) quantum entanglement. We will return to this in Section 3.3.

We now present an important class of completely positive maps. Consider the general map

$$\mathcal{L} : \hat{A} \longrightarrow \hat{A}' = \mathcal{L}(\hat{A}) , \tag{3.30}$$

with $\hat{A}$ and $\hat{A}'$ linear operators on the system Hilbert space. An important special case of such a map is given by

$$\mathcal{L} : \hat{A} \longrightarrow \hat{A}' = \sum_k \lambda_k \hat{B}_k \hat{A} \hat{B}_k^\dagger , \tag{3.31}$$

where the $\hat{B}_k$ are again linear operators. When $\lambda_k \geq 0$ for all k, the map in Eq. (3.31) is again a positive map. To prove this statement, note that

$$\langle \phi | \hat{B}_k \hat{A} \hat{B}_k^\dagger | \phi \rangle = ((\langle \phi | \hat{B}_k) \hat{A} (\hat{B}_k^\dagger | \phi \rangle)) \equiv \langle \phi' | \hat{A} | \phi' \rangle \geq 0 \tag{3.32}$$

for all $\hat{B}_k$ and $|\phi\rangle$ if $\hat{A}$ is non-negative. We then have

$$\langle \phi | \hat{A}' | \phi \rangle = \sum_k \lambda_k \langle \phi | \hat{B}_k \hat{A} \hat{B}_k^\dagger | \phi \rangle = \sum_k \lambda_k \langle \phi_k' | \hat{A} | \phi_k' \rangle . \tag{3.33}$$

The right-hand side of this equation is positive for all $\hat{B}_k, \lambda_k \geq 0$, and non-negative operators $\hat{A}$. Hence $\hat{A}'$ is a non-negative operator and $\mathcal{L}$ is positive. Furthermore, when $\lambda_k \geq 0$ for all k, such a map $\mathcal{L}$ is *completely* positive. To prove this statement, let $\mathcal{L}$ be a map on system 1 (henceforth denoted by $\mathcal{L}_1$) and consider a second system 2. Recall that $\mathcal{L}_1$ is completely positive if all its extensions $\mathcal{L}_1 \otimes \hat{\mathbb{I}}_2$ are positive. Define the extension $\mathcal{L}_{12} = \mathcal{L}_1 \otimes \hat{\mathbb{I}}_2$. System 2 can have arbitrary dimension, and may itself be composite. We thus have to show that $\mathcal{L}_{12}$ is positive. Let $\hat{A}_{12}$ be a non-negative operator on the composite system $1 + 2$:

$$\hat{A}_{12} = \sum_{j,k;l,m} a_{jk,lm} |\phi_j\rangle_1 |\psi_k\rangle_2 \langle \psi_m |_1 \langle \phi_l | , \tag{3.34}$$

with $a_{jk,lm} = a_{lm,jk}^*$, and define the operator $\hat{B}_i$ on system 1 as

$$\hat{B}_i = \sum_{p,q} b_{pq}^i |\phi_p\rangle \langle \phi_q | . \tag{3.35}$$

The map $\mathcal{L}_1 \otimes \hat{\mathbb{I}}_2$ is then given by the transformation

$$
\begin{aligned}
\hat{A}_{12} \to \hat{A}'_{12} &= \sum_i \lambda_i (\hat{B}_i \otimes \hat{\mathbb{I}}_2) A_{12} (\hat{B}_i \otimes \hat{\mathbb{I}}_2)^\dagger \\
&= \sum_i \lambda_i \sum_{p,q;r,s} \sum_{j,k;l,m} a_{jk,lm} b_{pq}^i |\phi_p\rangle_1 \langle \phi_q | \phi_j \rangle_1 |\psi_k\rangle_2 \langle \psi_m |_1 \langle \phi_l | \phi_r \rangle_1 \langle \phi_s | b_{sr}^{i*} \\
&= \sum_i \lambda_i \sum_{p,s} \sum_{j,k;l,m} a_{jk,lm} b_{pj} |\phi_p\rangle_1 |\psi_k\rangle_2 \langle \psi_m |_1 \langle \phi_s | b_{sl}^{i*} \\
&= \sum_i \lambda_i \sum_{j,k;l,m} a_{jk,lm} |\phi_j'\rangle_1 |\psi_k\rangle_2 \langle \psi_m |_1 \langle \phi_l' | ,
\end{aligned}
\tag{3.36}
$$

where we defined $|\phi'_j\rangle = \sum_p b_{pj}|\phi_p\rangle$. Since this is a convex sum over non-negative operators, the resulting operator is again non-negative and $\mathcal{L}$ is completely positive. This completes the proof.

Let us now return to the case of a single system and the states, operators, and maps defined on it. Consider a projection operator P defined by

$$P^\dagger = P \qquad \text{and} \qquad P^2 = P. \tag{3.37}$$

In terms of the states $\{|\phi_j\rangle\}$ this operator can be written as

$$P_j = |\phi_j\rangle\langle\phi_j|. \tag{3.38}$$

Suppose we have a set of projection operators $\{P_\mu\}$, with the states $|\mu\rangle$ not necessarily orthogonal. We can define a 'generalized projection operator' $\hat{E}_\nu$ as a weighted measure over this set:

$$\hat{E}_\nu = \sum_\mu \lambda_\mu^\nu P_\mu \, , \tag{3.39}$$

with $\lambda_\mu^\nu > 0$ and

$$\sum_\nu \hat{E}_\nu = \hat{\mathbb{1}} \, . \tag{3.40}$$

The operator E_k is called a 'positive operator-valued measure' or POVM for short. This can be generalised further by observing that

$$\lambda_\mu^\nu P_\mu = \alpha_\mu^\nu |\mu\rangle\langle\nu|\nu\rangle\langle\mu|\alpha_\mu^{\nu*} \, , \tag{3.41}$$

with $|\nu\rangle \in \{|\mu\rangle\}$ and $|\alpha_\mu^\nu|^2 = \lambda_\mu^\nu$. When we define the operator $\mathcal{A}$:

$$\mathcal{A} \equiv \sum_{\mu,\nu} \alpha_\mu^\nu |\mu\rangle\langle\nu| \tag{3.42}$$

we can write the POVM as

$$E_\nu = \sum_\mu \mathcal{A}_{\mu\nu}\mathcal{A}_{\mu\nu}^\dagger \, . \tag{3.43}$$

The operators $\mathcal{A}_{\mu\nu}$ are called 'Kraus operators'. We will use the techniques of superoperators, POVMs, and completely positive maps extensively throughout the book.

3.2 The fidelity

So far, we have developed the formalism that allows us to describe the evolution of quantum systems in the presence of information loss, which tends to manifest itself in quantum information processing as the occurrence of errors. A different question is how similar a given output state is to the ideal state without errors. This notion is quantified by the 'fidelity'. This concept is ubiquitous in quantum information processing, since it gives a numerical value to the quality of state-preparation procedures, quantum gates, etc. Equally important, the fidelity has an operational definition, which means that it does not rely on interpretation-dependent concepts.

The fidelity of a quantum state with respect to a reference state is the probability that the state is mistaken for the reference state in a measurement. For two pure states $|\psi\rangle$ and $|\psi_0\rangle$, this probability is the overlap of the two states in Hilbert space, or

$$F = |\langle \psi_0 | \psi \rangle|^2, \tag{3.44}$$

where we have chosen $|\psi_0\rangle$ as the reference state. Nevertheless, it is clear that this pure-state fidelity is symmetric in the states.

The fidelity of two pure states, although conceptually important, has limited applicability in practical problems. Usually the quantum state that is produced in a real procedure is mixed, and is represented by the density operator ρ. In this case, the fidelity with respect to a pure state follows by extending Eq. (3.44) using the definition Eq. (3.10)

$$F = \langle \psi_0 | \rho | \psi_0 \rangle = \text{Tr}[\rho \, |\psi_0\rangle \langle \psi_0|]. \tag{3.45}$$

Some authors define the fidelity as the square root of this number, but that precludes its interpretation given above. It also gives a misleadingly high value.

Finally, for completeness we define the fidelity for two mixed states ρ_1 and ρ_2:

$$F = \left[\text{Tr} \left(\sqrt{\rho_2^{1/2} \, \rho_1 \, \rho_2^{1/2}} \right) \right]^2. \tag{3.46}$$

This measure is again symmetric in ρ_1 and ρ_2, even though this is not obvious from the notation. It is not used very often, because for most applications the desired state is a pure state.

3.3 Entropy, information, and entanglement measures

One of the fundamental resources for quantum information processing is entanglement. It is therefore crucial that we are able to quantify the amount of entanglement in a system. There are many ways we can do this, and here we will introduce the most widely known techniques, namely the partial-transpose criterion, the von Neumann entropy, the concurrence, and the entanglement of formation.

3.3.1 Partial-transpose criterion

The partial-transpose criterion is an algorithm that can detect entanglement in the state of two systems. It does not give a numerical value of the entanglement, and can therefore not be considered a measure. Nevertheless, it can be a valuable tool in quantum information theory. It also provides a profound example of the theory of completely positive maps, developed in Section 3.1.

A density operator ρ of two systems $1 + 2$ is separable if and only if it can be written as

$$\rho = \sum_k p_k \rho_k^{(1)} \otimes \rho_k^{(2)} , \tag{3.47}$$

with $p_k \geq 0$ and $\sum_k p_k = 1$. The density operator $\rho_k^{(j)}$ is defined on system $j = 1, 2$. Consider again the transpose of an operator $\hat{A}$:

$$\mathcal{T} : \hat{A} \longrightarrow \hat{A}^T . \tag{3.48}$$

As we have seen, this is a trace-preserving, positive, but not completely positive map. We showed this by extending the map to the 'partial transpose' $\mathcal{T}_\wp \equiv \mathcal{T}_1 \otimes \hat{\mathbb{1}}_2$. The partial transpose is not positive on the composite system.

Under the partial transpose, the separable density operator from Eq. (3.47) will transform according to

$$\mathcal{T}_\wp : \rho \longrightarrow \rho' = \sum_k p_k \left(\rho_k^{(1)} \right)^T \otimes \rho_k^{(2)} . \tag{3.49}$$

However, $\mathcal{T}$ is positive and $(\rho_k^{(1)})^T$ is again a density operator. Therefore ρ' is another (separable) density operator. Now look at the eigenvalues of ρ': clearly, if ρ is separable, then ρ' has positive eigenvalues. Therefore, if ρ' has one or more *negative* eigenvalues, the original density operator ρ must have been entangled. This is the Peres–Horodecki partial-transpose criterion and is generally a necessary but not sufficient condition for separability.

Denoting a Hilbert space of dimension d by $\mathcal{H}_d$, it has been proved that for the Hilbert spaces $\mathcal{H}_2 \otimes \mathcal{H}_2$ and $\mathcal{H}_2 \otimes \mathcal{H}_3$ the partial-transpose criterion is both necessary and sufficient. In other words, ρ is separable *if and only if* the eigenvalues of its partial transpose ρ' are positive. For higher-dimensional Hilbert spaces this is no longer true. In that case there can exist density operators which are *not* separable, but for which the eigenvalues of ρ' are non-negative. Such states are said to exhibit 'bound entanglement'.

3.3.2 Von Neumann entropy

In classical physics, entropy is a measure of the disorder of a system. In statistical mechanics it is related (through Boltzmann's equation) to the number of microstates of a system that correspond to a given macrostate. The more microstates there are, the higher the entropy. An alternative viewpoint is that entropy is a measure of how much information we would gain if we could find out which of the microstates is giving rise to the macrostate. If there was only one microstate, no information would be gained, so the entropy is zero. More microstates give rise to higher entropies. Information theorists use exactly this interpretation of entropy to quantify how much memory it takes to represent a particular random variable associated with an information register or communication channel. This measure is called the 'Shannon entropy'. For a random variable A with probability distribution $\{p_a\}$ it is

$$H(A) = -\sum_a p_a \log_2 p_a. \tag{3.50}$$

This is a classical quantity, and we wish to define its quantum mechanical counterpart. In Section 3.1, we introduced the density matrix as a way of representing uncertainty in quantum states. But how much uncertainty is associated with a particular density matrix? We define the von Neumann entropy of a density operator ρ_A for system A as

$$S(A) = S(\rho_A) \equiv -\text{Tr}(\rho \log_2 \rho) = -\sum_i \lambda_i \log_2 \lambda_i, \tag{3.51}$$

where the λ_i are the eigenvalues of ρ_A. In order to make the definition complete, we must also assert that $0 \log_2 0 \equiv 0$ (or $0^0 = 1$).

Exercise 3.4: Prove the final equality in Eq. (3.51). (Hint: if matrices A and B are related by $A = e^B \equiv 1 + B + B^2/2 + ...$, then $B = \ln A$.)

Note that while $H(A)$ is a classical quantity, $S(A)$ is a quantum mechanical quantity, even though they are both numbers. They are closely related. As an example, a qubit density operator has eigenvalues x and $1 - x$, such that the von Neumann entropy becomes

$$S(\rho) = -x \log_2(x) - (1 - x) \log_2(1 - x) \equiv h(x). \tag{3.52}$$

This is identical to the Shannon entropy of a classical bit with probability distribution $\{x, 1 - x\}$. If ρ is a pure state, then it has one eigenvalue of unity (and the rest are zero), and the entropy is therefore zero. This is what we want, since there is no uncertainty built into pure states. For a maximally mixed state, $\rho = \hat{\mathbb{I}}/d$ (with d the dimension of the Hilbert space), it is easy to show that $S(\rho) = \log_2 d$, and this is the maximum possible value in a Hilbert space of that dimension.

The entropy is a useful measure of how mixed a particular state is. It is not a direct measure of entanglement in a system, but it can be used to at least partially characterize entanglement. Consider a two-qubit system that is in a pure state. If we use the partial trace in Eq. (3.18) to construct the density operator ρ for one of the two qubits, we find that if the two qubits are entangled, then ρ is mixed. If the two qubits are in a (pure) product state, then ρ is pure. The single-qubit entropy is therefore zero for an unentangled state and maximal for a maximally entangled state. It can also be proved that the entropy is monotonic: it is always non-negative, zero for separable states and maximal for maximally entangled states; the numerical value cannot be changed by 'local operations and classical communication' (LOCC) between the two subsystems; and the entropy is convex under information loss:

$$\sum_i p_i S(\rho_i) \geq S\left(\sum_i p_i \rho_i\right). \tag{3.53}$$

It makes no difference which of the two subsystems one chooses to look at; both give the same value of the entropy if together they are in a pure state. Therefore, the single-qubit entropy can be used to characterize entanglement, in this limited sense of a pure joint state. However, we would like to find a more general measure that would take any full density matrix of two qubits as its input and yield a numerical value which quantifies the amount of entanglement between the qubits.

3.3.3 Concurrence and entanglement of formation

The von Neumann entropy measure of entanglement for pure two-qubit states can be extended to the mixed case by first introducing the 'concurrence' C. Let ρ_{AB} denote the matrix representation of our two-qubit state in the computational basis $\{00, 01, 10, 11\}$ and ρ_{AB}^* is its complex conjugate. Then C is defined to be

$$C = [\max\{\beta_1 - \beta_2 - \beta_3 - \beta_4, 0\}], \tag{3.54}$$

where the β_i are the square roots of the eigenvalues (in decreasing order) of the matrix

$$\rho_{AB}\, \tilde{\rho}_{AB} = \rho_{AB}\, Y_A \otimes Y_B\, \rho_{AB}^*\, Y_A \otimes Y_B, \tag{3.55}$$

where Y_i is the Pauli Y operator acting on system i. The concurrence C is a monotonic entanglement measure in its own right, having a zero value for product states and a unity value for maximally entangled states. However, it does not reduce to the von Neumann entropy measure.

In contrast to the concurrence, the 'entanglement of formation' E_F corresponds exactly to the reduced von Neumann entropy in the case of pure two-qubit states. It is defined as

$$E_F(\rho) = h\left(\frac{1 + \sqrt{1 - C^2}}{2}\right), \tag{3.56}$$

where $h(x)$ is defined in Eq. (3.52). The entanglement of formation tells us how many Bell states we need on average to prepare the state ρ using some prescribed LOCC procedure.

3.3.4 Mutual information and the Holevo bound

We have introduced the Shannon entropy $H(A)$ for a classical random variable A, and the von Neumann entropy $S(\rho)$ for a quantum system in the state ρ. We can also introduce a 'joint entropy' $H(A, B)$ for two classical random variables A and B

$$H(A, B) = -\sum_{a,b} p(a, b) \log_2 p(a, b), \tag{3.57}$$

where $p(a, b)$ is the probability that $A = a$ and $B = b$. It is straightforward to extend this definition to a joint entropy $H(A, B, C, \ldots)$ of many random variables A, B, C, etc. The joint entropy measures the total uncertainty about the random variables. We can use the

Shannon entropy of two random variables A and B, and the joint entropy to define the 'mutual information' $H(A : B)$

$$H(A : B) = H(A) + H(B) - H(A, B). \tag{3.58}$$

It is a measure of how much information A and B have in common.

Quantum mechanically, we can also define the joint entropy and the mutual information, according to

$$S(A, B) = S(\rho_{AB}) = -\text{Tr}\left(\rho_{AB} \log_2 \rho_{AB}\right), \tag{3.59}$$

and

$$S(A : B) = S(A) + S(B) - S(A, B). \tag{3.60}$$

Usually, in quantum communication protocols, after the quantum mechanical part the communicating parties have strings of classical measurement outcomes (i.e., random variables), and therefore $S(A : B) = H(A : B) \equiv I(A : B)$ is the mutual information between A and B. In both the classical and the quantum case, $I(A : B)$ is symmetric.

For any (quantum) communication protocol, the mutual information between the two parties, Alice and Bob, is bounded. This is the celebrated 'Holevo bound', also known as 'Holevo's theorem'.

Holevo's theorem: With probability p_a Alice prepares a mixed state ρ_a chosen from a set of states $\{\rho_1, \ldots, \rho_N\}$. Bob performs an arbitrary (POVM) measurement on this state, with outcomes described by the random variable B. The mutual information between Alice and Bob is then bounded by

$$I(A : B) \leq S(\rho) - \sum_{a=1}^{N} p_a S(\rho_a), \tag{3.61}$$

where $\rho = \sum_a p_a \rho_a$.

The proof is given in Nielsen and Chuang (2000). This is a very powerful theorem, and we will use this bound in Chapter 5 to prove the security of quantum cryptographic protocols.

3.4 Correlation functions and interference of light

So far, we have concentrated on generally applicable measures of the quality of quantum states. However, these are not always the most useful measures when considering *optical coherence*. In particular, there exist quantum correlation functions that are more directly relevant for assessing typical results of optics experiments. To see this, let us first consider the famous Young's double-slit experiment, depicted in Fig. 3.2. Light impinges on the first screen, containing two slits that are placed close to each other. Due to Huygens' principle, the slits become the source of two diffracted optical waves, and a second (observation) screen displays the interference pattern resulting from the two waves. At the observation

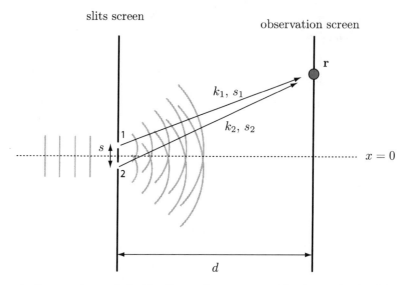

slits screen

observation screen

Fig. 3.2. A schematic diagram of Young's double-slit experiment. We consider the amplitude of the detected signal at a position **r** in the observation screen. The waves coming from the two slits have the same frequency ω, and their wavevectors have the same magnitude, $|\mathbf{k}| = |\mathbf{k}'| = k$. The distance between slit 1 or 2 and the point **r** is s_1 or s_2.

screen, light is reflected to the observer, who sees light and dark fringes. At the quantum level, individual photons are absorbed and re-emitted by the material of the screen. The simplest quantum model that allows us to capture this kind of process is to picture the observation screen as a set of individual atoms. Each atom has two levels, whose separation is equal to the energy of the photons in the original light beam.

The interaction Hamiltonian that describes how the atom and photon couple together will be properly derived in Chapter 7. Here we use the result of the derivation, and write

$$\mathcal{H}_I = -\mathbf{E}(\mathbf{r}, t) \cdot \mathbf{p}, \tag{3.62}$$

where **p** is the dipole operator that connects the ground and excited states of the atoms. If the atom is initially in the ground state, then the only possible transition is to the excited state. This increases the energy of the atom, so must be associated with a decrease in the energy of the field. The relevant associated field operator must therefore be the photon annihilation operator.

Let us consider a single optical mode, and construct its creation and annihilation operators $\hat{a}$ and $\hat{a}^\dagger$. In terms of this single mode, we can write the electric field operator in the form

$$\hat{E}(\mathbf{r}, t) = \hat{E}^+(\mathbf{r}, t) + \hat{E}^-(\mathbf{r}, t) \tag{3.63}$$

where

$$\hat{E}^+ \propto \hat{a} \, \exp[i(\omega t - \mathbf{k} \cdot \mathbf{r})]$$

$$\hat{E}^- \propto \hat{a}^\dagger \, \exp[-i(\omega t - \mathbf{k} \cdot \mathbf{r})]. \tag{3.64}$$

The detector absorbs a quantum of energy if the field undergoes the loss of a single photon, and the relevant matrix element for this process is therefore

$$M_{if} = \langle f | \hat{E}^+(\mathbf{r}, t) | i \rangle, \tag{3.65}$$

where $|i\rangle$ is the initial field state and $|f\rangle$ its final state. By 'Fermi's golden rule', the probability of the detector registering a photon is therefore proportional to $|M_{if}|^2$.

Exercise 3.5: By assuming a complete set of final states for the field, show that the square of the detection process matrix element can be written as

$$|M_{if}|^2 = \langle i | \hat{E}^-(\mathbf{r}, t)\hat{E}^+(\mathbf{r}, t) | i \rangle. \tag{3.66}$$

For a mixed initial state ρ of the field we can make a simple extension:

$$|M_{if}|^2 = \text{Tr}[\rho \hat{E}^-(\mathbf{r}, t)\hat{E}^+(\mathbf{r}, t)]. \tag{3.67}$$

This quantity is a measure of how well correlated the two operators $\hat{E}^+$ and $\hat{E}^-$ are at a position $\mathbf{r}$ and time t. We can generalize this, and define the 'first-order' quantum correlation function:

$$G^{(1)}(\mathbf{r}, \mathbf{r}', t, t') = \text{Tr}[\rho \hat{E}^-(\mathbf{r}, t)\hat{E}^+(\mathbf{r}', t')]. \tag{3.68}$$

This function is all we need to describe the interference that occurs in Young's experiment. To see this let us calculate it for a point $\mathbf{r}$ on the observation screen. We can now write the field operator as a superposition of two modes, one coming from each slit:

$$\hat{E}^+(\mathbf{r}, t) = \hat{E}_1^+(\mathbf{r}, t) + \hat{E}_2^+(\mathbf{r}, t). \tag{3.69}$$

With reference to Fig. 3.2, we see that the field at the screen originating from slit 1 is equal to the field that existed at the position of the slit a time $\tau = s_1/c$ earlier, with an amplitude that has fallen by a factor of $1/s_1$ for a spherical mode. Therefore

$$\hat{E}_i^+(\mathbf{r}, t) = \hat{E}^+\left(\mathbf{r}_i, t - \frac{s_i}{c}\right)\frac{1}{s_i}. \tag{3.70}$$

Here, $\mathbf{r}_i$ is the position vector for slit i. For the factors appearing in the denominator, we can assume that $s_1 \approx s_2 \approx R$. By substituting Eq. (3.70) into Eqs. (3.69) and Eq. (3.67) we find that

$$R^2|M_{if}|^2 = G^{(1)}(x_1, x_1) + G^{(1)}(x_2, x_2) + 2\text{Re}[G^{(1)}(x_1, x_2)] \tag{3.71}$$

where we have grouped $\mathbf{r}$ and t coordinates into a single spacetime variable x and $t_i \equiv t - s_i/c$. The first two terms on the right-hand side are autocorrelation functions, and so the phase terms in Eq. (3.64) are such that they cancel in $G^{(1)}$, which is then equal to a real number. On the other hand, the final term has different phases for the $\hat{E}^+$ and $\hat{E}^-$ parts and so can be written

$$G^{(1)}(x_1, x_2) = |G^{(1)}(x_1, x_2)|e^{i\omega(t_1 - t_2)} = |G^{(1)}(x_1, x_2)|e^{ik(s_1 - s_2)/c}. \tag{3.72}$$

We can capture the phase dependence more succinctly by defining the normalized correlation functions

$$g^{(1)}(x_1, x_2) \equiv \frac{G^{(1)}(x_1, x_2)}{\sqrt{G^{(1)}(x_1, x_1) G^{(1)}(x_2, x_2)}} = e^{ik(s_1 - s_2)/c}. \tag{3.73}$$

The positions of the maxima and minima of intensity are given entirely by the value of this function, $(k(s_1 - s_2)/c = 2n\pi$ for a maximum and $(2n + 1)\pi$ for a minimum). The 'visibility' of the interference pattern is given by

$$v = \frac{I_{max} - I_{min}}{I_{max} + I_{min}} = 2 \left(\frac{1}{\sqrt{G^{(1)}(x_1, x_1)}} + \frac{1}{\sqrt{G^{(1)}(x_2, x_2)}} \right)^{-1} \tag{3.74}$$

which takes its largest value of unity when $G^{(1)}(x_1, x_1) = G^{(1)}(x_2, x_2)$. To summarize, the first-order coherence function is sufficient to describe the results of simple interference experiments. However, as we will see next, we need to go further than this to fully capture the quantum nature of light.

3.5 Photon correlation measurements

We have so far explored experimental scenarios that can be understood by considering single-photon detection events. However, it is possible to tell a great deal more about the nature of a quantum field by making measurements that involve two or more photons. The first such experiment is due to Hanbury Brown and Twiss, who sent light into a 50:50 beam splitter and placed a detector in each of the two output modes, as shown in Fig. 3.3. The role of the beam splitter is to allow photons that are very closely separated in time to be detected independently. A single detector is not usually capable of resolving two or more photons that arrive simultaneously, so the two detectors are used to record the times of the arrival of different photons.[3] If we call these two times t and $t + \tau$, then such a set-up gives a measurement of the second-order quantum correlation function

$$G^{(2)}(t, t + \tau) = \text{Tr} \left[\rho \hat{E}^-(t) \hat{E}^-(t + \tau) \hat{E}^+(t + \tau) \hat{E}^+(t) \right]. \tag{3.75}$$

We have dropped the position dependence from the operators, since it is assumed that the measurement is sensitive to the photon correlation at a single position, namely that of the beam splitter. Once any transient effects of switching on the light source have settled, we do not expect $G^{(2)}$ to depend on absolute time t, and we may set $t = 0$ without loss of generality.

It will again be convenient to write $G^{(2)}$ in normalized form

$$g^{(2)}(\tau) = \frac{G^{(2)}(\tau)}{|G^{(1)}(0)|^2}, \tag{3.76}$$

---

[3] For more details on photon detectors, we refer the reader to Chapter 4.

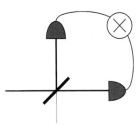

Fig. 3.3. A schematic diagram of the Hanbury, Brown, and Twiss apparatus. A 50:50 beam splitter is placed in the path of an incoming light beam and two detectors are placed in the two arms downstream. Photon counts are recorded as a function of time.

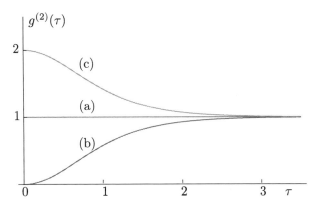

Fig. 3.4. Graph of $g^{(2)}(\tau)$. Curve (a) is expected for uncorrelated photons; (b) indicates photon anti-bunching; and (c) indicates photon bunching.

similar to the first-order correlation function. When photon arrival times are completely uncorrelated the second-order correlation function is a product of two first-order correlation functions $G^{(2)}(\tau) = |G^{(1)}(\tau)|^2$. Moreover, uncorrelated photon arrival times do not have a characteristic time scale τ, and we therefore expect $g^2(\tau) = 1$. Each photon acts as an independent particle, and Poisson statistics are therefore applicable. This is shown in curve (a) of Fig. 3.4. One of the states of light that produces this distribution is the coherent state $|\alpha\rangle$. We can use the relations in Eq. (3.64) to write

$$g^{(2)}(\tau) = \frac{\mathrm{Tr}[\rho\, \hat{a}^\dagger \hat{a}^\dagger \hat{a} \hat{a}]}{(\mathrm{Tr}[\rho\, \hat{a}^\dagger \hat{a}])^2}, \tag{3.77}$$

with $\rho = |\alpha\rangle\langle\alpha|$. Since $\hat{a}|\alpha\rangle = \alpha|\alpha\rangle$, it is clear that $g^{(2)}$ is independent of τ, and numerically equal to one. In this calculation we have assumed a continuous coherent light source, we ignored any decoherence, and we assumed that modes are not repopulated once photons are absorbed by the detectors.

Exercise 3.6: Show that for a number state ($\rho = |n\rangle\langle n|$) we expect

$$g^{(2)} = 1 - \frac{1}{n}. \tag{3.78}$$

For a mode with a single photon, we have $n = 1$, and $g^{(2)}$ therefore vanishes. This is a result that cannot be reproduced using a classical wave theory. Hence an observation of such a coincidence rate is a signature of the quantum nature of a light source. A $g^{(2)}$ value below one is an indicator of 'sub-Poissonian statistics': The variance of the photon number distribution is smaller than for a coherent state with a Poisson distribution. This is shown in curve (b) in Fig. 3.4.

In a real experiment, the coherence of the light will not be maintained for an infinite time, and the mode will be repopulated. Therefore we would expect a time dependence for $g^{(2)}$ such that for long enough times τ_l there is no correlation between the photon detector events, i.e. $g^{(2)}(\tau_l) = 1$. As we shall see later in Chapter 4, the quality of a single-photon source is measured by how small $g^{(2)}$ is at time $\tau = 0$. Since $g^{(2)}(0) < g^{(2)}(\tau_l)$, there is a tendency for photon-detection events to be spaced in time: photon arrival times are not random but are rather 'anti-bunched'.

Exercise 3.7: Show that $g^{(2)} = 2$ for the single-mode thermal state of frequency ω

$$\rho = \sum_{n=0}^{\infty} P_n |n\rangle \langle n| , \qquad (3.79)$$

with

$$P_n = (1 - \exp(-\hbar\omega/k_B T)) \exp(-n\hbar\omega/k_B T). \qquad (3.80)$$

This behaviour is shown in curve (c) in Fig. 3.4.

Again, in a real experiment we would eventually expect to return to $g^{(2)}(\tau_l) = 1$ for long enough τ_l. The enhanced coincidence count rate at time zero is associated with *super-Poissonian statistics* and $g^{(2)}(0) > g^{(2)}(\tau_l)$ indicates 'photon bunching'. Unlike anti-bunching, it is not a signature of a quantum light source, since it is also predicted by the classical theory of electromagnetic radiation.

3.6 References and further reading

The general theory of density operators can be found in most modern advanced textbooks on quantum mechanics. For the mathematical theory of POVMs, we refer the reader to Kraus (1983). The fidelity for mixed quantum states was analyzed by Jozsa (1994), and for a recent review of entanglement measures see Plenio and Virmani (2007). A treatment of coherence properties of the electromagnetic field and correlation functions can be found in Walls and Milburn (2008).

PART II

QUANTUM INFORMATION IN PHOTONS AND ATOMS

4

Photon sources and detectors

In optical quantum information processing, two of the most basic elements are the sources of quantum mechanical states of light, and the devices that can detect these states. In this chapter, we narrow this down to photon sources and photodetectors. We will describe first how detectors work, starting from abstract ideal detectors, via a complete description of realistic detectors in terms of POVMs, to a brief overview of current photodetectors. Subsequently, we will define what is a single-photon source, and how we can determine experimentally whether a source produces single photons or something else. Having laid down the ground rules, we will survey some of the most popular ways photons are produced in the laboratory. Finally, we take a look at the production of entangled photon sources and quantum non-demolition measurements of photons.

4.1 A mathematical model of photodetectors

Photodetectors are devices that produce a macroscopic signal when triggered by one or more photons. In the ideal situation, every photon that hits the detector contributes to the macroscopic signal, and there are no 'ghost' signals, or so-called dark counts. In this situation we can define two types of detector, namely the 'photon-number detector', and 'detectors without number resolution'.

First, the photon number detector is a (largely hypothetical) device that tells us how many photons there are in a given optical mode that is properly localized in space and time. This property is called 'photon-number resolution'. Every possible detection outcome (n) is associated with a POVM, $\hat{P}^{(n)}$, which in the case of an ideal photon number detector is a projection operator of the n-photon Fock state:

$$\hat{P}^{(n)} = |n\rangle \langle n| . \tag{4.1}$$

The probability of detecting n photons given an input state ρ is then given by the Born rule

$$p(n|\rho) = \text{Tr}\left[\hat{P}^{(n)}\rho\right] = \langle n| \rho |n\rangle . \tag{4.2}$$

It is also immediately clear that $\sum_n \hat{P}^{(n)} = \hat{\mathbb{I}}$, which ensures that the probabilities sum up to 1. These detectors are extremely powerful from a quantum information processing point of view, and there is a significant experimental effort to create such devices.

By contrast, the regular photodetector, also known as a 'bucket detector', is not capable of discriminating between different photon numbers: it can tell us only whether or not there were any photons present in the optical mode.[1] The macroscopic signal is typically an electric pulse, which is recorded on a data acquisition device, such as a computer. Historically, photon detectors (in the X-ray part of the spectrum) were often Geiger counters that made a clicking sound when they registered a photon. This has survived in the terminology, and we will generally denote the detector outcomes as a 'click' and 'no click'. Again, the POVMs associated with the detector outcomes are projection operators:

$$\hat{P}^{(\text{click})} = \sum_{n=1}^{\infty} \hat{P}^{(n)} = \sum_{n=1}^{\infty} |n\rangle \langle n| ,$$

$$\hat{P}^{(\text{no click})} = \hat{\mathbb{1}} - \hat{P}^{(\text{click})} = \hat{P}^{(0)} = |0\rangle \langle 0| . \qquad (4.3)$$

The requirement that $\hat{P}^{(\text{click})} + \hat{P}^{(\text{no click})} = \hat{\mathbb{1}}$ is satisfied by construction. The probability of recording a detector click given an input state ρ is

$$p(\text{click}|\rho) = \text{Tr}\left[\hat{P}^{(\text{click})}\rho\right] = \sum_{n=1}^{\infty} \langle n| \rho |n\rangle \qquad (4.4)$$

and $p(\text{no click}|\rho) = 1 - p(\text{click}|\rho)$, as required. As was already mentioned, these are idealized detector models. They do not include imperfections such as photon loss (non-unit efficiency of the detector), dark counts, afterpulsing, spatial non-uniformity, timing resolution and jitter, and non-uniform spectral response. Next, we will discuss the single-mode aspects of photon detectors, namely the detection efficiency and the dark-count rate.

4.1.1 Efficiency and dark counts

There is a conceptual subtlety when we want to discuss photon loss and dark counts in a photodetector: before the optical mode is detected, the state of the mode is typically not a number state, and we cannot speak of the number of photons before detection in a meaningful way. After detection we have a macroscopic signal indicating that a certain number of photons were detected, but the physical object we measured has been destroyed. So what do we mean when we say that we have lost a photon when beforehand the number of photons is undetermined, and afterwards there is no physical system left to compare the macroscopic signal with?

The key to this problem is to realize that the efficiency and the dark counts are properties of the detection system. This comprises not just the detector, but all the optics and free propagation from the source to the detector. Now suppose that we wish to make a photon number measurement on an arbitrary state ρ. We assume that the properties of the detector are independent of the state of the optical mode we wish to detect.[2] This allows us to choose

---

[1] Another defining property of bucket detectors is that they do not provide information about other degrees of freedom, such as where the photon hit the detection area.

[2] This is an approximation: for example, a bright enough source will induce saturation effects, or even damage the detector, leading to an effective change in detection efficiency. However, it is a very good approximation at the single-photon level.

a specific input state $|\psi\rangle$ without loss of generality:

$$|\psi\rangle = \sum_{n=0}^{\infty} c_n |n\rangle, \qquad (4.5)$$

where the c_n are normalized amplitudes. The detector is characterized by a set of POVMs $\{\hat{E}^{(n)}\}$, associated with all possible outcomes (n), that is completely general, and typically no longer a projection operator. The probability of finding the outcome n is then given by

$$p(n|\psi) = \text{Tr}\left[\hat{E}^{(n)} |\psi\rangle\langle\psi|\right] = \langle\psi|\hat{E}^{(n)}|\psi\rangle. \qquad (4.6)$$

The question is how we can include photon loss and dark counts in $\hat{E}^{(n)}$ in a well-defined and meaningful way.

If we can choose $|\psi\rangle$ as our input state without loss of generality, we can also choose a two-mode entangled state

$$|\psi'\rangle = \sum_{n=0}^{\infty} c_n |n\rangle_1 |n\rangle_2, \qquad (4.7)$$

on modes a_1 and a_2. This allows us to measure mode a_1 with our photodetector (finding outcome n), while mode a_2 is projected onto the output state $\rho_{\text{out}}^{(n)}$. This state will generally be mixed when $\hat{E}_1^{(n)}$ is a POVM:

$$\rho_{\text{out}}^{(n)} = d_{(n)}^{-1} \text{Tr}_1\left[\hat{E}_1^{(n)} |\psi'\rangle\langle\psi'|\right] \quad \text{and} \quad d_{(n)} = \text{Tr}_{12}\left[\hat{E}_1^{(n)} |\psi'\rangle\langle\psi'|\right]. \qquad (4.8)$$

This construction allows us to compare the detection outcome of mode a_1 with the state in mode a_2. We can 'measure' the output in mode a_2 with a hypothetical ideal photon number measurement. This yields the probability distribution of finding k photons in mode a_2, given that the realistic detector indicated n photons in mode a_1:

$$p(k|\rho_{\text{out}}^{(n)}) = \text{Tr}_2\left[\hat{P}_2^{(k)} \rho_{\text{out}}^{(n)}\right] = d_{(n)}^{-1} \text{Tr}\left[\hat{E}_1^{(n)} \hat{P}_2^{(k)} |\psi'\rangle\langle\psi'|\right]. \qquad (4.9)$$

It is clear that for $\hat{E}_1^{(n)} = \hat{P}_1^{(n)}$, the probability of finding n photons in mode a_2 is $p(k|\rho_{\text{out}}^{(n)}) = \delta_{nk}$. A general POVM $\hat{E}^{(n)}$ can be determined by choosing a specific form for the probability distribution $p(k|\rho_{\text{out}}^{(n)})$, as we will show below.

The probability $p(n|\rho_{\text{out}}^{(n)})$ is the fidelity of the output state $\rho_{\text{out}}^{(n)}$ with respect to the ideal state $|n\rangle$ that would have been created had the photon number detector been ideal. We call the fidelity

$$F^{(n)} \equiv p(n|\rho_{\text{out}}^{(n)}) = d_{(n)}^{-1} \langle\psi'| \left[\hat{E}_1^{(n)} \otimes |n\rangle_2\langle n|\right] |\psi'\rangle \qquad (4.10)$$

the 'confidence' in the detection outcome n. Having used the crutch of the secondary mode to define the detection efficiency, the dark counts, and the confidence, we can use these concepts in a counterfactual manner.

Detection efficiency

We now present a model for describing detectors that suffer only from photon loss (the dark counts are negligible). These are sometimes called 'lossy', or 'finite-efficiency' detectors. The physical mechanism of photon loss depends on the details of the actual detector that is used, but in general it is a very good approximation to say that each photon entering the detector has a certain probability η to trigger a count event. Classically, this would lead to a probability distribution of counts given the number of input photons. In order to construct the POVM associated with photon loss, we require that the probability distribution $p(k|\rho_{\text{out}}^{(n)})$ coincides with the corresponding classical probability distribution: take $k = n + m$ objects distributed over two containers labelled 'detected' and 'lost', with n objects detected and m lost. If the probability of an object going into the 'detected' container is η, then the probability to find n objects in the 'detected' container and m in the 'lost' container is

$$p(n, m) = \binom{n + m}{n} \eta^n (1 - \eta)^m. \tag{4.11}$$

This is the familiar binomial distribution. Given the input state $|\psi'\rangle$ in Eq. (4.7), we expect the probability distribution $p(n + m|\rho_{\text{out}}^{(n)})$ to be

$$p(n + m|\rho_{\text{out}}^{(n)}) = d_{(n)}^{-1} |c_{n+m}|^2 p(n, m). \tag{4.12}$$

The terms $|k\rangle_1$ in the superposition of $|\psi'\rangle$ with $k < n$ do not contribute to $p(n + m|\rho_{\text{out}}^{(n)})$ because there are not enough photons to trigger the detector (we assumed negligible dark counts). We will now use Eq. (4.9) and Eq. (4.12) to derive

$$
\begin{aligned}
p(n + m|\rho_{\text{out}}^{(n)}) &= d_{(n)}^{-1} \text{Tr}_{ab} \left[\hat{E}_1^{(n)} \hat{P}_2^{(n+m)} |\psi'\rangle\langle\psi'| \right] \\
&= d_{(n)}^{-1} \text{Tr}_1 \left[\hat{E}_1^{(n)} |c_{n+m}|^2 |n + m\rangle_1 \langle n + m| \right] \\
&= d_{(n)}^{-1} |c_{n+m}|^2 \langle n + m| \hat{E}_1^{(n)} |n + m\rangle \\
&= d_{(n)}^{-1} |c_{n+m}|^2 \binom{n + m}{n} \eta^n (1 - \eta)^m,
\end{aligned}
\tag{4.13}
$$

where the last equality follows from Eq. (4.12). The last two lines lead to the identification

$$p(n, m) = \langle n + m| \hat{E}^{(n)} |n + m\rangle, \tag{4.14}$$

from which we can deduce the POVM of a finite-efficiency photon number detector:

$$\hat{E}^{(n)} = \sum_{m=0}^{\infty} \binom{n + m}{n} \eta^n (1 - \eta)^m |n + m\rangle \langle n + m|. \tag{4.15}$$

Exercise 4.1: Show that $\sum_n \hat{E}^{(n)} = \hat{\mathbb{I}}$.

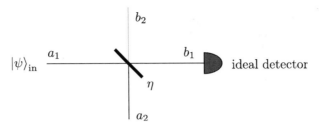

Fig. 4.1. **A model for detector loss using a beam splitter with transmission coefficient η. The reflected mode is traced over.**

This POVM was derived using only the requirement of compliance with a classical probability distribution. However, in many cases it is both convenient and instructive to have a physical model for detector loss. The commonly used model is provided by a beam splitter with transmission coefficient η placed in the optical mode (see Fig. 4.1). The reflected photons are considered lost, while the transmitted photons are detected with an ideal photon number detector. To show that this leads to the same POVM, consider the input state

$$|\psi\rangle = |k\rangle_1 |0\rangle_2, \tag{4.16}$$

on modes a_1 and a_2, and the beam-splitter transformation

$$\hat{a}_1^\dagger \rightarrow \sqrt{\eta}\,\hat{b}_1^\dagger + \sqrt{1-\eta}\,\hat{b}_2^\dagger. \tag{4.17}$$

Mode a_1 is the signal mode, which is transformed to mode b_1 by the beam splitter. The modes a_2 and b_2 are auxiliary modes. After the beam splitter, mode b_2 is traced out, since it is identified with the environment. This will lead to the following state of the optical modes:

$$
\begin{aligned}
|\phi\rangle\langle\phi| &= \frac{1}{k!}\hat{a}_1^{\dagger k}|0,0\rangle\langle 0,0|\hat{a}_1^k \\
&\rightarrow \frac{1}{k!}\left(\sqrt{\eta}\hat{b}_1^\dagger + \sqrt{1-\eta}\hat{b}_2^\dagger\right)^k |0,0\rangle\langle 0,0|\left(\sqrt{\eta}\hat{b}_1 + \sqrt{1-\eta}\hat{b}_2\right)^k \\
&= \frac{1}{k!}\sum_{l,m=0}^{k}\binom{k}{l}\binom{k}{m}\sqrt{\eta}^{2k-l-m}\sqrt{1-\eta}^{l+m}\,\hat{b}_1^{\dagger k-l}\hat{b}_2^{\dagger l}|0,0\rangle\langle 0,0|\hat{b}_1^{k-m}\hat{b}_2^m \\
&= \sum_{l,m=0}^{k} k!\sqrt{\frac{\eta^{2k-l-m}(1-\eta)^{l+m}}{(k-l)!\,l!\,(k-m)!\,m!}}\,|k-l,l\rangle\langle k-m,m| \\
&\equiv |\phi'\rangle\langle\phi'|. \tag{4.18}
\end{aligned}
$$

We can take the trace over mode b_2 to remove the environment from the state. This leads to the mixed single-mode state

$$\mathrm{Tr}_2\left(|\phi'\rangle\langle\phi'|\right) = \sum_{m=0}^{k}\binom{k}{m}\eta^{k-m}(1-\eta)^m |k-m\rangle\langle k-m|, \tag{4.19}$$

which is the state that results from preparing a state $|k\rangle$ and evolving it through a Poissonian loss mechanism. If the initial state is instead the two-mode number state $|\psi'\rangle = \sum_k c_k |k, k\rangle_{12}$, the resulting state after loss in mode a_1 becomes

$$\rho_{12} = \sum_{k=0}^{\infty} \sum_{m=0}^{k} |c_k|^2 \binom{k}{m} \eta^{k-m}(1-\eta)^m |k-m, k\rangle_{12}\langle k-m, k| . \tag{4.20}$$

Considering this two-mode photon-entangled state allows us to evaluate the probability of m photons in mode b_2 given a measurement outcome of n photons in mode b_1, similar to Eq. (4.9). First, we recall that we can write

$$\rho_{\text{out}}^{(n)} = d_{(n)}^{-1} \text{Tr}_1 \left[\hat{P}_1^{(n)} \rho_{12} \right] \quad \text{and} \quad d_{(n)} = \text{Tr}_{12} \left[\hat{P}_1^{(n)} \rho_{12} \right]. \tag{4.21}$$

This state has the same interpretation as the state in Eq. (4.8). Next we calculate the probability

$$
\begin{aligned}
p(n+m|\rho_{\text{out}}^{(n)}) &= d_{(n)}^{-1} \text{Tr}_{12} \left[\hat{P}_1^{(n)} \hat{P}_2^{(n+m)} \rho_{12} \right] \\
&= \sum_{i,j,k=0}^{\infty} \sum_{l=0}^{k} \frac{|c_k|^2}{d_{(n)}} \binom{k}{l} \eta^{k-l}(1-\eta)^l \delta_{i,n}\delta_{j,n+m}\delta_{i,k-l}\delta_{j,k} \\
&= d_{(n)}^{-1} |c_{n_m}|^2 \binom{n+m}{m} \eta^n (1-\eta)^m.
\end{aligned}
\tag{4.22}
$$

This is the same expression as Eq. (4.12), which means that the beam-splitter model for detector loss leads to the same POVM in Eq. (4.15). This model is constructed entirely with linear optical elements and perfect photon number detectors. It can therefore be very convenient in deriving properties of linear optical experiments with finite-efficiency detectors using the techniques introduced in Chapter 1.

From Eq. (4.15) we can find the POVM for photodetectors without photon number resolution (bucket detectors) by choosing $n = 0$. The POVM for finding no detector click is equivalent to the photon number POVM in the case when all photons are lost:

$$\hat{E}^{(\text{no click})} = \hat{E}^{(0)} = \sum_{m=0}^{\infty} (1-\eta)^m |m\rangle \langle m| . \tag{4.23}$$

The POVM for finding a detector click is

$$\hat{E}^{(\text{click})} = \hat{\mathbb{1}} - \hat{E}^{(\text{no click})} = \sum_{m=0}^{\infty} \left[1 - (1-\eta)^m \right] |m\rangle \langle m| . \tag{4.24}$$

In the situation where dark counts are negligible, this is a good approximation to the description of a real photodetector. Again, this finite-efficiency photodetector can be accurately modelled by a beam splitter with transmission coefficient η preceding an ideal photodetector.

Detector dark counts

Next, we consider the case of finite-efficiency detectors where the dark counts cannot be neglected. In the given time interval of detection, there will be some probability distribution $D(k)$ for the occurrence of k dark counts, such that

$$\sum_{k=0}^{\infty} D(k) = 1. \tag{4.25}$$

At this point, we do not assume any specific forms of this probability distribution: it can be a Poisson distribution, when all dark counts occur independently of each other; or it can be a thermal distribution, which may be more likely when the detector is sensitive to (and not properly shielded from) infrared wavelengths. In practice, it will be some multi-parameter distribution that is a result of a variety of physical mechanisms.

Classically, we again have a probability distribution in which objects are put in the 'detected' or 'lost' container, but now we have two types of objects the real objects (defined again via a two-mode entangled photon number state) and the false positives, or spurious objects. Let $n-k$ real and k spurious objects be deposited in the 'detected' container (giving a total of n detected objects), and let m objects be deposited in the 'lost' container. Including the binomial coefficient to count all possibilities this distribution can be realized, we have

$$p(n,m,k) = D(k) \binom{n-k+m}{m} \eta^{n-k}(1-\eta)^m \tag{4.26}$$

as the probability of finding n objects in the 'detected' container, given $n-k+m$ real objects and k spurious objects. We included a factor $D(k)$ since we assume that the probability of k spurious objects is independent of the number of real objects. There will be at most n spurious objects, and the probability distribution of finding n detected objects is therefore

$$p(n,m) = \sum_{k=0}^{n} D(k)p(n,m,k) = \sum_{k=0}^{n} D(k) \binom{n-k+m}{m} \eta^{n-k}(1-\eta)^m. \tag{4.27}$$

We can derive the POVM again via the relation $\langle n+m| \hat{E}^{(n)} |n+m\rangle = p(n,m)$. Using the following Ansatz for $\hat{E}^{(n)}$:

$$\hat{E}^{(n)} = \sum_{i=0}^{\infty} \sum_{j=0}^{n} f(n,i,j) |n+i-j\rangle \langle n+i-j| \tag{4.28}$$

we find (relabeling $j=k$ on the left-hand side):

$$\sum_{k=0}^{n} f(n,m,k) = \sum_{k=0}^{n} D(k) \binom{n-k+m}{m} \eta^{n-k}(1-\eta)^m. \tag{4.29}$$

This allows us to equate the expressions under the summation on both sides. Substituting this into Eq. (4.28) gives the POVM for the detector outcome 'n'

$$\hat{E}^{(n)} = \sum_{m=0}^{\infty} \sum_{k=0}^{n} D(k) \binom{n-k+m}{m} \eta^{n-k}(1-\eta)^m \lvert n+m-k \rangle \langle n+m-k \rvert, \qquad (4.30)$$

when the detector has efficiency η and dark count distribution $D(k)$.

We can choose any suitable probability distribution for the dark counts $D(k)$. One possibility is that the dark counts are distributed according to a Poisson distribution:

$$D(k) = \frac{\delta^k \exp(-\delta)}{k!}. \qquad (4.31)$$

Alternatively, the distribution may be thermal:

$$D(k) = \left(1 - e^{-\gamma}\right) \exp(-k\gamma), \qquad (4.32)$$

where γ is the characteristic parameter for a thermal distribution, typically $\gamma = \epsilon/k_B T$ with ϵ the corresponding energy value of the dark count, k_B Boltzmann's constant, and T the temperature. There may be a weighted sum or integral involved over all possible energy values. In many practical situations the dark-count rate is sufficiently low that the probability of having two dark counts is negligible. In that case we can choose the distribution

$$D(0) = 1 - \delta \quad \text{and} \quad D(1) = \delta. \qquad (4.33)$$

This gives the correct results up to order δ^2 in photon number detectors, but it significantly simplifies the POVM.

For bucket detectors the distribution in Eq. (4.33) is exact and completely general. The POVM for a bucket detector can again be found from the general photon number POVM by setting $\hat{E}^{(\text{no click})} = \hat{E}^{(0)}$:

$$\hat{E}^{(\text{no click})} = \hat{E}^{(0)} = D(0) \sum_{m=0}^{\infty} (1-\eta)^m \lvert m \rangle \langle m \rvert$$

$$\hat{E}^{(\text{click})} = \sum_{m=0}^{\infty} [1 - D(0)(1-\eta)^m] \lvert m \rangle \langle m \rvert. \qquad (4.34)$$

The dark counts are completely determined by a single parameter $D(0)$, which encapsulates all the physical mechanisms that may lead to the false positives.

Absolute detection efficiencies and dark currents

We can experimentally determine the 'absolute efficiency' of a detector as follows: consider the two-mode squeezed state in Eq. (1.235), which is a good approximation to the output state of a parametric downconverter (see Section 4.3.3). The critical property of this state is that the photons are created in pairs, i.e., when there are n photons in mode a_1, there are also n photons in mode a_2. This allows us to relate the count rates N_1 and N_2 in the detectors in modes a_1 and a_2. The measured count rate of detector $i = 1, 2$ is

$$N_i = \eta_i \tau_i N, \tag{4.35}$$

with η_i the detection efficiency of detector i, τ_i the transmittance from the source to the detector, and N the (unknown) rate of photon pairs produced by the source. The measured coincidence rate N_c is given by

$$N_c = \eta_1 \eta_2 \tau_1 \tau_2 N. \tag{4.36}$$

The detection efficiency of the detector in mode a_1 and mode a_2 can be expressed as

$$\eta_1 = \frac{N_c}{\tau_1 N_2} \quad \text{and} \quad \eta_2 = \frac{N_c}{\tau_2 N_1}. \tag{4.37}$$

The rates N_c, N_1, and N_2 are directly generated in the experiment, and τ_1 and τ_2 can be inferred by varying the distance from the detector to the source. Alternatively, τ_1 and τ_2 can be incorporated in the detection efficiency of the detection device. Since the measured efficiency of detector a_1 does not depend on the efficiency of detector a_2, and vice versa, this is an 'absolute measurement' of η_1 and η_2.

Similarly, the absolute dark current can be determined by measuring the count rate in the absence of a source.

4.2 Physical implementations of photodetectors

Now that we have developed a general theoretical model for photon detectors, we can discuss a number of physical implementations of both bucket detectors and photon number detectors. We first consider photomultiplier tubes and avalanche photodiodes, followed by a discussion on detector cascading. This is a technique that purports to turn many bucket detectors into photon number detectors. We conclude this section with photon number detectors based on superconducting devices.

4.2.1 Photomultiplier tubes and avalanche photodiodes

The most common photodetectors are 'photomultiplier tubes' (PMTs) and 'avalanche photodiodes' (APDs). They exploit the photoelectric effect to convert a photon into an electron, which is subsequently used in an electronic amplification process. In the case of a PMT, the electron is created at a photocathode in a vacuum tube. It is then accelerated towards a 'dynode' that is held at a positive voltage with respect to the photocathode. When the electron hits the dynode, its kinetic energy releases several more electrons, which in turn are accelerated towards a second dynode held at a positive voltage with respect to the first. This process repeats several times, until finally, a macroscopic number of electrons hit the anode, where they are detected as a current spike, giving the 'click' of the detector (see Fig. 4.2).

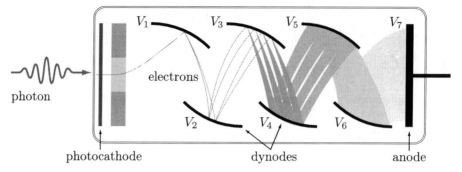

Fig. 4.2. A photomultiplier tube. A single photon produces an electron that is accelerated to a dynode at positive voltage V_1. The dynode will release several electrons, which in turn are accelerated to the dynode at $V_2 > V_1$, and so on. At the anode, the electrons create a macroscopic current spike.

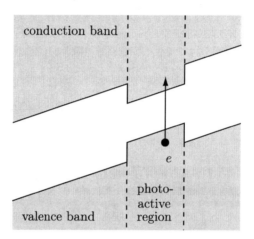

Fig. 4.3. Energy structure of an avalanche photodiode. An electron is promoted from the valence band to the conduction band and causes an avalanche of secondary electron–hole pairs. This leads to a macroscopic signal.

In an APD, the incoming photon promotes an electron from the valence band to the conduction band of a semiconductor. The gap left in the valence band is called a 'hole' and acts as a particle in its own right, and has a positive charge.[3] A large potential difference is applied across the semiconductor, which causes the electron to accelerate towards the positive end (and the hole to the negative end). In doing so, the accelerating particles promote other valence electrons to the conduction band, creating further electron–hole pairs. In turn, these pairs get accelerated and produce more electrons and holes, which eventually becomes an avalanche. When the current becomes macroscopic in size, the electronics of the APD registers a current spike, giving the 'click' of the detector (see Fig. 4.3). The avalanche must be stopped actively by reversing the potential across the semiconductor. The effect of

---

[3] We will return to holes in Chapter 11.

this reversal is that the detector has a period immediately after a detection event in which it cannot register incoming photons. This is called the 'dead time', and typically lasts for a few nanoseconds.

In quantum optics, APDs are mostly used. These devices generally do not have number-resolving capabilities (although there are exceptions), and it is not difficult to see why: two photons will not create a conduction electron at exactly the same time, and the second electron will therefore be just another conduction electron in the avalanche, rather than the seed of an avalanche. Other properties of APDs that can have an effect on the quality of quantum information processing are the spatial uniformity (or lack thereof) of the detection area, time resolution and jitter, afterpulsing, and the spectral response of the detector:

Spatial uniformity: The active area of a detector may have non-uniformities that affect properties such as the detection efficiency $\eta(\mathbf{r})$, which then becomes a function of position on the detection area $\mathbf{r} = (x, y)$. This can be decomposed in terms of the Hermite–Gaussian or Laguerre–Gaussian modes $\eta(\mathbf{r}) = \sum_{nm} \eta_{nm} u_{nm}(\mathbf{r})$ introduced in Section 1.3, such that each incoming transverse mode $u_{nm}(\mathbf{r})$ has a constant detection efficiency η_{nm}. The detector will be triggered with probability η_{nm} by photons in these transverse modes. When the transverse mode information is not recorded in the detector, the modes have to be included in the POVM description of the detector.

Time resolution and jitter: A photon wave packet is often several nanoseconds in duration, but the time resolution of APDs may be as short as tens of picoseconds. In a very practical sense, the detector collapses the 'wave function' of the photon in the temporal domain. Consequently, there will be a temporal variation in the moment when the detector clicks, but this is due to the nature of the wave packet, rather than the structure of the detector. There is, however, also a detector-specific timing noise, called 'jitter'. This is mainly due to thermal electronic noise in the APD, and typically follows a Gaussian distribution.

Spectral response: In addition to the spatial uniformity of the detection area, the detection efficiency depends on the frequency of the incoming light. Clearly, the efficiency of the detector will drop to zero when the wavelength becomes too long for the photoelectric effect to occur. In general, there will be a spectral band over which the detection efficiency is approximately constant. The characteristics of this band are determined by the physics of the detector.

Afterpulsing: A detector click that was triggered by a true photon (as defined above in the two-mode setting) can be followed by one or more spurious detector clicks. This is called afterpulsing. Just like dark counts, these pulses can lead to false positives. A specific example of afterpulsing is so-called hot carrier luminescence, where the cascade produces photons in the optical range. These photons can enter the interferometer and reflect back into the detectors (Ulu *et al.*, 2000).

All these effects must be taken into account when designing protocols, since they may compromise the fidelity of the quantum information processes.

As an example, we will now determine the POVM for a photodetector with spatial non-uniformity of the detection efficiency. The detection efficiency $\eta(\mathbf{r})$ is a function of the position **r** on the active detection area. What is the POVM for a detector click in such a photodetector? Since we assume that the detector has no number-resolving capability, we choose to calculate the POVM for 'no click' $\hat{E}^{(0)}$, and calculate the POVM for a click via the relation $\hat{E}^{(1)} = \hat{\mathbb{I}} - \hat{E}^{(0)}$. We have already established that for a transverse mode $u_{nm}(\mathbf{r})$, the detection efficiency η_{nm} is constant. The POVM for this mode is therefore

$$\hat{E}_{nm}^{(0)} = \sum_{k=0}^{\infty} (1 - \eta_{nm})^k |k\rangle \langle k| . \tag{4.38}$$

Since each mode labeled by n and m independently responds to the incoming light field, the total POVM for no click in the photodetector is

$$\hat{E}^{(0)} = \bigotimes_{n,m} \hat{E}_{nm}^{(0)}. \tag{4.39}$$

It is usually infeasible to calculate an analytic form for $\hat{E}^{(0)}$. Fortunately, this is hardly ever necessary. Many quantum information protocols require the photon to be strictly in a single mode, in which case the effective POVM reduces again to Eq. (4.38). Only when transverse mode functions are used can variations in η_{nm} become problematic. In that case it is usually sufficient to consider only a few modes.

4.2.2 Photon number resolution: detector cascading

A conceptually straightforward way in which regular photodetectors (such as PMTS or APDS) can be used to achieve single-photon resolution, i.e., to build a photon number detector, is by 'cascading', shown in Fig. 4.4. In a detector cascade, the incoming mode is divided equally over a large number of modes, which are all detected. If the number of modes is much larger than the typical number of photons in the input mode, the probability for two photons entering the same detector becomes arbitrarily small.

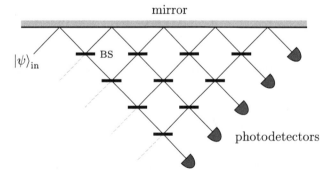

Fig. 4.4.   A detector cascade with 50:50 beam splitters (BS). A photon in the input mode is equally likely to end up in any of the photodetectors.

A particular N-port detector cascade is given by the discrete Fourier transform

$$U_{jk} = \frac{1}{\sqrt{N}} \exp\left[2\pi i \frac{(j-1)(k-1)}{N} \right], \qquad (4.40)$$

which leads to the transformation of the mode operators

$$\hat{b}_j = \sum_{k=1}^{N} U_{jk} \hat{a}_k \quad \text{and} \quad \hat{b}_j^\dagger = \sum_{k=1}^{N} U_{jk}^* \hat{a}_k^\dagger. \qquad (4.41)$$

A photon entering any input port has an equal probability of exiting any output port. We next derive an expression for the POVMs of the detector cascade, including the probability of two photons in the same output mode, detector efficiencies, and dark counts. For simplicity, we assume that all the detectors in the cascade are identical photodetectors.

We follow the same strategy as in Section 4.1. First, we construct the input state we wish to detect on mode a_1. Then, we extend this to a two-mode entangled state, including the hypothetical mode a_0

$$|\psi'\rangle = \sum_{n=0}^{\infty} c_n |n, n\rangle_{10}, \qquad (4.42)$$

which allows us to talk about the counterfactual probability that the detector indicated n photons, while mode a_1 contained k photons:

$$p(k|\rho_{\text{out}}^{(n)}) = \text{Tr}_{10}\left[\hat{E}_1^{(n)} \hat{P}_0^{(k)} |\psi'\rangle\langle\psi'| \right]. \qquad (4.43)$$

The classical probability of distributing k identical objects over N containers is the well-known multinomial distribution:

$$p_{\text{cascade}}(k, \vec{n}) = \frac{k!}{n_1! \cdots n_N!} p_1^{n_1} \cdots p_N^{n_N}, \qquad (4.44)$$

where p_i is the probability of putting the object in the ith container, and $\vec{n} = (n_1, \ldots, n_N)$ is the N-tuple of the number of objects in each container. For an arbitrary N-tuple $\vec{n}$ we define $|\vec{n}|$ as the number of non-zero n_i, and $\|\vec{n}\| = \sum_i n_i$ is the total number of objects. In the case of the detector cascade $p_i = 1/N$, and the probability of n input photons triggering n detectors is given by

$$p_{\text{cascade}}(n|n) = \sum_{|\vec{n}|=n} p_{\text{cascade}}(n, \vec{n}) = \binom{N}{n} \frac{n!}{N^n}. \qquad (4.45)$$

This probability approaches 1 in the limit $N \to \infty$, as required.

For the more general case we return to the quantum input state ρ after the application of the N-port in Eq. (4.40). The probability of finding the photon distribution $\vec{n}$ in the output modes, given a distribution $\vec{k} = (k_1, \ldots, k_N)$ in the input modes is ($\rho = |\vec{k}\rangle\langle\vec{k}|$)

$$p_{\text{cascade}}(\vec{n}|\vec{k}) = \langle\vec{n}| U \rho U^\dagger |\vec{n}\rangle = \left| \langle\vec{n}| U |\vec{k}\rangle \right|^2. \qquad (4.46)$$

For a detector cascade, we have $\vec{k} = (k, 0, \ldots, 0)$. This can be evaluated in terms of multi-dimensional Hermite polynomials, but that is beyond the scope of this book. For more information see Kok and Braunstein (2001). The confidence of a detector outcome given an input state $|\psi\rangle = \sum_k c_k |k\rangle$ is

$$F^{(n)} = \frac{|c_n|^2 p_{\text{cascade}}(n|n)}{\sum_{m=0}^{\infty} |c_{n+m}|^2 p_{\text{cascade}}(n|n+m)}. \tag{4.47}$$

The probability of an n-fold detector coincidence in detectors with efficiency η, given n input photons is

$$p_{\text{cascade}}(n|n) = \binom{N}{n} \frac{\eta^n n!}{N^n}. \tag{4.48}$$

This probability approaches η^n as $N \to \infty$. Detector cascading therefore works well only with high-efficiency photodetectors.

There are several experimental implementations of detector cascades, and the one that is most closely related to the theoretical model of the detector cascade given here is the 'visible light photon counter' (VLPC), in which the active area of the detector is divided into many independent active regions (see e.g., Waks *et al.*, 2003 and 2004). Each region acts as a separate photodetector, with its own efficiency, dark-count rate, and detector dead time. Rather than using many beam splitters to divide the beam over the detectors, the VLPC relies on the fact that each individual detection area is much smaller than the beam size, and the photon wave packet randomly collapses onto a single area. The probability that two photons collapse onto the same area is small, and roughly follows the distribution for the detector cascades derived above. Here, N is the number of active areas in the VLPC.

Time multiplexing

Rather than randomly distributing the photons in an input mode over N spatial output modes in a detector cascade, the photons may be distributed randomly over N time intervals. This is called 'time multiplexing', and to date there have been two distinct implementations of this idea.

The time multiplexer shown in Fig. 4.5a is the temporal equivalent of the detector cascade described in the previous section. An incoming photon encounters a fibre coupler that acts

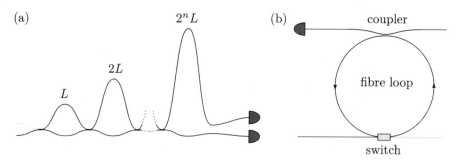

Fig. 4.5. Time-multiplexing detector cascade: (a) Fitch *et al.* (2003) and Achilles *et al.* (2003); (b) Thé and Ramos (2007).

as a 50:50 beam splitter. The two possibilities for the photon are to travel to a second 50:50 fibre coupler via the short path or the long path with length L. The short path has negligible length. This process repeats again at the next beam splitter, but now the long path has twice the length ($2L$). After n iterations, the most recent long path has length $2^n L$. An incoming photon will travel a distance that is a superposition of all possible paths through the multiplexer. Each path can be uniquely described by an n-bit string: when a photon is reflected at the kth beam splitter into the long arm, the bit string records a 1 in the $(n-k)^{\text{th}}$ place. If the photon is transmitted into the short path, the bit string records a 0. Since each delay line has a length that is a power of two with respect to L, each possible path has a unique number, corresponding to its length. This is translated into a unique time of arrival for each path. When n is sufficiently large, it is very unlikely that two incoming photons take exactly the same path, and will therefore result in two separate clicks in the detectors. Each path is equally likely. The probability distribution for the detected photons given a number of input photons is therefore the same as that of the detector cascade described above, although by its very nature the time multiplexer is considerably slower than the spatial detector cascade.

A slightly different version of a multiplexer is shown in Fig. 4.5b. The incoming photons are stored in a fibre loop by means of an optical switch. Weakly coupled to the loop is another fibre that leads to a photodetector. Every time the wave packet passes through the fibre coupler, there is a small probability that a photon is transferred to the fibre and into the detector. The difference between this model and the previous multiplexer is that here the beam is not split off randomly, but rather a small amplitude is leaked out of the fibre loop many times in a row, like grains of sand falling slowly through your fingers. We need a cut-off in time, after which we declare that the number of detector clicks equals the number of photons in the incoming mode. If the fibre loop has length L, the cut-off time is $\tau = NL/v$, where v is the speed of light in the fibre, and N is the number of revolutions. To maximize the confidence in the detection outcome, the fibre coupler should have a transmission coefficient of $1/N$. In the language of cascades, this corresponds to a single mode passing through N beam splitters with reflection coefficient $1/N$. If the previous cascade had a tree structure, this cascade has a linear structure.

4.2.3 Superconducting photon-number detectors

A particularly elegant solution to the problem of achieving photon-number resolution is the use of the superconducting-to-Ohmic transition in areas and nanowires. There are two basic mechanisms, called 'transition edge sensors' (TES) and 'superconducting nanowire detectors' (SND). We will briefly describe both mechanisms, and compare current state-of-the-art implementations. We finish this section with an overview of some experimental implementations of photon-number resolving detectors.

In a TES, the temperature of the active area is held at the point where the material is at the edge of the transition from the superconducting state to the Ohmic state (see Fig. 4.6). The active material is still superconducting, but any rise in temperature will take it into the transition to the Ohmic regime. An incoming photon will be absorbed in the material

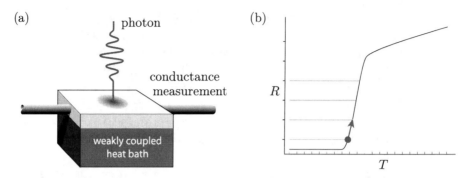

The transition edge of a superconducting material, as used in a TES photon-number resolving detector. R is the resistivity of the active detection area, and T is the temperature. The transition between the superconducting and Ohmic state is nearly linear, which allows for photon-number detection up to a certain number of photons.

and creates a small rise in temperature. This will cause the resistance of the material to rise very steeply at the transition edge, creating a measurable change in the current through the material. Since the transition edge is almost linear, a second absorbed photon will increase the resistance of the material by the same amount. The measured resistivity is then proportional to the number of detected photons. When too many photons are absorbed by the detection area, the transition edge is saturated and the detector loses its photon-number resolution.

The quantum efficiency of the TES can be made very high by making the active area sufficiently thick, such that most incoming photons will be absorbed. However, after a detection event the active area must again be cooled to the initial temperature. If the active area is too thick, this will take a long time, resulting in a drop in the repetition rate. A detection efficiency of $\eta = 88\%$ and a repetition rate of 50 kHz have been reported (Rosenberg *et al.*, 2005). This detector operates at a temperature around 100 mK.

In an SND a single superconducting nanowire is meandering such that it creates a number of active two-dimensional regions, each of which can act as a single-photon detector. In this sense, the detection statistics follows that of the detector cascade. When a photon is absorbed by the wire it creates a hotspot in which the superconductivity breaks down. The supercurrent must flow around the hotspot, and therefore becomes most dense in the region adjacent to the hotspot. If the bias current is strong enough, the supercurrent will exceed the critical current in the dense areas, and superconductivity is lost. This creates a resistive barrier. The total resistivity of the nanowire is proportional to the number of resistive barriers in the wire, and therefore to the number of photons that created the barriers.

The SND has a number of advantages over the TES, and one important drawback (Divochiy *et al.*, 2008). The advantages are its extremely low dark-count rate (0.15 Hz), and the very high repetition rate (80 MHz). It also operates at a temperature that is an order of magnitude higher than the TES, which makes an important practical difference in implementation. The downside of the SND is its low detection efficiency ($\eta = 2\%$), which is partly due to the difficulty of filling a two-dimensional surface with a one-dimensional wire. This prevents

Table 4.1.   Important properties of some number-resolving photodetectors

Detector[a]	R (kHz)	D (Hz)	η (%)	λ (nm)	N_{max}	T (K)
APDC[b]	20	1.6×10^8	33	1 064	1 024	246
PMT[c]	670	400	< 7	523	4	293
VLPC[d]	15	2×10^4	85	543	10	6–7
TM[e]	10	–	66	700–800	8–16	293
TES[f]	50	400	88	1550	< 10	< 0.1
SND[g]	80 000	0.15	2	1300	4	2

[a] Here, R is the repetition rate, D the dark count rate, η the detection efficiency, λ the frequency, N_{max} the maximum detectable photon number, and T the operating temperature.
[b] Avalanche photodiode cascades (Jiang *et al.*, 2007).
[c] Photomultipliers (Zambra *et al.*, 2004).
[d] Visible light photon counters (Waks *et al.*, 2003, 2004).
[e] Time multiplexing (Achilles *et al.*, 2003, and Fitch *et al.*, 2003; no dark count rates reported).
[f] Superconducting transition edge sensors (Rosenberg *et al.*, 2005).
[g] Superconducting nanowire detectors (Divochiy *et al.*, 2008).

the SND from reconstructing photon-number distributions with high fidelity. Nevertheless, the large repetition rate compensates for this in the bit rate that can be achieved in quantum communication. Both TES and SND operate at telecom wavelengths.

Finally, in Table 4.1 we list some important properties of a number of experimental detectors with photon-number resolution. The repetition rate R gives a bound on the possible quantum information processing speed, and the dark-count rate D tends to have an immediate effect on the fidelity of quantum information processing. The efficiency η is the total collection and detection efficiency of the detector, not just the quantum efficiency. The maximum detectable photon number N_{max} can in some cases be very high, but is meaningful only when η is reasonably high as well.

4.3 Single-photon sources

In this section we discuss the criteria for good single-photon sources. It is important to remember that the behaviour of a single-photon source is determined by the way it triggers photodetectors. We describe the Hanbury Brown and Twiss experiment and the Hong–Ou–Mandel experiment, and show how they provide diagnostic tools for single-photon sources. Subsequently, we discuss some of the most widely used sources in single-photon experiments, such as weak coherent states and photons from heralded parametric downconversion. We end the section with an overview of triggered quantum structures, and more exotic photon sources. The physics of generating photons with atom-like structures will be described in more detail in Chapters 7 and 11.

Intuitively, we can immediately formulate two requirements for single-photon sources. First, whenever we 'push the button' on our source to create a single-photon state, the device must indeed produce *one and only one* photon. That is, when placed in front of an ideal detector, triggering the source should always result in the detector indicating the presence of a single photon. Second, the photon should be *pure*. By this, we mean that the source creates a single photon in a discrete single optical mode. If the source produces a single photon in an incoherent mixture of modes (e.g., due to entanglement of the photon with another system), the quantum interference that can be achieved with these states is limited. Note, however, that this may not be such a problem for quantum cryptographic applications.

There is a third requirement that is important for applications such as quantum computing with single photons, linear optics, and photodetection: the source should be reproducible to the extent that two independent sources produce identical photons. To a large extent, these requirements are captured by the Hanbury Brown and Twiss and Hong–Ou–Mandel experiments.

4.3.1 Hanbury Brown and Twiss, and Hong–Ou–Mandel

The results of Hanbury Brown and Twiss, and Hong–Ou–Mandel experiments give an indication of the quality of single-photon sources. Furthermore, these experiments can be performed for *any* physical realization of a single-photon source, and therefore provide a good metric for the quality of the sources, as well as an excellent way to compare different single-photon sources. Indeed, we can use these experiments to *define* how a perfect single-photon source should behave. Together with the source efficiency, they give all the relevant information for quantum information processing.

The Hanbury Brown and Twiss (HBT) experiment tells us how well the input field is described by a single-photon state, regardless of the purity of this state. It constitutes a measurement of $g^{(2)}(\tau)$ of a single input mode. In the previous chapter, we have seen that a single mode a_1 with n photons gives rise to a value

$$g^{(2)} = 1 - \frac{1}{n}. \tag{4.49}$$

Therefore, a single-photon input state should lead to $g^{(2)} = 0$. Since $g^{(2)}$ is proportional to $\mathrm{Tr}(\rho\,\hat{a}_1^\dagger\hat{a}_1^\dagger\hat{a}_1\hat{a}_1)$, the observable associated with the measurement of $g^{(2)}$ is proportional to $\hat{a}_1^\dagger\hat{a}_1^\dagger\hat{a}_1\hat{a}_1$. There is no detector that is described directly by this operator, but it is not too hard to devise one that is constructed from simple optical components. Consider two modes b_1 and b_2, and photon-number detectors described by $\hat{b}_1^\dagger\hat{b}_1$ and $\hat{b}_2^\dagger\hat{b}_2$. We can multiply the measurement result of the two detectors, which forms the new observable $\hat{b}_1^\dagger\hat{b}_1\hat{b}_2^\dagger\hat{b}_2$. We can translate this into a measurement that depends on $\hat{a}_1^\dagger\hat{a}_1^\dagger\hat{a}_1\hat{a}_1$ by transforming modes b_1 and b_2 into modes a_1 and a_2, where a_2 is required by unitarity of the transformation. In particular, we can choose the 50:50 beam-splitter transformation

$$\hat{b}_1 \to \frac{\hat{a}_1 + \hat{a}_2}{\sqrt{2}} \quad \text{and} \quad \hat{b}_2 \to \frac{\hat{a}_1 - \hat{a}_2}{\sqrt{2}}. \tag{4.50}$$

(a)   (b)

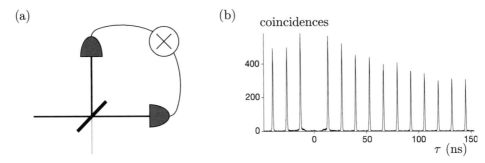

Fig. 4.7. The Hanbury Brown and Twiss experiment. (a) The experimental set-up consists of a 50:50 beam splitter (BS) with one input mode (the second input is held at vacuum). (b) A typical $g^{(2)}$ plot for a single-photon source, adapted by permission from Macmillan Publishers Ltd: Nature, Santori *et al.* (2002).

Similarly, the normalization factor $\mathrm{Tr}(\rho\,\hat{a}_1^\dagger\hat{a}_1)^2$ in $g^{(2)}$ is proportional to the photon number in mode a_1. If we choose the vacuum input state in mode a_2, we obtain $\hat{n}_{\mathrm{in}} = \hat{n}_1 + \hat{n}_2$, and the measurement of the photocurrents in modes b_1 and b_2 suffice to determine $g^{(2)}$. This set-up is given in Fig. 4.7a.

Whenever the input mode contains only a single photon, pure or mixed, it is easy to see that coincidence counts in modes b_1 and b_2 can never occur. Indeed, this is consistent with Eq. (4.49). In addition, if the input states have non-zero amplitude for components $|n\rangle$ with $n \geq 2$, then the coincidence rate will never be zero. Therefore, a zero coincidence rate $\hat{n}_1\hat{n}_2$, given a non-zero total photocurrent $\hat{n}_1 + \hat{n}_2$, indicates that the input field was in a single-photon state. Experimental data for a periodically pumped photon source is shown in Fig. 4.7b. At zero time delay between the detection windows in modes b_1 and b_2 no coincidence counts are found. The question remains whether the single-photon input state is pure. To determine this, we need to perform another experiment.

Exercise 4.2: Show that the experimental set-up in Fig. 4.7a allows a measurement of $g^{(2)}$.

The Hong–Ou–Mandel (HOM) experiment can be used to determine the purity and the distinguishability of two single-photon states. The set-up involves a single 50:50 beam splitter, and single-photon input states in each input mode (see Fig. 4.8a). If we label the (spatial) input modes a_1 and a_2, the mode transformation of the beam splitter can be written

$$\hat{b}_1(k) = \frac{\hat{a}_1(k) + \hat{a}_2(k)}{\sqrt{2}} \quad \text{and} \quad \hat{b}_2(k) = \frac{\hat{a}_1(k) - \hat{a}_2(k)}{\sqrt{2}}, \qquad (4.51)$$

where $k = \omega_k/c$ indicates the wave number. The input states are two single-photon states

$$|1_f\rangle_1 = \int dk\, f(k)\, \hat{a}_1^\dagger(k)\, |0\rangle_1 \quad \text{and} \quad |1_g\rangle_2 = \int dk'\, g(k')\, \hat{a}_2^\dagger(k')\, |0\rangle_2\,. \qquad (4.52)$$

We have suppressed the polarization degree of freedom, assuming that the two input states have the same polarization. However, it is straightforward to put the polarization back in

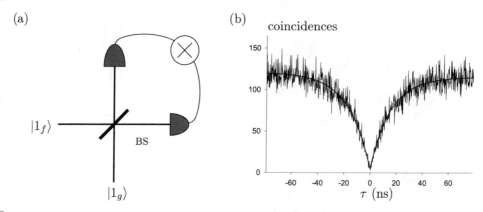

Fig. 4.8. The Hong–Ou–Mandel effect. (a) The experimental set-up consists of a 50:50 beam splitter (BS) with two single-photon input states. The output ports are detected with standard photodetectors. (b) A typical Hong–Ou–Mandel 'dip' in the coincidence counts between the two detectors, adapted from Lounis *et al.* (2000) with permission from Elsevier.

by adding the label λ to f and $\hat{a}_1$, and the label λ' to g and $\hat{a}_2$. The functions f and g are normalized such that

$$\int_{-\infty}^{\infty} |f(k)|^2 \, dk = \int_{-\infty}^{\infty} |g(k)|^2 \, dk = 1. \tag{4.53}$$

We are interested in the coincidence counts in the two detectors, which is determined by a second-order correlation function C:

$$C = \int \left\langle \hat{a}_1^\dagger(t)\hat{a}_2^\dagger(t') \, \hat{a}_1(t)\hat{a}_2(t') \right\rangle dt \, dt'. \tag{4.54}$$

We assume that C is time-averaged over a period that is much longer than the carrier frequency of the photons. The output state before the photodetectors is

$$|\psi_{\text{out}}\rangle = U \, |\psi_{\text{in}}\rangle, \tag{4.55}$$

where U is the transformation of the beam splitter in Eq. (4.51), and $U \, |0\rangle = |0\rangle$. We can then write the output state as

$$\begin{aligned}
|\psi_{\text{out}}\rangle &= \int dk\,dk' f(k)g(k') \left[U\hat{a}_1^\dagger(k)U^\dagger \, U\hat{a}_2^\dagger(k')U^\dagger \right] U \, |0\rangle \\
&= \frac{1}{2} \int dk\,dk' f(k)g(k') \left[\hat{a}_1^\dagger(k) + \hat{a}_2^\dagger(k) \right] \left[\hat{a}_2^\dagger(k') - \hat{a}_1^\dagger(k') \right] |0\rangle \\
&= \frac{1}{2} \int dk\,dk' f(k)g(k') \left[|1\rangle_{1k} \, |1\rangle_{2k'} - |1\rangle_{1k} \, |1\rangle_{1k'} \, |0\rangle_2 \right. \\
&\quad \left. + |1\rangle_{2k} \, |1\rangle_{1k'} - |0\rangle_1 \, |1\rangle_{2k} \, |1\rangle_{2k'} \right].
\end{aligned} \tag{4.56}$$

Using this output state, we can calculate C, and we find

$$C = \frac{1}{2} - \frac{1}{2} \int dk\, dk'\, f(k) f^*(k') g(k') g^*(k). \qquad (4.57)$$

When the two functions f and g are identical, the coincidence counts reduce to zero, $C = 0$. This is the famous Hong–Ou–Mandel dip, shown in Fig. 4.8b. The shape of the dip depends on the mechanism of the source described by the functions f and g, and is often modeled as a Gaussian or a Lorentzian.

Exercise 4.3: Using Eq. (1.117) and averaging over t and t', verify Eq. (4.57).

When the single-photon states are in a mixed state described by a density operator ρ, the HOM dip will never become zero, even when the two input states are identical ($\rho_1 = \rho_2$). This can be seen by considering the example

$$\rho = p \left| 1_f \right\rangle \left\langle 1_f \right| + (1 - p) \left| 1_g \right\rangle \left\langle 1_g \right|, \qquad (4.58)$$

where f and g may denote orthogonal mode functions (for example different time bins or different Hermite–Laguerre modes). When both input photons are in this state, the output will be a statistical mixture of four components. With probability p^2 the contribution to the coincidence counts will be from $\left| 1_f \right\rangle_1$ and $\left| 1_f \right\rangle_2$, which gives the maximum HOM dip. Similarly, with probability $(1-p)^2$ the states $\left| 1_g \right\rangle_1$ and $\left| 1_g \right\rangle_2$ contribute a maximum HOM dip. However, with probability $2p(1-p)$ two completely distinguishable photons interfere on the beam splitter and give no dip at all. Since the HOM experiment requires a statistical ensemble to show the dip, the ratio between the minimum and the maximum count rate (taken when the delay between the photons is large enough to distinguish them) is $2p(1-p)$. A complete dip (no coincidence counts at zero time delay) therefore tells us that the single-photon states are pure.

In conclusion, a maximal HOM dip can be reached only when the input states are pure indistinguishable single photons. Any multi-photon component in the input states will show up as a coincidence count at zero time delay. The full HOM dip is therefore a good indicator for good single-photon sources. There are two things it cannot tell us, however. First, it does not reveal the efficiency of the photon source. In order to determine this, a calibrated count rate must be established (taking detector efficiency and dark counts into account). Second, when the dip is non-maximal, the experiment does not tell us how much of the coincidence rate is due to impurity of the sources, and how much is due to multi-photon components in the input states. In order to determine this, an HBT experiment must be performed.

Finally, since both HBT and HOM experiments operate in a post-selected fashion, the figures of merit $g^{(2)}(0)$ and C do not distinguish sources that produce the state $\left| 1 \right\rangle \left\langle 1 \right|$ and the state $\rho_s = (1 - \eta_s) \left| 0 \right\rangle \left\langle 0 \right| + \eta_s \left| 1 \right\rangle \left\langle 1 \right|$. Whenever the latter state is collapsed onto the vacuum, no spurious coincidences in the detectors occur to produce false positives. It is tempting to call η_s the *efficiency* of the source, but this is correct only in a post-selected set-up. When ρ_s is freely propagating we cannot make ontological statements about the optical mode that elevate the number basis over all others, for to do so is to commit the PEF (see Section 3.1). The value of η_s can be measured with calibrated detectors, using the procedure leading to Eq. (4.37). The quality of a single-photon source Q_{SP} can therefore be defined as

$$Q_{SP} = \eta_s (1 - 2C) \left[1 - g^{(2)}(0) \right], \tag{4.59}$$

where $1 - 2C$ is the visibility of the HOM experiment, and $g^{(2)}(0)$ is the value of the HBT experiment at zero time delay. However, rather than their product Q_{SP}, it is often much more instructive to quote the three separate values, since these reveal how the source performs in different circumstances.

4.3.2 Weak coherent states and post-selection

We have mentioned that a single-mode laser operating far above threshold is described by a coherent state

$$|\alpha\rangle = e^{-|\alpha|^2/2} \sum_{n=0}^{\infty} \frac{\alpha^n}{\sqrt{n!}} |n\rangle . \tag{4.60}$$

For any non-zero value of α, the coherent state has non-zero amplitude for each photon number. We can calculate the probability of finding n photons in the coherent state as

$$p(n|\alpha) \equiv |\langle n|\alpha\rangle|^2 = \frac{|\alpha|^{2n} e^{-|\alpha|^2}}{n!}. \tag{4.61}$$

This is a Poissonian probability distribution. When the laser is highly attenuated ($|\alpha|^2 \ll 1$ and $\exp(-|\alpha|^2) \simeq 1$), the probability of two photons in a pulse (or time interval) becomes very small. Indeed, due to the Poissonian photon statistics of the coherent state, when a single photon appears with probability $p = |\alpha|^2$, the two-photon contribution has probability $p^2/2$. Can we use a highly attenuated laser as an approximation to a single-photon source?

The answer to this is a cautious 'yes'. When a weak coherent source is used as a single-photon source, the empty pulses will not trigger the detector, whereas the pulses with one or more photons do (assuming reasonably good photodetectors). Even though the rate of photon generation is small, the rate of two-photon generation and higher is even smaller and visibilities in single-photon experiments can be made arbitrarily close to one (at the expense of the data-acquisition rate). This is called *post-selection*.

Post-selection of weak coherent states can fail in a dramatic way when more than one weak coherent state is used in an interferometry experiment. Suppose an N-port interferometer has two weak coherent input states operating as approximations to single-photon sources, and a set of photodetectors in the output modes. Post-selection requires that the detectors must indicate the presence of two photons if a detection event is to be accepted (for this example we do not care whether the detectors are ideal, number-resolving, or bucket detectors). The probability that both weak coherent input states result in a single photon triggering a detector is p^2. The data-acquisition rate of the interferometry experiment will be dominated by this factor. However, the probability that one of the weak coherent sources produces two photons in the detectors, while the other produces nothing is $p^2/2$. There are two ways this failure mode can happen (either source 1 or source 2 produces the two photons), so this rate is of the order p^2. *If* the experiment is set up in such a way that the detector signatures do not tell us whether we had two weak coherent sources yielding a single photon each or if one produced two photons, then the error rate is of the same order as the data-acquisition rate, and the weak coherent sources are no longer good single-photon sources. The type of experiment determines whether post-selected weak coherent states are a good approximation to single-photon sources.

As an example, we will now show that an optical pulse in a weak coherent state fails the HOM test. First, we consider the coincidence rate in a HOM experiment given two perfectly distinguishable input photons. Distinguishability means that the photons occupy different optical modes, labelled 1 and 2. If the input modes are a_1, a_2, and b_1, b_2, and the output modes are c_1, c_2, and d_1, d_2, the transformations of a 50:50 beam splitter are

$$\hat{a}_i^\dagger \rightarrow \frac{\hat{c}_i^\dagger + \hat{d}_i^\dagger}{\sqrt{2}} \quad \text{and} \quad \hat{b}_i^\dagger \rightarrow \frac{\hat{c}_i^\dagger - \hat{d}_i^\dagger}{\sqrt{2}}, \tag{4.62}$$

with $i = 1, 2$. The input state $|1, 0; 0, 1\rangle_{a_1,a_2;b_1,b_2}$ then transforms into

$$|1, 0; 0, 1\rangle_{a_1,a_2;b_1,b_2} \rightarrow \frac{1}{2} \left(|1, 1; 0, 0\rangle - |1, 0; 0, 1\rangle + |0, 1; 1, 0\rangle - |0, 0; 1, 1\rangle \right). \tag{4.63}$$

The detectors in modes c and d do not indicate whether photons in mode 1 or 2 were detected, they just indicate whether a photon was present in either mode (c or d). This state therefore yields a coincidence rate of 50%, which is the 'base rate' in that no interference has occurred in the experiment.

Second, we compare this situation with the case where the input states of the HOM experiment are two weak coherent states. Since these states are used as approximations to single-photon states, the phase of the coherent state is unimportant, and phase stability is not required. Furthermore, the finite time interval that constitutes a detection coincidence is limited by the temporal resolution of the detector, which is typically much longer than the period of the optical frequency ω. This means that the phases of the two weak coherent input states are randomized with respect to each other, and we can model the weak coherent state as an incoherent Poissonian mixture:

$$\int_0^{2\pi} dt \, |\alpha\rangle \langle\alpha| = e^{-|\alpha|^2} \int_0^{2\pi} dt \sum_{n,m=0}^{\infty} \frac{\alpha^{*n} \alpha^m \, e^{i(n-m)\omega t}}{\sqrt{n!m!}} |m\rangle \langle n|$$

$$= e^{-|\alpha|^2} \sum_{n=0}^{\infty} \frac{|\alpha|^2}{n!} |n\rangle \langle n| \,. \tag{4.64}$$

The two input states of the HOM experiment are both given by this randomized weak coherent state, and in contrast to the previous set-up, the mode-matching on the beam splitter is perfect. That is, the photons are perfectly indistinguishable, and we take $a_1 = a_2 = a$, etc.

The input state is then given by

$$\rho_{\text{in}} \propto |0,0\rangle \langle 0,0| + p \left(|1,0\rangle \langle 1,0| + |0,1\rangle \langle 0,1| \right)$$

$$+ \frac{p^2}{2} \left(|2,0\rangle \langle 2,0| + |0,2\rangle \langle 0,2| + 2|1,1\rangle \langle 1,1| \right) + O(p^3). \qquad (4.65)$$

If we post-select on events where we see two photons in the output modes (by using either number-resolving detectors or detector cascades) the normalized part of the input state responsible for these events is

$$\rho_{\text{in}} = \frac{1}{4} |2,0\rangle \langle 2,0| + \frac{1}{4} |0,2\rangle \langle 0,2| + \frac{1}{2} |1,1\rangle \langle 1,1|. \qquad (4.66)$$

Notice that the state no longer depends on p. If we define $|\psi^{\pm}\rangle = (|2,0\rangle \pm |0,2\rangle)/\sqrt{2}$, the beam splitter transforms the input state into the output state

$$\rho_{\text{out}} = \frac{1}{2} |\psi^-\rangle \langle \psi^-| + \frac{1}{4} |\psi^+\rangle \langle \psi^+| + \frac{1}{4} |1,1\rangle \langle 1,1|. \qquad (4.67)$$

The states $|\psi^{\pm}\rangle$ do not give rise to coincidence counts, and all the coincidences are due to the state $|1,1\rangle$. The coincidence rate for this state is therefore 25%.

Compared to the HOM experiment with distinguishable input photons, the coincidence rate is reduced by a factor 2. This is the *classical* limit to the HOM experiment, since we have shown it to be true for coherent states that are often regarded as the classical solutions to the field equations of motion. Deviations from the Poisson distribution can reduce the coincidence rate, and the input state becomes more like a single-photon state.

4.3.3 Heralded parametric downconversion

Another practical way of producing high-quality single-photon sources is via 'parametric downconversion' (PDC). In this system a crystal with a large $\chi^{(2)}$ nonlinearity is pumped with a strong laser at a frequency ω_P. There is a small probability (proportional to $\chi^{(2)}$) that photons from the pump laser are absorbed by the crystal and re-emitted as *two* photons with frequencies ω_S and ω_I (see Fig. 4.9). The indices S and I stand for 'signal' and 'idler', which are the traditional names for the output modes. The interaction Hamiltonian governing this process can be written as

$$\mathcal{H} \propto e^{i\theta} \chi^{(2)} \hat{a}_{\mathbf{k}_P, \omega_P} \hat{a}^{\dagger}_{\mathbf{k}_S, \omega_S} \hat{a}^{\dagger}_{\mathbf{k}_I, \omega_I} + e^{-i\theta} \chi^{(2)} \hat{a}^{\dagger}_{\mathbf{k}_P, \omega_P} \hat{a}_{\mathbf{k}_S, \omega_S} \hat{a}_{\mathbf{k}_I, \omega_I}, \qquad (4.68)$$

where we have introduced a phase θ to be general. Next, we note that the pump is very bright, and is therefore very well described by a classical coherent state $\hat{a}_{\mathbf{k}_P, \omega_P} \to \alpha_P$. Collecting all factors in a single complex number $\xi \propto \alpha_P \chi^{(2)} e^{i\theta}$, we write the Hamiltonian as

$$\mathcal{H}_{\text{PDC}} = \xi \, \hat{a}^{\dagger}_{\mathbf{k}_S, \omega_S} \hat{a}^{\dagger}_{\mathbf{k}_I, \omega_I} + \xi^* \, \hat{a}_{\mathbf{k}_S, \omega_S} \hat{a}_{\mathbf{k}_I, \omega_I}. \qquad (4.69)$$

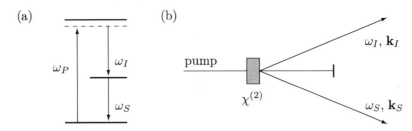

Fig. 4.9. The principle of parametric downconversion. (a) In a nonlinear crystal a photon from the pump can be absorbed and re-emitted into two photons. (b) The two photons emerge from the crystal in two different modes.

This is the generator of two-mode squeezing, introduced in Section 1.5.2. The operator for the pump has become a *parameter* in the Hamiltonian, and the photons from the pump are converted down to two photons with lower energy. Hence the name 'parametric downconversion'.

The output state of a two-mode squeezer is given in Eq. (1.235), and choosing $\tanh |\xi| = \lambda$ we can write

$$|\text{PDC}(\lambda, \phi)\rangle = \frac{1}{\sqrt{1 - \lambda^2}} \sum_{n=0}^{\infty} \lambda^n e^{in\phi} |n\rangle_S \otimes |n\rangle_I. \tag{4.70}$$

The parameter λ tends to be very small, of the order 10^{-2}. The two lowest-order contributions to the output state are therefore

$$|\text{PDC}(\lambda)\rangle = \sqrt{1 - \lambda^2} |0, 0\rangle + \lambda |1, 1\rangle_{SI} + O(\lambda^2), \tag{4.71}$$

where we have set $\phi = 0$ for convenience. We can place a photodetector in one of the output modes, say the signal mode. When dark counts are negligible, a detector click will then herald the presence of a photon in the idler mode. Moreover, since λ is small, this will be an almost ideal single-photon state with a negligible two-photon contribution.

Exercise 4.4: Given a click in a lossy photodetector (without dark counts) in the signal mode, calculate the output state in the idler mode of a parametric downconverter.

In practice, filtering of the modes can be used to make the signal and idler modes extremely pure, and both HBT and HOM experiments are easily performed in the lab. However, the efficiency of PDC as a single-photon source is very low. Theoretical proposals exist in which a large array ($\sim 10^4$) of PDCs are used in parallel to produce a single photon with unit probability. However, this is a rather large overhead for a single deterministic single-photon source, and is unlikely as a practical solution for the foreseeable future. Alternatively, a single PDC can be used to create a heralded photon, which is subsequently switched into a fibre loop, where it can be stored for a reasonable amount of time until it is needed. In this case switching losses are the dominant source of errors (see Pittman *et al.*, 2002).

Table 4.2.   Important properties of some single-photon sources, after Lounis and Orrit (2005).

Source[a]	ES	γ (MHz)	τ (ns)	FL	R (MHz)	ME	T (K)
PDC[b]	sharp line	10^3	10^{-3}	Yes	10^{-3}	Yes	300
CA[c]	sharp line	10	15	Yes	0.1	No	10^{-6}
IT[d]	sharp line	10	15	Yes	0.1	No	10^{-3}
LTM[e]	ZPL/broad	$30/10^3$	4	Yes/No	100	No	1–10
RTM[f]	broad band	10^6	4	No	100	No	300
CC[g]	broad band	10^6–10^7	10	No	10	No	300
QD[h]	sharp line	10^3	0.3	Yes	1 000	Yes	1–30
NC[i]	broad line	$3 \cdot 10^4$	30	No	10	No	300

[a] Here, ES is the emission spectrum, γ is the linewidth, τ is the emission lifetime, FL stands for 'Fourier-limited', R is the emission rate, ME stands for 'multiple excitations', and T is the operating temperature.

[b] Parametric downconversion (Lounis and Orrit, 2005).

[c] Cold atom (Lounis and Orrit, 2005).

[d] Ion in trap (McKeever *et al.*, 2004).

[e] Molecule at low temperature (Basché *et al.*, 1992, and Brunel *et al.*, 1999).

[f] Molecule at room temperature (Lounis and Moerner, 2000, and Treussart *et al.*, 2002).

[g] Colour centre at room temperature (Gruber *et al.*, 1997).

[h] III-V quantum dot (Brokmann *et al.*, 2004).

[i] II-VI nanocrystal (Santori *et al.*, 2002).

4.3.4 Other single-photon sources

Both weak coherent states and heralded parametric downconversion suffer from multi-photon contributions to the wave function. In the case of the weak coherent state we have seen that this leads to the classical limit of the HOM experiment. Photons produced in heralded parametric downconversion can perform very well in HBT and HOM experiments, since the second-order term proportional to $O(\lambda^2)$ can be made arbitrarily small. However, this means that the source efficiency $|\lambda|^2$ must also be small, and the data-acquisition rate low. The reason for these multi-photon contributions is that the number of quantum excitations in the source is ill defined. In a weak coherent state the laser has an *average* number of photons, and in parametric downconversion we cannot tell how many photons are being converted in the crystal.

An obvious way around this problem is to take a single system that acts as the photon emitter. When a photon is emitted, the system needs to be reloaded before a second photon can be emitted. Examples of such systems are atoms and ions in cavities or traps (or both), colour centres in materials such as diamond, quantum dots in microcavities, and single molecules. In Chapter 7 we will describe how single photons can be produced in atom-like structures in cavities, and in Chapter 11 we describe single-photon emitters in the solid state, such as quantum dots and colour centres in diamond.

Some important properties of single-photon sources are given in Table 4.2. One such property is the emission spectrum, which indicates whether the source produces a broad range of frequencies, or whether it produces a narrow line in the frequency spectrum. Similarly, the emission lifetime indicates the time interval in which most of the energy of the source is concentrated. The linewidth and lifetimes are often Lorentzian or Gaussian functions, for which a meaningful measure can be given. When the lifetime and the linewidth are related via a Fourier transform, the source is said to be 'Fourier-limited'. This means that the output state of the source is a pure state, and is therefore a critical property when we need a maximum HOM dip (i.e., no coincidences in the detectors). The emission rate is an important consideration for any scalable quantum information processing protocol.

4.4 Entangled photon sources

There are important applications in optical quantum information processing where entangled photon sources are required. Rather than applying entangling gates to two single-photon states, which will be addressed in Chapter 6, it may be more feasible to create sources of entangled photons. In this section, we give a brief overview of these devices.

4.4.1 Post-selected parametric downconversion

In the previous section we have seen how the output of a parametric downconverter is naturally described by two-mode squeezing. In this section we treat parametric downconversion in a bit more detail, and show how it can be used to create maximally entangled photon pairs (at least in some post-selected way).

The signal and idler modes have frequencies ω_S, ω_I and wave vectors $\mathbf{k}_S$, $\mathbf{k}_I$, respectively. However, how are the values of these quantities determined? There are so-called 'phase-matching' conditions inside the crystal that need to be satisfied by the process. They amount to energy and momentum conservation of the process:

$$\omega_P = \omega_S + \omega_I \quad \text{and} \quad \mathbf{k}_P = \mathbf{k}_S + \mathbf{k}_I. \tag{4.72}$$

These conditions hold only in the limit of infinite crystal size. For finite crystals, the phase-matching conditions will be approximate, and the delta functions must be replaced by squared sinc functions. Here, we will use the delta functions for simplicity.

We can write the broadband Hamiltonian as an integral over all signal and idler wave vectors, and include the phase-matching conditions as delta functions. The interaction Hamiltonian in the parametric approximation was given by Eq. (4.69), and the broadband Hamiltonian becomes

$$\mathcal{H}_{\text{PDC}} = \int d\mathbf{k}_S \, d\mathbf{k}_I \, d\omega_S \, d\omega_I \, \delta(\mathbf{k}_P - \mathbf{k}_S - \mathbf{k}_I)\delta(\omega_P - \omega_S - \omega_I)$$
$$\times \left[\xi \, \hat{a}^\dagger_{\mathbf{k}_S,\omega_S} \hat{a}^\dagger_{\mathbf{k}_I,\omega_I} + \xi^* \, \hat{a}_{\mathbf{k}_S,\omega_S} \hat{a}_{\mathbf{k}_I,\omega_I} \right]. \tag{4.73}$$

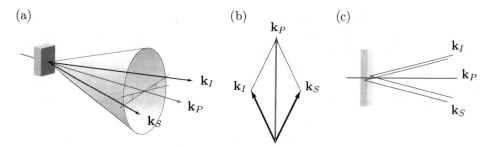

Fig. 4.10. The output modes of PDC form cones that satisfy the phase-matching conditions of Eq. (4.72): (a) the output cone for a single crystal; (b) phase-matching conditions; and (c) double crystal for entangled photon-pair generation.

The squeezing parameter ξ generally depends on the wave vectors and frequencies as well. The delta functions remove the integration over, say, $\mathbf{k}_S$ and ω_S, but there is still an integral over $\mathbf{k}_I$. The effect of this is that by symmetry the downconverted photons form 'cones' (see Fig. 4.10). The experimenter can select specific $\mathbf{k}_S$ and $\mathbf{k}_I$ by placing a screen with pinholes in the output cone. When the two pinholes are placed correctly, the number of photons passing through one pinhole is identical to the number of photons passing through the other pinhole. The output states are therefore superpositions of $|n\rangle_S |n\rangle_I$, as we found earlier.

Apart from the frequency and the wave vectors, we have the polarization as a degree of freedom. In the above set-up, the downconversion crystal can be cut along certain axes such that the two modes $\mathbf{k}_S$ and $\mathbf{k}_I$ have identical polarization. This is called 'type-I parametric downconversion' (type-I PDC). When the crystal is cut such that the output modes have orthogonal polarization, we speak of 'type-II' PDC. We can create polarization-entangled photon pairs as follows: we place two thin crystals, cut for type-II PDC, on top of each other. Since they are very thin, the output cones from the crystals almost perfectly overlap each other. The crystals are rotated around the pump axis such that the first crystal produces photons according to $\hat{a}^\dagger_{H,\mathbf{k}_S} \hat{a}^\dagger_{V,\mathbf{k}_I}$, and the second crystal produces photons according to $\hat{a}^\dagger_{V,\mathbf{k}_S} \hat{a}^\dagger_{H,\mathbf{k}_I}$. Here, H and V are linear polarization directions. Since the thickness of the crystals is much shorter than the coherence length of the pump pulse, the two terms add coherently. Including a π phase shift and a $\pi/2$ polarization rotation in the signal (or idler) mode, the Hamiltonian becomes proportional to the generator

$$\hat{K}_+ = \hat{a}^\dagger_{H,\mathbf{k}_S} \hat{a}^\dagger_{V,\mathbf{k}_I} - \hat{a}^\dagger_{V,\mathbf{k}_S} \hat{a}^\dagger_{H,\mathbf{k}_I}. \tag{4.74}$$

Using $S(\xi) = \exp(\xi \hat{K}_+ + \xi^* \hat{K}_-)$, the state created by this PDC set-up is (assuming ξ real)

$$S(\xi) |0\rangle = \frac{1}{\cosh|\xi|} \sum_{n=0}^{\infty} \tanh^n|\xi| \frac{\hat{K}_+^n}{n!} |0\rangle = \frac{1}{\cosh|\xi|} \sum_{n=0}^{\infty} \tanh^n|\xi| |\Phi_n\rangle$$

$$\equiv \sqrt{1-\lambda^2} \sum_{n=0}^{\infty} \lambda^n |\Phi_n\rangle, \tag{4.75}$$

where we have defined $\lambda = \tanh |\xi|$, and

$$|\Phi_n\rangle = \frac{1}{\sqrt{n+1}} \sum_{m=0}^{n} (-1)^m |m, n-m\rangle_S |n-m, m\rangle_I . \tag{4.76}$$

This state includes the horizontal and the vertical component of each mode. It is easy to see that $\langle\Phi_n|\Phi_m\rangle = \delta_{nm}$, since both states have different total photon numbers. When $\lambda \ll 1$, the resulting state is approximately

$$S(\xi)|0\rangle \simeq \sqrt{1-\lambda^2}|0,0\rangle + \frac{\lambda}{\sqrt{2}}(|H,V\rangle - |V,H\rangle) + O(\lambda^2). \tag{4.77}$$

In other words, there is a small probability of creating a maximally entangled state in the polarization of two photons. Such a state can be produced with parametric downconversion, where typically $\lambda \approx 10^{-2}$. This is how most entangled photon states are produced in practice.

Parametric downconversion is not the only way to create entangled photon pairs. Similar to single-photon sources, we can construct a system that must be reloaded in order to create a second entangled photon pair, thus suppressing unwanted multi-pair production. A level system for a physical device that can produce entangled photons is shown in Fig. 4.11. There are four levels: one ground state $|g\rangle$, one upper excited state $|e\rangle$, and two non-degenerate (possibly meta-stable) intermediate states $|s_1\rangle$ and $|s_2\rangle$. The system cannot decay directly from $|e\rangle$ to $|g\rangle$, for example, because of a selection rule. Instead, the only possible decay mechanisms are via the intermediate states. If the decay path is via $|s_1\rangle$, the system creates a single photon with polarization L and frequency ω_1, followed almost immediately by a second photon with polarization R and frequency ω_2. This state can be written as $|L_1, R_2\rangle$. By contrast, if the decay takes the path via $|s_2\rangle$, the two-photon state that is produced is

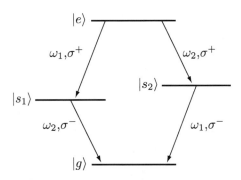

Fig. 4.11.   A level system for creating two entangled photons in frequency and polarization. The direct transition from $|e\rangle$ to $|g\rangle$ is forbidden owing to an angular-momentum selection rule. The intermediate states $|s_1\rangle$ and $|s_2\rangle$ are relatively short-lived. The system can take two paths from $|e\rangle$ to $|g\rangle$, namely via $|s_1\rangle$ or $|s_2\rangle$. In the first case the system produces two photons with polarization L and R, and frequencies ω_1 and ω_2, respectively: $|L_1, R_2\rangle$. The second path gives rise to the two-photon state $|R_1, L_2\rangle$. Since the system does not record which path was taken, the state of the emitted photons is $(|L_1, R_2\rangle + |R_1, L_2\rangle)/\sqrt{2}$, that is, an entangled state.

$|R_1, L_2\rangle$. When the system does not record in any way which path has been taken, the two states are superposed coherently, and the output state of the device is

$$|\Psi_{\text{out}}\rangle = \frac{|L_1, R_2\rangle + |R_1, L_2\rangle}{\sqrt{2}}. \tag{4.78}$$

Note that this state is entangled in both frequency and polarization. This is sometimes called 'hyper-entanglement', and it may or may not be a desirable feature. For example, hyper-entanglement is an essential ingredient in some protocols for optical quantum computing (see, e.g., Yoran and Reznik, 2003). However, for the purpose of producing clean polarization entangled states the frequency entanglement may be disastrous, since a trace operation over the frequency degree of freedom destroys all entanglement.

4.5 Quantum non-demolition photon detectors

In addition to photodetectors (which destroy the photons), and sources (which create photons), one can imagine a device that measures the number of photons in a mode without destroying them. Such a device would be both a detector *and* a source. These are called 'quantum non-demolition' (QND) measurements. They do not have to operate in the photon-number basis; any orthonormal basis will do. An ideal QND measurement projects the measured mode onto the eigenstate corresponding to the eigenvalue that was inferred from the measurement outcome. It is therefore a perfect von Neumann measurement.

One way to implement a QND measurement of photon number is using a cross-Kerr nonlinearity, described by the Hamiltonian

$$\mathcal{H}_K = \hbar \kappa \, \hat{a}_1^\dagger \hat{a}_1 \, \hat{a}_2^\dagger \hat{a}_2, \tag{4.79}$$

acting on modes a_1 and a_2. This symmetric Hamiltonian induces a phase shift in either mode that depends on the intensity of the field in the other mode. When the interaction κ is strong enough, such a device can be used to create an optical switch for photons. Under the interaction $\mathcal{H}_K t$, the mode $\hat{a}_1$ transforms into

$$\begin{aligned}
\hat{a}_1(t) &= \exp\left(-\frac{i}{\hbar}\mathcal{H}_K t\right) \hat{a}_1 \, \exp\left(\frac{i}{\hbar}\mathcal{H}_K t\right) \\
&= \hat{a}_1 - i\tau[\hat{a}_1^\dagger\hat{a}_1, \hat{a}_1]\hat{a}_2^\dagger\hat{a}_2 + \frac{(-i\tau)^2}{2!}\left[\hat{a}_1^\dagger\hat{a}_1, [\hat{a}_1^\dagger\hat{a}_1, \hat{a}_1]\right](\hat{a}_2^\dagger\hat{a}_2)^2 + \dots \\
&= \hat{a}_1 \, e^{i\tau \, \hat{a}_2^\dagger\hat{a}_2}, \tag{4.80}
\end{aligned}$$

where $\tau = \kappa t$, and t is the duration of the interaction. When one mode, say a_1, is the signal mode, and a_2 is a probe mode, the presence of n photons in mode a_1 induces a phase shift $\phi = n\tau$ in mode a_2. Such a phase, when it is sufficiently large, can be measured with a Mach–Zehnder interferometer, as shown in Fig. 4.12. We will discuss the optical implementation of $\mathcal{H}_K$ in Chapter 8.

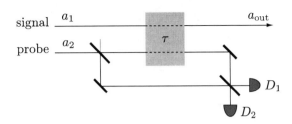

Fig. 4.12. Schematic of an optical QND measurement for a single photon in a_1. The cross-Kerr nonlinearity induces a phase shift τ in the probe mode a_2, dependent on the presence or absence of a photon in the signal mode a_1. The phase can be measured with a Mach–Zehnder interferometer. The detection signature in D_1 and D_2 indicates whether there is a photon in mode a_{out}.

The QND measurement can now be implemented by combining the cross-Kerr nonlinearity with a Mach–Zehnder interferometer, as shown in Fig. 4.12. From Eq. (1.152) in Chapter 1, the difference between the measured intensities in detectors D_1 and D_2 in a Mach–Zehnder interferometer can be written as

$$\Delta I = I_{D_1} - I_{D_2} = \cos\phi \left(\hat{a}_2^\dagger \hat{a}_2 + \hat{a}_2'^\dagger \hat{a}_2' \right) + \sin\phi \left(\hat{a}_2^\dagger \hat{a}_2' + \hat{a}_2'^\dagger \hat{a}_2 \right), \qquad (4.81)$$

and therefore gives an estimation of the phase ϕ. When the secondary input mode a_2' of the Mach–Zehnder interferometer is the vacuum, this reduces to

$$\Delta I = \cos\tau \, \hat{a}_2^\dagger \hat{a}_2. \qquad (4.82)$$

In a single-shot measurement with a *single* photon in mode a_2, the only way to gain full knowledge about the presence or absence of a photon in the output mode a_{out} is if the absence of photons in a_1 produces a click in one detector (e.g., D_1), and the presence of a photon in a_1 produces a click in D_2. The phase shift that is needed for such a switch in ΔI is $\tau = \pi$. However, such large cross-Kerr phase shifts are impossible to achieve (see Appendix 3).

On the other hand, when the probe field is in a coherent state $|\alpha\rangle$, the intensity difference is $\Delta I = \cos\tau \, |\alpha|^2$. The average photon number $|\alpha|^2$ may be very large, and we can distinguish between smaller intensity differences. The precision of such a measurement is given by the variance of ΔI, $\mathrm{Var}(\Delta I)$, and we require that the change in intensity difference is (much) larger than $\mathrm{Var}(\Delta I)$. In practice, τ will be much smaller than one, and we can evaluate how large $|\alpha|$ must be given a value of τ (see Exercise 4.5). Another advantage of using a bright coherent state is that n-photon components in the state of the signal mode a_1 induce a phase shift $n\tau$ in the probe mode. When $n\tau < \pi$, this may induce a measurable intensity difference (depending on the variance), and the set-up can distinguish more than just the single-photon state from the vacuum. Finally, when more exotic states are used in the Mach–Zehnder interferometer (possibly using the secondary input port), the number of photons may in principle be measured deterministically for much smaller values of τ. We will describe these advanced phase-estimation techniques in more detail in Chapter 13.

Exercise 4.5: Show that in order to construct a deterministic photon-number QND measurement with a coherent probe state, we must choose $|\alpha| \gg \tau^{-2}$.

When we take into account various practical imperfections, such as detection inefficiencies, dark currents, photon loss in the Kerr medium, etc., the QND measurement is no longer a perfect von Neumann measurement, and the question is how the output state in mode a_{out} is affected. In general, the output state will no longer be a pure state, even if the input state in a_1 was pure. The QND measurement can therefore no longer be modeled as a projection, but instead we need to model it as a completely positive map. The techniques presented in Sections 3.1 and 4.1 can be used to evaluate these maps explicitly.

The use of Kerr nonlinearities is not essential for constructing a QND measurement. For example, a single-photon QND experiment was performed by Nogues *et al.* (1999), using the strong atom–photon interactions of microwave cavity quantum electrodynamics. If one is willing to sacrifice the deterministic character, single-photon QND measurements can also be implemented with passive linear-optical elements and auxiliary single-photon sources (Kok *et al.*, 2002b). Traditionally, QND measurements have been considered for the quadrature operators of the electromagnetic field. For an introduction to this type of QND measurement, see for example Chapter 14 of Walls and Milburn (2008).

4.6 References and further reading

Most of the references to the primary literature regarding photon detectors and sources are contained in the captions of Table 4.1 and Table 4.2. An overview of state-of-the-art photon-number detectors was given by Divochiy *et al.* (2008), and a recent comprehensive review of single-photon sources is Lounis and Orrit (2005). Entangled photons have been generated with parametric downconversion for years (Kwiat *et al.*, 1995), and recently other sources have emerged. Kuzmich *et al.* (2003) produced entangled photons in atomic ensembles, and photonic Bell pairs have been created with solid-state devices by Stevenson *et al.* (2006), Akopian *et al.* (2006), and Hafenbrak *et al.* (2007).

5 Quantum communication with single photons

After the preceding four introductory chapters, we are finally ready to discuss optical quantum information processing. While in principle quantum communication can be implemented with atoms or electrons, in practice the implementation of choice for long-distance quantum communication will almost certainly be optical. In this chapter, we will develop some of the key topics in quantum communication with single photons, and we will discuss continuous-variable quantum communication in Chapter 8. In some ways, quantum communication is the most technologically advanced part of quantum information processing. Most protocols in this chapter, such as teleportation and cryptography, have been convincingly demonstrated in the lab, and there are already several commercial organizations that sell cryptographic systems based on quantum mechanical principles. In particular, the promise of secure communication with quantum cryptography is one of the driving forces behind the development of single-photon sources and detectors. In this chapter, we first construct several optical representations of the qubit, including polarized photons and time-bin encoding. Since the single-photon implementation of a qubit is central to optical quantum information processing, this constitutes a large part of the chapter. In Section 5.2 we discuss quantum teleportation and entanglement swapping with single photons. We also present a method for entanglement distillation using entanglement swapping. In Section 5.3 we introduce decoherence-free subspaces, which can be used to reduce noise in the transmission of quantum information, and even establish communication channels in the absence of shared reference frames. Finally, in Section 5.4 we analyze quantum cryptography with single photons. We will give a proof of security for BB84 and EKERT91, and show how multi-photon sources can be used to increase the data transmission via decoy states.

5.1 Photons as information carriers

There are several representations in which a single photon can carry a single qubit of quantum information. The advantage of using photons is that single-qubit operations can be implemented with high precision. The most common implementation is the 'dual-rail representation', where a single photon can be in a superposition of two optical modes. These modes can be either spatial modes, polarization, or orthogonal transverse mode functions. The dual-rail implementation is contrasted with the 'single-rail implementation', where the qubit is carried by a single optical mode, and the qubit degree of freedom is the photon number. This representation is much less common, since the corresponding single-qubit operations do not preserve photon number and are therefore very hard to implement in practice. We will concentrate here on the dual-rail implementation, and consider spatial modes, polarization, and 'orbital angular-momentum' (OAM) states of the photon. Finally, we will discuss time-bin encodings, in which the computational basis states of the qubit are associated with a photon in two different time intervals.

At this point, we recall our discussion in Chapter 1 about optical modes. In most books the electromagnetic field is quantized in terms of plane waves that are extended to infinity in all directions. The advantage of this approach is that the quantization becomes relatively simple in terms of wave vectors. On the other hand, true plane-wave modes are not physical, and any real, localized optical mode must be a superposition of the unphysical plane waves. In general we can then talk only of approximate wave vectors. Nevertheless, the localized physical modes are also true modes in the sense that their creation and annihilation operators obey the bosonic commutation relations. In the remainder of this book, we take optical modes to mean *physical* modes that are suitably localized.[1] In addition, the modes are labelled with an index for polarization, and a continuous variable for the (approximate) wave vector. We will often omit the explicit wavevector dependence when its continuous character is not important to the discussion.

5.1.1 Spatial dual-rail qubits

We now define the single-photon representation of a qubit. The computational basis for the dual-rail implementation on two spatial modes a_1 and a_2 is denoted by

$$|0\rangle = \hat{a}_1^\dagger |0,0\rangle_{12} = |1,0\rangle_{12} \quad \text{and} \quad |1\rangle = \hat{a}_2^\dagger |0,0\rangle_{12} = |0,1\rangle_{12}, \tag{5.1}$$

which are defined as the eigenstates of the Pauli Z matrix. We have suppressed the polarization degree of freedom (or any other degree of freedom, for that matter) in the creation operator, since it is assumed to be the same for all operators. The mode functions associated with the operators $\hat{a}_j$ can be assumed to be suitably localized. The modes a_1 and a_2 may have the same direction and frequency, in which case the index serves to distinguish two parallel, non-overlapping beams. In Section 1.3 we have seen that this situation is well defined in the paraxial approximation, and suitably described by Gaussian mode functions.

Single-qubit operations can be written as mode transformations on a_1 and a_2. These are the generalized beam-splitter and phase-shift transformations. In Eq. (1.146) we saw that the general Hamiltonian of a two-mode transformation is

$$\mathcal{H}(\theta, \varphi) = \theta\, e^{i\varphi}\, \hat{a}_1^\dagger \hat{a}_2 + \theta\, e^{-i\varphi}\, \hat{a}_2^\dagger \hat{a}_1, \tag{5.2}$$

which leads to a (Bogoliubov) transformation on the creation and annihilation operators

$$\hat{a}_1 \rightarrow \cos\theta\, \hat{a}_1 - i e^{i\varphi} \sin\theta\, \hat{a}_2,$$
$$\hat{a}_2 \rightarrow -i e^{-i\varphi} \sin\theta\, \hat{a}_1 + \cos\theta\, \hat{a}_2. \tag{5.3}$$

---

[1] In Chapter 7 we will show how to derive the Jaynes–Cummings Hamiltonian using plane waves, but will again see the necessity of introducing physical, localized modes.

The parameter θ describes the reflectivity of the beam splitter, while φ determines the phase shift imparted on the reflected mode by the beam splitter. Physically, this phase can be implemented by using a suitable coating, or by using an additional phase shift.

When the creation operators associated with Eq. (5.3) are applied to the vacuum state, we obtain the qubit transformations

$$|0\rangle = \hat{a}_1^\dagger |0,0\rangle_{12} \rightarrow \cos\theta \, |1,0\rangle_{12} + ie^{-i\varphi} \sin\theta \, |0,1\rangle_{12}$$
$$= \cos\theta \, |0\rangle + ie^{-i\varphi} \sin\theta \, |1\rangle,$$
$$|1\rangle = \hat{a}_2^\dagger |0,0\rangle_{12} \rightarrow ie^{i\varphi} \sin\theta \, |1,0\rangle_{12} + \cos\theta \, |0,1\rangle_{12}$$
$$= ie^{i\varphi} \sin\theta \, |0\rangle + \cos\theta \, |1\rangle. \tag{5.4}$$

Using suitable values for θ and φ, it is straightforward to define the eigenstates of the Pauli X and Y matrices:

$$|+\rangle \equiv \frac{1}{\sqrt{2}} \left(|1,0\rangle_{12} + |0,1\rangle_{12} \right) \quad \text{and} \quad |-\rangle \equiv \frac{1}{\sqrt{2}} \left(|1,0\rangle_{12} - |0,1\rangle_{12} \right),$$
$$|\circlearrowright\rangle \equiv \frac{1}{\sqrt{2}} \left(|1,0\rangle_{12} + i|0,1\rangle_{12} \right) \quad \text{and} \quad |\circlearrowleft\rangle \equiv \frac{1}{\sqrt{2}} \left(|1,0\rangle_{12} - i|0,1\rangle_{12} \right). \tag{5.5}$$

All the qubit rotations are contained in Eq. (5.3), and can therefore be implemented with beam splitters and phase shifters.

Some transformations are not rotations, however. The Hadamard gate has determinant minus one ($\det H = -1$), and can therefore not be reduced to a rotation and implemented with a generalized beam splitter of the form in Eq. (5.3). Instead, this transformation is obtained by using a mode-dependent phase shift in addition to a single-qubit rotation. For a phase π (which is sufficient) this operator is the Pauli Z matrix.

Exercise 5.1: How do we implement a Hadamard transformation?

5.1.2 Polarization qubits

Another natural single-photon qubit degree of freedom is polarization, since the generators of polarization transformations are also Pauli matrices. The polarization degree of freedom for a mode with wave vector $\mathbf{k}$ is given by the vectors $\boldsymbol{\epsilon}_\lambda(\mathbf{k})$, with λ indicating one of two possible orthogonal polarization states. Two often-used orthogonal polarization directions are the horizontal and vertical directions with respect to some reference frame:

$$\epsilon_H = \begin{pmatrix} 1 \\ 0 \end{pmatrix} \quad \text{and} \quad \epsilon_V = \begin{pmatrix} 0 \\ 1 \end{pmatrix}, \tag{5.6}$$

where we have suppressed the $\mathbf{k}$ dependence since it is common to all polarization vectors in this discussion. Note that we have also omitted the third dimension of the vector $\boldsymbol{\epsilon}_\lambda$. We are working in the Coulomb gauge, in which the polarization is perpendicular to the direction of propagation ($\epsilon_\lambda \cdot \mathbf{k} = 0$), so the third term of ϵ_λ is always zero. We next show that the polarization vectors span a two-dimensional complex vector space (a Hilbert space), and may therefore be used to define a qubit degree of freedom.

First, the two vectors ϵ_H and ϵ_V can be considered eigenvectors of the Pauli Z matrix with eigenvalues $+1$ and -1, respectively. Similarly, we can define diagonal polarization

$$\epsilon_D = \frac{1}{\sqrt{2}} \begin{pmatrix} 1 \\ 1 \end{pmatrix} \quad \text{and} \quad \epsilon_A = \frac{1}{\sqrt{2}} \begin{pmatrix} 1 \\ -1 \end{pmatrix}, \tag{5.7}$$

where D and A indicate 'diagonal' and 'anti-diagonal', respectively. Since the diagonal and anti-diagonal polarizations are related to the horizontal and vertical polarization according to

$$\epsilon_D = \frac{\epsilon_H + \epsilon_V}{\sqrt{2}} \quad \text{and} \quad \epsilon_A = \frac{\epsilon_H - \epsilon_V}{\sqrt{2}}, \tag{5.8}$$

these vectors can be considered eigenvectors of the Pauli X matrix. Finally, the polarization vectors can have complex components. In particular, circular polarization is defined as

$$\epsilon_L = \frac{1}{\sqrt{2}} \begin{pmatrix} 1 \\ i \end{pmatrix} \quad \text{and} \quad \epsilon_R = \frac{1}{\sqrt{2}} \begin{pmatrix} 1 \\ -i \end{pmatrix}. \tag{5.9}$$

It is not difficult to verify that these are the eigenvectors of the Pauli Y matrix. Since these six vectors are identified with the eigenvectors of the Pauli matrices, they form the three orthogonal axes in a Bloch sphere. Formally, any complex 2×2 unitary matrix rotates a polarization vector ϵ_λ to another vector $\epsilon_{\lambda'}$.

If a photon with polarization ϵ_H in mode a is created by the creation operator $\hat{a}_H^\dagger$, and a photon in the same spatial mode with polarization ϵ_V is created by the creation operator $\hat{a}_V^\dagger$, we can immediately work out the Bogoliubov transformation for a creation operator that creates a photon in an arbitrary polarization $\epsilon_\lambda = (\alpha, \beta)$ with $|\alpha|^2 + |\beta|^2 = 1$. Since $\epsilon_\lambda = \alpha \epsilon_H + \beta \epsilon_V$, we have

$$\hat{a}_\lambda^\dagger = \alpha \, \hat{a}_H^\dagger + \beta \, \hat{a}_V^\dagger. \tag{5.10}$$

We will use this correspondence between the polarization vectors and the mode operators throughout the book, without always explicitly mentioning it. It is a consequence of the direct relation between mode functions and mode operators, developed in Chapter 1. We now define the horizontal and vertical polarization states as the computational basis:

$$|0\rangle \equiv \hat{a}_H^\dagger |0,0\rangle_{HV} = |1,0\rangle_{HV} = |H\rangle,$$

$$|1\rangle \equiv \hat{a}_V^\dagger |0,0\rangle_{HV} = |0,1\rangle_{HV} = |V\rangle. \tag{5.11}$$

The remaining eigenstates of the Pauli operators X and Y for the qubit follow automatically from this identification

$$|+\rangle \equiv |D\rangle \quad \text{and} \, |-\rangle \equiv |A\rangle,$$

$$|\circlearrowleft\rangle \equiv |L\rangle \quad \text{and} \, |\circlearrowright\rangle \equiv |R\rangle, \tag{5.12}$$

where we have implicitly introduced modes a_D, a_A, a_L, and a_R.

Physically, polarization rotations are implemented with quarter- and half-wave plates. These plates are pieces of material with a birefringent property, where light with different polarization travels at different speeds through the material. The birefringence of a uniaxial material is characterized by a single optical axis. If the field is linearly polarized, such that the **E** field is oscillating along this axis, the propagation velocity takes a different value. We will assume in this discussion that the velocity along the axis v_f is faster than the velocity perpendicular to the axis v_s. When the polarization is orthogonal to this, the propagation velocity is minimal. The phase shift induced by a dielectric is given by $\phi_i = v_i k l / c$, where v_i denotes the velocity of propagation in the respective direction ($i = f, s$), k is the wave number, and l is the thickness of the dielectric wave plate. To induce a relative phase shift of π between the polarization along the two perpendicular directions, we can choose l such that

$$l = \frac{\lambda}{2} \frac{c}{v_f - v_s}, \tag{5.13}$$

where λ is the wavelength of the light. This is a 'half-wave plate', and is shown schematically in Fig. 5.1. The larger the birefringent effect ($v_f \gg v_s$), the thinner the material can be, and fewer photon losses may occur due to absorption in the dielectric. Note also that the wave plate is wavelength-dependent.

The optical axis of the material is oriented in the plane of the wave plate. Now suppose that this axis makes an angle θ with the horizontal axis. A single photon with horizontal polarization ϵ_H is then created by $\hat{a}_H^\dagger$:

$$\hat{a}_H^\dagger = \cos\theta \, \hat{a}_F^\dagger + \sin\theta \, \hat{a}_S^\dagger, \tag{5.14}$$

where the subscripts F and S denote the linear polarization along and perpendicular to the optical axis, respectively. The half-wave plate induces a π phase shift in the slow axis:

$$\hat{a}_F^\dagger \to \hat{a}_F^\dagger \quad \text{and} \quad \hat{a}_S^\dagger \to -\hat{a}_S^\dagger. \tag{5.15}$$

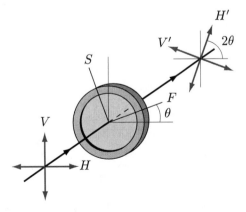

Fig. 5.1. The effect of a 'half-wave plate' (HWP) on a linearly polarized beam of light. Axes parallel and perpendicular to the optical axis of the half-wave plate are indicated by F and S, respectively. A spatial rotation over an angle θ of the wave plate induces a polarization rotation over an angle 2θ.

Horizontally polarized light therefore becomes

$$\hat{a}_H^\dagger = \cos\theta\,\hat{a}_F^\dagger + \sin\theta\,\hat{a}_S^\dagger \rightarrow \cos\theta\,\hat{a}_F^\dagger - \sin\theta\,\hat{a}_S^\dagger. \tag{5.16}$$

We can express $\hat{a}_F^\dagger$ and $\hat{a}_S^\dagger$ in terms of $\hat{a}_H^\dagger$ and $\hat{a}_V^\dagger$, which leads to

$$\hat{a}_H^\dagger \rightarrow \cos\theta\left(\cos\theta\,\hat{a}_H^\dagger - \sin\theta\,\hat{a}_V^\dagger\right) - \sin\theta\left(\sin\theta\,\hat{a}_H^\dagger + \cos\theta\,\hat{a}_V^\dagger\right)$$
$$= \cos 2\theta\,\hat{a}_H^\dagger - \sin 2\theta\,\hat{a}_V^\dagger. \tag{5.17}$$

Unitarity of the operation of the half-wave plate immediately gives us the expression for the vertical polarization

$$\hat{a}_V^\dagger \rightarrow \sin 2\theta\,\hat{a}_H^\dagger + \cos 2\theta\,\hat{a}_V^\dagger. \tag{5.18}$$

Therefore, a rotation over an angle θ of the half-wave plate results in a polarization rotation over an angle 2θ. In the Bloch sphere, this corresponds to a rotation around the y axis through the states $|\circlearrowleft\rangle$ and $|\circlearrowright\rangle$, generated by the Pauli Y matrix

$$U_{\mathrm{HWP}}(\theta) = U_Y(2\theta) = \exp\left(2i\theta Y\right) = \begin{pmatrix} \cos 2\theta & -\sin 2\theta \\ \sin 2\theta & \cos 2\theta \end{pmatrix}, \tag{5.19}$$

in the $\{\epsilon_H, \epsilon_V\}$ polarization basis.

Similarly, a quarter-wave plate is a birefringent material that is cut such that the linear polarization perpendicular to the optical axis is retarded by a relative phase $\pi/2$. If the incoming light has left-handed circular polarization ϵ_L:

$$\hat{a}_L^\dagger = \frac{1}{\sqrt{2}}\left(\hat{a}_H^\dagger + i\hat{a}_V^\dagger\right), \tag{5.20}$$

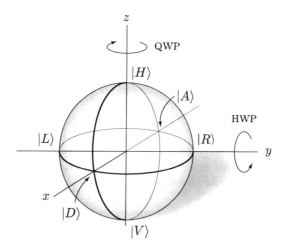

Fig. 5.2.   The Bloch sphere for a single-photon polarization qubit. The effects of a 'half-wave plate' (HWP) and a 'quarter-wave plate' (QWP) are indicated as rotations around the y and the z axis, respectively.

then the quarter-wave plate will change the polarization of, say, the vertically polarized wave, and the polarization becomes

$$\hat{a}_L^\dagger = \frac{1}{\sqrt{2}}\left(\hat{a}_H^\dagger + i\hat{a}_V^\dagger\right) \rightarrow \frac{e^{-i\pi/4}}{\sqrt{2}}\left(\hat{a}_H^\dagger + ie^{i\pi/2}\hat{a}_V^\dagger\right) = e^{-i\pi/4}\hat{a}_A^\dagger. \qquad (5.21)$$

A quarter-wave plate can therefore change circular polarization into linear polarization, and vice versa. In the Bloch sphere, this corresponds to a rotation around the axis through the states $|H\rangle$ and $|V\rangle$, generated by the Pauli Z matrix

$$U_{\text{QWP}} = \exp\left(-\frac{i\pi}{4}Z\right) = \exp\left(-i\pi/4\right)\begin{pmatrix} 1 & 0 \\ 0 & i \end{pmatrix}, \qquad (5.22)$$

in the ϵ_H-ϵ_V polarization basis.

An arbitrary polarization rotation can be decomposed into three rotations around two specific non-parallel axes. One possible decomposition would be a rotation α around the y axis, $\exp(-i\alpha Y)$, followed by a β rotation around the x axis, $\exp(-i\beta X)$, and finally another rotation around the y axis, this time of angle γ, $\exp(-i\gamma Y)$. The y-axis rotations are implemented by half-wave plates oriented at suitable angles $\alpha/2$ and $\gamma/2$, as described in Eq. (5.15). The x-axis rotation can be implemented using two quarter-wave plates (rotated over an angle $\pi/2$ with respect to each other) and a half-wave plate physically rotated over an angle $\beta/2$:

$$\exp(-i\beta X) = \exp\left(i\frac{\pi}{4}Z\right)\exp(-i\beta Y)\exp\left(-i\frac{\pi}{4}Z\right)$$

$$= \begin{pmatrix} 1 & 0 \\ 0 & -i \end{pmatrix}\begin{pmatrix} \cos\beta & -\sin\beta \\ \sin\beta & \cos\beta \end{pmatrix}\begin{pmatrix} 1 & 0 \\ 0 & i \end{pmatrix}$$

$$= \begin{pmatrix} \cos\beta & -i\sin\beta \\ -i\sin\beta & \cos\beta \end{pmatrix}. \qquad (5.23)$$

An arbitrary rotation in the Bloch sphere can therefore be written in terms of half-wave plates and quarter-wave plates as

$$U(\alpha,\beta,\gamma) = U_{\text{HWP}}\left(\frac{\gamma}{2}\right)U_{\text{QWP}}^\dagger U_{\text{HWP}}\left(\frac{\beta}{2}\right)U_{\text{QWP}}U_{\text{HWP}}\left(\frac{\alpha}{2}\right)$$

$$= \exp(-i\gamma Y)e^{i\pi Z/4}\exp(-i\beta Y)\,e^{i\pi Z/4}\exp(-i\alpha Y). \qquad (5.24)$$

The adjoint of the quarter-wave plate transformation corresponds to the interchange of the fast and slow axes, in other words a rotation over $\pi/2$. This decomposition of an arbitrary rotation into rotations about the y and x axes is not unique, and half-wave and quarter-wave plates may also be used to implement different decompositions of qubit rotations.

Frequently, experiments do not need arbitrary polarization rotations. In that case the experimenter can often make do with fewer quarter- and half-wave plates. There are also other optical elements available that may be used to change the polarization of a light beam.

These elements are generally variations of the mechanism described here. It is worth noting that even though we treated the half- and quarter-wave plates quantum mechanically, they can be understood completely classically in electrodynamics. However, it is convenient for the applications in this book to give the quantum mechanical treatment.

Finally, a single-photon qubit in the polarization implementation can be translated deterministically into a single-photon qubit on two spatial modes with linear optical elements. The physical set-up again involves a birefringent material: when polarized light is incident at a non-normal angle to the surface of a birefringent crystal with an optical axis at an angle $< 90°$ with the surface of the crystal, the angle of transmission will be different for orthogonally polarized beams. This allows us to separate horizontal and vertical polarization spatially, and it is the principle behind 'polarizing beam splitters'. The mode transformation of a polarizing beam splitter may have the following form:

$$\hat{a}_{H,1} \to \hat{b}_{H,1} \quad \text{and} \quad \hat{a}_{V,1} \to \hat{b}_{V,2}$$

$$\hat{a}_{H,2} \to \hat{b}_{H,2} \quad \text{and} \quad \hat{a}_{V,2} \to \hat{b}_{V,1}, \tag{5.25}$$

where $\hat{a}$ and $\hat{b}$ denote the input and the output modes of the polarizing beam splitter. In this case the horizontally polarized light is always transmitted, while the vertically polarized light is always reflected. It is straightforward to construct the Hamiltonian that gives rise to this transformation.

To transform polarization qubits into qubits on two spatial modes, we split the two polarization modes into two spatial modes using a polarizing beam splitter. A suitable polarization rotation in one of the spatial modes can then be used to align the polarization of one mode with the other, thus effectively removing this degree of freedom. Of course, it is still there; we just choose not to encode any information in this degree of freedom. Since all steps in this procedure are unitary, we can easily reverse them, and translate spatial qubits into polarization qubits. The two representations are therefore equivalent.

5.1.3 Orbital angular-momentum qubits

We now turn our attention to the degrees of freedom associated with the transverse mode functions. The Hermite–Gaussian (HG) and Laguerre–Gaussian (LG) transverse mode functions were constructed as solutions of the paraxial wave equation. We will show that the HG and LG mode functions of order one (for a beam propagating in the z direction) can be used to encode a qubit degree of freedom. The orthonormal mode functions of order one are

$$u_{0,1}^{\text{HG}}(k; \mathbf{r}) = \frac{-1}{s^4(z)} \frac{\sqrt{2}\, y}{\sqrt[4]{\pi}} \exp\left(ikz - i\omega_k t - \frac{x^2 + y^2}{s^2(z)} \right),$$

$$u_{1,0}^{\text{HG}}(k; \mathbf{r}) = \frac{-1}{s^4(z)} \frac{\sqrt{2}\, x}{\sqrt[4]{\pi}} \exp\left(ikz - i\omega_k t - \frac{x^2 + y^2}{s^2(z)} \right), \tag{5.26}$$

where $s^2(z)$ was defined in Eq. (1.96). These two mode functions can be represented by a two-dimensional vector

$$u_{0,1}^{\text{HG}}(k;\mathbf{r}) = \begin{pmatrix} 1 \\ 0 \end{pmatrix} \quad \text{and} \quad u_{1,0}^{\text{HG}}(k;\mathbf{r}) = \begin{pmatrix} 0 \\ 1 \end{pmatrix}, \tag{5.27}$$

by virtue of their orthonormality.

When we rotate the Hermite–Gaussian modes over an angle θ around the z axis, we obtain another set of orthonormal modes. The rotation is given by

$$x' = \cos\theta\, x + \sin\theta\, y$$
$$y' = -\sin\theta\, x + \cos\theta\, y, \tag{5.28}$$

and the rotated Hermite–Gaussian modes become

$$u_{0,1}^{\text{HG}}(k;\mathbf{r}') = \cos\theta\, u_{0,1}^{\text{HG}}(k;\mathbf{r}) + \sin\theta\, u_{0,1}^{\text{HG}}(k;\mathbf{r})$$
$$u_{1,0}^{\text{HG}}(k;\mathbf{r}') = -\sin\theta\, u_{0,1}^{\text{HG}}(k;\mathbf{r}) + \cos\theta\, u_{0,1}^{\text{HG}}(k;\mathbf{r}), \tag{5.29}$$

where we used that $x^2+y^2 = x'^2+y'^2$. These mode functions are linear superpositions of the Hermite–Gaussian functions in Eq. (5.26), and therefore again solutions to the paraxial wave equation. In the special case of $\theta = \pi/4$ we obtain the Hermite–Gaussian modes at $45°$:

$$u_{0,1}^{\overline{\text{HG}}}(k;\mathbf{r}) = \frac{1}{\sqrt{2}}\begin{pmatrix} 1 \\ 1 \end{pmatrix} \quad \text{and} \quad u_{1,0}^{\overline{\text{HG}}}(k;\mathbf{r}) = \frac{1}{\sqrt{2}}\begin{pmatrix} 1 \\ -1 \end{pmatrix}. \tag{5.30}$$

It is clear that the HG mode functions of order one are analogous to the polarization states ϵ_H and ϵ_V of the previous section, and the HG modes rotated over $45°$ are analogous to ϵ_D and ϵ_A. We therefore construct the following superposition to complete the analogy:

$$\frac{u_{0,1}^{\text{HG}}(k;\mathbf{r}) + iu_{0,1}^{\text{HG}}(k;\mathbf{r})}{\sqrt{2}} \quad \text{and} \quad \frac{u_{0,1}^{\text{HG}}(k;\mathbf{r}) - iu_{0,1}^{\text{HG}}(k;\mathbf{r})}{\sqrt{2}}. \tag{5.31}$$

These are the Laguerre–Gaussian modes of order one:

$$u_{0,1}^{\text{LG}}(k;\mathbf{r}) = \frac{u_{0,1}^{\text{HG}}(k;\mathbf{r}) + iu_{0,1}^{\text{HG}}(k;\mathbf{r})}{\sqrt{2}},$$
$$u_{0,-1}^{\text{LG}}(k;\mathbf{r}) = \frac{u_{0,1}^{\text{HG}}(k;\mathbf{r}) - iu_{0,1}^{\text{HG}}(k;\mathbf{r})}{\sqrt{2}}, \tag{5.32}$$

with

$$u_{0,1}^{\text{LG}}(k;\mathbf{r}) = \frac{-1}{s^4(z)}\frac{re^{i\phi}}{\sqrt[4]{\pi}}\exp\left(ikz - i\omega_k z - \frac{r^2}{s^2(z)}\right),$$
$$u_{0,-1}^{\text{LG}}(k;\mathbf{r}) = \frac{-1}{s^4(z)}\frac{re^{-i\phi}}{\sqrt[4]{\pi}}\exp\left(ikz - i\omega_k z - \frac{r^2}{s^2(z)}\right). \tag{5.33}$$

Consequently, we can define the remaining orthogonal pair of vectors as

$$u_{0,1}^{\text{LG}}(k;\mathbf{r}) = \frac{1}{\sqrt{2}}\begin{pmatrix}1\\i\end{pmatrix} \quad \text{and} \quad u_{0,-1}^{\text{LG}}(k;\mathbf{r}) = \frac{1}{\sqrt{2}}\begin{pmatrix}1\\-i\end{pmatrix}. \tag{5.34}$$

Following the procedure for polarization qubits, we use these three orthogonal pairs of transverse mode functions to define the eigenstates of the three Pauli matrices.

Exercise 5.2: Show that the modes in Eq. (5.32) are the Laguerre–Gaussian modes of order one.

Now that we have constructed a two-dimensional complex Hilbert space from the transverse mode functions of order one, we can represent a qubit by a single photon that is supported on these modes. Again, this is completely analogous to the polarization qubit. First, the modes $u_{nm}^{\text{HG}}(k;\mathbf{r})$ with $n,m = 0,1$ are associated with the annihilation operators $\hat{a}_{nm}$, such that for single photons we can define

$$|0\rangle \equiv |\text{HG}_{01}\rangle = \hat{a}_{01}^{\dagger}|0,0\rangle_{01,10} = |1,0\rangle_{01,10},$$
$$|1\rangle \equiv |\text{HG}_{10}\rangle = \hat{a}_{10}^{\dagger}|0,0\rangle_{01,10} = |0,1\rangle_{01,10}, \tag{5.35}$$

as the eigenstates of a Pauli Z matrix, and hence the computational basis states. The single-photon eigenstates of the Pauli X matrix can then be written as

$$|+\rangle \equiv |\overline{\text{HG}}_{01}\rangle = \frac{\hat{a}_{01}^{\dagger} + \hat{a}_{10}^{\dagger}}{\sqrt{2}}|0,0\rangle_{01,10},$$
$$|-\rangle \equiv |\overline{\text{HG}}_{10}\rangle = \frac{\hat{a}_{01}^{\dagger} - \hat{a}_{10}^{\dagger}}{\sqrt{2}}|0,0\rangle_{01,10}. \tag{5.36}$$

Finally, the eigenstates of the Pauli Y matrix can be constructed as

$$|\circlearrowleft\rangle \equiv |\text{LG}_{1,0}\rangle = \hat{b}_{1,0}^{\dagger}|0,0\rangle_{1,-1} = |1,0\rangle_{1,-1},$$
$$|\circlearrowright\rangle \equiv |\text{LG}_{-1,0}\rangle = \hat{b}_{-1,0}^{\dagger}|0,0\rangle_{1,-1} = |0,1\rangle_{1,-1}, \tag{5.37}$$

where we used the Bogoliubov transformation for transforming the creation and annihilation operators of HG mode functions into those of LG mode functions via

$$\hat{b}_{1,0} = \frac{\hat{a}_{1,0} + i\hat{a}_{0,1}}{\sqrt{2}} \quad \text{and} \quad \hat{b}_{-1,0} = \frac{\hat{a}_{1,0} - i\hat{a}_{0,1}}{\sqrt{2}}. \tag{5.38}$$

This means that the state space of a single photon with transverse mode functions of order one is again a Bloch sphere (see Fig. 5.3). The remaining questions regarding the optical implementation of the transverse mode functions as a qubit are then threefold: (1) Can we create single photons in transverse modes of order one? (2) Can we implement arbitrary qubit transformations? and (3) Can we detect which transverse mode the single photon is in?

Starting with a regular Gaussian beam $u_0^{\text{LG}}(k,\mathbf{r})$, we can create a beam that has a higher-order transverse mode function using a hologram, shown in Fig. 5.4. We first review the basic

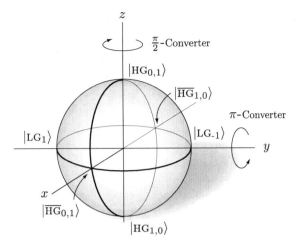

The Bloch sphere for a single-photon qubit with transverse-mode shapes of order one. The effects of the lenses are indicated as rotations around the *y* and the *z* axis, respectively.

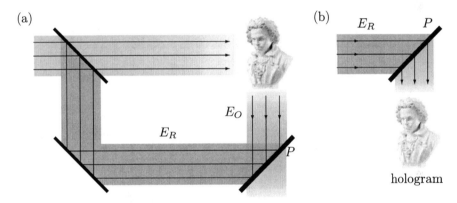

A hologram is produced in two stages. (a) A scattered wave and a reference wave produce an interference pattern on a plate *P*; (b) illumination of *P* with a second reference wave produces the scattered light, retrieving the image of the scatterer.

principle behind holography, and then show how this can be used to create *any* transverse mode function, and in particular the order-one LG mode.

A hologram is produced by recording the interference pattern of light scattered from an object with a coherent reference beam. When the interference pattern is subsequently illuminated with the reference beam, in well-defined directions the diffracted light will have the phases and amplitudes of the original scattered beam, 'showing' the viewer the object. We treat the light classically. Suppose that the electric field of the reference beam is given by

$$E_R = \exp(i\mathbf{k} \cdot \mathbf{r}), \tag{5.39}$$

and the light scattered from the object is given by

$$E_O = \sum_j E_j \exp(i\mathbf{k}_j \cdot \mathbf{r}). \tag{5.40}$$

For simplicity, we assume that the light remains polarized in one direction, and we normalized the amplitude of the reference beam. Also note that the scattered light propagates in a different direction from the reference beam (assume that all $\mathbf{k}_j$ propagate approximately in the same direction). The interference pattern between the two beams is then given by

$$|E_R + E_O|^2 = 1 + I_O + \sum_j \left(E_j \exp\left[i(\mathbf{k}_j - \mathbf{k}) \cdot \mathbf{r}\right] + E_j^* \exp\left[-i(\mathbf{k}_j - \mathbf{k}) \cdot \mathbf{r}\right] \right), \tag{5.41}$$

where I_O is the intensity of the scattered light. This pattern can be recorded as a phase image, and when we illuminate the interference pattern again with the reference beam, the scattered light takes the form

$$e^{i\mathbf{k}\cdot\mathbf{r}}|E_R + E_O|^2 = e^{i\mathbf{k}\cdot\mathbf{r}} + I_O\, e^{i\mathbf{k}\cdot\mathbf{r}} + \sum_j \left(E_j e^{i\mathbf{k}_j \cdot \mathbf{r}} + E_j^* e^{-i(\mathbf{k}_j - 2\mathbf{k})\cdot\mathbf{r}} \right). \tag{5.42}$$

We now notice several things: the first two terms propagate in the direction of the reference beam, and the last term (inside the brackets) is a diffracted wave propagating in the direction $\sim \mathbf{k}_j - 2\mathbf{k}$. For a suitable choice of $\mathbf{k}$ and $\mathbf{k}_j$, these can be separated from the third term, which remarkably depends only on the scattered light. Since all the phase relations are preserved, an observer looking into this scattered light sees the object in true 3D.

We can use this technique to create any transverse mode function, and in particular LG mode functions. First, we calculate the interference pattern of the wave with transverse mode function $u_{l,+l}^{(\text{LG})}(k, \mathbf{r})$, with a reference plane wave. Let the LG wave propagate in the z direction, and the plane wave has wave vector $\mathbf{k} = k(\sin\theta, 0, \cos\theta)$. In the $z = 0$ plane, this pattern becomes

$$\Phi(x, y, 0) = l \arctan\left(\frac{y}{x}\right) + kx\sin\theta, \tag{5.43}$$

where l is the 'orbital angular-momentum' (OAM) quantum number. We can produce a phase plate according to the specifications of $\Phi(x, y, 0)$, which produces a structure shown in Fig. 5.5 for $l = 2$. By virtue of the holographic technique presented above, when we illuminate such a plate at the same angle as the original reference beam, light with the desired transverse mode function is scattered into the mode propagating in the z direction. Since the probability of scattering into the desired mode is less than one, this procedure can be used for creating single-photon OAM ('orbital angular-momentum') qubits only in post-selected fashion.

The second question is how to implement arbitrary qubit rotations with optical elements. First, in order to construct the physical implementation of a rotation around the y axis in the Bloch sphere we define the HG transverse modes in some basis $\mathbf{r}' = (x', y')$, with

$$x' = \cos\theta\, x - \sin\theta\, y,$$
$$y' = \sin\theta\, x + \cos\theta\, y. \tag{5.44}$$

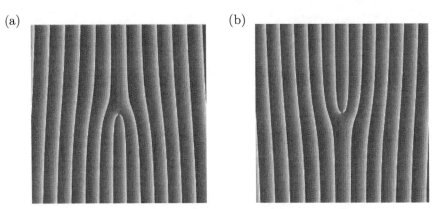

(a)   (b)

Fig. 5.5.   Defects in phase holograms can induce angular momentum. (a) The hologram that increases the orbital angular momentum by 2; (b) the hologram that decreases the orbital angular momentum by 2.

This means that we can write

$$u_{0,1}^{\text{HG}}(k;\mathbf{r}') = \begin{pmatrix} \cos\theta \\ -\sin\theta \end{pmatrix} \quad \text{and} \quad u_{1,0}^{\text{HG}}(k;\mathbf{r}') = \begin{pmatrix} \sin\theta \\ \cos\theta \end{pmatrix}, \tag{5.45}$$

in the basis of the transverse mode functions $u_{0,1}^{\text{HG}}(k;\mathbf{r})$ and $u_{0,1}^{\text{HG}}(k;\mathbf{r})$. The annihilation operators corresponding to the HG modes in the primed (rotated) coordinate system are

$$\hat{a}_{0,1}' = \cos\theta\,\hat{a}_{0,1} - \sin\theta\,\hat{a}_{1,0}$$
$$\hat{a}_{1,0}' = \sin\theta\,\hat{a}_{0,1} + \cos\theta\,\hat{a}_{1,0}. \tag{5.46}$$

This is completely analogous to the polarization case, as it should be: every orthogonal mode function has its own annihilation operator. In the case of the polarization, we had an element (the half-wave plate) that introduced a minus sign in one polarization state ϵ_S relative to the orthogonal polarization ϵ_F. Due to the correspondence between mode functions (in this case polarization) and the mode operators, we obtained the Bogoliubov transformation

$$\hat{a}_F \rightarrow \hat{a}_F \quad \text{and} \quad \hat{a}_S \rightarrow -\hat{a}_S. \tag{5.47}$$

Eq. (5.15) showed how this transformation is sufficient to create a rotation in the Bloch sphere around the y axis. Similarly, we imagine that we have a physical device that performs the following transformation on the HG mode function in the $\mathbf{r}'$ frame:

$$u_{0,1}^{\text{HG}}(k;\mathbf{r}') \rightarrow u_{0,1}^{\text{HG}}(k;\mathbf{r}') \quad \text{and} \quad u_{1,0}^{\text{HG}}(k;\mathbf{r}') \rightarrow -u_{1,0}^{\text{HG}}(k;\mathbf{r}'). \tag{5.48}$$

This then means that we have the transformation rule for the mode operators:

$$\hat{a}_{0,1}' \rightarrow \hat{a}_{0,1}' \quad \text{and} \quad \hat{a}_{1,0}' \rightarrow -\hat{a}_{1,0}', \tag{5.49}$$

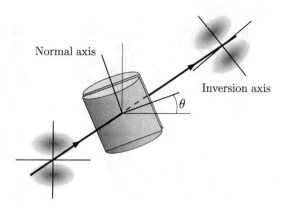

Fig. 5.6. (a) Two cylindrical lenses with focal length $f_{x'}$ a distance $2f_{x'}$ apart. This system will cause inversion in the x' direction: $x' \to -x'$. (b) The two lenses are rotated with respect to the xy frame by an angle θ to induce a rotation over 2θ around the y axis in the Bloch sphere.

where the prime again indicates that we consider the rotated HG mode functions. The primed basis can be considered an eigenbasis for the device that induces this transformation, and whatever the device will be, it is clear that it will operate similarly to the half-wave plate.

When we compare the transformation in Eq. (5.48) with the functional form of $u_{0,1}^{\text{HG}}(k; \mathbf{r}')$ and $u_{1,0}^{\text{HG}}(k; \mathbf{r}')$ in Eq. (5.26), we see that it is equivalent to the spatial inversion

$$x' \to -x' \quad \text{and} \quad y' \to y'. \tag{5.50}$$

This can be implemented with *cylindrical* lenses, which have curvature in only one direction (in this case the x' direction). In other words, the focal length $f_{x'}$ has some finite value f, but the focal length $f_{y'} \to \infty$. When we combine two identical cylindrical lenses and place them a distance $d = 2f$ apart, the x'-coordinate in the transverse field mode will be inverted, while the y'-coordinate remains unaffected (see Fig 5.6). Rotating the cylindrical lens around the z axis (the direction of propagation) over an angle θ induces a rotation

$$\hat{a}_{0,1}' = \cos 2\theta \, \hat{a}_{0,1} - \sin 2\theta \, \hat{a}_{1,0}$$
$$\hat{a}_{1,0}' = \sin 2\theta \, \hat{a}_{0,1} + \cos 2\theta \, \hat{a}_{1,0}, \tag{5.51}$$

which amounts to a rotation around the y axis in the Bloch sphere. The pair of cylindrical lenses in Fig. 5.6 is called a π-converter, since it imparts a phase shift $e^{i\pi} = -1$ on the optical wave along the inversion axis. It is the transverse-mode analog of the half-wave plate for polarization (see Fig. 5.7a).

To generate arbitrary qubit rotations, we also need the analog of the quarter-wave plate. Such a device must be able to convert order-one HG modes to modes with orbital angular momentum, or LG modes of order one. From the mode operator transformations we see that this involves a $\pi/2$ phase shift (a factor i), hence we call this device a $\pi/2$-converter. It can also be implemented with a set of cylindrical lenses. However, this time the two lenses (with focal length f in the x direction) are separated by a distance $d = \sqrt{2}f$. In

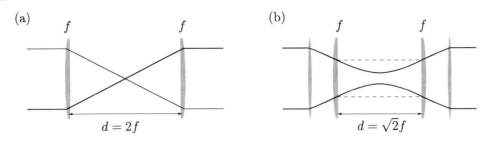

(a) f f (b) f f

$d = 2f$ $d = \sqrt{2}f$

Fig. 5.7. Transverse-mode converters: (a) the π-converter, and (b) the $\pi/2$-converter. The cylindrical lenses have focal length f in the x direction, and are placed a distance d apart. In (b) the incoming mode must be made to converge by a second set of lenses.

addition, changing d means that we have to add another set of (cylindrical) lenses to ensure that the incoming beam converges correctly onto the mode converter, as shown in Fig. 5.7. Mathematically, the description of the $\pi/2$-converter is rather involved, and we refer the reader to the original paper by Beijersbergen *et al.* (1993) for the details.

We considered the cylindrical lens system in the regime of geometrical optics (or ray optics). This means that we have made the implicit assumption that the wave vector of the beam k is *much* larger than $2f_{x'}/w^2$, with w the linear dimension of the spot size. This is a fairly good approximation in most experimental conditions, but it has severe implications for the miniaturization of optical circuits using single-photon OAM qubits. Other methods for implementing unitary transformations for single-photon OAM qubits are using Dove prisms, which act similarly to the π-converter, and regular beam splitters and phase shifters (see Zou and Mathis, 2005).

Finally, we have to implement a way to perform a single-qubit measurement with optical elements. Given arbitrary single-qubit transformations for the OAM qubit, we have to construct a qubit detector that can make reliable measurements in a single basis only (typically the computational basis). The combination of a fixed measurement basis and arbitrary qubit rotations then allows for arbitrary effective measurements. To make a (theoretically) deterministic measurement in the basis $\{|\mathrm{HG}_{01}\rangle, |\mathrm{HG}_{10}\rangle\}$, it is sufficient to find an interferometer that spatially separates the two modes $u_{0,1}^{\mathrm{HG}}(k, \mathbf{r})$ and $u_{1,0}^{\mathrm{HG}}(k, \mathbf{r})$. The resulting two spatial modes are then sent into a regular photodetector that is insensitive to the transverse-mode functions.

The spatial separation can be implemented using the so-called fractional Fourier transform $\mathcal{F}_f^{(x)}(\theta)$ (FRFT). The FRFT is an operator whose eigenfunctions are the Hermite polynomials, and the eigenvalues are phase shifts proportional to θ:

$$\mathcal{F}_f^{(x)}(\theta)[H_n(x)] = e^{in\theta} H_n(x). \tag{5.52}$$

Optically, this operator can be implemented using a dielectric material that has a variable index of refraction $n(\mathbf{r})$ in the transverse direction $\mathbf{r} = (x, y)$. In particular, $n(\mathbf{r})$ is a quadratic function in the x direction, but constant in the y direction:

$$n(\mathbf{r}) = n_0 - n_2 x^2. \tag{5.53}$$

This behaviour is independent of the intensity of the field, and works equally well for photons as for laser beams. When we wish to distinguish between the modes $u_{0,1}^{\mathrm{HG}}(k,\mathbf{r})$ and $u_{1,0}^{\mathrm{HG}}(k,\mathbf{r})$, we need the FRFT in the x and y directions. However, since the mode functions are simple products of the Hermite polynomials, the FRFT in the x direction acts independently of the FRFT in the y direction, and we have

$$\mathcal{F}_f^{(x)}(\theta)\mathcal{F}_f^{(y)}(\phi)\left[H_n(x)\,H_m(y)\right] = e^{in\theta+im\phi}H_n(x)\,H_m(y). \qquad (5.54)$$

The FRFT in the y direction is implemented using a dielectric with $n(\mathbf{r}) = n_0 - n_2 y^2$, and no variation in the x direction.

A Mach–Zehnder interferometer can be used as a switch between output modes, dependent on the relative phase shift in the two arms of the interferometer. If one input mode is in an unknown (coherent or incoherent) superposition of single-photon states $|\mathrm{HG}_{01}\rangle$ and $|\mathrm{HG}_{10}\rangle$, and the second input mode is in the vacuum state $|0\rangle$, then a dielectric medium described by Eq. (5.53) can be used to apply a phase shift to $|\mathrm{HG}_{01}\rangle$, but not to $|\mathrm{HG}_{10}\rangle$ (or vice versa). For the appropriate phase difference θ, we have a perfectly deterministic switch, as shown in Fig. 5.8. We can measure the output modes of the interferometer with photodetectors, resulting in a deterministic measurement of $|\mathrm{HG}_{01}\rangle$ and $|\mathrm{HG}_{10}\rangle$, which therefore serves as a measurement in the computational basis $\{|0\rangle,|1\rangle\}$ for the OAM qubit. A technique for measuring the OAM of a single photon using Dove prisms was proposed by Leach *et al.* (2002).

Exercise 5.3: Calculate the value of θ for the Mach–Zehnder interferometer in Fig. 5.8 to act as a separator of transverse modes into different spatial modes.

The interferometric set-up serves another purpose, besides providing an optical implementation for the measurement of an OAM qubit. Using single-qubit rotations to transform the photons from the modes $u_{nm}^{\mathrm{HG}}(k,\mathbf{r})$ to $u_{1,1}^{\mathrm{LG}}(k,\mathbf{r})$, and reducing the angular momentum to zero, we have (in principle) a completely unitary method to translate between spatial (dual-rail) single-photon qubits and OAM qubits. By extension we therefore also have a complete optical-unitary transformation between the OAM qubit and the single-photon polarization qubit with passive optical elements. This allows us to extend the results for linear optical-mode transformations (the Bogoliubov transformations) for dual-rail or polarization qubits to OAM qubits. In particular, the difficulties that we will encounter for information processing with single-photon polarization qubits (such as deterministically entangling two photons) will also apply to single-photon OAM qubits.

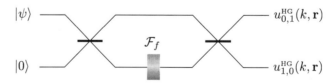

Fig. 5.8. Separating the transverse-mode functions $u_{0,1}^{\mathrm{HG}}(k,\mathbf{r})$ and $u_{1,0}^{\mathrm{HG}}(k,\mathbf{r})$ into spatial-mode functions by means of a graded index of refraction that acts as the FRFT $\mathcal{F}_f$. When the output modes of the interferometer are sent to regular photodetectors, this set-up is a qubit measurement in the basis $\{|\mathrm{HG}_{01}\rangle,|\mathrm{HG}\rangle_{10}\} \equiv \{|0\rangle,|1\rangle\}$.

Finally, unlike the spin degree of freedom for a photon (the polarization), which can be at most ± 1, the orbital angular momentum of a single photon can in principle have an arbitrary large (integer) value. In addition, the technique of using mode converters to implement qubit rotations can be extended to higher-order mode functions, and the FRFT can be used to spatially separate higher-order mode functions as well. This means that the transverse-mode functions can be used to encode *qudits*, i.e., quantum systems with d basis states. This has been demonstrated experimentally for second-order transverse modes by Langford *et al.* (2004).

5.1.4 Time-bin encoding

The last type of single-photon qubit we consider here is the so-called time-bin qubit, in which the photon can occupy two consecutive time intervals. The computational basis states are $|0\rangle \equiv |e\rangle = \hat{b}^\dagger_{\mu=e} |0\rangle$ and $|1\rangle \equiv |\ell\rangle = \hat{b}^\dagger_{\mu=\ell} |0\rangle$, for the 'early' and 'late' time intervals, respectively. The time-bin operators $\hat{b}_\mu$ and $\hat{b}^\dagger_\mu$ were defined in Section 1.4, and particularly in Eq. (1.141).

There are no passive linear optical interferometers that translate time-bin qubits to polarization qubits, but we can still construct a linear Bogoliubov transformation that relates the time-bin qubit unitarily to the polarization qubit. For this to work, we need fast optical switches. The interferometric device that translates between time-bin and polarization qubits is shown in Fig. 5.9. The optical switch acts on the incoming beam, and operates in the interval between the early time bin and the late time bin. When switched on, it induces a polarization rotation over $90°$. Suppose that the incoming photon is polarized in the vertical direction. When it is in the early time bin, the switch does nothing, and the photon is reflected in the polarizing beam splitter and takes the long path through the interferometer. If, however, the photon is in the late time bin, the switch has turned on a polarization rotation, and the photon is transmitted through the polarizing beam splitter, such that it takes the shortest path. The second bin catches up with the first at the second polarizing beam splitter, and the result is a single photon in the output mode with a polarization state that is dependent on the path it took in the circuit. A linear superposition of the time-bin qubit then translates perfectly (in principle) to a coherent superposition in polarization. The circuit

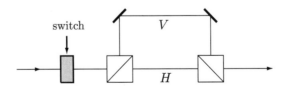

Fig. 5.9. Time-bin to polarization circuit. The single photon in state $|V\rangle$ is sent into the polarizing beam splitter such that it is reflected, and takes the longer path. Between the two time bins, the optical switch induces a polarization rotation over $90°$, sending the late qubit in the second time bin into the shorter optical path. The second polarizing beam splitter recombines the two paths into a single spatial mode.

can be run in reverse to translate polarization qubits (and by extension arbitrary dual-rail qubits) into time-bin qubits.

In practice, optical switches such as these can be quite lossy. The question then becomes why one would go to such lengths to implement qubits this way. The answer is that time-bin qubits have beneficial properties when single photons are used for quantum communication: the decoherence of time-bin qubits is much lower than for polarization qubits when the communication takes place in optical fibres. Tension in fibres changes the propagation velocity of light, and any asymmetric tension on a fibre induces a different propagation velocity for different polarizations, which in turn introduces an unwanted qubit rotation.[2] When the qubit is sent through a commercial fibre over long distances, this qubit rotation is unknown, and the result is decoherence of the quantum information that is being communicated. Time-bin qubits have the advantage that the decoherence mechanism in the transmission through a fibre is much weaker, which results in a higher fidelity of quantum communication protocols.

5.2 Quantum teleportation and entanglement swapping

Having spent quite some time on the definition of single photons as qubit carriers, we are now ready to put them to use. We will use exclusively polarization qubits in the remainder of this chapter, but the techniques of the previous section show how this can be extended to general dual-rail qubits, including OAM qubits and time-bin qubits.

5.2.1 Two-photon entanglement

The first application we consider here is quantum teleportation. In Chapter 2, we introduced the quantum teleportation protocol for qubits, and one may expect that we can replace every $|0\rangle$ with $|H\rangle$, and every $|1\rangle$ with $|V\rangle$, and find the optical implementation. Unfortunately, it is not that straightforward. Whereas we have seen that individual qubit rotations and measurements can be implemented optically in a deterministic manner, the quantum teleportation protocol requires two-qubit operations that we have not considered thus far. In fact, general two-qubit operations cannot be implemented deterministically with the linear optical devices that are described by linear Bogoliubov transformations. Assuming that we have high-quality single-photon sources, this presents us with two problems. First, we need to create maximally entangled Bell states for teleportation:

$$\left|\Psi^{\pm}\right\rangle_{12} = \frac{1}{\sqrt{2}} \left(|HV\rangle \pm |VH\rangle\right) \quad \text{and} \quad \left|\Phi^{\pm}\right\rangle_{12} = \frac{1}{\sqrt{2}} \left(|HH\rangle \pm |VV\rangle\right), \qquad (5.55)$$

---

[2] This is, in fact, exactly how polarization rotations are implemented when the single-photon qubits are kept in a fibre as opposed to the free-space implementation discussed earlier. Mathematically, this is analogous to the quarter- and half-wave plates, in which we had an optical symmetry axis.

where $|H\rangle_j = \hat{a}_{jH}^{\dagger} |0\rangle_j$ and $|V\rangle_j = \hat{a}_{jV}^{\dagger} |0\rangle_j$, with $j = 1, 2$. Second, Alice needs the ability to perform a measurement in the Bell basis. We will postpone the problem of creating Bell states to the next chapter, where we will discuss in detail how to implement two-qubit gates for single-photon qubits, and circumvent the problem by assuming that we can use the output of a suitably engineered parametric downconverter. Instead, we will discuss why creating entanglement in a deterministic fashion is difficult, and how this leads to a partial, or probabilistic, optical implementation of the Bell measurement.

The physical reason why it is difficult to entangle two single-photon polarization qubits is that photons do not interact directly with each other. Mathematically, this is due to the linearity of the Bogoliubov transformations. Consider the Hong–Ou–Mandel effect, in which the state of two identical photons after a 50:50 beam splitter is given by

$$|1, 1\rangle \rightarrow \frac{|2, 0\rangle - |0, 2\rangle}{\sqrt{2}}. \tag{5.56}$$

This is clearly an entangled state, since we cannot write it as a product of $|0\rangle$ and $|2\rangle$

$$\frac{1}{\sqrt{2}} (|2, 0\rangle_{12} - |0, 2\rangle_{12}) \neq (\alpha |0\rangle_1 + \beta |2\rangle_1) (\alpha' |0\rangle_2 + \beta' |2\rangle_2). \tag{5.57}$$

However, on the level of the mode operators, there is no entanglement:

$$\hat{a}_1^{\dagger} \hat{a}_2^{\dagger} \rightarrow \left(\frac{\hat{b}_1^{\dagger} + \hat{b}_2^{\dagger}}{\sqrt{2}} \right) \left(\frac{\hat{b}_1^{\dagger} - \hat{b}_2^{\dagger}}{\sqrt{2}} \right) = \frac{1}{2} \left(\hat{b}_1^{\dagger 2} - \hat{b}_2^{\dagger 2} \right). \tag{5.58}$$

It is straightforward to verify that these operators give rise to Eq. (5.56) when acting on the vacuum. The entanglement in the state arises from the cancellation of cross terms in the special product of the operators $\hat{b}_1^{\dagger}$ and $\hat{b}_2^{\dagger}$ in Eq. (5.58), and from the fact that repeated application of the creation operator leads to orthonormal number states.

By contrast, consider two polarization qubits. The entangling operation we are interested in requires a transformation of the form

$$|HH\rangle \rightarrow \frac{1}{\sqrt{2}} (|HH\rangle + |VV\rangle) \tag{5.59}$$

or something that is equivalent. In terms of the creation operators, this requires a transformation

$$\hat{a}_{1H}^{\dagger} \hat{a}_{2H}^{\dagger} \rightarrow \hat{b}_{1H}^{\dagger} \hat{b}_{2H}^{\dagger} + \hat{b}_{1V}^{\dagger} \hat{b}_{2V}^{\dagger}, \tag{5.60}$$

which takes a product of two creation operators to an inseparable product

$$\hat{b}_{1H}^{\dagger} \hat{b}_{2H}^{\dagger} + \hat{b}_{1V}^{\dagger} \hat{b}_{2V}^{\dagger} \neq \left(\sum_{\lambda=H,V} U_{1\lambda} \hat{b}_{1\lambda}^{\dagger} + U_{2\lambda} \hat{b}_{2\lambda}^{\dagger} \right) \left(\sum_{\lambda=H,V} V_{1\lambda} \hat{b}_{1\lambda}^{\dagger} + V_{2\lambda} \hat{b}_{2\lambda}^{\dagger} \right). \tag{5.61}$$

Here, U and V are unitary matrices. The terms in brackets are Bogoliubov transformations (see Section 1.2.2), and correspond to the linear optical transformation in Eq. (1.157). In other words, the right-hand-side in Eq. (5.61) is exactly what you get with passive linear-optical devices such as beam splitters, phase shifters, polarizers, etc. In turn, this means that passive linear-optical elements alone cannot create deterministic maximal entanglement in dual-rail photonic qubits.

The reverse situation is also true by virtue of unitarity: we cannot take a maximally entangled input state of two dual-rail photonic qubits and turn it into a separable state. The question remains, however, of whether we can perform a complete deterministic Bell measurement. We will give a simple example of a probabilistic Bell measurement based on linear optics and photon counting, and use this to construct an implementation of the quantum teleportation and entanglement swapping of polarization qubits.

5.2.2 Photonic Bell measurements

Consider the simple interferometer shown in Fig. 5.10. Modes a_1 and a_2 are the two spatial input modes of a 50:50 beam splitter, which induces a mode transformation

$$\hat{a}_{1\lambda} = \frac{1}{\sqrt{2}} \left(\hat{b}_{1\lambda} + \hat{b}_{2\lambda} \right) \quad \text{and} \quad \hat{a}_{2\lambda} = \frac{1}{\sqrt{2}} \left(\hat{b}_{1\lambda} - \hat{b}_{2\lambda} \right), \quad (5.62)$$

where λ indicates the polarization degree of freedom. The beam splitter does not affect the polarization. In the output modes b_1 and b_2 of the beam splitter we place polarization-sensitive photodetectors. If such detectors are not available, we can separate the polarization modes into two spatial modes with a polarizing beam splitter, and detect the four resulting output modes with polarization-insensitive detectors. We will now investigate what state (and therefore what detector signature) the different Bell states give rise to after the beam splitter.

Consider the singlet state $\left| \Psi^- \right\rangle = (|HV\rangle - |VH\rangle)/\sqrt{2}$. In terms of creation operators acting on the vacuum, this is

$$\left| \Psi^- \right\rangle = \frac{1}{\sqrt{2}} \left(\hat{a}_{1H}^\dagger \hat{a}_{2V}^\dagger - \hat{a}_{1V}^\dagger \hat{a}_{2H}^\dagger \right) |0\rangle . \quad (5.63)$$

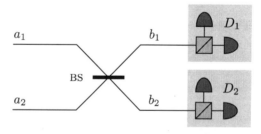

Fig. 5.10. Probabilistic Bell measurement with one 50:50 beam splitter (BS) and two polarization-sensitive detectors D_1 and D_2. The detectors each consist of a polarizing beam splitter and two photodetectors. Only the $|\Psi^\pm\rangle_{12}$ Bell states can be identified unambiguously in this set-up. The $|\Phi^\pm\rangle_{12}$ Bell states give an ambiguous detector signature.

The mode transformations in Eq. (5.62) then turn this into

$$|\Psi^-\rangle \rightarrow \frac{1}{2\sqrt{2}} \left[\left(\hat{b}_{1H}^\dagger + \hat{b}_{2H}^\dagger \right) \left(\hat{b}_{1V}^\dagger - \hat{b}_{2V}^\dagger \right) - \left(\hat{b}_{1V}^\dagger + \hat{b}_{2V}^\dagger \right) \left(\hat{b}_{1H}^\dagger - \hat{b}_{2H}^\dagger \right) \right] |0\rangle$$

$$= \frac{1}{\sqrt{2}} \left(\hat{b}_{1V}^\dagger \hat{b}_{2H}^\dagger - \hat{b}_{1H}^\dagger \hat{b}_{2V}^\dagger \right) |0\rangle = |\Psi^-\rangle. \tag{5.64}$$

The singlet state is invariant under a 50:50 beam splitter! The other Bell states give rise to the following states:

$$|\Psi^+\rangle = \frac{1}{\sqrt{2}} \left(|HV, 0\rangle + |0, HV\rangle \right)$$

$$|\Phi^+\rangle = \frac{1}{2} \left(|2H, 0\rangle - |0, 2H\rangle + |2V, 0\rangle - |0, 2V\rangle \right)$$

$$|\Phi^-\rangle = \frac{1}{2} \left(|2H, 0\rangle - |0, 2H\rangle - |2V, 0\rangle + |0, 2V\rangle \right), \tag{5.65}$$

where $|2H\rangle = \hat{b}_{1H}^{\dagger 2} |0\rangle / \sqrt{2}$, and $|HV\rangle = \hat{b}_{1H}^\dagger \hat{b}_{1V}^\dagger |0\rangle$, etc.

Exercise 5.4: Verify Eqs. (5.65).

All terms in all three output states in Eq. (5.65) have two photons either in spatial mode b_1 or mode b_2, whereas the singlet state has one photon in mode b_1 and one in mode b_2. A detection coincidence in D_1 and D_2, regardless of polarization, therefore tells us that a $|\Psi^-\rangle$ was detected without ambiguity. Similarly, the state $|\Psi^+\rangle$ gives rise to two detector clicks in the *same* output mode b_1 or b_2, but with *different* polarization. No other Bell state gives this detector signature, and we can uniquely identify the $|\Psi^+\rangle$ input state as well.

The $|\Phi^\pm\rangle$ input states, on the other hand, are problematic. They do give rise to orthogonal output states (as required by unitarity), but the orthonormality comes in the phases rather than the photon number of the modes. Photodetectors are insensitive to the phases of the different terms in the superposition (phase and photon number are conjugate variables), and this set-up therefore cannot tell the difference between $|\Phi^+\rangle$ and $|\Phi^-\rangle$. In this set-up (with ideal, lossless detectors) we can therefore unambiguously identify two out of the four Bell states.

Is this the best we can do? Are there other interferometric set-ups that allow us to tell all the Bell states apart deterministically? It turns out that we can choose to detect other Bell states, but we can identify at most two Bell states unambiguously in deterministic set-ups like the one described above. The proof of this is given by Lütkenhaus *et al.* (1999), and is a bit too long to include here. The assumptions of the proof are that we may use any linear optical elements, classical feed-forward, and perfect number-resolving photodectection, and we require perfect distinguishability between all four Bell states. This still leaves open the case where we can distinguish between all Bell states with arbitrary high (but not perfect) success probability. Indeed, if we are willing to admit a small error probability, we can distinguish between all four Bell states. We will return to this in Appendix 2.

5.2.3 Teleportation of single photons

We can finally turn our attention to quantum teleportation and entanglement swapping. In Fig. 5.11 we show the schematic set-up. For simplicity, assume that all detectors are photon-number resolving. Two 'parametric downconverters' (PDCs) are used to create entangled photon pairs, as described in Section 4.4. In the case of teleportation, the PDC that creates entangled photon pairs in modes a_1 and a_2 is operated as a heralded single-photon source, while the other PDC creates the entanglement channel for teleportation. The polarization rotation θ_1 in mode a_1, together with the polarization-sensitive detector D_1, is used to create arbitrary pure single-photon polarization qubit states in mode a_2. For entanglement swapping, both PDCs are used as entanglement channels. The Bell measurement detects modes a_2 and a_3, and the outcome of a successful Bell measurement determines the corrective polarization rotation θ_4. The detector D_4 in mode a_4 is used to verify the teleported state. This experiment was performed in 1997 by Bouwmeester *et al.*

Suppose that the PDCs create photon pairs in the singlet state of polarization:

$$\left|\Psi^-\right\rangle = \frac{1}{\sqrt{2}}\left(|HV\rangle - |VH\rangle\right). \tag{5.66}$$

Strictly speaking, this is wrong. As shown in Section 4.4, a downconverter never creates states with definite photon number, but rather a coherent superposition of different numbers of photon pairs. When the probability of creating a photon pair is small, most of the time there will be no photon pairs created, sometimes a single pair is created, and with even lower probability two or more pairs are created. However, in the set-up of Fig. 5.11, we know that photons have been produced in the PDCs, since they trigger all four detectors in

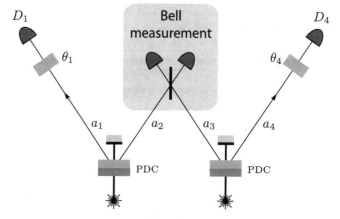

Fig. 5.11. Optical set-up for post-selected quantum teleportation and entanglement swapping. The 'parametric downconverters' (PDCs) create entangled photon pairs, and the polarization rotations θ_1 and θ_4, together with the polarization-sensitive detectors D_1 and D_4, can be chosen such that the measurement statistics are consistent with teleportation or entanglement swapping. The Bell measurement is implemented as shown in Fig. 5.10.

the experiment (ignoring detector dark counts). So we can use Eq. (5.66), provided we take into account the appropriate detection signatures.

The Bell measurement and the polarization rotations θ_1 and θ_4 apply to different modes, and we can therefore perform the Bell measurement first, and manipulate the polarization afterwards. Writing the output of the two PDCs according to Eq. (5.66) in the Bell basis for modes a_1, a_4, and a_2, a_3 gives

$$
\begin{aligned}
\left|\Psi^-\right\rangle_{12}\left|\Psi^-\right\rangle_{34} = {} & \frac{1}{2}\left|\Psi^+\right\rangle_{23}\left|\Psi^+\right\rangle_{14} - \frac{1}{2}\left|\Psi^-\right\rangle_{23}\left|\Psi^-\right\rangle_{14} \\
& - \frac{1}{2}\left|\Phi^+\right\rangle_{23}\left|\Phi^+\right\rangle_{14} + \frac{1}{2}\left|\Phi^-\right\rangle_{23}\left|\Phi^-\right\rangle_{14}.
\end{aligned}
\tag{5.67}
$$

If only the singlet state is identified in the Bell measurement, this leads to the singlet state $\left|\Psi^-\right\rangle_{14}$ in modes a_1 and a_4. The detectors D_1 and D_4 should therefore find measurement statistics consistent with the singlet state. This is indeed what was found in the experiment by Pan *et al.* (1998), and the set-up can therefore be interpreted as entanglement swapping. If the Bell measurement is used to distinguish both $\left|\Psi^-\right\rangle$ and $\left|\Psi^+\right\rangle$, extra operations may have to be applied to mode a_4 (or mode a_1), in accordance with the rules for quantum teleportation.

When mode a_1 is measured in a particular polarization basis set by θ_1, the state in mode a_2 before the Bell measurement becomes

$$
\begin{aligned}
\frac{|HV\rangle_{12} - |VH\rangle_{12}}{\sqrt{2}} \rightarrow {} & \frac{|H\rangle_1}{\sqrt{2}}\left(\cos\theta_1\,|V\rangle_2 + \sin\theta_1\,|H\rangle_2\right) \\
& + \frac{|V\rangle_1}{\sqrt{2}}\left(-\cos\theta_1\,|H\rangle_2 + \sin\theta_1\,|V\rangle_2\right).
\end{aligned}
\tag{5.68}
$$

The measurement of an H- or V-polarized photon in D_1 therefore produces a single-photon qubit in the state $\cos\theta_1\,|V\rangle_2 + \sin\theta_1\,|H\rangle_2$ or $\cos\theta_1\,|H\rangle_2 - \sin\theta_1\,|V\rangle_2$, respectively, in mode a_2. This can be interpreted as the state preparation of the single-photon input qubit in mode a_2. After the Bell measurement, the state is $\left|\Psi^-\right\rangle_{14}$, and a measurement of mode a_1 produces the state in mode a_4 instead of mode a_2. We can interpret this as quantum teleportation. Note that modes a_1 and a_4 never interacted. Moreover, the set-up in Fig. 5.11 makes it clear that entanglement swapping and quantum teleportation are fundamentally the same thing, because the teleported qubit may be itself part of an entangled state.

All output modes are detected in the set-up in Fig. 5.11, and as a consequence there is no freely propagating teleported photon. We call this 'post-selected quantum teleportation'. Some form of post-selection is always necessary in a proof-of-principle experiment in order to verify that teleportation has occurred with high fidelity. We have seen that by using Eq. (5.66) as an approximation of the output of a PDC, we also need the measurement outcome of D_1 to signal that there was indeed an entangled photon pair in the PDC on modes a_1 and a_2, and the detector D_4 indicates an entangled photon pair in modes a_3 and a_4. However, modes a_2 and a_3 are always subjected to a Bell measurement, and a natural question is to ask whether the heralding of photons in a successful Bell measurement is sufficient to ensure that there are always two entangled photons, one in mode a_1 and one in mode a_4, even when we do not use D_1 and D_4. In other words, can we prepare a freely propagating, or 'event-ready' Bell state in this way?

Unfortunately, this is not possible. The reason is that the probability of finding one entangled photon pair in each PDC is similar to the probability of finding two pairs in one PDC, and nothing in the second. Even with ideal photon-number resolving detectors the Bell measurement cannot always distinguish between these two cases. If one PDC creates a double pair in, say, modes a_1 and a_2, while the other creates no pairs, the two photons trigger D_1. The photons in mode a_2 can produce the right Bell measurement signature in the detectors, e.g., a photon in D_2 and one in D_3 (recall that this constitutes a projection onto the singlet state). If detector D_1 cannot distinguish between one and two photons, then without using detector D_4 we cannot distinguish between a true teleportation event with a teleported photon in mode a_4 and a spurious event where mode a_4 is in the vacuum state. If the detector D_1 indicates a click, we may *think* that the input state in mode a_2 is a single photon in the polarization state $|\psi(\theta_1)\rangle$, and conditioned on a successful Bell measurement we expect a teleportation event with high fidelity. However, the *actual* state of mode a_4 before detection has the form

$$\rho = p\,|\psi(\theta_1)\rangle\,\langle\psi(\theta_1)| + (1-p)\,|0\rangle\,\langle0|, \tag{5.69}$$

where p is comparable to $(1-p)$. Without further post-selection we cannot say that the state $|\psi(\theta_1)\rangle$ was teleported with high fidelity and efficiency p, since that would be elevating one basis in Hilbert space over all others, thus committing the 'partition ensemble fallacy' (see Section 3.1). There is therefore a real difference between post-selected teleportation and non-post-selected teleportation. Whether this presents a practical problem very much depends on the application that the teleportation protocol is used for. For example, if it is used in a photon-number conserving experiment, the spurious events will always reveal themselves in the end. But if it is used in conjunction with states that do not have a definite photon number (such as, for example, coherent states), it is generally impossible to tell the spurious events from real teleportation. Currently, the vast majority of photon interference experiments operate in post-selected fashion.

5.2.4 Entanglement distillation

Entanglement swapping can be used in a distillation procedure that consumes two photon pairs with degraded entanglement, and produces a photon pair that is more strongly entangled than the initial pairs. We assume that the entire procedure is operated in post-selected fashion, which simplifies our description while still dealing with an interesting problem. In the discussion below, we follow the argument of Bose *et al.* (1999b).

Consider a protocol in which we wish to distribute maximally entangled photon pairs, say the state $|\Phi^+\rangle$, between two distant communicating parties, Alice and Bob. One of the photons in the singlet will travel to Alice, and the other will travel to Bob, e.g., via an optical fibre. Generally, there will be absorption in the fibre, and we saw in Chapter 4 that this can be modelled by a 'beam splitter' in each fibre mode. Let the transmission coefficient be parametrized by an angle θ such that a single photon with polarization λ is transformed according to

$$|1\rangle_\lambda |0\rangle_l \rightarrow \cos\theta |1\rangle_\lambda |0\rangle_l + \sin\theta |0\rangle_\lambda |1\rangle_l, \tag{5.70}$$

where l denotes the lost mode. If the loss is polarization-independent, and Alice and Bob post-select their detection events on both finding a photon in their detector set-up, the measurement statistics are determined by the (unnormalized) state

$$\rho = \frac{\cos^4\theta}{2} \left(|HH\rangle \langle HH| + |HH\rangle \langle VV| + |VV\rangle \langle HH| + |VV\rangle \langle VV| \right), \tag{5.71}$$

where the renormalization factor $\cos^4\theta$ is the probability of Alice and Bob both finding a photon in their detectors. The state ρ can be found by projecting the loss modes onto the vacuum state. Note also that the coherence between the photons is preserved in this noise model: ρ describes a pure state. This can work only in the post-selected model, because without post-selection we cannot make the projection of the loss mode onto the vacuum.

In the above model of lossy transmission, no entanglement degradation occurs in the post-selected data, but the data rate is reduced by a factor $\cos^4\theta$. However, we can modify the noise model such that now the loss is polarization-dependent. We assume that the loss is stronger for vertically polarized light. We can simplify this model (without loss of generality) by incorporating the losses for horizontally polarized light in the data rate, and then the transmitted state of light (in the post-selected set-up) can be treated as a lossy channel where only the vertically polarized light experiences loss. After projection of the loss mode onto the vacuum, the measurement statistics are consistent with the input state

$$\frac{|HH\rangle + \cos^2\theta\, |VV\rangle}{\sqrt{1 + \cos^4\theta}} \equiv \cos\phi\, |HH\rangle + \sin\phi\, |VV\rangle = |\psi(\phi)\rangle. \tag{5.72}$$

In other words, the measurement statistics are no longer those of maximally entangled photons, and any quantum communication protocols that Alice and Bob wish to engage in will be noisy.

Using two copies of $|\psi(\phi)\rangle$, the entanglement between Alice and Bob can be increased via entanglement swapping. When the initial state is $|\psi(\phi)\rangle_{12} \otimes |\psi(\phi)\rangle_{34}$, and an ideal Bell measurement is performed on modes a_2 and a_3, the output states are given by

$$|\psi_{\text{out}}\rangle \propto \frac{\cos^2\phi\, |HH\rangle \pm \sin^2\phi\, |VV\rangle}{\sqrt{\cos^4\phi + \sin^4\phi}}$$

$$|\psi_{\text{out}}\rangle = |\Psi^\pm\rangle, \tag{5.73}$$

depending on the Bell measurement outcome. In other words, for two outcomes of a Bell measurement, the remaining entangled state is a maximally entangled Bell state. The other two states are less entangled than the input state $|\psi(\phi)\rangle$. We can implement this Bell measurement optically with a 50:50 beam splitter and polarization-sensitive photodetectors. We must ensure that the Bell measurement distinguishes the two states that lead to the successful distillation events.

Exercise 5.5: Show that the failure probability for this distillation protocol is $\frac{1}{2}(\cos^4\phi + \sin^4\phi)$, and calculate the concurrence of $|\psi(\phi)\rangle$ and $|\psi_{\text{out}}\rangle$.

5.3 Decoherence-free subspaces for communication

The entanglement distillation procedure discussed above is probabilistic, with a maximum success probability of 1/2 in the case of perfect photodetection. A natural question is whether there are deterministic protocols that can help protect quantum information in communication protocols. Without employing the full machinery of quantum error-correction codes, there are indeed techniques that protect against some noise in a deterministic fashion. When there is a source of decoherence in the communication channel that has a particular mathematical form, we can often choose an encoding that is insensitive to this type of decoherence. The cost of this encoding is that each logical qubit must be implemented with a number of physical qubits. The Hilbert space of the physical qubits is then split into two subspaces, each associated with one logical qubit state. When the logical qubit is no longer sensitive to the particular type of decoherence, these are called 'decoherence-free subspaces' (DFSs). We will now give an explicit example of a DFS.

Suppose that we can send multiple photons in different modes simultaneously through a communication channel, for example an optical fibre. To be specific, assume that the photons have different frequencies, and that the propagation of different frequency modes through the fibres is about equal. As usual, we take the qubit degree of freedom to be the polarization of the photon. A photon that propagates through the fibre can be subjected to an unwanted qubit transformation U, due to stresses in the fibre. This transformation is unitary, since the fibre does not generally retain a memory of the state of the photon passing through. Moreover, U may change reasonably rapidly over time, making it impractical to correct for it by calibration. The transmitted qubits are therefore no longer pure, since we must in some mathematical sense average over all possible U.

When many photons are transmitted through the fibre, and if the transmission time is shorter than the changes in U, each photon experiences *almost the same* decoherence process U. Symbolically, we write this as

$$\left|\Psi_{\text{in}}^{(N)}\right\rangle \rightarrow \left|\Psi_{\text{out}}^{(N)}\right\rangle = U^{\otimes N}\left|\Psi_{\text{in}}^{(N)}\right\rangle. \tag{5.74}$$

Let θ parameterize $U = U(\theta)$. When two (distinguishable) single-photon qubits in modes a_1 and a_2 with initial state $|\psi\rangle_{12}$ travel through the fibre, the output state becomes

$$\rho_{\text{out}} \propto \int d\theta\, p(\theta)\, U(\theta) \otimes U(\theta)\, |\psi\rangle_{12} \langle\psi|\, U^\dagger(\theta) \otimes U^\dagger(\theta), \tag{5.75}$$

where $p(\theta)$ is a probability distribution over θ. We will now show how we can define decoherence-free subspaces in the Hilbert space $\mathscr{H}$ of the two qubits. We are seeking two subspaces $\mathscr{V}_0 \subset \mathscr{H}$ and $\mathscr{V}_1 \subset \mathscr{H}$ such that

$$\mathscr{V}_0 \cup \mathscr{V}_1 = \mathscr{H} \quad \text{and} \quad \mathscr{V}_0 \cap \mathscr{V}_1 = \varnothing. \tag{5.76}$$

The two subspaces $\mathcal{V}_0$ and $\mathcal{V}_1$ are used to encode the logical qubit as follows. By construction, any state $|\psi(0)\rangle \in \mathcal{V}_0$ remains in $\mathcal{V}_0$ under the transformation $U \otimes U$. We can write this symbolically as

$$U \otimes U \, \mathcal{V}_0 U^\dagger \otimes U^\dagger = \mathcal{V}_0. \tag{5.77}$$

If we write $\mathcal{V}_0 \oplus \mathcal{V}_1 = \mathcal{H}$, it is straightforward to prove that $\mathcal{V}_1$ is also invariant under $U \otimes U$. First note that $\mathcal{H}$ is invariant under $U \otimes U$. We then have

$$U \otimes U \, (\mathcal{V}_0 \oplus \mathcal{V}_1) \, U^\dagger \otimes U^\dagger = U \otimes U \, \mathcal{V}_0 \, U^\dagger \otimes U^\dagger \oplus U \otimes U \, \mathcal{V}_1 \, U^\dagger \otimes U^\dagger$$
$$= \mathcal{V}_0 \oplus U \otimes U \, \mathcal{V}_1 \, U^\dagger \otimes U^\dagger$$
$$= \mathcal{V}_0 \oplus \mathcal{V}_1. \tag{5.78}$$

In the last line we have used $\mathcal{H} = \mathcal{V}_0 \oplus \mathcal{V}_1$ is invariant under $U \otimes U$, and we therefore find

$$(U \otimes U) \, \mathcal{V}_1 \, (U^\dagger \otimes U^\dagger) = \mathcal{V}_1. \tag{5.79}$$

To encode a logical qubit in these subspaces means that we have to find a state $|\psi(0)\rangle \in \mathcal{V}_0$ and a state $|\psi(1)\rangle \in \mathcal{V}_1$, and use these as the computational basis of the logical qubit. After transmission, a measurement of the logical qubit in the computational basis boils down to projecting the state of the two physical qubits onto the subspace $\mathcal{V}_0$ or $\mathcal{V}_1$. It is not obvious a priori that the invariant subspaces $\mathcal{V}_0$ and $\mathcal{V}_1$ exist, since the transformation $U \otimes U$ is rather general. However, we will show by explicit construction that there exist such states. In fact, we will prove something slightly stronger, namely that there is a subspace $\mathcal{V}_0$ that is one-dimensional.

Any single-qubit unitary U can be decomposed into rotations generated by the Pauli operator X and rotations generated by Z. Starting with X, we can formulate the invariance requirement for a state $|\psi\rangle$ as

$$U_X \otimes U_X |\psi\rangle = \exp\left[-i\theta(X_1 \otimes \hat{\mathbb{1}}_2 + \hat{\mathbb{1}}_1 \otimes X_2)\right] |\psi\rangle = e^{i\varphi} |\psi\rangle. \tag{5.80}$$

This is satisfied when

$$(X_1 \otimes \hat{\mathbb{1}}_2 + \hat{\mathbb{1}}_1 \otimes X_2) |\psi\rangle = 0. \tag{5.81}$$

When we write $|\psi\rangle$ in the most general form

$$|\psi\rangle = c_{00} |00\rangle + c_{01} |01\rangle + c_{10} |10\rangle + c_{11} |11\rangle, \tag{5.82}$$

the requirement in Eq. (5.81) reduces to

$$c_{00} = -c_{11} \quad \text{and} \quad c_{01} = -c_{10}. \tag{5.83}$$

Similarly, for rotations generated by Z_1 and Z_2, the requirement becomes

$$(Z_1 \otimes \hat{\mathbb{1}}_2 + \hat{\mathbb{1}}_1 \otimes Z_2) |\psi\rangle = 0, \tag{5.84}$$

which leads to the restriction $c_{00} = c_{11} = 0$. There are no additional restrictions we need to impose, since all rotations can be decomposed into rotations generated by X and Z, and the invariant one-dimensional subspace $\mathcal{V}_0$ is therefore spanned by the singlet state

$$|\psi(0)\rangle = \frac{|01\rangle - |10\rangle}{\sqrt{2}}. \tag{5.85}$$

The complementary subspace $\mathcal{V}_1$ is then spanned by the triplet states

$$\mathcal{V}_1 = \mathrm{Span}\left(|00\rangle, |11\rangle, \frac{|01\rangle + |10\rangle}{\sqrt{2}}\right). \tag{5.86}$$

The two subspaces can be evolved unitarily by the receiving party, so the qubit measurement is not restricted to the computational basis. However, these generally require two-qubit entangling operations, which are difficult to implement.

Apart from unknown polarization rotations in optical fibres, this technique can be used when the communicating parties lack a common reference frame. The polarization of a photon is defined with respect to an external reference frame, and both parties in the communication must have access to the same reference frame if their measurement statistics are to exhibit the right correlations. It so happens that for N qubits, a rotated reference frame induces a transformation $U^{\otimes N}$, which is exactly the problem considered above. This means that our encoding is also capable of establishing a communication channel in the absence of a shared frame of reference. This application of decoherence-free subspaces is reviewed in detail by Bartlett *et al.* (2007).

5.4 Quantum cryptography

In this section we give a brief description how the standard protocols introduced in Section 2.2 (BB84 and EKERT91) can be implemented with single-photon qubits. Recall that in BB84, Alice sends a series of qubits (in this case single photons) randomly prepared in X and Z eigenstates, and Bob measures the X and Z operators. If the preparation and detection observables coincide, Alice and Bob share a secret random bit. In EKERT91, Alice and Bob share Bell pairs, and both measure their qubits randomly in the X or Z basis. When their measurement bases coincide, they know that their measurement outcomes are perfectly correlated, and they again share a secret random bit. In both cases Alice and Bob sacrifice part of the key to test the correlation between their keys. Any eavesdropping permitted by the laws of quantum mechanics will show up as reduced correlations, and results in abortion of the communication channel. Since neither protocol requires two-qubit gates or (deterministic) Bell measurements, the implementation is straightforward when good single-photon or entangled photon sources are available. We first prove the security of EKERT91, and then in a sequence of steps translate this to BB84. We end this section with a consideration of multiphoton states (such as coherent states), and introduce so-called decoy states that can be used to increase the data rate of the communication channel.

Before we start the security proof of BB84 and EKERT91, let us recall the purpose of quantum cryptography, and equally important, the issues that it does *not* address. First, the quantum mechanical part of quantum cryptography is the 'key distribution' (QKD). Once Alice and Bob share a secret key, they can classically encode a message by adding the key to the message (addition modulo two, if the key and the message are in binary). Both BB84 and EKERT91 provide a provably secure means to distribute the key. That is, an eavesdropper who tries to intercept the key during the QKD stage will leave an easily detectable tell-tale sign. However, an encrypted communication channel is only as secure as its weakest component, and in practice undetected eavesdropper attacks may still be possible on parts of the protocol other than the QKD component. This is an important problem that occupies many people in the security community. Second, QKD is not an authentication protocol. In other words, neither BB84 nor EKERT91 can guarantee Alice that the person she shares the key with is who she thinks he is. There exist both classical and quantum authentication protocols, but we will restrict our discussion to quantum key distribution.

5.4.1 Security of Ekert91

While BB84 is the older QKD protocol, it turns out that the security proof of EKERT91 is easier. This can in turn be used to prove the security of BB84 via an elegant argument by Lo and Chau (1999) and Shor and Preskill (2000). We first give a strict definition of what constitutes secure communication, and show that near-maximally entangled pairs shared between two parties give rise to the possibility of secure key distribution. Such pairs can be distilled from noisy entangled pairs via entanglement distillation or by employing quantum error correction on the communication channel. We then reduce the protocol based on Bell pairs to the BB84 protocol.

What does it mean for a protocol to be secure? We could require that an eavesdropper (Eve) obtains no information at all about the key shared by Alice and Bob, but this is too restrictive: in any practical situation, there will be noise induced by the channel between Alice and Bob. As a matter of principle, they cannot tell whether this was due to unrecoverable information loss to the environment, or whether Eve replaced the noisy communication channel by a less noisy channel and induced the additional noise by eavesdropping. The security of a 'quantum key distribution' (QKD) protocol must therefore put a bound on the amount of information Eve can obtain.

Definition: A QKD protocol between Alice and Bob is *secure* if for some parameters $a, b > 0$, any eavesdropping strategy that respects the laws of quantum physics either results in abortion of the protocol, or the protocol is not aborted it succeeds with probability $1 - O(2^{-a})$ *and* the mutual information of Alice and Bob with the environment is smaller than 2^{-b}.

The protocol parameters a and b are chosen by Alice and Bob, and the mutual information between the joint system of Alice and Bob and the environment puts a bound on the amount of information that Eve can obtain about the distributed key.

Suppose that Alice and Bob attempt to share n Bell states $|\Phi_n^+\rangle = 2^{-n/2}(|00\rangle + |11\rangle)^{\otimes n}$ using some imperfect channel. The fidelity of the noisy state ρ that they *actually* shared is given by

$$F = \langle \Phi_n^+ | \rho | \Phi_n^+ \rangle. \tag{5.87}$$

The success probability of the QKD protocol can be identified with the fidelity, since F is interpreted as the probability that the state ρ is mistaken for the state $|\Phi_n^+\rangle$. Next, we prove that

$$F > 1 - 2^{-a} \quad \Rightarrow \quad S(\rho) < [(2n+a)\ln 2 + 1]2^{-a} + O(2^{-2a}), \tag{5.88}$$

where $S(\rho) = -\text{Tr}(\rho \ln \rho)$ is the von Neumann entropy of ρ. If $F > 1 - 2^{-a}$, then the largest eigenvalue of ρ must be larger than $1 - 2^{-a}$. The entropy of ρ can then be written as

$$S(\rho) < -(1 - 2^{-a})\ln(1 - 2^{-a}) - \sum_i \lambda_i \ln \lambda_i, \tag{5.89}$$

where $\sum_i$ denotes the sum over the remaining eigenvalues λ_i of ρ. This term is maximal when $\lambda_i = \lambda_j$ for all i and j, or

$$\sum_i \lambda_i \ln \lambda_i = 2^{-a} \ln \frac{2^{-a}}{2^{2n} - 1}, \tag{5.90}$$

from which Eq. (5.88) follows after some algebra. We can choose a value of b that scales at most logarithmically with n, and write $S(\rho) < 2^{-b}$, with

$$b = a - \log_2 [(2n+a)\ln 2 + 1], \tag{5.91}$$

which is logarithmic in n.

The bound on the entropy of the n shared entangled pairs can be related to the *mutual information* between the n pairs and the environment via Holevo's theorem. We write AB for the system of qubits shared by Alice and Bob, and E denotes the environment (including Eve). The mutual information between the system AB and the environment is then

$$I(AB : E) \leq S(\rho). \tag{5.92}$$

In the worst-case scenario, the information $I(AB : E)$ leaked from Alice and Bob's n entangled pairs is all captured by Eve's probe system. The bound on the entropy $S(\rho)$ in Eq. (5.88) therefore limits the amount of information Eve can obtain about the distributed key. Alice and Bob can estimate the fidelity F by sacrificing part of the shared entangled pairs. If they find that $F > 1 - 2^{-a}$, they can be confident that their communication channel is secure within the parameters a and b. This constitutes a proof of the EKERT91 protocol.

As noted before, in practice the fidelity of the shared entangled pairs will be degraded by various channel imperfections, which may or may not be due to an eavesdropper. In particular, the fidelity easily drops below the threshold value of $1 - 2^{-a}$. However, rather than dismissing this channel, Alice and Bob can introduce entanglement distillation or quantum error correction into the QKD protocol. This reduces the number of shared entangled pairs, but increases the fidelity and lowers the information held by the environment below the security threshold of 2^{-b}.

5.4.2 Security of BB84

So far, we have considered the establishment of k Bell pairs with high fidelity for the purpose of secure QKD in EKERT91. In a series of steps that demonstrably will not increase the mutual information between Alice and Bob on the one hand, and the environment on the other, we now reduce EKERT91 to BB84. The security of the former then implies the security of the latter.

First, we observe that Alice does not have to delay the measurement of her qubits until she has sent their entangled partners to Bob. She can measure her qubits at any point after the creation of the Bell pairs, since both her and Bob's remaining part of the original EKERT91 protocol (without entanglement distillation) is strictly local. In the BB84 protocol, Alice must use two random variables in order to determine which states to send to Bob: one for the determination of the bases, and one for the determination of the basis states once a basis has been chosen. Suppose Alice creates the entangled state

$$|\Phi^+\rangle = \frac{|00\rangle + |11\rangle}{\sqrt{2}} = \frac{|++\rangle + |--\rangle}{\sqrt{2}}, \qquad (5.93)$$

and measures randomly in the X basis or in the Z basis. Due to the nature of entanglement, her measurement outcomes are completely random, but the remaining state (that is sent to Bob) is maximally correlated to her measurement outcome. This procedure can therefore play the role of the random variable determining the basis states. We are using the inherent probabilistic nature of quantum mechanical measurements as the random-number generator.

Next, the secure QKD protocol proceeds as follows. We define the 'key qubits' as the qubits that are used to share the random key, and the 'check qubits' as the qubits that are used in eavesdropper detection. In EKERT91 we admitted entanglement distillation or quantum error correction to establish Bell pairs that were sufficiently pure for the required level of security. Since Alice has already measured her part of the entangled pairs, entanglement distillation is not an option, and we must use quantum error correction. Alice encodes k qubits in a stabilizer code of n key qubits that corrects d errors. In particular, she encodes her key qubits by choosing random eigenvalues for the stabilizer generators of the code. She then mixes the n key qubits with the n check qubits, and sends everything to Bob, who stores the qubits in a quantum memory and publicly acknowledges receipt of the qubits. Alice then announces her random stabilizer eigenvalues, which qubits are the key bits and which are the check bits, and the values of the check bits. Bob compares the check bits, and decides whether the channel is secure or not. He then measures the stabilizer generators and uses Alice's announcement of her choice of stabilizer generator eigenvalues to decipher the key.

This general procedure requires that Bob stores his qubits in a quantum memory, and the measurement of the stabilizer generators generally requires the type of control provided by a quantum computer. Therefore, one may think that the security of this protocol is conditional on having perfect quantum memory and a sufficiently powerful quantum computer. However, there are quantum error-correction codes, such as the CSS codes, which do not need either a quantum memory or a quantum computer. The protocol then proceeds by Bob measuring the incoming photons in the appropriate bases, followed by a comparison of the check qubits. This is formally identical to BB84. The combined error rate of bit flips and phase flips in the limit $n \to \infty$ is 11%.

5.4.3 Multi-photon states and decoy states

So far, we have assumed that the photon sources are near perfect: true single-photon sources in the case of BB84, and entangled photon pairs for EKERT91. However, this is a rather strong technological requirement. As we discussed in Section 4.3, a less challenging method for producing a source that, under certain conditions, can be used as a single-photon source is to attenuate a laser such that the probability of two or more photons being emitted is small. Of course, such a source has a large vacuum contribution, and as a result the data rate is reduced significantly. In addition, there is always a finite probability of measuring more than one photon in an attenuated laser source; we found previously in Eq. (4.61) that the probability of measuring n photons is given by the Poisson distribution

$$p(n|\alpha) = \frac{|\alpha|^{2n} e^{-|\alpha|^2}}{n!}, \tag{5.94}$$

where α is the complex amplitude of the coherent laser state.

Let us now consider how an eavesdropper may exploit multi-photon events. First of all, since the components used by Alice and Bob are not perfect, Eve may be able to hide her presence in the noise. The typical imperfections that can hide Eve are channel loss between Alice and Bob, the fact that Alice's source will often produce no photons, and that Bob's detectors are inefficient. Hence only a fraction (or *yield Y*) of the total number of pulses sent by Alice is ever detected by Bob. On the other hand, for the purposes of a security proof we grant Eve complete technological power within the boundaries of quantum mechanics and she therefore has access to a lossless channel to both Alice and Bob. Imagine that Eve can non-destructively measure how many photons are in each pulse sent by Alice. This is in principle allowed by quantum mechanics, using a perfect QND detector (see Section 4.5). If she measures a single photon, she blocks that signal completely. If she measures two or more photons, she keeps one and sends the rest on to Bob. This kind of attack is called a 'photon number-splitting strategy'.

The probability of more than one photon in each pulse is given by $p_{multi} = 1 - e^{-|\alpha|^2}(|\alpha|^2 + 1)$, and if this is greater than Y, Eve can disguise her activity completely. We summarize this in the security condition

$$Y > p_{multi}. \tag{5.95}$$

Put another way, this condition states that the total number of pulses detected by Bob is greater than the total number of pulses that can be attacked by Eve. For realistic, very lossy channels, Y is small and Alice is forced to reduce $|\alpha|$ in order to satisfy this condition (in this limit, we can write $|\alpha|^2 < \sqrt{2Y}$). This severely reduces the efficacy of the protocol, since only a small fraction $|\alpha|^2 \ll 1$ of the pulses generated by Alice can be used by Bob. In the remainder of this section we will discuss a way to overcome this restriction.

Imagine that Alice has access to two different attenuated laser sources, characterized by different values of α. One will act as the signal source (α_s) and the other as a 'decoy-state' source (α_d). As the nomenclature suggests, only the signal source will ultimately be used to transmit the key; the decoy source replaces the signal for a random selection of transmitted pulses in order to foil Eve's attempted espionage. The protection by decoy states works, because Eve's strategy sends an abnormally high number of multiphoton pulses to Bob. Therefore, if $|\alpha_d|^2 > |\alpha_s|^2$, the proportion of decoy-state pulses getting through to Bob is higher than the proportion transmitted by Alice. Eve's attack can then be straightforwardly inferred by Bob when Alice publicly announces which of her pulses were decoys and which were signals.

To discuss the idea more thoroughly, let us first define y_n to be the yield for an n-photon state, which must be independent of whether that state is part of a signal or decoy pulse. We then find the yields Y_s and Y_d of the decoy and signal pulses:

$$Y_d = \sum_n p(n|\alpha_d)\, y_n,$$

$$Y_s = \sum_n p(n|\alpha_s)\, y_n. \tag{5.96}$$

In fact, by running a series of decoy pulses with a whole set of values of α_d, it is possible for Alice and Bob to calculate exactly which values of p_n characterize their communication channel. Any attempt by Eve to intercept the message and keep some information herself can then immediately be detected by Alice and Bob. This means that Alice and Bob can use any value of α_s for their signal, with full confidence that a photon number-splitting attack will always be detected.

If the yield for a single-photon state is given by η, then in a simple model one would expect that the probability of an n-photon Fock state making it through the channel would be $1 - (1 - \eta)^n$, which is approximately ηn in the small η limit. The condition for security without decoys in the photon number-splitting attack is then $\eta > |\alpha|^2/2$, so $|\alpha|^2$ must be chosen to be 2η at most, leading to $Y_s < 2\eta^2$. On the other hand, with decoy states there is no such restriction, and Alice is free to increase the intensity of her source until $Y_s \approx 1$, a possible improvement in data-transmission rates by several orders of magnitude.

5.5 References and further reading

In this chapter we have discussed only a fraction of quantum communication with single photons. We have indicated where the straightforward implementation of communication protocols breaks down due to physical restrictions posed by the implementation. The holographic technique for creating transverse-mode shapes was pioneered by Heckenberg *et al.* (1992), and the detection of both the polarization and the orbital angular momentum of a single photon using interferometric techniques was proposed by Leach *et al.* (2004). Post-selected teleportation with photons was demonstrated experimentally by Bouwmeester *et al.* (1997) and Boschi *et al.* (1998), and entanglement swapping was demonstrated by Pan *et al.* (1998). See also Kok and Braunstein (2000) for a detailed discussion about post-selection in optical protocols. For general entanglement distillation we refer to Bennett *et al.* (1996). The concept of decoherence-free subspaces were introduced by Zanardi and Rasetti (1997), and demonstrated experimentally in two-photon systems by 2004. Details of the security proof of BB84 can be found in Lo and Chau (1999), Nielsen and Chuang (2000), Shor and Preskill (2000), and Gottesman and Preskill (2003). The theory of decoy states was developed by Hwang (2003) and Lo *et al.* (2005).

6
Quantum computation with single photons

In the previous chapter we developed some of the basic aspects of quantum information processing with single photons as qubits. Apart from noting the obvious benefit of using light for quantum communication, we identified some difficulties in manipulating quantum information that is encoded in photons. In particular, it is difficult to construct two-qubit gates for photonic qubits. In this chapter we will have to face this difficulty head-on in our discussion of quantum computation with single photons and linear-optical elements. The possibility of a quantum computer based on single photons, linear optics, and photon counting was demonstrated in a landmark paper by Knill, Laflamme, and Milburn in 2001. Their protocol is commonly referred to as the 'KLM protocol'. In subsequent years, the KLM protocol has been dramatically slimmed down in terms of complexity and the necessary resources to create the universal set of quantum gates. For pedagogical reasons we give the streamlined version here, and we briefly describe the original KLM protocol in Appendix 2. We will make extensive use of the results in Section 1.4 about mode transformations, and Section 2.3 about cluster-state quantum computing. We start this chapter, however, with a description of linear-optical networks, and how they fail as deterministic quantum computers. In Section 6.2 we discuss the principle of post-selection, and how this can be used to create probabilistic gates with deterministic feed-forward control. Having developed these techniques, we show in Section 6.3 how to build linear-optical quantum computers using so-called fusion gates, both in the one-way model and in the circuit model. In Section 6.4 we show how these models can be made tolerant against photon loss, and in Section 6.5 we consider more general errors and discuss fault-tolerance thresholds for quantum computing with linear optics and single photons.

6.1 Optical N-port interferometers and scalability

In Chapter 1 we stated that *any* unitary transformation $U(N)$ on N optical modes can be realized in practice using $O(N^2)$ beam splitters and phase shifters. This raises a pressing question: a quantum computer relies on implementing unitary transformations, so why can we not build a quantum computer based on such an N-port interferometer? The unitary transformation on n qubits in a quantum computer is a $2^n \times 2^n$ matrix. The interferometer $U(N)$ must therefore have $N = 2^n$ optical modes, which is exponentially large in the number of qubits. This unitary is implemented with at most $N(N-1)/2$ beam splitters, which is also exponentially large in the number of qubits. In other words, such an implementation would generally need an exponentially large amount of resources, and any exponential speed-up that the quantum computer offers will be offset by the necessary resources.

6.1.1 Optical simulation of a quantum computer

Even though the N-port interferometer offers only a semi-classical simulation of a linear-optical quantum computer, it is instructive to show how it is implemented, and where it fails to be a real quantum computer. Each input (and output) mode in the N-port corresponds to a bit *string*, rather than a qubit. A single photon in a particular mode indicates that the state of the computer is given by the bit string corresponding to that mode. Since the photon can be in a coherent superposition of modes, the state of the computer can be a superposition as well. This is different from our usual qubit representation, where each qubit is represented by a single photon in two optical modes (spatial, polarization, etc.). In this simulation there is only one photon entering the input mode associated with the bit string $|00\ldots0\rangle$, and exiting with high probability in the output mode that corresponds to the correct outcome of the computation.

We construct the optical elements shown in Fig. 6.1 that are used to build the simulator. The arbitrary phase gate, the Hadamard, and the cx gate suffice to construct any unitary transformation. The (arbitrary) phase gate is defined by the transformation

$$U(\phi) = \begin{pmatrix} 1 & 0 \\ 0 & e^{i\phi} \end{pmatrix}. \tag{6.1}$$

Suppose that we want to apply the phase gate to qubit i at some stage in the computation. We implement this gate in the N-port by inserting an optical phase shift ϕ in each mode corresponding to the bit strings where the value of qubit i is 1. Therefore, to implement a single phase gate we need to insert $N/2 = 2^{n-1}$ phase shifters in the interferometer. This is an exponential growth in the number of optical elements needed for only one gate. Similarly, a Hadamard gate acting on this qubit can be implemented with a 50:50 beam splitter and two $-\pi/2$ phase shifters (see Fig. 6.1):

$$|1,0\rangle \rightarrow \frac{|1,0\rangle + |0,1\rangle}{\sqrt{2}} \quad \text{and} \quad |0,1\rangle \rightarrow \frac{|1,0\rangle - |0,1\rangle}{\sqrt{2}}. \tag{6.2}$$

In the case of many qubits, the beam splitter must be applied to any pair of modes that corespond to the bit strings $|s_i, 0, s_j\rangle$ and $|s_i, 1, s_j\rangle$ for all partial bit strings s_i, s_j. Again, we

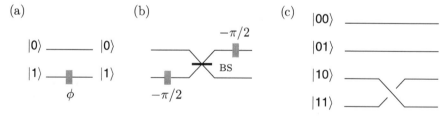

Fig. 6.1. Elements of a classical simulation of a linear-optical quantum computer. Horizontal lines correspond to bit strings, rather than single qubits. (a) The arbitrary phase gate ϕ is generated by an optical phase shift; (b) the Hadamard gate is generated by a 50:50 beam splitter (BS) with appropriate phase corrections; and (c) the cx gate is a simple mode swap. The broken line indicates that there is no interaction between the crossing modes.

need of the order of 2^{n-1} optical elements. It is now clear where the exponential increase in resources comes from in this simulation: each quantum gate must be applied an exponentially large number of times. The last gate we discuss here is the cx gate, which is also the simplest gate. It corresponds to a simple swap of the modes carrying $|10\rangle$ and $|11\rangle$.

Exercise 6.1: How many mode swaps need to be implemented for every cx gate in an N-port, where $N = 2^n$?

Finally, there is another reason why the single-photon N-port construction we have presented here can only be a semi-classical simulation of a quantum computer: the equations of motion for a single photon are exactly the classical Maxwell equations. In particular, rather than using a single photon as the input of the N-port, we could have used a classical coherent light pulse. Whenever the single photon ends up in (predominantly) one output mode, the classical coherent pulse will also end up in that mode. Another way of saying this is that single-photon interference is essentially the same as classical interference. Genuine quantum computing with single photons and linear optics therefore relies critically on higher-order interference, such as the Hong–Ou–Mandel effect.

6.2 Post-selection and feed-forward gates

The previous section demonstrates that our construction of a quantum computer based on linear optics and single photons must be amended. In particular, rather than doing a many-qubit computation with a *single* photon, we encode each individual qubit in a single photon with two modes. This dual-rail representation was introduced in detail in Chapter 5. We constructed a complete set of single-qubit gates, based on a spatial dual-rail representation, polarization, orbital angular momentum, and time-bin qubits. We also showed that we cannot create deterministic two-qubit gates, nor can we perform a deterministic Bell measurement. These are enormous obstacles to scalable quantum computation. In this section we will first give a theoretical construction of a deterministic two-photon cz gate using nonlinear optical elements, and argue that this cannot be created in practice. Next, we show that we can create 'probabilistic' two-qubit entangling gates when we allow the use of so-called ancilla qubits and photodetection. These gates can then in principle be used for the creation of cluster states. An alternative technique is gate teleportation, which historically led to the klm protocol.

6.2.1 A cz gate based on Kerr nonlinearities

Conceptually, it is relatively straightforward to come up with a photon–photon interaction that can be used to create a cz gate. In Chapter 4 we encountered the cross-Kerr nonlinearity governed by the interaction Hamiltonian

$$\mathcal{H}_K = \hbar\kappa \, \hat{a}_1^\dagger \hat{a}_1 \, \hat{a}_2^\dagger \hat{a}_2 \tag{6.3}$$

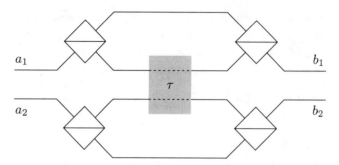

Fig. 6.2. The cz gate based on the cross-Kerr nonlinearity. Each qubit mode is split into two modes with a polarizing beam splitter, and the modes with vertical polarization are sent into the Kerr medium. The logical state then transforms into $e^{ijk\tau} |j, k\rangle$, with $j, k = 0, 1$. When $\tau = \pi$, this amounts to the cz gate.

on modes a_1 and a_2. This interaction leads to the mode transformations

$$\hat{a}_1(t) = \hat{a}_1 \, e^{i\kappa t \, \hat{a}_2^\dagger \hat{a}_2} \quad \text{and} \quad \hat{a}_2(t) = \hat{a}_2 \, e^{i\kappa t \, \hat{a}_1^\dagger \hat{a}_1} \tag{6.4}$$

where t is the interaction time. We saw that for sufficiently large $\tau = \kappa t$ the cross-Kerr nonlinearity can act as a photon switch. This suggests that such a nonlinearity can be used in quantum computation with single-photon qubit states.

The construction of the cz gate is as follows (see Fig. 6.2): consider two single-photon qubits in the polarization encoding, such that each input a_j consists of two polarization modes $a_{j,H}$ and $a_{j,V}$. The output modes are denoted by b_1 and b_2, also with a polarization degree of freedom. We assume that the logical basis is given by $|0\rangle = |H\rangle$ and $|1\rangle = |V\rangle$. The input qubits are each split into two spatial modes with horizontal and vertical polarization. The modes of vertical polarization are sent into the interaction region described by $\mathcal{H}_K$. We can now calculate the effect of this device on the two-qubit computational basis.

If both qubits are in the state $|00\rangle$, the corresponding photons are both horizontally polarized, and neither will go through the interaction region. The output therefore remains unchanged. Alternatively, when one of the photons is vertically polarized (that is, when the qubit state is $|01\rangle$ or $|10\rangle$), the phase due to the interaction $\mathcal{H}_K$ is

$$\begin{aligned}
|01\rangle &= |H, V\rangle = \hat{a}_{1,H}^\dagger \hat{a}_{2,V}^\dagger |0, 0\rangle \\
&\rightarrow e^{-i\tau \hat{a}_{1,V}^\dagger \hat{a}_{1,V}} \, \hat{a}_{1,H}^\dagger \hat{a}_{2,V}^\dagger |0, 0\rangle = e^{-i\tau \hat{a}_{1,V}^\dagger \hat{a}_{1,V}} |H, V\rangle = |H, V\rangle \\
&= |01\rangle \,, \tag{6.5}
\end{aligned}$$

where we have used that the photon number in mode $a_{1,V}$ is zero. A similar reasoning leads to $|10\rangle \rightarrow |10\rangle$. Finally, if both qubits are in the state $|11\rangle$, the interaction of the modes leads to the transformation

$$\hat{a}_{1,V} \hat{a}_{2,V} \rightarrow \hat{a}_{1,V} \hat{a}_{2,V} \, e^{i\tau (\hat{a}_{1V}^\dagger \hat{a}_{1V} + \hat{a}_{2V}^\dagger \hat{a}_{2V} - 1)} \,. \tag{6.6}$$

It is left as an exercise to the reader to verify that

$$|11\rangle \to e^{-i\tau} |11\rangle . \tag{6.7}$$

If we choose $\tau = \pi$, and the path lengths for the horizontal and vertical modes are kept equal, the device in Fig. 6.2 implements a deterministic CZ gate on single-photon polarization qubits.

Exercise 6.2: Verify Eq. (6.7).

Unfortunately, it is physically impossible to create cross-Kerr nonlinearities of this magnitude ($\tau = \pi$) and at the same time keep the noise low enough for quantum information processing purposes. In Appendix 3 we show why this is fundamentally so. This means that we have to find a different way to implement two-qubit gates for single photons. Most of the rest of this chapter is devoted to demonstrating how we can build probabilistic two-qubit gates for single photons, and how we can use them to construct an efficient quantum computation.

6.2.2 A probabilistic CZ gate for single-photon qubits

We will now construct a probabilistic CZ gate with linear optics, ancilla photons, photodetection, and feed-forward processing. The construction consists of two stages: we first build a CZ gate using linear optics and a black-box operation, and then we give a linear-optical implementation of the black box.

The CZ gate based on the black-box operation has four optical input modes: each qubit has two optical modes, corresponding to the logical states $|0\rangle$ and $|1\rangle$. The $|1\rangle$ modes of the two qubits are the input modes of a Mach–Zehnder interferometer. In each arm of the interferometer we insert a black-box operation (see Fig. 6.3). The black box acts on a single optical mode, and has the following effect on photon-number states:

$$|\psi\rangle = c_0 |0\rangle + c_1 |1\rangle + c_2 |2\rangle \xrightarrow[\text{black box}]{} c_0 |0\rangle + c_1 |1\rangle - c_2 |2\rangle , \tag{6.8}$$

where the c_i are arbitrary amplitudes. Its action on higher photon-number states is not defined. A quick examination of this operation tells us that the black box cannot just be a linear phase shift implemented by a unitary operator $\exp(i\phi\hat{n})$, with $\hat{n}$ the number operator

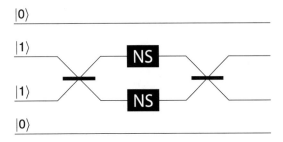

Fig. 6.3. The CZ gate for two single-photon qubits in the dual-rail representation.

on the mode. For a linear phase shift to produce a shift of -1 in the state $|2\rangle$, the phase of $|1\rangle$ must be $\pm i$. We therefore call the black-box operation the 'nonlinear sign', or NS gate. A better name would be 'nonlinear phase gate'.

Now, let us consider what happens to the various input states in the computational basis. The $|00\rangle$ state corresponds to two photons, both in the modes that do not interact with the beam splitter (the top and bottom lines in Fig. 6.3). We therefore have the transformation $|00\rangle \to |00\rangle$. The next two input states we consider are $|01\rangle$ and $|10\rangle$: only one photon enters the first beam splitter, in either input mode. The NS gate does not affect the optical states $|0\rangle$ and $|1\rangle$, so the second beam splitter simply reverses the effect of the first, and the transformation becomes

$$|01\rangle \to |01\rangle \quad \text{and} \quad |10\rangle \to |10\rangle \ .$$

The nontrivial operation of the gate happens when there are two photons impinging on the first beam splitter, one in each input mode. This is the two-qubit state $|11\rangle$. The input state will transform into the following output state:

$$|1, 1\rangle \to \frac{|2, 0\rangle - |0, 2\rangle}{\sqrt{2}} \ . \tag{6.9}$$

This state will pick up an overall minus sign due to the NS gate, before it will be returned to the state $|1, 1\rangle$ by the second beam splitter. Consequently, the transformation of the photon states is $|1, 1\rangle \to -|1, 1\rangle$, which translates in a transformation of the computational basis state $|11\rangle$:

$$|11\rangle \to -|11\rangle \ . \tag{6.10}$$

This amounts to a CZ gate on the two input qubits.

How can the NS gate be implemented with linear optics and projective measurements? It is worth going through this in some detail. We want to use only photon-number states, passive linear optics, and (perfect) photodetection. As we mentioned before in Chapter 1, the most general linear-optical network is the N-port, which can be drawn as Fig. 6.4. Without

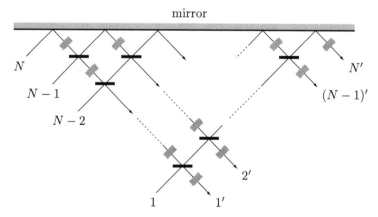

Efficient construction of an arbitrary N-port with beam splitters and phase shifters.

loss of generality, we can choose the lowest input mode in the figure as the input mode of the NS gate (indicated in the figure). The other modes represent ancillas. This way the main mode couples to the ancilla modes with just one beam splitter. Conservation of photon number means that if the number of ancilla photons is not equal to the number of detected photons (assuming perfect photodetectors), then we must add or subtract photons in the main mode. A nonlinear phase shift in the main mode does not alter the photon number, and we must therefore register all the input ancilla photons in the photodetectors. The resulting gate operation is probabilistic because there is a non-zero chance that we obtain the wrong detection outcome.

The simplest N-port that could possibly be turned into an NS gate is a 2-port with an n-photon ancilla state $|\psi\rangle_1 |n\rangle_2$, where $|\psi\rangle$ is defined in Eq. (6.8). The mode transformation is a 2×2 unitary matrix, and we can write it as a matrix equation

$$\begin{pmatrix} \hat{a}_1^\dagger \\ \hat{a}_2^\dagger \end{pmatrix} = \begin{pmatrix} u & v \\ -v^* & u^* \end{pmatrix} \begin{pmatrix} \hat{b}_1^\dagger \\ \hat{b}_2^\dagger \end{pmatrix}, \tag{6.11}$$

where $\hat{a}_j^\dagger$ are the input mode operators, $\hat{b}_j^\dagger$ the output mode operators, and $|u|^2 + |v|^2 = 1$.

We want the NS gate to succeed with an appreciable probability (we need two of them for the CZ gate to work), and we will therefore first determine the success probability of this 2-port protocol. For this, consider the vacuum contribution in the input state $|\psi\rangle$. The 2-port transforms this as

$$c_0 |0, n\rangle = c_0 \frac{\hat{a}_2^{\dagger n}}{\sqrt{n!}} |0, 0\rangle \rightarrow c_0 \frac{\left(-v^* \hat{b}_1^\dagger + u^* \hat{b}_2^\dagger\right)^n}{\sqrt{n!}} |0, 0\rangle = u^{*n} c_0 |0, n\rangle + \dots \tag{6.12}$$

Since $c_0 |0\rangle$ must remain untouched by the NS gate, the term u^{*n} on the right-hand side must be the success probability amplitude of the gate, leading to $p_{\text{success}} = |u|^{2n}$. This becomes very low, very quickly, as n increases. We therefore set $n = 1$ in the remainder of this calculation ($n = 0$ would lead to a deterministic gate, and we already know that is impossible).

Exercise 6.3: Show that for $n = 1$ the relevant term in the output state (before detection) is given by

$$u^* c_0 |0, 1\rangle + \left(|u|^2 - |v|^2\right) c_1 |1, 1\rangle + u \left(|u|^2 - 2|v|^2\right) c_2 |2, 1\rangle . \tag{6.13}$$

The NS gate then requires that

$$\frac{\left(|u|^2 - |v|^2\right)}{u^*} = 1 \quad \text{and} \quad \frac{u}{u^*} \left(|u|^2 - 2|v|^2\right) = -1 . \tag{6.14}$$

Show that the equations in (6.14) cannot be solved for u and v.

This exercise shows that we cannot build an NS gate with one photon in one ancilla mode. It is a bit more difficult to show that this is true for any photon number $|n\rangle$ in the ancilla mode of a 2-port.

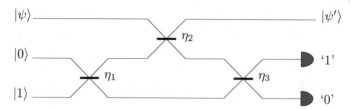

Fig. 6.5. The NS gate based on three variable beam splitters and single-photon detection.

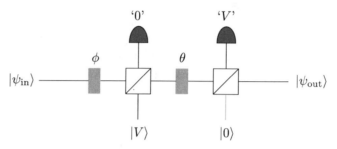

Fig. 6.6. A polarization-based NS gate with polarization rotations.

The next attempt to build a simple NS gate must involve a 3-port. We have seen that increasing photon number decreases the success probability, and we therefore assume a single photon in one of the ancilla modes, and vacuum in the other mode. The set-up is shown in Fig. 6.5.

Exercise 6.4: Show that the gate in Fig. 6.5 works if

$$\eta_1^2 = \eta_3^2 = \frac{1}{4 - 2\sqrt{2}} \quad \text{and} \quad \eta_2^2 = 3 - 2\sqrt{2}, \tag{6.15}$$

and show that the success probability of this NS gate is 1/4.

The NS gate makes use of beam splitters with non-standard transmission amplitude (η_1 and η_2). In Chapter 1, we showed that the beam splitter is formally equivalent to a polarization rotation, and we are going to make use of this in the next example. Fig. 6.6 shows a polarization-based NS gate acting on the input mode in horizontal polarization. It is therefore still a single-mode NS gate. The two elements labeled ϕ and θ are polarization rotations over angles ϕ and θ, respectively. The first polarizing beam splitter has one vertically polarized ancilla photon in the secondary input port, and no photons are detected in the output port. The second polarizing beam splitter has no ancilla input photons, and the secondary output must detect a single vertically polarized photon. We will now show that this interferometer acts as an NS gate when the detector signature is $(0, V)$.

The input state is horizontally polarized, and can be written as

$$|\psi_{\text{in}}\rangle = c_0 |0\rangle_H + c_1 |1\rangle_H + c_2 |2\rangle_H = \left(c_0 + c_1 \hat{a}_H^\dagger + \frac{c_2}{\sqrt{2}} \hat{a}_H^{\dagger 2} \right) |0\rangle . \tag{6.16}$$

The first polarization rotation yields the transformation $\hat{a}_H^\dagger \to \cos\phi\,\hat{a}_H^\dagger + \sin\phi\,\hat{a}_V^\dagger$. After detecting no photons in the secondary output mode of the first polarizing beam splitter, the state of the optical modes just before the element θ is

$$\left(c_0 + c_1 \cos\phi\,\hat{a}_H^\dagger + \frac{c_2}{\sqrt{2}}\cos^2\phi\,\hat{a}_H^{\dagger 2}\right)\hat{a}_V^\dagger\,|0\rangle\ .$$

The second polarization rotation is given by

$$\hat{a}_H^\dagger \to \cos\theta\,\hat{a}_H^\dagger + \sin\theta\,\hat{a}_V^\dagger \quad \text{and} \quad \hat{a}_V^\dagger \to -\sin\theta\,\hat{a}_H^\dagger + \cos\theta\,\hat{a}_V^\dagger\ . \tag{6.17}$$

Detecting a single vertically polarized photon at the second output port then yields the final output state

$$|\psi_{\text{out}}\rangle = c_0\cos\theta\,|0\rangle + c_1\cos\phi\cos 2\theta\,|1\rangle + c_2\cos^2\phi\cos\theta(1-3\sin^2\theta)\,|2\rangle \tag{6.18}$$

in the horizontally polarized mode. If this is to implement an NS gate, the following equations must hold:

$$\frac{\cos\phi\cos 2\theta}{\cos\theta} = 1 \quad \text{and} \quad \cos^2\phi\left(1 - 3\sin^2\theta\right) = -1\ . \tag{6.19}$$

The success probability for this gate is $\cos^2\theta$. To solve for ϕ and θ, we first eliminate $\cos\phi$, and then we substitute $w = \cos^2\theta$. We find

$$p_{\text{success}} = w = \frac{21 - 7\sqrt{2}}{49} \simeq 0.227\ . \tag{6.20}$$

This leads to rotation angles $\theta \approx 61.5°$ and $\phi \approx 150.5°$. The success probability of this gate is slightly smaller than the previous NS gate.

Exercise 6.5: Verify Eq. (6.18) and solve for ϕ and θ.

The NS gates operate properly when a specific number of photons is detected in some auxiliary modes, and no photons in others. The detectors must therefore not only have photon-number discriminating capabilities, they must also have a near-perfect detection efficiency. If the detection efficiency is lower than one, there is a probability that some photons may have entered a mode where no photons should be, without triggering the detector. This lowers the fidelity of the gate, leading to potential qubit loss and dephasing. In Chapter 4 we saw that it is extremely difficult to make detectors that have near-perfect detection efficiency. We will address this problem in Section 6.3.

6.2.3 Improving success probabilities of optical gates

What is the maximum success probability of the NS gate? There is a simple proof due to Knill (2003) that limits the maximum success probability of a linear-optical NS gate to one half.

The proof reveals an interesting property of optical interferometers, so we will describe it in some detail. There are two stages: first, we show that we can transform a two-photon input state $|1, 1\rangle$ into a state $|2, 0\rangle$ with passive linear-optics and use of a single NS gate. In the second stage we prove that the average photon number in each output mode is smaller than or equal to one. Reconciliation of these two facts then requires that the success probability of the NS gate cannot exceed one half.

For the first stage, we write $|1, 1\rangle = \hat{a}_1^\dagger \hat{a}_2^\dagger |0, 0\rangle$ and we mix the two modes on a beam splitter with transmission amplitude $\cos(\pi/8)$. The mode transformations are

$$\hat{b}_1^\dagger = \cos(\pi/8)\,\hat{a}_1^\dagger + \sin(\pi/8)\,\hat{a}_2^\dagger \quad \text{and} \quad \hat{b}_2^\dagger = -\sin(\pi/8)\,\hat{a}_1^\dagger + \cos(\pi/8)\,\hat{a}_2^\dagger. \quad (6.21)$$

The output state is then

$$|\psi_{\text{out}}\rangle = -\sqrt{2}\cos(\pi/8)\sin(\pi/8)\,|2, 0\rangle + \left[\cos^2(\pi/8) - \sin^2(\pi/8)\right]|1, 1\rangle$$
$$+ \sqrt{2}\cos(\pi/8)\sin(\pi/8)\,|0, 2\rangle\,. \quad (6.22)$$

Applying the NS gate to mode a_1 changes the sign of the first term in the superposition, and we can write

$$|\psi_{\text{out}}\rangle \to \sqrt{2}\cos(\pi/8)\sin(\pi/8)\,|2, 0\rangle + \left[\cos^2(\pi/8) - \sin^2(\pi/8)\right]|1, 1\rangle$$
$$+ \sqrt{2}\cos(\pi/8)\sin(\pi/8)\,|0, 2\rangle$$
$$= \frac{1}{2}\left[\sin(\pi/4)\,\hat{b}_1^{\dagger 2} + 2\cos(\pi/4)\,\hat{b}_1^\dagger\hat{b}_2^\dagger + \sin(\pi/4)\,\hat{b}_2^{\dagger 2}\right]|0, 0\rangle$$
$$= \frac{1}{\sqrt{2}}\left(\frac{\hat{b}_1^\dagger + \hat{b}_2^\dagger}{\sqrt{2}}\right)^2 |0, 0\rangle_{12} = \frac{\hat{c}_1^{\dagger 2}}{\sqrt{2}}\,|0, 0\rangle_{12} = |2, 0\rangle_{12}\,. \quad (6.23)$$

The NS gate therefore lets us combine two photons into the same mode. No linear-optical unitary mode transformation can accomplish this deterministically, because the transformation $\hat{a}_i^\dagger\hat{a}_j^\dagger \to \hat{b}_k^{\dagger 2}$ requires that two input modes are mapped to the same output mode. This requires two identical columns in the unitary matrix U that describes the transformation, and therefore $\det U = 0$. However, unitary matrices must have $|\det U| = 1$.

For the second stage of the proof, suppose that we have N optical modes. A selection of these modes is occupied by one photon, while the remaining modes are in the vacuum state. We can write the input state as

$$|\psi\rangle = \hat{a}_1^\dagger \dots \hat{a}_k^\dagger |0\rangle_{1,\dots,N}\,, \quad (6.24)$$

where we have k input photons in the first k modes (we may have to relabel the modes to get the state in this form). We now use this state as the input of an N-port that implements a unitary transformation U, and calculate the average photon number of an arbitrary output mode j:

$$\langle \hat{n}_j \rangle = \langle \psi | \, U^\dagger \hat{a}_j^\dagger \hat{a}_j U \, | \psi \rangle = \langle \psi | \, U^\dagger \hat{a}_j^\dagger U \, U^\dagger \hat{a}_j U \, | \psi \rangle$$

$$= \langle \psi | \left(\sum_{m=1}^{N} u_{jm}^* \hat{a}_m^\dagger \right) \left(\sum_{n=1}^{N} u_{jn} \hat{a}_n \right) | \psi \rangle = \sum_{m,n=1}^{N} u_{jm}^* u_{jn} \langle \psi | \, \hat{a}_m^\dagger \hat{a}_n \, | \psi \rangle$$

$$= \sum_{m=1}^{N} |u_{jm}|^2 \leq 1 \,, \tag{6.25}$$

where we used the relation $U^\dagger \hat{a}_j U = \sum_{n=1}^{N} u_{jn} \hat{a}_n$, and the inner product in the second-to-last line is a Kronecker delta δ_{mn} because $|\psi\rangle$ has definite photon number in each mode. The final inequality follows from the unitarity of U. This tells us that on average we can never measure more than one photon in the output modes of an arbitrary unitary N-port if the input modes had at most one photon each.

The NS gate is probabilistic, and each application of the gate gives a 'success' or 'failure' indication. If we ignore this indication, the state of the output mode $\hat{c}$ in Eq. (6.23) becomes a mixture:

$$\rho = p \, |2\rangle \langle 2| + (1 - p)\rho' \,, \tag{6.26}$$

where p is the success probability of the NS gate, and ρ' may or may not contain the projector $|2\rangle \langle 2|$. The average photon number in mode a is then

$$\langle \hat{n}_a \rangle = \text{Tr}\left[\hat{n}_a \rho \right] = 2p + q \,, \tag{6.27}$$

with q the contribution from ρ'. Since $\langle \hat{n}_a \rangle \leq 1$ and $q \geq 0$, we obtain

$$p \leq \frac{1}{2} \,. \tag{6.28}$$

This is not a tight bound. Subsequent work by Scheel and Lütkenhaus (2004) and Eisert (2005) has shown that the maximum success probability of the NS gate is one quarter, and the construction above is therefore optimal.

When we apply the probabilistic NS gate, we are hoping for a specific detection signature indicating 'success'. In this signature we include the possibility that we still have to apply a *deterministic* passive linear-optical operation. When we get a different detection signature, there are two possibilities:

(i) The detection signature indicates that with more ancilla photons and photodetection the NS gate can still be made to work;

(ii) The detection signature indicates a catastrophic error, and the gate has failed beyond repair. This is the case when the detected number of photons is larger than the number of ancilla photons. Such a signature can happen only when there was a photon in the input state, thus revealing information about this state.

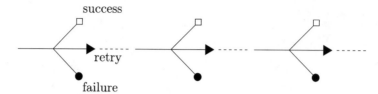

Fig. 6.7. Schematic structure of increasing the success probability of the NS gate via feed-forward.

In the event of possibility (i), we can apply another post-selection linear-optical network designed to salvage the gate. The detection signature of this second round again has three possible outcomes, namely 'success', 'failure', or 'retry' in the case of a correctable failure. We can repeat this procedure until we obtain either a catastrophic failure or a successful gate operation (see Fig. 6.7).

We can use this construction to put increasingly tighter bounds on the success probability of the feed-forward NS gate. The probability of a catastrophic failure typically depends on the input state, and we therefore have to consider the *maximal* failure rate. This occurs when $c_0 = c_1 = 0$, and $c_2 = 1$, and determines a failure rate of $p_{\text{failure}} \approx 0.61$. This is larger than 1/2, the minimum failure bound for general optical gates we found before. We can sum the catastrophic error probabilities at each iteration, but due to the number of parameters this is very difficult and only feasible numerically. The result indicates that the catastrophic failure rate approaches 2/3 (Scheel *et al.*, 2006). The improvement in gate probability seems very modest (after one iteration we obtain $p_{\text{success}} \approx 0.28$, which is only marginally larger than one quarter). The prospects of feed-forward gates are therefore not very promising, and in the next section we will encounter different gates that are simpler and have a higher initial success probability.

6.2.4 A cx gate that consumes entanglement

Previously, we have seen how we can construct entangling gates like the cz and the cx gate with single-photon states as ancillas. The success probability of these gates is relatively low: 1/16 without feed-forward, and towards 1/9 with feed-forward. In this section we investigate whether *entangled* ancilla states can help increase the success probability. We find that it does, and the success probability rises to 1/4.

To see how entanglement can be useful, consider the following hand-waving argument: its use makes sense only if we send one part of the entangled photon pair into an interferometer with the control qubit, and the other part into a second interferometer with the target qubit, because we need to establish (or remove) entanglement between the control and the target qubit. This will be our general approach. We will first construct a formal entanglement-assisted 'completely positive' (CP) map that induces a probabilistic cx gate, and second, we will give an optical implementation.

Suppose we use the following Bell state on modes 1 and 2 as our entangled resource

$$|\Phi^+\rangle_{12} = \frac{1}{\sqrt{2}} \left(|HD\rangle_{12} + |HA\rangle_{12} + |VD\rangle_{12} - |VA\rangle_{12} \right), \tag{6.29}$$

where we have written the second qubit in the diagonal polarization basis. The qubit input states are

$$|\psi\rangle_c = \alpha |H\rangle + \beta |V\rangle \quad \text{and} \quad |\phi\rangle_t = \gamma |D\rangle + \delta |A\rangle , \tag{6.30}$$

where c and t denote 'control' and 'target', respectively. In the computational basis, the joint state can be written as

$$\alpha(\gamma + \delta) |HH\rangle_{ct} + \alpha(\gamma - \delta) |HV\rangle_{ct} + \beta(\gamma + \delta) |VH\rangle_{ct} + \beta(\gamma - \delta) |VV\rangle_{ct} .$$

We will now show how the following two maps will create a cx gate:

$$\mathcal{M}_1 = |H\rangle \langle HH| + |V\rangle \langle VV| \quad \text{and} \quad \mathcal{M}_2 = |D\rangle \langle DD| + |A\rangle \langle AA| . \tag{6.31}$$

The bras in $\mathcal{M}_1$ ($\mathcal{M}_2$) act on the modes 1 and c (2 and t) and the ket is c (or t). These are CP maps, rather than proper projectors in Hilbert space, because they are not idempotent: $\mathcal{M}_i \neq \mathcal{M}_i^2$. First, $\mathcal{M}_1$ gives

$$\mathcal{M}_1 |\psi\rangle_c \left|\Phi^+\right\rangle_{12} = \alpha |HD\rangle_{c2} + \alpha |HA\rangle_{c2} + \beta |VD\rangle_{c2} - \beta |VA\rangle_{c2} . \tag{6.32}$$

Subsequently, the second map gives

$$\begin{aligned}
|\psi_{\text{out}}\rangle_{ct} &= \mathcal{M}_2 \mathcal{M}_1 |\psi\rangle_c \left|\Phi^+\right\rangle_{12} |\phi\rangle_t \\
&= \alpha\gamma |HD\rangle_{ct} + \beta\gamma |VD\rangle_{ct} + \alpha\delta |HA\rangle_{ct} - \beta\delta |VA\rangle_{ct} .
\end{aligned} \tag{6.33}$$

When we write this in the computational basis we obtain

$$\alpha(\gamma + \delta) |HH\rangle_{ct} + \alpha(\gamma - \delta) |HV\rangle_{ct} + \beta(\gamma - \delta) |VH\rangle_{ct} + \beta(\gamma + \delta) |VV\rangle_{ct} .$$

The two maps therefore induce a swap between $|VH\rangle$ and $|VV\rangle$, while doing nothing to $|HH\rangle$ and $|HV\rangle$. This is a cx gate. It is also not hard to see that if we replace D and A with H and V, respectively, two $\mathcal{M}_1$ maps will implement a cz gate.

Exercise 6.6: Verify that the two maps $\mathcal{M}_1$ and $\mathcal{M}_2$ operate as advertised.

How do we implement the maps $\mathcal{M}_1$ and $\mathcal{M}_2$? To answer this question we note a couple of things: first, each map takes two input photons and produces one output photon. We therefore need to measure at least one photon. Second, we may assume initially that there are no ancilla photons (other than the entangled photon pair), since each added photon has the tendency to lower the success probability on account of creating more possible output states.

Now consider the map $\mathcal{M}_1 = |H\rangle \langle HH| + |V\rangle \langle VV|$. This operation means that when the two input photons have equal polarization in the computational basis $\{H, V\}$, they will enter two different output modes, one of which will be measured. This is a description of a polarizing beam splitter rotated to separate horizontal and vertical polarization, where one output mode is detected. Without detection, the action of the polarizing beam splitter is

$$\alpha |HH\rangle + \beta |HV\rangle + \gamma |VH\rangle + \delta |VV\rangle$$
$$\rightarrow \alpha |H, H\rangle + \beta |HV, 0\rangle + \gamma |0, VH\rangle + \delta |V, V\rangle , \tag{6.34}$$

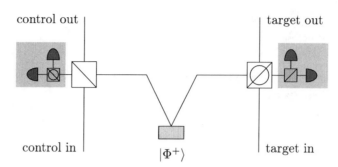

control out   target out

control in $|\Phi^+\rangle$ target in

Fig. 6.8.   The entanglement-assisted cx gate.

where we have added commas to separate the spatial modes in the second line to avoid confusion. We see that post-selecting on a single measured photon in one mode will ensure that there is a single photon in the second mode with exactly the same polarization. Note also that zero or two photons in one mode indicate a failure of the map.

We have to be careful, because a polarization-insensitive photodetector will erase the polarization information of the detected photon, and thus cause the output state to be mixed. On the other hand, measuring a horizontally polarized photon (say) will create the map $|H\rangle\langle HH|$, which is not what we want. We have to measure the photon in a conjugate (or mutually unbiased) polarization basis $\{D, A\}$ (diagonal polarization) or $\{L, R\}$ (circular polarization). We follow the literature and choose D and A.

The map $\mathcal{M}_2$ works in exactly the same way as $\mathcal{M}_1$, except that we have to swap the roles of $\{H, V\}$ and $\{D, A\}$: the polarizing beam splitter is rotated to separate diagonal polarization, while the photodetector measures horizontally and vertically polarized photons. The two maps are then combined with the entangled ancilla state to form the linear-optical interferometer shown in Fig. 6.8. The gate succeeds when each photodetector detects exactly one photon. Counting all possibilities, it is not too difficult to see that the overall success probability is 1/4, a great improvement over gates that do not consume entanglement.

There is one more subtlety in the operation of this gate. Given a successful event (one detected photon), the photodetector in $\mathcal{M}_1$ ($\mathcal{M}_2$) can indicate D or A (H or V). Unitarity requires that these two orthogonal detector results produce orthogonal output states of the gate. Indeed, we need to apply a (deterministic) single-photon operation depending on the measurement outcome.

Exercise 6.7: Prove that the linear-optical cx based on entangled resources succeeds with probability 1/4. What are the single-photon operations for each measurement outcome?

6.3 Building quantum computers with probabilistic gates

Having introduced probabilistic two-photon entangling gates like the cz gate and the cx gate, we can in principle use them to create universal cluster states and perform any quantum computation. However, the gates introduced so far are not the most economical in creating cluster states, and when we consider realistic photodetectors we find that they are very hard to make loss-tolerant since they rely in part on *not* counting photons. Furthermore, it is not necesary to have a cx or a cx gate if you want to construct a cluster state. Any entangling operation that gives you the right result will do. In this section we introduce 'fusion gates', and show how they can be used to create cluster states efficiently. We also show how a simple qubit encoding allows the fusion gates to be used in the circuit model for quantum computing.

6.3.1 Fusion gates and four-photon GHZ states

The two maps $\mathcal{M}_1$ and $\mathcal{M}_2$ that make up the fixed components of the cx gate can be used to create multi-photon entangled states. If the two photons in modes 2 and 3 are part of an entangled pair, the resulting output photons from the map $\mathcal{M}_1$ form a three-photon GHZ state:

$$\mathcal{M}_1 \left(|HH\rangle_{12} + |VV\rangle_{12}\right) \left(|HH\rangle_{34} + |VV\rangle_{34}\right) = |HHH\rangle_{134} + |VVV\rangle_{134} \, , \qquad (6.35)$$

where we detected a single photon with polarization D in mode 2, and we suppressed the normalization. The map $\mathcal{M}_1$ fused the two entangled photon pairs at the cost of a single photon. We therefore rename $\mathcal{M}_1$ as a 'fusion' gate. In particular, it is the first of two fusion gates, and we will therefore call it the 'type-I fusion gate' $\mathcal{F}_I$:

$$\mathcal{F}_I : \begin{cases} \text{detector outcome } D: |H\rangle \langle HH| + |V\rangle \langle VV| \, , \\ \text{detector outcome } A: |H\rangle \langle HH| - |V\rangle \langle VV| \, . \end{cases} \qquad (6.36)$$

The second fusion gate $\mathcal{F}_{II}$ (type-II) is based on the map $\mathcal{M}_2$ with the modification that *both* output modes are measured with a photodetector in the $\{H, V\}$ basis. When the two detectors each indicate the presence of a single photon the gate is successful. There are four possible detector outcomes that indicate a successful gate operation: $(H, H), (H, V), (V, H)$, and (V, V). When both detectors give the same outcome (H, H) or (V, V), the resulting output state is found by taking the inner product of the input state with $\langle HV| + \langle VH|$, while opposite polarization leads to an output state that is found by the inner product with $\langle HH| + \langle VV|$:

$$\mathcal{F}_{II} : \begin{cases} \text{detector outcome } (H, H) \text{ or } (V, V): \langle HV| + \langle VH| \, , \\ \text{detector outcome } (H, V) \text{ or } (V, H): \langle HH| + \langle VV| \, . \end{cases} \qquad (6.37)$$

Clearly, this gate removes two photons from the input state, and can therefore not be used to produce larger entangled states when the input states are entangled photon pairs. However, when we apply $\mathcal{F}_{II}$ to two photons that are each part of a three-photon GHZ state (say, the output of $\mathcal{F}_I$), then we can create the following four-photon GHZ state:

$$\mathcal{F}_{II} \left(|HHH\rangle + |VVV\rangle\right) \left(|HHH\rangle + |VVV\rangle\right) = \begin{cases} |HHHH\rangle + |VVVV\rangle \\ |HHVV\rangle + |VVHH\rangle \, , \end{cases}$$

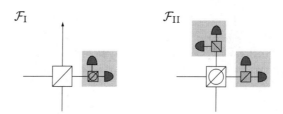

Fig. 6.9. Two fusion gates $\mathcal{F}_\mathrm{I}$ and $\mathcal{F}_\mathrm{II}$. The type-I fusion gate $\mathcal{F}_\mathrm{I}$ is a polarizing beam splitter rotated to separate the $\{H, V\}$ direction, followed by a single-photon detection in the basis $\{D, A\}$ in only one output mode, and the type-II fusion gate $\mathcal{F}_\mathrm{II}$ is a polarizing beam splitter oriented to the $\{D, A\}$ direction, followed by single-photon detection in the $\{H, V\}$ basis in each output mode.

depending on the detection signature. We have again suppressed normalization, and assumed that the fusion gate acted on one photon in the first superposition and one in the second. The two fusion gates are shown in Fig. 6.9. There is a third fusion gate $\mathcal{F}'_\mathrm{II}$ that acts similarly to $\mathcal{F}_\mathrm{II}$, but in the basis $\{D, A\}$, rather than $\{H, V\}$. We will use this gate in Section 6.3.3.

Exercise 6.8: Construct the dual-rail equivalent of the two polarization-based fusion gates.

The type-II fusion gate consumes more photons than the type-I fusion gate, which begs the question: why use type-II at all? It turns out that $\mathcal{F}_\mathrm{II}$ has two important practical advantages over $\mathcal{F}_\mathrm{I}$. First, the failure mode of $\mathcal{F}_\mathrm{II}$ less catastrophic when we want to create cluster states. We will discuss this in the next section. Second, the type-II gate still works in the presence of lossy sources and detectors. We demonstrate this with a practical example.

So far, we have assumed that our photodetectors are perfect. In other words, we can always tell exactly how many photons there were present in a particular detected mode. In practice, this is of course not the case at all. There will be losses in the detector, and often the detector gives only a binary output (click/no-click) without any additional information about photon number (see Chapter 4 for more details). In addition, we may assume that our entangled resource photons are always created in pairs, but sometimes pair-production may fail altogether (see, e.g., parametric downconversion). Remarkably, we can still create high-fidelity four-photon GHZ states when we choose the right configuration of $\mathcal{F}_\mathrm{I}$ and $\mathcal{F}_\mathrm{II}$ gates.

In order for $\mathcal{F}_\mathrm{II}$ to be helpful at all, we must first create two three-photon GHZ gates with two $\mathcal{F}_\mathrm{I}$ gates, each acting on two entangled photon pairs (shown in Fig. 6.10). There are several ways in which the creation of the three-photon GHZ state can fail, even though the detector signature in $\mathcal{F}_\mathrm{I}$ indicates success. First, not every source of photon pairs may in fact create a pair. If the source efficiency is denoted by η_s, the output of the source ρ_s can be written as

$$\rho_s = \eta_s \left| \Phi^+ \right\rangle \left\langle \Phi^+ \right| + (1 - \eta_s) \left| 00 \right\rangle \left\langle 00 \right| . \tag{6.38}$$

The probability that both sources produce an entangled pair is η_s^2 (assuming all sources are identical). We accept a certain event only when we find a detector click, which means

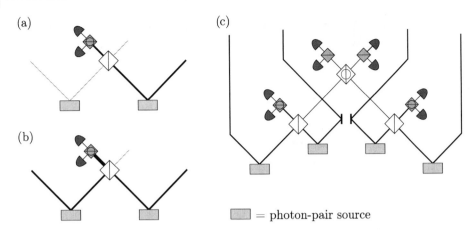

Fig. 6.10. Failure modes of the three-photon GHZ state factory. (a) One of the sources may fail to produce an entangled photon pair; (b) one of the photons may not be detected by the photodetectors; and (c) taking the possibly 'empty' optical modes and applying a type-II fusion gate to them ensures that a high-fidelity four-photon GHZ state is created. This construction assumes that the dark counts in the detectors and higher-order pair production in the sources are negligible.

that *at least*[1] one source must have produced an entangled photon pair. If the other source did not create a photon pair, then the output mode of $\mathcal{F}_I$ will be empty (the vacuum state). Second, even if both sources produce an entangled photon pair, the natural evolution of the polarizing beam splitter may send two photons into the detected modes. If the detector has detection efficiency η_d, then there is a probability $2\eta_d(1 - \eta_d)$ that only one of the two photons triggers a detector. As a result, we accept the event, even though the output of $\mathcal{F}_I$ is again empty.

Finite-source and detection efficiency will cause empty output modes in the fusion gates. Whenever the output mode of $\mathcal{F}_I$ is *not* empty (conditioned on a successful detection signature), the fusion gate has worked correctly. For this it is essential that the detector dark counts and the multiple pair-production of the sources are negligible. This places severe restrictions on parametric downconversion as the source of entangled photon pairs. Given these caveats, we can now apply the type-II fusion gate to the output modes of the two type-I fusion gates. Since two photons must be detected in $\mathcal{F}_{II}$, whenever one of the output modes of the two $\mathcal{F}_I$ is empty, $\mathcal{F}_{II}$ cannot give a two-fold detection coincidence, and therefore cannot give the required detection signature. The overall success probability p of creating a loss-tolerant four-photon GHZ state is

$$p = \frac{(\eta_s \eta_d)^4}{8},\qquad (6.39)$$

since we have four successfully operating sources, four detected photons, and three fusion gates with intrinsic success probability of 1/2.

---

[1] We assume that detector dark counts are negligible. This is often a very good approximation, but it is important to remember that in general dark counts will give us false positives, which directly affect the fidelity of these operations.

Exercise 6.9: Show that the creation of four-photon GHZ states with the fusion gates described above requires at most bucket detectors.

6.3.2 Optical cluster-state generation

The four-photon GHZ states that are created according to the loss-tolerant procedure in the previous section are locally equivalent (or, more specifically, LC-equivalent) to cluster states. In this section, we will explore how the two fusion gates connect two arbitrary cluster states to make a large cluster state. This procedure, together with the recipe to create small GHZ states, is sufficient to create universal cluster states for quantum computing. In the final part of this section we show how failure of the fusion gates affects the clusters.

Since our aim is to join two clusters by applying a fusion gate to a single qubit from each cluster, we first write an arbitrary cluster state $|C\rangle$ explicitly as a single qubit that is entangled with the rest of the cluster. Recall that a cluster state is obtained by preparing a collection of qubits in the $|+\rangle$ state, followed by CZ gates between the qubit pairs that require a connection in the graph. Suppose the cluster state consists of N qubits, and we wish to apply the fusion gate to qubit 1. The cluster state $|C\rangle$ is obtained from the state $|+\rangle_1 |\psi\rangle_{2..N}$ by applying CZ gates between qubit 1 and all its neighbours in $|C\rangle$:

$$|C\rangle = \prod_{k \in n(1)} CZ_{1k} |+\rangle_1 |\psi\rangle_{2..N}. \tag{6.40}$$

Here, $n(1)$ is the neighbourhood of qubit 1 in the state $|C\rangle$. Given that the CZ gate CZ_{1k} is by definition a Pauli Z operation on qubit k when qubit 1 is in the state $|1\rangle$, we can write the cluster state as

$$|C\rangle = |0\rangle_1 |\psi\rangle_{2..N} + |1\rangle_1 \prod_{k \in n(1)} Z_k |\psi\rangle_{2..N}, \tag{6.41}$$

where we have suppressed the overall normalization factor. We now consider two cluster states $|C_A\rangle$ and $|C_B\rangle$:

$$|C_A\rangle = |0\rangle_1 |\psi_A\rangle_{2..N} + |1\rangle_1 \prod_{j \in n(1)} Z_j |\psi_A\rangle_{2..N}$$

$$|C_B\rangle = |0\rangle_{1'} |\psi_B\rangle_{2'..M'} + |1\rangle_{1'} \prod_{j \in n(1')} Z_j |\psi_B\rangle_{2'..M'}. \tag{6.42}$$

We can then apply the fusion gates $\mathcal{F}_{\mathrm{I}}$ and $\mathcal{F}_{\mathrm{II}}$ to qubits 1 and $1'$.

First, we consider the effect of $\mathcal{F}_{\mathrm{I}}$ on qubits 1 and $1'$. The operation can be written as

$$\mathcal{F}_{\mathrm{I}} = |0\rangle_{1\,11'}\langle 00| \pm |1\rangle_{1\,11'}\langle 11|, \tag{6.43}$$

where the relative sign ($\pm$) depends on the measurement outcome. This leads to

$$\mathcal{F}_{\mathrm{I}} |C_A, C_B\rangle = |0\rangle_1 |\psi_A\rangle_{2..N} |\psi_B\rangle_{2'..M'} \pm |1\rangle_1 \prod_{j \in n(1,1')} Z_j |\psi_A\rangle_{2..N} |\psi_B\rangle_{2'..M'}, \tag{6.44}$$

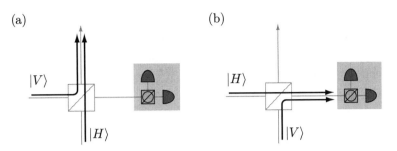

Fig. 6.11. **Failure modes of $\mathcal{F}_I$ with perfect detectors. (a) Zero photons are detected: the input modes were in the state $|V, H\rangle$; (b) two photons are detected: the input modes were in the state $|H, V\rangle$. This is a measurement in the computational basis.**

where $n(1, 1') = n(1) \cup n(1')$ is the union of the neighbourhoods of qubit 1 and $1'$, respectively. When the measurement outcome of $\mathcal{F}_I$ indicates that the minus sign is actuated, we can apply a corrective Z operation to qubit 1. It is clear that the resulting state is again a cluster state, since it takes the same form as Eq. (6.41).

The fusion gate $\mathcal{F}_I$ succeeds at most only half the time, so it is important to know what happens when the gate fails. Assuming perfect photodetectors, the measurement outcomes that indicate gate failure are 'no photons' and 'two photons'. In the latter case, the two photons must have opposite polarization, as was determined in Eq. (6.34). First, if no photons are detected, then both photons must be in the output mode. This can happen only when one incoming photon is horizontally polarized, and the other vertically. Moreover, we know exactly which photon has what polarization (see Fig. 6.11). Therefore, finding no photons is an effective single-photon qubit measurement in the computational basis $\{H, V\}$. Second, when two photons are detected there is again only one possible two-photon state that causes both photons to enter the output mode that leads to the detector. Hence a two-photon detector outcome is also equivalent to a measurement in the computational basis. As a result, the failure of $\mathcal{F}_I$ results in a Z measurement of the two qubits it is acting on. Recall from Chapter 2 that the Z measurement of a qubit in a cluster state breaks all the bonds of that qubit with the rest of the cluster. When we are trying to increase the size of the cluster, this a rather bad way to fail.

Next consider the effect of a type-II fusion gate $\mathcal{F}_{II}$ on two qubits that are part of a cluster. We can deduce from Eq. (6.42) that a successful application of $\mathcal{F}_{II}$ yields the state

$$|\psi_A\rangle_{2..N} |\psi_B\rangle_{2'..M'} + \prod_{j \in n(1,1')} Z_j |\psi_A\rangle_{2..N} |\psi_B\rangle_{2'..M'} ,$$

$$\text{or} \quad |\psi_A\rangle_{2..N} \prod_{j \in n(1')} Z_j |\psi_B\rangle_{2'..M'} + \prod_{j \in n(1)} Z_j |\psi_A\rangle_{2..N} |\psi_B\rangle_{2'..M'} . \quad (6.45)$$

It is not so easy to see directly that these are again cluster states, so we will use the stabilizer formalism to prove that $\mathcal{F}_{II}$ produces cluster states. This approach also has the advantage that we can see exactly what bonds the type-II fusion gate creates.

Suppose that before the application of $\mathcal{F}_{II}$, the qubits 1 and $1'$ are connected to several other qubits in their respective clusters. We will consider an arbitrary neighbour i of qubit 1 and j from qubit $1'$, and study how their stabilizers change under $\mathcal{F}_{II}$. Since our result is true for two arbitrary neighbours, it is true for all neighbours. The stabilizer generator for qubits i, 1, $1'$, and j can be written as

$$S_i = \mathcal{A}\, X_i Z_1$$
$$S_1 = \mathcal{B}\, Z_i X_1$$
$$S_{1'} = X_{1'} Z_j\, \mathcal{C}$$
$$S_j = Z_{1'} X_j\, \mathcal{D}, \qquad (6.46)$$

where $\mathcal{A}$, $\mathcal{B}$, $\mathcal{C}$, and $\mathcal{D}$ are products of Pauli Z operations on the neighbouring qubits that are not i, 1, $1'$, or j. A successful application of $\mathcal{F}_{II}$ projects the two qubits onto the state $|HH\rangle + |VV\rangle$ or $|HV\rangle + |VH\rangle$. We consider here the case where the projection is onto $|HH\rangle + |VV\rangle$ and leave the alternative as an exercise. The stabilizer generators for this state are $X_1 X_{1'}$ and $Z_1 Z_{1'}$. Since the fusion gate removes the two qubits it acts upon, we have to find two new stabilizers for qubits i and j that commute with the stabilizer of the projected fusion state. We find these by multiplying $S_i S_j$ with $Z_1 Z_{1'}$, and $S_1 S_{1'}$ with $X_1 X_{1'}$:

$$S_i' = \mathcal{A}\, X_i X_j\, \mathcal{D} \quad \text{and} \quad S_j' = \mathcal{B}\, Z_i Z_j\, \mathcal{C}. \qquad (6.47)$$

We can transform these stabilizer generators into the standard form by applying a Hadamard operation to either qubit i or qubit j. Applying, say, H_j then yields the new generators

$$\bar{S}_i = \mathcal{A}\, X_i Z_j\, \mathcal{D} \quad \text{and} \quad \bar{S}_j = \mathcal{B}\, Z_i X_j\, \mathcal{C}. \qquad (6.48)$$

This means that $\mathcal{F}_{II}$ creates a bond between qubit i and qubit j. When there are multiple neighbours of qubit 1 and $1'$, one of the neighbouring qubits k inherits all the links with the neighbours of qubits 1 and $1'$, while the other qubits inherit both their own neighbours *and* all the neighbours from k.

Exercise 6.10: Construct the stabilizer generators $\bar{S}_i$ and $\bar{S}_j$ when $\mathcal{F}_{II}$ projects onto $|HV\rangle + |VH\rangle$.

Assuming again perfect photodetectors, the type-II fusion gate fails when we find two photons in the same output mode of the polarizing beam splitter. This can happen only when one photon was in the polarization state $|D\rangle$ and the other in $|A\rangle$. Which detector finds the two photons determines which photon was in $|D\rangle$ and which one was in $|A\rangle$. A failed $\mathcal{F}_{II}$ gate therefore acts as an X measurement on each photon. From Chapter 2 we recall that an X measurement of a qubit in a cluster does *not* break the bonds but rearranges the bonds among the neighbours of the measured qubit. In particular, in a linear cluster the X measurement shortens the cluster by two qubits, while redundantly encoding one of the neighbours of the measured qubit. When we try to fuse clusters, the gate failure of $\mathcal{F}_{II}$ is therefore considerably less catastrophic than gate failure of $\mathcal{F}_I$. Linear-optical quantum computing now proceeds by creating the necessary cluster states with fusion gates, and performing single-qubit measurement on the photons. These measurements are polarization measurements (although any dual-rail representation works), and can be performed deterministically, up to detector inefficiencies. In Section 6.4 we consider finite-detection efficiencies.

6.3.3 Photonic quantum computing in the circuit model

The one-way model of quantum computing using cluster states is a very elegant way to perform quantum computation with probabilistic entangling gates, but it is not the only way. When we choose the right encoding of a logical qubit, we can still do quantum computing in the circuit model. The idea behind this model is to make a large multi-qubit encoding for each logical qubit, such that the fusion gates can be used to implement probabilistic gates without destroying the logical qubit when the gate fails. In this section we will introduce the encoding that makes this possible, and then we construct the necessary circuit elements.

The computational basis states for the logical qubits are defined as n-qubit states

$$|0\rangle^{(n)} = \frac{|+\rangle^{\otimes n} + |-\rangle^{\otimes n}}{\sqrt{2}} \quad \text{and} \quad |1\rangle^{(n)} = \frac{|+\rangle^{\otimes n} - |-\rangle^{\otimes n}}{\sqrt{2}}. \tag{6.49}$$

These are even and odd 'parity states', respectively. The parity of a bit string is even when the string contains an even number of 1s, and it is odd if the string contains an odd number of 1s. Classically, a bit string always has a well-defined parity. In quantum mechanics, however, qubit states can be superpositions of different parity states (for example, the three-qubit GHZ state $|000\rangle + |111\rangle$ does not have a well-defined parity).

Exercise 6.11: Show that $|0\rangle^{(n)}$ is a uniform superposition of all even parity qubit strings, and that $|1\rangle^{(n)}$ is a uniform superposition of all odd parity qubit strings.

Exercise 6.12: Construct an operator that corresponds to the parity observable for n qubits.

Having defined $|0\rangle^{(n)}$ and $|1\rangle^{(n)}$ as even and odd parity states, we can introduce a recursive construction of the states:

$$|0\rangle^{(n+1)} = \frac{|0\rangle^{(n)} |0\rangle + |1\rangle^{(n)} |1\rangle}{\sqrt{2}} \quad \text{and} \quad |1\rangle^{(n+1)} = \frac{|0\rangle^{(n)} |1\rangle + |1\rangle^{(n)} |0\rangle}{\sqrt{2}}. \tag{6.50}$$

From this construction it is immediately clear that a measurement in the computational basis $\{|0\rangle, |1\rangle\}$ of one qubit will not destroy the encoding, but instead will reduce the size n of the encoding by one. For example, when a qubit is in a state $\alpha |0\rangle^{(n+1)} + \beta |1\rangle^{(n+1)}$, measurement of a qubit in the computational basis would give $\alpha |0\rangle^{(n)} + \beta |1\rangle^{(n)}$ for outcome 0 and $\alpha |1\rangle^{(n)} + \beta |0\rangle^{(n)}$ for outcome 1. It is clear from Eq. (6.50) that a bit-flip X of a single qubit in the logical qubit encoding will induce a bit-flip $X^{(n)}$ of the logical qubit.

How do we apply single-qubit operations to these encoded logical qubits? The Pauli $Z^{(n)}$ operation must induce the unitary transformation $|0\rangle^{(n)} \rightarrow |0\rangle^{(n)}$ and $|1\rangle^{(n)} \rightarrow -|1\rangle^{(n)}$. Since $|0\rangle^{(n)}$ has an even number of 1s, and $|1\rangle^{(n)}$ has an odd number of 1s, we can make a $Z^{(n)}$ operation by applying Z operations to *all* n qubits in the encoding. The next single-qubit operation we consider is an arbitrary rotation around the x axis in the Bloch sphere of the encoded qubit. Such a rotation is generated by

$$U_X^{(n)}(\theta) = \cos\theta \, \hat{\mathbb{1}} - i \sin\theta \, X^{(n)} \,. \tag{6.51}$$

We have already seen that a bit-flip X on a single qubit in the encoding acts as a bit-flip $X^{(n)}$ on the encoded qubit. We can therefore implement an arbitrary rotation around the x axis by performing a single-qubit rotation on any of the qubits in the encoding:

$$U_X(\theta) = \cos\theta \, \hat{\mathbb{1}} - i \sin\theta \, X \,. \tag{6.52}$$

The single-qubit gates $Z^{(n)}$ and $U_X^{(n)}(\theta)$ are not universal for quantum computing. In order to generate a universal set of gates we have to add two more gates, for example the single-qubit phase gate $\Phi = U_Z^{(n)}(\pi/4)$ and the two-qubit CX gate. However, these cannot be implemented deterministically with linear-optical elements, and we therefore have to construct them from the fusion operators. It is not intuitively obvious that the fusion gates should be capable of this, but we will show that it is indeed the case by explicit construction.

Before we construct the two remaining gates, we need to understand better how the fusion gates act on the encoded qubits. We use the variation $\mathcal{F}'_{\text{II}}$ instead of $\mathcal{F}_{\text{II}}$. The gate $\mathcal{F}'_{\text{II}}$ works exactly like $\mathcal{F}_{\text{II}}$, but in the basis $\{D, A\}$, rather than $\{H, V\}$. The main practical difference between the two gates is that the failure mode of $\mathcal{F}'_{\text{II}}$ amounts to a computational basis (Z) measurement $\{H, V\}$, rather than a measurement in the X basis. Whereas a Z measurement is bad news for cluster-state generation, an X measurement of a single photon collapses the parity state $|0\rangle^{(n)}$, as can be seen from Eq. (6.49), and is therefore catastrophic. The action of a successful $\mathcal{F}'_{\text{II}}$ is

$$\mathcal{F}'_{\text{II}} : \begin{cases} \text{detector outcome } (D, D) \text{ or } (A, A): \langle HH| + \langle VV| \,, \\ \text{detector outcome } (D, A) \text{ or } (A, D): \langle HH| - \langle VV| \,, \end{cases} \tag{6.53}$$

while failure results in a measurement in the computational basis. This reduces the encoding from size n to $n - 1$.

How do we create the parity encoding of the logical qubit? We first use the type-I and type-II fusion gates $\mathcal{F}_{\text{I}}$ and $\mathcal{F}_{\text{II}}$ to create the four-photon GHZ state from the previous section, given by $|HHHH\rangle + |VVVV\rangle \equiv |0000\rangle + |1111\rangle$. When we apply a Hadamard operation to each photonic qubit, we obtain the state $|+{+}{+}{+}\rangle + |-{-}{-}{-}\rangle = |+\rangle^{\otimes 4} + |-\rangle^{\otimes 4}$. (Incidentally, this state can also be created with $\mathcal{F}'_{\text{II}}$, so only one version of the type-II gate is required for the circuit model protocol.) By definition, this is the logical computational basis state $|0\rangle^{(4)}$. We can make smaller states by measuring photons in the $\{H, V\}$ basis and applying a bit flip to one photon in the remaining state, depending on the measurement outcome. Next, we can again use Eq. (6.50) to show that a successful type-II fusion gate $\mathcal{F}'_{\text{II}}$ applied to one photon from each logical qubit $|0\rangle^{(n+1)}$ and $|0\rangle^{(m+1)}$ creates the state

$$\mathcal{F}'_{\text{II}} : \; |0\rangle^{(n+1)} |0\rangle^{(m+1)} \rightarrow |0\rangle^{(n+m)} \,. \tag{6.54}$$

Depending on the detector outcome, it may be necessary to apply a phase flip $Z^{(n+m)}$ to the photons in $|0\rangle^{(n+m)}$. Similarly, when the input photons of $\mathcal{F}'_{\text{II}}$ are part of the logical qubits $|1\rangle^{(n+1)}$ and $|0\rangle^{(m+1)}$, a successful type-II fusion operation creates $|1\rangle^{(n+m)}$. We can summarize these two results by defining an arbitrary pure encoded qubit

$$|\psi\rangle^{(n+1)} = \alpha |0\rangle^{(n+1)} + \beta |1\rangle^{(n+1)} \,. \tag{6.55}$$

The successful fusion operator $\mathcal{F}'_{\text{II}}$ then creates

$$\mathcal{F}'_{\text{II}} : \ |\psi\rangle^{(n+1)} |0\rangle^{(m+1)} \to |\psi\rangle^{(n+m)} \,, \tag{6.56}$$

subject to a phase-flip $Z^{(n+m)}$ given the measurement outcome.

Exercise 6.13: Verify Eq. (6.56).

We now show how we can create the phase gate $\Phi^{(n)}$. The entire protocol is shown in Fig. 6.12. First, we apply Φ to a single qubit in an $n+1$ encoded qubit $|\psi\rangle^{(n+1)}$; second, we prepare another encoded qubit in the state $|0\rangle^{(m+1)}$ and we send one of its qubits into $\mathcal{F}'_{\text{II}}$, together with the qubit we applied the Φ gate to; finally, we perform a parity measurement (which is a computational basis measurement in this encoding) and apply a corrective bit-flip and/or phase shift to the output qubit. The output state is then $\Phi^{(m)} |\psi\rangle^{(m)}$.

When the fusion gate fails, the participating qubits are measured in the computational basis and hence are removed from the encoding. The gate Φ can therefore be attempted again, but now on $|\psi\rangle^{(n)}$ with ancilla qubit $|0\rangle^{(m)}$. For sufficiently large encoding n and m, the gate succeeds before the encoding runs out of steam. We can choose the size of the ancilla $|0\rangle^{(m)}$ such that on average the output qubit has the same level of encoding as the input qubit.

The cx gate is implemented as shown in Fig. 6.13. The control and target qubits are encoded in n and m level qubits, respectively, and we require one ancilla qubit $|0\rangle^{(k)}$. First, we use a type-I fusion gate on a qubit from the control and a qubit from the ancilla. The resulting qubit is fed into an $\mathcal{F}'_{\text{II}}$ gate, together with a qubit from the target. Depending on the parity of the control qubit, a corrective X operation must be applied to the remaining ancilla and target qubits. The ancilla becomes the new control qubit. When the fusion gates fail, the encoding of the qubits is lowered but no quantum information is destroyed. The level of encoding must be chosen such that the overall failure probability falls well below the fault-tolerance threshold.

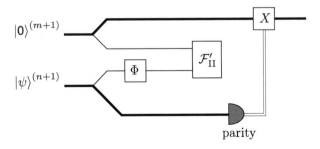

Fig. 6.12. Using the fusion gate $\mathcal{F}'_{\text{II}}$ and the phase gate Φ to create the $\Phi^{(n)}$ gate.

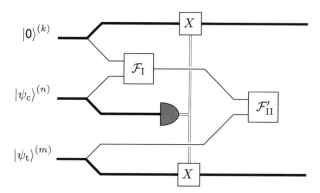

Fig. 6.13.   Using the fusion gates $\mathcal{F}_I$ and $\mathcal{F}'_{II}$ to create the cx gate.

Exercise 6.14: Verify that the Φ and cx gates operate in the manner shown in Figs. 6.12 and 6.13.

6.4 Photon counting and quantum memories

In addition to the errors that occur due to the natural failure modes of the fusion gates, there are also errors in the computation that arise because of noise or other physical imperfections, and we need to protect against these errors as well. One of these imperfections that is typical for implementations using single photons is loss, either due to inefficient sources or detectors.

6.4.1 The tree encoding

We look first at protection from photon loss in the one-way model of quantum computation with single photons and linear-optical elements. The situation is initially as shown in Fig. 6.14a. Suppose that a particular qubit (the middle bottom) must be measured in the basis $M(\pm\alpha)$ as part of a quantum algorithm. However, there will be some probability that the photon carrying the qubit will be lost, due to source or detection inefficiency. Since $M(\pm\alpha)$ is part of the algorithm,[2] the actual sign of α is not determined until just before the measurement is made. We cannot therefore perform this measurement at the cluster state-preparation stage, but we must do the loss protection in-line.

To protect against photon loss, we will 'plant a tree': we will create a (large) tree structure at the site of the qubit that is to be measured in $M(\pm\alpha)$, as shown in Fig. 6.14b. The trunk of the tree consists of two qubits, both of which will be measured in the X basis at the state-preparation stage. A highly connected cluster state results, shown in Fig. 6.14c. This encoding must be repeated for every qubit in the cluster, which will generally lead to an

---

[2] For brevity, we will denote both the basis and the observable by $M(\pm\alpha)$.

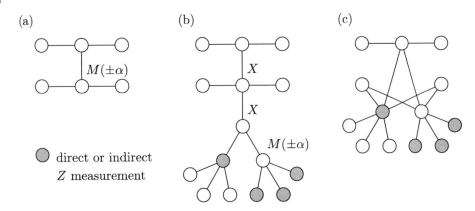

(a)

$M(\pm\alpha)$

⬤ direct or indirect
Z measurement

(b)

X

X

$M(\pm\alpha)$

(c)

Fig. 6.14. Tree encoding for protection against photon loss. (a) A particular algorithm requires that we measure a certain qubit in the cluster in the basis $M(\pm\alpha)$. The photon carrying the qubit may be lost. (b) We replace the qubit by a tree, and perform the two X measurements at the state-preparation stage. (c) The encoded cluster is now highly connected. If the measurement $M(\pm\alpha)$ in (b) is successful, then Z measurements of the shaded qubits will yield the cluster in (a). If $M(\pm\alpha)$ fails due to photon loss, we can attempt it again on the other qubit on the same level in the tree. The lost qubit can still be projected onto the computational basis by measuring the next level in the tree.

even higher degree of connectivity. Here, we will consider the tree encoding of only this one qubit, in order to keep the figures readable. Since the two adjacent X measurements occur at the state-preparation process, we can assume that they are always successful, and below we will freely switch between Fig. 6.14b and 6.14c when discussing the tree encoding.

When the $M(\pm\alpha)$ measurement in Fig. 6.14b succeeds, the shaded qubits must be measured in the Z basis. Recalling that a Z measurement breaks all the bonds, it is easy to see that successful Z measurements turn Fig. 6.14c back into 6.14a. When the $M(\pm\alpha)$ measurement in Fig. 6.14b fails, we can retry on another qubit at the same (horizontal) level. We may have several branches coming off the lowest qubit labelled X, which gives us many opportunities to get the $M(\pm\alpha)$ measurement done. However, every lost photon will cause local decoherence in the cluster. Somehow, we have to measure the lost qubit in the computational basis (the Z basis), and this is where the rest of the tree comes in.

We can perform an *indirect* Z measurement when we find that a qubit is lost. Consider Fig. 6.15, and suppose that the black qubit is lost. We wish to retroactively measure this qubit in the computational basis. This seems an impossible task, until we realize that a successful measurement of the stabilizer generator associated with a qubit directly below the lost qubit (qubit 1 in Fig. 6.15) will determine the computational basis state of the lost qubit. The cluster state $|\Psi\rangle$ was prepared as an eigenstate of stabilizer generators, one of which is

$$S_1 = Z_{\text{lost}}X_1Z_2Z_3Z_4 \quad \text{with} \quad S_1|\Psi\rangle = |\Psi\rangle . \tag{6.57}$$

This equation means that if we measure X on qubit 1 and Z on its neighbours (including the not-yet-lost photon), the product of the five measurement outcomes (±1) will be $+1$. Now

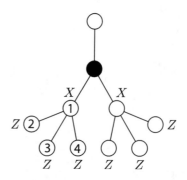

Fig. 6.15. Suppose the black qubit is lost. Its value in the computational basis can be inferred by measuring the remainder of the stabilizer generator of a neighbouring qubit.

suppose the photon is lost. An X measurement of qubit 1, together with Z measurements of qubits 2–4 will give outcomes ± 1. When we multiply these outcomes, we obtain either $+1$ or -1. Assuming no other errors have occurred, this means that a Z measurement of the lost qubit would have yielded $+1$ or -1, respectively, perfectly correlated with the actual measurement outcomes. The tree structure allows us to make a *counterfactual measurement* of the lost photon! This is very counterintuitive, and a direct consequence of the entanglement that was present in the cluster state before the photon was lost.

We have just seen that we can do an indirect Z measurement on a particular photon, even when it is lost. However, the measurement involves several other photon-detection events, any of which may fail due to photon loss. However, only one of the X measurements in Fig. 6.15 must succeed, because the remaining (lower) Z measurements connected to the successfully X-measured qubit can again be either direct or indirect measurements. This leads to contradictory requirements on the tree structure: on the one hand, we want many qubits at the level of the $M(\pm\alpha)$ and X measurements, so we can repeat the measurement often, and photon loss may be high. On the other hand, we want to minimize the number of (direct or indirect) Z measurements we need to do. It turns out that the maximum permissible photon loss is $\eta = 50\%$. Physically, this limit can be understood as follows: suppose that the photons are not lost, but are captured by your evil twin. If you and your twin can both do the quantum computation with half of the qubits, you can do the computation twice. This can be used to violate the no-cloning theorem, since the two systems can generate perfect identical output states based on a single (possibly unknown) input state of the computer. Therefore, it is impossible to recover from photon loss if the losses are 50% or more. Numerical simulations have shown that the total number of qubits in a tree scales as a polylog function of the loss rate. For more details, see Varnava *et al.* (2006).

When the photon loss becomes considerable (but still less than 50%), the tree structure that is necessary to protect against loss becomes very large. Consequently, phase errors become important, and the trees lose their advantage. Other error-correction codes must therefore complement the tree encoding. Nevertheless, the tree encoding is a wonderful example of the counterintuitive power of cluster states and the stabilizer formalism.

6.4.2 The redundant encoding

In Section 6.3, we showed how a parity encoding, together with two fusion operators $\mathcal{F}_I$ and $\mathcal{F}'_{II}$, can in principle be used to perform quantum computing in the circuit model. In this section, we show how an extra layer of encoding, the so-called redundant encoding, can be used to protect the quantum computation from photon loss. Again, we assume that there are no other errors. First, we determine how to create the redundant encoding with fusion gates. Second, we show how to encode a universal set of quantum gates in this encoding, and finally, we discuss the loss threshold in this error model.

We have already encountered a three-qubit redundant encoding in Chapter 2, where we called it a 'majority code': $|\bar{0}\rangle = |000\rangle$ and $|\bar{1}\rangle = |111\rangle$. Here, we assume that the basic qubits are parity-encoded (see Eq. (6.49)), which yields the redundant encoding

$$|\bar{0}\rangle = |0\rangle_1^{(n)} \otimes \cdots \otimes |0\rangle_q^{(n)} \equiv |0\rangle^{(n,q)} \quad \text{and}$$

$$|\bar{1}\rangle = |1\rangle_1^{(n)} \otimes \cdots \otimes |1\rangle_q^{(n)} \equiv |1\rangle^{(n,q)} . \tag{6.58}$$

A logical qubit can then be written as

$$|\psi\rangle^{(n,q)} = \alpha |0\rangle^{(n,q)} + \beta |1\rangle^{(n,q)} . \tag{6.59}$$

This state can be constructed efficiently from $|\psi\rangle^{(n)}$ (see Eq. (6.55)) with the following resource:

$$|0\rangle^{(n,q)} + |1\rangle^{(n,q)} = |0\rangle_1^{(n)} \otimes \cdots \otimes |0\rangle_q^{(n)} + |1\rangle_1^{(n)} \otimes \cdots \otimes |1\rangle_q^{(n)} . \tag{6.60}$$

We can obtain the redundantly encoded qubit $|\psi\rangle^{(n,q)}$ from $|\psi\rangle^{(n)}$ and this resource state by applying a fusion gate $\mathcal{F}'_{II}$ to a photon in $|\psi\rangle^{(n)}$ and a photon in the resource state. After a successful fusion operation (which may take several attempts), the photons supporting the original qubit $|\psi\rangle^{(n)}$ are subjected to a parity measurement[3]

$$|\psi\rangle^{(n)} |0\rangle^{(n,q)} = \left[\alpha \left(|0\rangle^{(n)} |0\rangle + |1\rangle^{(n)} |1\rangle \right) + \beta \left(|0\rangle^{(n)} |1\rangle + |1\rangle^{(n)} |0\rangle \right) \right]$$

$$\otimes \left(|0\rangle |0\rangle^{(n,q-1)} + |1\rangle |1\rangle^{(n,q-1)} \right)$$

$$\xrightarrow[\mathcal{F}'_{II}]{} |0\rangle^{(n)} \left(\alpha |0\rangle^{(n,q-1)} + \beta |1\rangle^{(n,q-1)} \right)$$

$$+ |1\rangle^{(n)} \left(\alpha |1\rangle^{(n,q-1)} + \beta |0\rangle^{(n,q-1)} \right). \tag{6.61}$$

Depending on the outcome, a bit-flip in the redundant encoding will be applied, and the state $|\psi\rangle^{(n,q)}$ is created. The resource state in Eq. (6.60) can be created from $|0\rangle_1^{(n)} \otimes \cdots \otimes |0\rangle_q^{(n)}$ by applying a Hadamard on the first parity-encoded qubit, followed by $q-1$ cx gates between qubit 1 and the remaining qubits.

---

[3] The notation suggests that each parity qubit has a level n encoding, but this is not necessary. Each parity qubit i, with $i \in \{1, \ldots, q\}$ can have a different level of encoding n_i. However, so as not to burden the notation too much, we will use the superscript (n,q), where n is understood as a vector $n = (n_1, \ldots, n_q)$.

Exercise 6.15: Construct the Pauli operators in the parity encoding and the redundant encoding. How many single-photon operations are necessary to implement each operator?

How do we apply a universal set of quantum gates at the level of the redundant encoding? In the parity encoding we constructed the Pauli gates, arbitrary rotations around the x axis in the Bloch sphere $U_X^{(n)}(\theta)$, the phase gate $\Phi^{(n)}$, and the CX gate. These last two are probabilistic gates. In the redundant encoding, on the other hand, we construct the universal set consisting of the Pauli gates, the CX gate, a rotation $U_X^{(n,q)}(\pi/4)$, and arbitrary rotations around the z axis $U_Z^{(n,q)}(\varphi)$

$$U_X^{(n,q)}\left(\frac{\pi}{4}\right) = \frac{1}{\sqrt{2}}\begin{pmatrix} 1 & -i \\ -i & 1 \end{pmatrix} \quad \text{and} \quad U_Z^{(n,q)}(\varphi) = \begin{pmatrix} e^{-i\varphi} & 0 \\ 0 & e^{i\varphi} \end{pmatrix}. \tag{6.62}$$

The gate $U_Z^{(n,q)}(\varphi)$ is shown in Fig. 6.16. The bundle of fat lines denotes the redundant encoding. The auxiliary state is given by $|0\rangle^{(n+1)}$. As before, the detection of the parity-encoded qubit in the figure is a parity measurement, which comprises a computational basis measurement of all photons in the computational basis. When a single photon is lost in this measurement, the parity of the qubit is unknown. The redundant encoding ensures that the measurement can be repeated on another parity qubit in the code.

The gate $U_X^{(n,q)}(\frac{\pi}{4})$ is shown in Fig. 6.17, where the auxiliary state is given by $|\text{aux}\rangle^{(n,q)} = |0\rangle^{(n,q)} + |1\rangle^{(n,q)}$. The single-qubit operation is performed on a single photon in the redundant encoding, after which a fusion gate is applied that links the photon to the auxiliary state. The parity qubit associated with the single-photon operation is measured with a standard parity measurement, while the rest is measured in the $X^{(n,q)}$ basis $\{|+\rangle^{(n,q)}, |-\rangle^{(n,q)}\}$. At the single-photon level, the diagonal basis amounts to a successful measurement of a single photon in the X basis in *each* parity encoding. The encoding n must be large enough to allow the diagonal measurement, but not so large that the success probability of a parity measurement becomes too small.

Finally, we consider the CX gate between two redundantly encoded qubits. This is simply a series of q CX gates on the parity encoding that forms the basis of the redundant encoding. When photon loss occurs, we remove that particular parity qubit by a measurement in the $X^{(n)}$ basis.

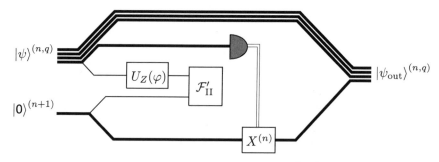

Fig. 6.16. Using the fusion gate $\mathcal{F}_{\text{II}}'$ and a single-photon qubit rotation $U_Z(\varphi)$ to create the gate $U_Z^{(n,q)}(\varphi)$ in the redundant encoding.

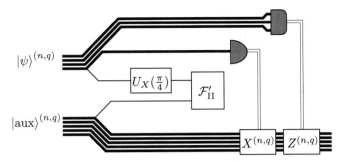

Fig. 6.17. Using the fusion gate $\mathcal{F}'_{II}$ and a single-photon qubit rotation $U_X(\frac{\pi}{4})$ to create the gate $U_X^{(n,q)}(\frac{\pi}{4})$ in the redundant encoding. The auxiliary state is $|\text{aux}\rangle^{(n,q)} = |0\rangle^{(n,q)} + |1\rangle^{(n,q)}$.

As was the case for the tree encoding, there are conflicting requirements in the parity and redundant encodings. On the one hand, we need a large n to ensure that the $X^{(n,q)}$ measurement is nearly deterministic, but at the same time n must be small enough for a parity measurement to succeed with considerable success probability. This translates again into a maximum loss probability. Several thresholds for different situations have been established numerically. For example, the loss threshold for simple propagation is about 18% for the photon sources, detectors, and fibre attenuation. When one or more components have lower loss (e.g. sources and attenuation), the remaining components (in this case the detectors) typically have higher loss thresholds. Similarly, the loss threshold for the cx gate is about 10%. For more details, see Hayes *et al.* (2008).

6.4.3 Quantum memory for single photons

In both the cluster-state and the circuit model of linear-optical quantum computation with single photons, an essential part of the protocol is the adaptation of subsequent qubit operations depending on measurement outcomes. While the measurements are being performed, the remaining photons that will be subjected to gates depending on the measurement outcomes must be stored somewhere. This means that we will always need a quantum memory to store the single-photon qubit.

The most common quantum memory is a delay line in free space or in an optical fibre. While practical at the level of proof-of-principle experiments, the space requirements of delay lines mean that they will not be the solution for a large-scale optical quantum computer. It is more likely that a scalable delay line will take the form of a microcavity with a high Q-factor (see, e.g., Armani *et al.*, 2003). Alternatively, single-photon qubits could be stored in atoms in cavities, or in atomic ensembles. We will return to this topic in Chapters 7 and 10.

6.5 Threshold theorem for linear-optical quantum computing

In practice, photon loss is by far the dominant error in optical implementations of quantum computing. Other errors, such as dephasing and depolarization, occur much less frequently.

However, when a sufficiently large number of logical operations are made, the probability of a depolarization (or dephasing) error becomes arbitrarily close to one, and we have to protect our computation against these errors. In particular, tree structures and the redundant encoding require a very large number of photons per logical qubit when the loss rate is appreciable. In the tree encoding, an error rate of 1% per gate with detectors of $\eta = 80\%$ efficiency requires between 200 and 300 qubits in a tree. Consequently, depolarization and dephasing are no longer negligible, and we need to find a threshold for a more general error model.

We consider the cluster-state model introduced in Section 6.3.2. This model uses two-photon maximally entangled Bell states as input states, so we assume that we have a suitable source for these states. Furthermore, we can apply deterministic single-qubit operations in the form of linear-optical elements, and we have some kind of single-photon quantum memory. We can also apply the two fusion gates $\mathcal{F}_\mathrm{I}$ and $\mathcal{F}_\mathrm{II}$, which operate in a probabilistic fashion. Finally, we assume that we have polarization-sensitive photodetectors that can distinguish between zero, one, and two photons. This may seem a departure from the realistic photodetectors that indicate whether or not any photons were present (the so-called bucket detectors), but at this time no threshold for optical cluster-state quantum computing with bucket detectors is known. Since bucket detectors provide less information than the simple number-resolving photodetectors considered here, the threshold is expected to be worse with bucket detectors.

The computation proceeds in time steps of length τ, and each qubit is subjected to exactly one operation per time step. This can be a single-qubit operation, a fusion gate, detection, or just a single-photon memory operation in which no logical operation is applied to the qubit. The error model is characterized by two parameters, γ and ϵ. The loss probability per photon per cycle τ is denoted by γ, and it occurs at each step in the computation. After the Bell-state preparation, the loss channel is applied to both qubits independently. Similarly, the loss channel is applied before single-qubit operations, the memory channel, and before the photon enters a fusion gate. Finally, the loss channel is applied just before photodetection. Clearly, γ is used to parameterize not only the detector efficiency, but also memory loss and any other losses.

The second parameter ϵ characterizes the probability of a single depolarizing error. It can occur both in the single-qubit components (single-qubit gates, quantum memory and photon measurement), as well as in the two-qubit components (Bell-state preparation and fusion gates). In the single-qubit components, the depolarizing channel introduces a random Pauli gate X, Y, or Z with probability $\epsilon/3$, while the probability of no error is $1 - \epsilon$. The Kraus operators for the single-qubit components are therefore

$$A_0 = \sqrt{1 - \epsilon}\,\hat{\mathbb{I}} \quad \text{and} \quad A_i = \sqrt{\frac{\epsilon}{3}}\,\sigma_i, \tag{6.63}$$

with $\sigma_i \in \{X, Y, Z\}$. The two-qubit components are subjected to depolarizing noise with total probability ϵ as well: after the Bell-state preparation and before a fusion gate, with probability $(1 - \epsilon)$ no error will occur, and with probability $\epsilon/15$ a two-qubit error such as

$X \otimes \hat{\mathbb{I}}$ or $X \otimes Y$ will occur. The Kraus operators for the two-qubit components are

$$A_{00} = \sqrt{1-\epsilon}\,\hat{\mathbb{I}} \otimes \hat{\mathbb{I}} \quad \text{and} \quad A_{ij} = \sqrt{\frac{\epsilon}{15}}\,\sigma_i \otimes \sigma_j \quad (i,j) \neq (0,0)\,. \qquad (6.64)$$

It was shown numerically by Dawson *et al.* (2006) that the fault-tolerant threshold for this error model is $\gamma < 3 \cdot 10^{-3}$ and $\epsilon < 10^{-4}$, comparable to thresholds for other models of quantum computation. Recently, Raussendorf and Harrison (2007) proved a general threshold theorem for two-dimensional cluster states, allowing an error probability of 0.75%.

6.6 References and further reading

In this chapter, we showed how to implement a quantum computer with linear-optical elements and single-photon qubits. There remain many challenges in this approach, mainly related to the relatively low success probabilities of the entangling gates, and the high photon-loss rates in sources and detectors, compared to fault-tolerant thresholds. In addition, due to the feed-forward nature of post-selection quantum gates and the natural speed of propagation of photons, any quantum computer based on linear optics requires quantum memories. We will return to this issue in Chapters 7 and 10.

The classical simulation of quantum computing with linear optics was considered by Cerf *et al.* (1998), and Adami and Cerf (1999), before it was proved that scalable quantum computing is possible with linear optics, single photons, and photodetection (Knill, Laflamme, and Milburn, 2001) (KLM). See Kok *et al.* (2007) for a general review on linear-optical quantum computing with photonic qubits. Many probabilistic two-photon gates have been proposed, most notably by KLM, but also by Ralph *et al.* (2002a) and Rudolph and Pan (2001). The general theory of multi-mode unitary gates was developed by Scheel *et al.* (2003), Lapaire *et al.* (2003), and van Loock and Lütkenhaus (2004). Many gates have been demonstrated experimentally, for example the two-photon CX gate proposed by Ralph *et al.* (2002b) was implemented by O'Brien *et al.* (2003), and a four-photon cluster state was created by Walther *et al.* (2005). Many more experiments have since continued to improve the technology of linear-optical quantum information processing with single photons and photodetection. The maximum success probability of two-photon gates was studied by Knill (2003), Scheel and Lütkenhaus (2004), and Eisert (2005). The evolution of linear-optical quantum computing with single photons can be traced from the KLM protocol and the introduction of the fusion gates by Pittman *et al.* (2001), via Yoran and Reznik (2003), and Nielsen (2004), to the optical cluster-state generation with fusion gates due to Browne and Rudolph (2005), and the circuit model due to Gilchrist *et al.* (2007). Photon loss tolerance using tree structures in cluster states was introduced by Varnava *et al.* (2006), and loss tolerance in the circuit model was constructed by Hayes *et al.* (2008). Finally, fault-tolerant thresholds were derived by Dawson *et al.* (2006), and Raussendorf and Harrington (2007).

7

Atomic quantum information carriers

We have so far spoken almost exclusively of photons and linear-optical elements, and seen just how powerful those two components can be for information processing. They provide unbreakable cryptographic tools, and allow for efficient quantum computing. However, many more possibilities become available when we allow photons to interact with atoms and solid matter in a quantum mechanical way. In particular, a quantum memory, the principal difficulty for linear-optical quantum computing, can be created. In this chapter we will take the first steps towards a full understanding of a photon's interaction with atoms. We will show how to describe the interactions within a system consisting of photons and few-level atoms and show how this interaction can be manipulated and exploited to provide quantum information processors based on both atomic and photonic qubits. We will also show that photon emission from atoms can degrade the quantum information contained within atoms, and we will present a formalism to model this effect. We begin with a general discussion of atom–photon interactions.

7.1 Atomic systems as qubits

Let us first consider an electron in an isolated atom. It is bound there by the Coulomb force due to the charge distribution of all the other electrons and the nucleus. The potential that describes this coupling is given by $V(\mathbf{r})$. The electron will of course also have kinetic energy, so that the time-independent Schrödinger equation that determines its wave function $\psi(\mathbf{r})$ takes the form

$$\mathcal{H}\,\psi(\mathbf{r}) = \left[-\frac{\hbar^2}{2m}\nabla^2 + V(\mathbf{r}) \right] \psi(\mathbf{r}), \tag{7.1}$$

where m is the mass of the electron. We would like to calculate how this atomic electron is affected by the application of an electromagnetic field. As defined in Eq. (1.1), the electric and magnetic fields can be described by a vector potential operator $\mathbf{A}(\mathbf{r}, t)$ and a scalar potential operator $\Phi(\mathbf{r}, t)$:

$$\mathbf{E}(\mathbf{r}, t) = -\nabla\Phi(\mathbf{r}, t) - \frac{\partial\mathbf{A}(\mathbf{r}, t)}{\partial t}, \tag{7.2}$$

$$\mathbf{B}(\mathbf{r}, t) = \nabla \times \mathbf{A}(\mathbf{r}, t). \tag{7.3}$$

Given the vectorial character of the electromagnetic field, the Schrödinger equation is no longer of the form in Eq. (7.1). The momentum operator is modified to include a term proportional to the vector potential operator, $-i\hbar\nabla \rightarrow -i\hbar\nabla + e\mathbf{A}(\mathbf{r}, t)$, and the potential $V(\mathbf{r})$ gains a term involving the scalar potential operator $V(\mathbf{r}) \rightarrow V(\mathbf{r}) - e\Phi(\mathbf{r}, t)$. The Hamiltonian can then be written as

$$\mathcal{H}_e = \frac{1}{2m}\left[-i\hbar\nabla - e\mathbf{A}(\mathbf{r}, t)\right]^2 + V(\mathbf{r}) - e\Phi(\mathbf{r}, t). \tag{7.4}$$

This Hamiltonian of an electron in a general electromagnetic field is exact. We can simplify it by choosing a convenient gauge for $\mathbf{A}$ and Φ. Since we will be interested in optical interactions, we choose again the Coulomb gauge, in which $\nabla \cdot \mathbf{A} = 0$ and $\Phi = 0$. This leads to

$$\mathcal{H}_e = -\frac{\hbar^2}{2m}\nabla^2 + \frac{ie\hbar}{m}\mathbf{A} \cdot \nabla + \frac{e^2}{2m}A^2 + V, \tag{7.5}$$

where we have suppressed the functional dependence of $\mathbf{A}$ and V on $\mathbf{r}$ and t in our notation, and $A^2 \equiv |\mathbf{A}|^2$.

We will be interested in the regime where the field is not too strong, which allows us to neglect the term proportional to A^2. In addition, we may use the time-dependent form of the momentum operator $-i\hbar\nabla \rightarrow m\,d\mathbf{r}/dt$, yielding

$$\mathcal{H}_e \approx -\frac{\hbar^2}{2m}\nabla^2 - e\,\mathbf{A}(\mathbf{r}, t) \cdot \frac{d\mathbf{r}}{dt} + V(\mathbf{r}). \tag{7.6}$$

We now make the so-called dipole approximation, i.e., that the electromagnetic field has no spatial variation across the extent of our atom. Since optical frequencies have wavelengths between 400 and 700 nm, this is an extremely good approximation for atoms with a typical size of 0.1 to 1 nm. This assumption allows us to choose a specific form for our vector potential in the Coulomb gauge

$$\mathbf{A}(\mathbf{r}, t) = \frac{\mathbf{B}(t) \times \mathbf{r}}{2}. \tag{7.7}$$

The second term in the Hamiltonian in Eq. (7.6) can then be rewritten using

$$\mathbf{A}(\mathbf{r}, t) \cdot \frac{d\mathbf{r}}{dt} = \frac{d(\mathbf{A}(\mathbf{r}, t) \cdot \mathbf{r})}{dt} - \frac{d\mathbf{A}(\mathbf{r}, t)}{dt} \cdot \mathbf{r}, \tag{7.8}$$

which, using Eq. (7.7) becomes

$$\mathbf{A}(\mathbf{r}, t) \cdot \frac{d\mathbf{r}}{dt} = -\frac{d\mathbf{A}(\mathbf{r}, t)}{dt} \cdot \mathbf{r}. \tag{7.9}$$

By definition, the negative time derivative of the vector potential is the electric field, and our final, approximate Hamiltonian for an electron in an electromagnetic field is

$$\mathcal{H}_e = -\frac{\hbar^2}{2m}\nabla^2 + e\,\mathbf{E}(t) \cdot \mathbf{r} + V(\mathbf{r}). \tag{7.10}$$

The electric dipole operator for an electron is defined as $\mathbf{p} = -e\mathbf{r}$, and so the principal effect of the application of a field to an atomic electron is the interaction of the electric dipole with the electric field, which modifies the isolated Hamiltonian through the term

$$\mathcal{H}_d = -\mathbf{E}(t) \cdot \mathbf{p}. \tag{7.11}$$

In the next section, we will explore the effect of optical radiation on atoms, which can be calculated by choosing an oscillatory form for $\mathbf{E}$.

7.1.1 Manipulating atomic states with classical light

We now consider the simplest nontrivial atomic system, with only two accessible energy levels for the electron. Both levels are eigenstates of an isolated atomic Hamiltonian such as that in Eq. (7.1), and we will label them $|g\rangle$ and $|e\rangle$, for 'ground' and 'excited' state, respectively. The frequency separation between these states is $\omega_e - \omega_g = \omega_0$. In due course we will interpret such a system as a qubit with $|g\rangle = |0\rangle$ and $|e\rangle = |1\rangle$. We are interested in the situation where the system interacts with an electromagnetic field that oscillates with (optical) frequency ω. If the optical field is propagating in direction $\mathbf{n}$, then its polarization $\boldsymbol{\epsilon}$ must be perpendicular to $\mathbf{n}$. We can describe the field by

$$\mathbf{E} = E_0 \left[\boldsymbol{\epsilon} \exp(i\omega t - i\mathbf{k} \cdot \mathbf{n}\, r) + \boldsymbol{\epsilon}^* \exp(-i\omega t + i\mathbf{k} \cdot \mathbf{n}\, r]) \right], \qquad (7.12)$$

where $r = |\mathbf{r}|$, and E_0 is the (real) amplitude of the field. Without loss of generality, we can assume that our two-level system is positioned at the origin: $|\mathbf{r}| = r = 0$. Using our definition of the dipole $\mathbf{p} \equiv -e\mathbf{r}$, the coupling matrix element M_{eg} between $|e\rangle$ and $|g\rangle$ due to the Hamiltonian in Eq. (7.11) can be written as

$$\begin{aligned}
M_{eg} &= \langle e| \mathcal{H}_{\mathrm{d}} |g\rangle \\
&= E_0 e \langle e| \mathbf{r} |g\rangle \cdot \left[\boldsymbol{\epsilon} \exp(i\omega t) + \boldsymbol{\epsilon}^* \exp(-i\omega t) \right] \\
&= e E_0\, \mathbf{r}_{eg} \cdot \boldsymbol{\epsilon}\, e^{i\omega t} + e E_0\, \mathbf{r}_{eg} \cdot \boldsymbol{\epsilon}^*\, e^{-i\omega t}.
\end{aligned} \qquad (7.13)$$

Similarly, $M_{ge} = M_{eg}^*$, and since $\mathbf{r}$ is an operator with odd parity, the diagonal matrix elements vanish: $M_{ee} = M_{gg} = 0$. Therefore, in matrix form in the basis $\{|g\rangle, |e\rangle\}$ we can write the total Hamiltonian as

$$\mathcal{H}_e(t) = \begin{pmatrix} -\hbar\omega_0/2 & e E_0 \mathbf{r}_{eg}^* \cdot (\boldsymbol{\epsilon} e^{i\omega t} + \boldsymbol{\epsilon}^* e^{-i\omega t}) \\ e E_0 \mathbf{r}_{eg} \cdot (\boldsymbol{\epsilon} e^{i\omega t} + \boldsymbol{\epsilon}^* e^{-i\omega t}) & \hbar\omega_0/2 \end{pmatrix}. \qquad (7.14)$$

It is difficult to solve the Schrödinger equation exactly for this Hamiltonian, since $\mathcal{H}_e(t)$ is time-dependent. There is, however, a powerful procedure and approximation that can be used to remove the time dependence. First, we make a unitary transformation $\tilde{U}(t)$ into a frame that rotates at the same frequency as the optical field.

$$\tilde{U}(t) = \begin{pmatrix} e^{i\omega t/2} & 0 \\ 0 & e^{-i\omega t/2} \end{pmatrix}. \qquad (7.15)$$

Since there is a time dependence to $\tilde{U}(t)$, we must go back to the Schrödinger equation to derive the Hamiltonian in this new rotating frame. In the laboratory frame, we would have written

$$\mathcal{H}_e(t) |\psi(t)\rangle = -i\hbar \frac{d}{dt} |\psi(t)\rangle. \qquad (7.16)$$

Using $\left|\psi'(t)\right\rangle = \tilde{U}(t)\left|\psi(t)\right\rangle$, or equivalently $\left|\psi(t)\right\rangle = \tilde{U}^\dagger(t)\left|\psi'(t)\right\rangle$, we find

$$\mathcal{H}_e(t)\tilde{U}^\dagger(t)\left|\psi'(t)\right\rangle = -i\hbar\tilde{U}^\dagger(t)\frac{d\left|\psi'(t)\right\rangle}{dt} - i\hbar\frac{d\tilde{U}^\dagger(t)}{dt}\left|\psi'(t)\right\rangle. \tag{7.17}$$

Multiplying from the right by $\tilde{U}(t)$ and rearranging yields

$$\left(\tilde{U}(t)\mathcal{H}_e(t)\tilde{U}^\dagger(t) + i\hbar\tilde{U}(t)\frac{d\tilde{U}^\dagger(t)}{dt}\right)\left|\psi'(t)\right\rangle = -i\hbar\frac{d\left|\psi'(t)\right\rangle}{dt}. \tag{7.18}$$

This equation has exactly the same form as our original Schrödinger equation but now with the effective Hamiltonian

$$\mathcal{H}'_e = \left(\tilde{U}\mathcal{H}_e\tilde{U}^\dagger + i\hbar\tilde{U}\frac{d\tilde{U}^\dagger}{dt}\right), \tag{7.19}$$

where we have dropped the explicit time dependence for notational simplicity. The Hamiltonian in Eq. (7.14) is therefore transformed into

$$\mathcal{H}'_e(t) = \begin{pmatrix} -\hbar(\omega_0 - \omega)/2 & eE_0\mathbf{r}^*_{eg}\cdot(\boldsymbol{\epsilon}^* + \boldsymbol{\epsilon}e^{2i\omega t}) \\ eE_0\mathbf{r}_{eg}\cdot(\boldsymbol{\epsilon} + \boldsymbol{\epsilon}^*e^{-2i\omega t}) & \hbar(\omega_0 - \omega)/2 \end{pmatrix}. \tag{7.20}$$

The off-diagonal Hamiltonian elements now have two terms. The first is time-independent and describes the component of the oscillating field that precesses in the same direction as the coherence of the atom. The other term is time-dependent and describes counter-propagation of the two precessions; it therefore oscillates rapidly. We can assume that it has little effect on the atom dynamics so long as the external field is close to resonance with the transition from the ground state to the excited state, i.e. when $(\omega - \omega_0) \ll \omega$, and $E_0 \ll \omega$. This is the 'rotating-wave approximation' (RWA), and leads to

$$\mathcal{H}'_e = \begin{pmatrix} -\frac{\hbar}{2}(\omega_0 - \omega) & eE_0(\mathbf{r}_{eg}\cdot\boldsymbol{\epsilon})^* \\ eE_0(\mathbf{r}_{eg}\cdot\boldsymbol{\epsilon}) & \frac{\hbar}{2}(\omega_0 - \omega) \end{pmatrix}. \tag{7.21}$$

We can simplify the notation by defining the 'detuning' $\nu \equiv \omega_0 - \omega$ and the 'Rabi frequency' $\Omega \equiv 2eE_0(\mathbf{r}_{eg}\cdot\boldsymbol{\epsilon})/\hbar$. The Hamiltonian then takes on the standard form

$$\mathcal{H}'_e(t) = \frac{\hbar}{2}\begin{pmatrix} -\nu & \Omega^* \\ \Omega & \nu \end{pmatrix}. \tag{7.22}$$

A full justification of the RWA and higher-order corrections to it is made later in this section using Floquet theory.

Solving the Schrödinger equation for a two-level system in the presence of a detuned field is now a simple time-independent eigenvalue problem, which we solve in the usual way. The eigenvalues are

$$\lambda_\pm = \pm\frac{\hbar}{2}\sqrt{\nu^2 + \Omega^2}, \tag{7.23}$$

with corresponding eigenvectors

$$|\lambda_+\rangle = \cos\frac{\theta}{2}|e\rangle + \sin\frac{\theta}{2}|g\rangle\,, \tag{7.24}$$

$$|\lambda_-\rangle = -\sin\frac{\theta}{2}|e\rangle + \cos\frac{\theta}{2}|g\rangle\,, \tag{7.25}$$

where the mixing angle θ is defined as

$$\theta = \arctan\left(\frac{\Omega}{\nu}\right). \tag{7.26}$$

If the field is resonant with the atomic transition, the detuning becomes zero ($\nu = 0$), and so $\theta = \pi/2$. The eigenstates $|\lambda_\pm\rangle$ are then equal superpositions of $|g\rangle$ and $|e\rangle$. These states are important and so are assigned unique symbols, namely

$$|+\rangle = \frac{1}{\sqrt{2}}(|g\rangle + |e\rangle)$$

$$|-\rangle = \frac{1}{\sqrt{2}}(|g\rangle - |e\rangle)\,, \tag{7.27}$$

with eigenvalues $\lambda_\pm = \pm\hbar\Omega/2$. Compare these states to the eigenstates of the Pauli operator X in Eq. (2.4) as equal superpositions of $|0\rangle$ and $|1\rangle$.

We can use the Bloch sphere picture to explore the time evolution, as shown in Fig. 7.1a. We know that the natural evolution of a quantum system in this picture is a precession around the axis that connects the two eigenstates. If $|g\rangle$ is the North Pole and $|e\rangle$ is the South Pole, then the two states in Eq. (7.27) are on the equator. An initial ground state therefore precesses towards the equator and eventually all the way to $|e\rangle$. This motion is repeated periodically and the resultant oscillations, which are depicted to the right of Fig. 7.1, are called 'Rabi oscillations'. If the field is slightly off resonance, such that $\nu \neq 0$, Eq. (7.25) tells us that the line connecting the eigenstates now tilts slightly away from the equator, as shown in Fig. 7.1b. In this case the oscillations become incomplete: the population does not go all the way from $|g\rangle$ to $|e\rangle$. The oscillations also have higher frequency, since the energy difference between eigenstates is larger. Finally, for a highly off-resonant field, the eigenstates start to approach $|g\rangle$ and $|e\rangle$ and the oscillations become very shallow and very fast. This situation is depicted in Fig. 7.1c.

If we now imagine that $|g\rangle$ and $|e\rangle$ represent qubit state $|0\rangle$ and $|1\rangle$ we can recast the time evolution under an optical field in terms of single-qubit gate operations. For example, if the field is resonant with the atomic transition, the atom–field Hamiltonian is simply

$$\mathcal{H}'_e = \frac{\hbar\Omega}{2}X\,, \tag{7.28}$$

with X the usual Pauli X operator. In this language, the time-evolution operator of the system becomes

$$U(t) = \exp\left(-\frac{i}{\hbar}\frac{\hbar\Omega}{2}Xt\right) = U_X\left(\frac{\Omega t}{2}\right)\,, \tag{7.29}$$

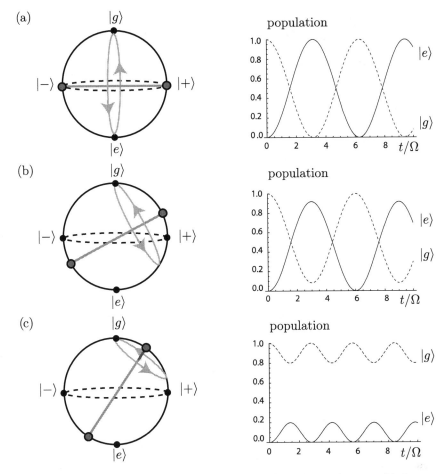

Fig. 7.1. On the left-hand side are Bloch sphere representations of the $|g\rangle$, $|e\rangle$ subspace, with the eigenstates in each of three cases shown as large grey dots. The populations of ground and excited states as a function of time are shown on the right in units of Ω. The initial state is the ground state $|g\rangle$. (a) The resonant case $v = 0$; (b) slightly off-resonant situation with $v/\Omega = 0.3$; and (c) far off-resonant case of $v/\Omega = 2$.

where $U_X(\theta)$ is the single-qubit rotation of angle 2θ around the x axis of the Bloch sphere, which we encountered first in Chapter 2 in Eq. (2.9). The angle θ can straightforwardly be varied by changing the time for which the laser is applied to the atom. Once the laser is switched off, there is no time evolution of the qubit.[1]

In Section 5.1, we found that rotations of any angle around two different axes of the Bloch sphere are needed to perform an arbitrary single-qubit rotation. The second axis could be

---

[1] Throughout this discussion we implicitly describe the qubits in the rotating frame. In the laboratory frame there would be a constant build-up of phase between the two states $|0\rangle$ and $|1\rangle$. However, this phase is always known and can be taken account of at the end of any calculation.

generated simply by detuning the optical field from resonance, but a more practical method
will be explored in the exercise below.

Exercise 7.1: In practice, it can be easier to obtain a second axis by varying the phase of
the optical field rather than its frequency. That is, an extra factor ζ enters Eq. (7.12):

$$\mathbf{E} = E_0 \left[\boldsymbol{\epsilon} \exp(i\zeta + \omega t - i\mathbf{k} \cdot \mathbf{n}\, r) + \boldsymbol{\epsilon}^* \exp(-i\zeta - \omega t + i\mathbf{k} \cdot \mathbf{n}\, r) \right] . \tag{7.30}$$

Show that changing the phase ζ of a resonant pulse rotates the axis connecting eigenstates
in the equatorial plane of the Bloch sphere by an angle ζ. When $\zeta = \pi/2$, show that the
time evolution of the qubit is equivalent to a rotation around the y axis of the Bloch sphere,
i.e., U_Y in Eq. (2.9).

A potential control problem is spontaneous emission of a photon from the excited state,
a process we will discuss in Section 7.3. Three-level systems offer an alternative possible
system that can allow us to overcome this difficulty by preventing population of optically
excited states, a topic we will return to later in this section.

7.1.2 Floquet theory and the rotating-wave approximation

In the discussion above, we made the particularly useful rotating-wave approximation,
which allowed us to cast the Hamiltonian of an interacting atom-oscillating field system
in a time-independent form. However, our justification of this step was not rigorous and
we will here show how to put it on a firmer footing, as well as derive corrections to it.
The basis of this approach was developed in the nineteenth century by Gaston Floquet,
a French mathematician who studied the properties of linear differential equations with
periodic coefficients. The Schrödinger equation

$$\mathcal{H}_e(t)\,\psi(t) = i\hbar \frac{d}{dt}\psi(t) , \tag{7.31}$$

with $\mathcal{H}_e(t)$ defined in Eq. (7.14), falls in this class of equation. Floquet showed that this
equation must have solutions of the form

$$\psi(t) = e^{-iqt} \sum_j \phi_j(t) , \tag{7.32}$$

where q is a real number and the ϕ_j are periodic functions with the same period as that of
the Hamiltonian. The label j runs over each dimension of the Hilbert space we are working
in, and so in our case it can be g or e. The periodic functions can be expanded in a Fourier
series:

$$\phi_j(t) = \sum_n \phi_j^n e^{in\omega t} . \tag{7.33}$$

Each matrix element of the Hamiltonian, labelled $\mathcal{H}_{kl}$, can also be expanded in a Fourier
series

$$\mathcal{H}_{kl}(t) = \sum_m e^{im\omega t} \mathcal{H}_{kl}^m , \tag{7.34}$$

where k and l again label the dimensions of the Hilbert space. Substituting the forms Eqs. (7.32), (7.33), and (7.34) into Eq. (7.31) gives

$$\hbar \sum_{n,j} (q - n\omega) e^{i(n\omega - q)t} \phi_j^n = \sum_{j,l,m,k} e^{i[(m+k)\omega - q]t} \mathcal{H}_{jl}^m \phi_l^k . \qquad (7.35)$$

Now let us consider one particular term on the left-hand side of the above equation, characterized by particular values j' and n'. It has a time dependence $e^{i(n'\omega - q)t}$ and we can therefore equate this term with the sum of all the terms on the right-hand side with this same time dependence. Such terms have $m + k = n'$ and so

$$\hbar \left(q - n'\omega \right) \phi_{j'}^{n'} = \sum_{l,k} \mathcal{H}_{j'l}^{n'-k} \phi_l^k . \qquad (7.36)$$

The set of equations generated by writing Eq. (7.36) for each value of j' and n' reduces to the matrix equation

$$\hbar(q\hat{\mathbb{I}} - \mathcal{D})\boldsymbol{\phi} = \mathcal{H}_e \boldsymbol{\phi} , \qquad (7.37)$$

where $\boldsymbol{\phi}$ is a vector whose components are labelled by the two indices n' and j', and which has a dimension d given by the product of the original Hilbert space dimension with the number of Fourier components of interest. Here, $\mathcal{H}$ is the $d \times d$ Hamiltonian matrix, and $\mathcal{D}$ is a diagonal matrix with elements $n'\omega$. Finding the solutions is an eigenvalue problem:

$$(\mathcal{H}_e + \hbar\mathcal{D}) \boldsymbol{\phi} = \hbar q \, \boldsymbol{\phi} . \qquad (7.38)$$

For the specific case we are interested in, given by Eq. (7.14), the problem simplifies significantly. There are only two Fourier components on the off-diagonal of the Hamiltonian (at frequencies $\pm\omega$), and only one on the diagonal (the constant zeroth-order component). Thus, the so-called 'Floquet Hamiltonian' $\mathcal{H}^F = \mathcal{H}_e + \hbar\mathcal{D}$ takes the following matrix form

$$\mathcal{H}^F = \hbar \begin{pmatrix} & \vdots & \vdots & \vdots & \vdots & \vdots & \vdots & \\ \cdots & \omega_g - \omega & 0 & 0 & \Lambda/2 & 0 & 0 & \cdots \\ \cdots & 0 & \omega_e - \omega & \Omega/2 & 0 & 0 & 0 & \cdots \\ \cdots & 0 & \Omega^*/2 & \omega_g & 0 & 0 & \Lambda/2 & \cdots \\ \cdots & \Lambda^*/2 & 0 & 0 & \omega_e & \Omega/2 & 0 & \cdots \\ \cdots & 0 & 0 & 0 & \Omega^*/2 & \omega_g + \omega & 0 & \cdots \\ \cdots & 0 & 0 & \Lambda^*/2 & 0 & 0 & \omega_e + \omega & \cdots \\ & \vdots & \vdots & \vdots & \vdots & \vdots & \vdots & \end{pmatrix}, \qquad (7.39)$$

where the two coupling parameters are defined as $\Omega = 2eE_0 \mathbf{r}_{eg} \cdot \boldsymbol{\epsilon}/\hbar$ and $\Lambda = 2eE_0 \mathbf{r}_{eg}^* \cdot \boldsymbol{\epsilon}/\hbar$. The basis states used to express the Floquet Hamiltonian, which are characterized by the two quantum numbers $j \in \{g, e\}$, and the Fourier index s, are 'Floquet states' $|j, s\rangle$. The power of Floquet theory is now evident: we have recast a time-dependent problem as a time-independent one that looks identical to the eigenvalue problem for solving a time-independent Schrödinger equation. The transformation is exact. Moreover, all of the methods developed for solving time-independent problems, such as perturbation theory, time-evolution operators, etc., can be used in this case as well.

The Floquet Hamiltonian has a high degree of symmetry. Indeed, it is clear that its eigenvalues must fall into two sets, labelled 1 and 2, each of which corresponds to a ladder of values: $q_{1(2),p} = \lambda_{1(2)} + p\omega$ (p integer). We choose the $\lambda_{1,2}$ (arbitrarily) to be the two eigenvalues with the smallest absolute value. The eigenvectors have a similar symmetry: if the eigenvector corresponding to eigenvalue $q_{\xi,r}$ is called $|q_{\xi,r}\rangle$, then the series of eigenstates must satisfy

$$\langle j,s| q_{\xi,r}\rangle = \langle j,s+t| q_{\xi,r+t}\rangle . \tag{7.40}$$

Substitution of the eigenstate solutions to the Floquet Hamiltonian back into the full (time-dependent) form for the wave functions, Eq. (7.32), reveals that there are really only two inequivalent solutions; each eigenvalue on the same ladder gives a wave function of identical form. Thus we need consider only the two solutions corresponding to the two eigenvalues $\lambda_{1,2}$.

We are now in a position to understand the RWA rigourously. If the coupling terms, Λ and Ω, are small compared to the driving frequency ω of the field, we can use perturbation theory to simplify Eq. (7.39). Consider the near-resonant case, where the detuning $\nu \equiv \omega_e - \omega_g - \omega$ is small ($\ll \omega$). Both the detuning and laser couplings can be considered small parameters, so that the Floquet Hamiltonian splits up into a block diagonal form that is a ladder of effectively uncoupled two-dimensional matrices. Each matrix gives an equivalent pair of solutions, which can be found by applying first-order degenerate perturbation theory. We only need to diagonalize one such matrix, for example in the basis $|j,s\rangle \in \{|e,0\rangle, |g,1\rangle\}$, which yields

$$\mathcal{H}^F = \frac{\hbar}{2} \begin{pmatrix} \omega_e & \Omega \\ \Omega^* & \omega_e - 2\nu \end{pmatrix}. \tag{7.41}$$

Up to a constant factor on the diagonal, this is precisely the same matrix as we found in Eq. (7.22). Its eigenvalues are

$$\lambda_\pm = \frac{\hbar}{2}\left(\omega_e - \nu \pm \sqrt{\nu^2 + \Omega^2}\right), \tag{7.42}$$

with corresponding eigenvectors

$$|\lambda_+\rangle = \cos\frac{\theta}{2}|e,0\rangle + \sin\frac{\theta}{2}|g,1\rangle \tag{7.43}$$

$$|\lambda_-\rangle = -\sin\frac{\theta}{2}|e,0\rangle + \cos\frac{\theta}{2}|g,1\rangle , \tag{7.44}$$

and $\theta = \arctan(\Omega/\nu)$. In order to get a proper time-dependent solution of the Schrödinger equation back from these solutions we must substitute into Eq. (7.32), using Eq. (7.33). Thus, $|\lambda_+\rangle$ and $|\lambda_-\rangle$ are then equivalent to full wavefunction solutions:

$$|\psi_+(t)\rangle = e^{-i\lambda_+ t/\hbar}(\cos\theta |e\rangle + \sin\theta e^{i\omega t}|g\rangle) \tag{7.45}$$

$$|\psi_-(t)\rangle = e^{-i\lambda_- t/\hbar}(-\sin\theta |e\rangle + \cos\theta e^{i\omega t}|g\rangle). \tag{7.46}$$

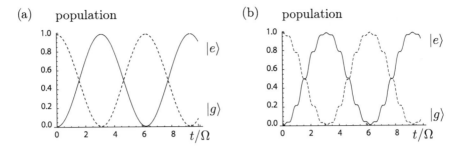

If the driving strength becomes sufficiently large, the RWA breaks down and the simple Rabi oscillation picture is no longer accurate. (a) For $\Omega = \Lambda = \omega/500$ the RWA holds; (b) for $\Omega = \Lambda = \omega/5$ the RWA breaks down and higher-frequency oscillations appear.

These are precisely the same as those we obtained from our simpler analysis in Eq. (7.25) when you take into account that Eq. (7.25) was obtained in the frame rotating at frequency ω. Our new analysis, however, puts the RWA on a firmer footing and allows us to predict when it is no longer valid.

If, in Eq. (7.39), the inequality $\omega \gg \Lambda, \Omega$ is *not* satisfied, our simple two-component solutions are no longer valid. Rather, there will be significant mixing of states with an energy difference $\omega_g - \omega_e - \omega$, and for stronger driving there will be coupling to even more energetically distant states. This means that our simple, single, Rabi oscillation is no longer a good description of the behaviour of our two-level atomic system. Extra higher-frequency oscillations also become important. This is shown in Fig. 7.2.

7.1.3 Three-level systems

We conclude our discussion of classically driven atoms by introducing a third level that has a similar energy to the ground level, and which can also be excited optically to the higher energy state $|e\rangle$. The two lower-lying levels might occur in a real system for spin half electrons, or for an electron coupled to the nucleus through the hyperfine interaction. We will not discuss their origin here, rather keeping the discussion general, but we will consider specific systems later in the book.

Let us label our two lower levels as $|0\rangle$ and $|1\rangle$, in anticipation of their eventual iden-tification as a qubit. The scheme is depicted in Fig. 7.3a. If both transitions $|0\rangle \leftrightarrow |e\rangle$ and $|1\rangle \leftrightarrow |e\rangle$ are optically active, we have a so-called Λ-configuration; if only one is optically active, we have an L-configuration. In angular frequency units, the Hamiltonian for a Λ-system can be written in a matrix form using the basis $\{|0\rangle, |1\rangle, |e\rangle\}$:

$$\mathcal{H}_\Lambda(t) = \hbar \begin{pmatrix} 0 & 0 & \Omega_1 \cos \omega_1 t \\ 0 & \delta & \Omega_2 \cos \omega_2 t \\ \Omega_1 \cos \omega_1 t & \Omega_2 \cos \omega_2 t & \omega \end{pmatrix}. \tag{7.47}$$

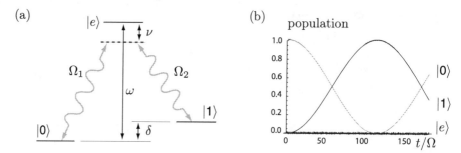

Fig. 7.3. (a) The three-level system is coupled as shown with coupling strength Ω_1 and Ω_2. Each laser is here detuned by the same amount from the transition that it addresses. (b) When the detuning is sufficiently large, oscillations between $|0\rangle$ and $|1\rangle$ occur through the coupling to $|e\rangle$, though $|e\rangle$ itself is hardly populated at all. This kind of oscillation is called a 'Raman transition'. The parameters here are $\Omega_1 = \Omega_2 = \Omega$, $\nu = 20\Omega$, $\delta = 10\Omega$, and $\omega = 500\Omega$. A full numerical calculation of Eq. (7.47) is shown, with an initial state $|0\rangle$.

Here, ω_1 and ω_2 are the frequencies of the transitions $|0\rangle \leftrightarrow |e\rangle$ and $|1\rangle \leftrightarrow |e\rangle$, respectively, and Ω_1 and Ω_2 are the respective coupling strengths of the laser addressing the transitions. We are assuming that each transition can be addressed individually and that there is no crosstalk between the two. This condition can be achieved through frequency selectivity or an angular-momentum selection rule (see Chapter 11). We can choose $|0\rangle$ as the zero energy level, and $|1\rangle$ lies at an energy $\hbar\delta$ above it. Similarly, the energy associated with $|e\rangle$ is $\hbar\omega$.

Let us now make a unitary transformation similar to the one in Eq. (7.15) that took us into the rotating frame earlier

$$\tilde{U}(t) = \begin{pmatrix} 1 & 0 & 0 \\ 0 & e^{-i(\omega_1 - \omega_2)t} & 0 \\ 0 & 0 & e^{-i\omega_1 t} \end{pmatrix}. \tag{7.48}$$

Using the Hamiltonian transformation in Eq. (7.19) gives an effective Hamiltonian

$$\mathcal{H}'_\Lambda(t) = \frac{\hbar}{2} \begin{pmatrix} 0 & 0 & \Omega_1(1 + e^{2i\omega_1 t}) \\ 0 & 2(\delta + \omega_2 - \omega_1) & \Omega_2(1 + e^{2i\omega_2 t}) \\ \Omega_1(1 + e^{-2i\omega_1 t}) & \Omega_2(1 + e^{-2i\omega_2 t}) & 2(\omega - \omega_1) \end{pmatrix}. \tag{7.49}$$

We again make the RWA, which finally gives us the simple, time-independent form of the Hamiltonian

$$\mathcal{H}'_\Lambda = \frac{\hbar}{2} \begin{pmatrix} 0 & 0 & \Omega_1 \\ 0 & 2(\nu_1 - \nu_2) & \Omega_2 \\ \Omega_1 & \Omega_2 & 2\nu_1 \end{pmatrix}, \tag{7.50}$$

where $\nu_1 = \omega - \omega_1$ and $\nu_2 = \omega - \omega_2 - \delta$.

Let us consider the situation in which the two lasers are equally detuned from their respective transitions, i.e., $\nu_1 = \nu_2 \equiv \nu$. If we work in the limit where the detunings are themselves large compared with the coupling strengths, i.e., $\nu \gg \{\Omega_1, \Omega_2\}$, we can treat the problem using degenerate perturbation theory. The excited state $|e\rangle$ remains an approximate eigenstate, and an effective two-dimensional Hamiltonian can be written for the other two states, giving

$$\mathcal{H}'_{\Lambda,2D} = -\frac{1}{4\nu} \begin{pmatrix} \Omega_1^2 & \Omega_1\Omega_2 \\ \Omega_1\Omega_2 & \Omega_2^2 \end{pmatrix}. \tag{7.51}$$

Straightforward diagonalization yields the two eigenstates that can again be parameterized by a single angle θ, as in Eq. (7.25). The angle becomes

$$\theta = \arctan\left(\frac{2\Omega_1\Omega_2}{\Omega_2^2 - \Omega_1^2}\right), \tag{7.52}$$

and the two eigenstates have energy separation $\Delta = (\Omega_1^2 + \Omega_2^2)/4\nu$. This kind of induced coupling between two levels via a higher third state is called a 'Raman transition'.

Varying the coupling strengths allows a whole range of values of θ, which allows us to vary the amplitude and frequency of the Rabi oscillations and so perform arbitrary single-qubit gate operations. The Rabi frequency when $\Omega_1 = \Omega_2 = \Omega$ is $\Delta = \Omega^2/2\nu$, which is the situation represented in Fig. 7.3. We conclude that in a Λ-system it is possible to perform any unitary operation on the qubit subspace $\{|0\rangle, |1\rangle\}$, by the fast switching of optical fields coupling to level $|e\rangle$, even though that level is never really populated. The two-qubit states need not be coupled directly to each other, and can be chosen to interact only very weakly with electromagnetic radiation. They can therefore have very low spontaneous-emission rates and represent a very robust qubit.

Finally, we return to Eq. (7.47), but this time we will analyze the Hamiltonian for the Λ-system differently. We make the same unitary transformation, Eq. (7.48), as well as the RWA, and again assume that $\nu_1 = \nu_2 = \nu$. This gives

$$\mathcal{H}'_\Lambda = \frac{1}{2} \begin{pmatrix} 0 & 0 & \Omega_1 \\ 0 & 0 & \Omega_2 \\ \Omega_1 & \Omega_2 & 2\nu \end{pmatrix}. \tag{7.53}$$

At this point, instead of considering the limit of large detuning, we make no further approximations and specifically do not use perturbation theory. Since there is no direct coupling between $|0\rangle$ and $|1\rangle$, and each has the same diagonal energy, the Hamiltonian in Eq. (7.53) always has one eigenvector of the form

$$|\psi(\theta)\rangle = \cos\theta\,|0\rangle - \sin\theta\,|1\rangle, \tag{7.54}$$

where θ is again a mixing angle.

Exercise 7.2: Convince yourself that the Hamiltonian in Eq. (7.53) must have an eigenvector of the form of Eq. (7.54). Show that the mixing angle is $\theta = \arctan(\Omega_1/\Omega_2)$.

By varying the ratio $R = \Omega_1/\Omega_2$ we can change the relative amplitudes of $|0\rangle$ and $|1\rangle$ in this eigenstate. Let us imagine a situation in which we begin with both lasers switched off, and allow the system to relax to its ground state, $|0\rangle$. We now turn on the second laser slowly,

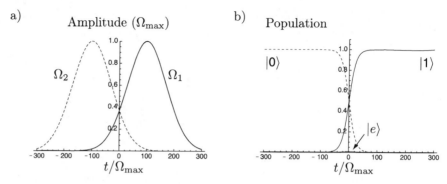

a) Amplitude (Ω_{max})   b) Population

Fig. 7.4. Full numerical simulation of the Hamiltonian in Eq. (7.47) for the STIRAP protocol. (a) Pulse sequence for the STIRAP protocol. Both lasers are resonant with the transition that they address ($\nu_1 = \nu_2 = 0$), and each has the Gaussian amplitude profile shown (Ω_{max} is the same for both pulses). The Ω_2 pulse is applied first to prepare the eigenstate structure of the Hamiltonian correctly, and then Ω_1 is increased as Ω_2 decreases. (b) The result of sequence (a) is to transfer population coherently from $|0\rangle$ to $|1\rangle$ with almost no population ever leaking into $|e\rangle$.

such that $\Omega_2 \neq 0$, while keeping the first laser off, $\Omega_1 = 0$. Then $R = 0$ and $|0\rangle$ remains an eigenstate. Now imagine slowly ramping down the power in the second laser, while increasing it in the first laser, and causing θ to increase till eventually it becomes $\theta = \pi/2$ as $\Omega_2 \to 0$. At this stage our eigenstate is $|1\rangle$. If the change in the laser power happens sufficiently slowly, the state of the system will follow the eigenstate $|\psi(\theta)\rangle$. This process is called 'adiabatic following'. Our argument shows that it is possible to coherently move from $|0\rangle$ to $|1\rangle$, while *never populating* $|e\rangle$. This process is known as 'stimulated Raman adiabatic passage' (STIRAP) and offers an alternative method of single-qubit manipulation without populating a vulnerable excited state. Fig. 7.4 shows the STIRAP pulse sequence and population change. It demonstrates that in contrast to regular Raman coupling the duration of laser pulses is not critical for performing quantum coherent control using STIRAP.

7.2 The Jaynes–Cummings Hamiltonian

We have so far discussed how a classical field interacts with an atom, and we now proceed with the quantization of the field. We will see that much of the behaviour discussed in the previous section has a quantum counterpart. In Section 7.1.1, we introduced a monochromatic electric field

$$\mathbf{E} = E_0 \left(\boldsymbol{\epsilon} \exp(i[\omega t - \mathbf{k} \cdot \mathbf{n}r]) + \boldsymbol{\epsilon}^* \exp(-i[\omega t - \mathbf{k} \cdot \mathbf{n}r]) \right). \tag{7.55}$$

To quantize this field, we recall from Chapter 1 that a general classical field can be expressed in terms of the vector potential as follows:

$$\mathbf{A}(\mathbf{x}, t) = \sum_\lambda \int d\mathbf{k} \left[\frac{A_\lambda(\mathbf{k})\boldsymbol{\epsilon}_\lambda(\mathbf{k})e^{i\mathbf{k}\cdot\mathbf{r}-i\omega_\mathbf{k}t}}{\sqrt{(2\pi)^3 2\omega_\mathbf{k}}} + \frac{A_\lambda^*(\mathbf{k})\boldsymbol{\epsilon}_\lambda^*(\mathbf{k})e^{-i\mathbf{k}\cdot\mathbf{r}+i\omega_\mathbf{k}t}}{\sqrt{(2\pi)^3 2\omega_\mathbf{k}}} \right], \tag{7.56}$$

where $A_\lambda(\mathbf{k})$ is the complex amplitude of the component with wavevector $\mathbf{k}$ and polarization λ. We assume the free-space dispersion relation $c^2|\mathbf{k}|^2 = \omega_\mathbf{k}^2$. From this classical expression, in Eq. (1.27) we constructed the quantum field

$$\hat{\mathbf{A}}(\mathbf{r}, t) = \sum_{\lambda=1}^{2} \int d\mathbf{k} \sqrt{\frac{\hbar}{\varepsilon_0}} \left[\boldsymbol{\epsilon}_\lambda(\mathbf{k}) \, \hat{a}_\lambda(\mathbf{k}) u(\mathbf{k}; \mathbf{r}, t) + \boldsymbol{\epsilon}_\lambda^*(\mathbf{k}) \, \hat{a}_\lambda^\dagger(\mathbf{k}) u^*(\mathbf{k}; \mathbf{r}, t) \right], \qquad (7.57)$$

where $\hat{a}_\lambda(\mathbf{k})$ and $\hat{a}_\lambda^\dagger(\mathbf{k})$ are annihilation and creation operators for a photon in the modes $u(\mathbf{k}; \mathbf{r}, t)$ and $u^*(\mathbf{k}; \mathbf{r}, t)$, respectively. For plane waves, we have

$$u(\mathbf{k}; \mathbf{r}, t) = \frac{e^{i\mathbf{k}\cdot\mathbf{r} - i\omega_\mathbf{k} t}}{\sqrt{(2\pi)^3 2\omega_\mathbf{k}}}, \qquad (7.58)$$

where the wave vector is denoted by $\mathbf{k}$ and the polarization by λ.

In order to simplify the derivation of the Jaynes–Cummings Hamiltonian, we will consider the interaction of a single discrete mode of the field with an atom centred at position $\mathbf{r} = 0$. From Eq. (7.11) we know that the interaction term in the Hamiltonian in the dipole approximation is $\mathcal{H}_d = e\hat{\mathbf{E}} \cdot \mathbf{r}$. Again, $\mathbf{E}$ is assumed to be constant over the extent of the atom and we can use $\hat{\mathbf{E}} = -\partial_t \hat{\mathbf{A}}$ to calculate it. Since $\hat{\mathbf{E}}$ then has explicit time dependence, $\mathcal{H}_d$ is given in the Heisenberg picture. However, we would like to work in the Schrödinger picture, where the time dependence is contained in the wave functions, not the operators. We transform to this picture by evaluating the dipole coupling Hamiltonian at $t = 0$. We therefore use $\hat{\mathbf{E}}(t = 0, \mathbf{r} = 0)$ to obtain

$$\mathcal{H}_d = \sum_{\lambda=1}^{2} \int d\mathbf{k} \sqrt{\frac{e^2 \hbar \omega_\mathbf{k}}{2(2\pi)^3 \varepsilon_0}} \left[\mathbf{r} \cdot \boldsymbol{\epsilon}_\lambda(\mathbf{k}) \, \hat{a}_\lambda(\mathbf{k}) + \mathbf{r} \cdot \boldsymbol{\epsilon}_\lambda^*(\mathbf{k}) \, \hat{a}_\lambda^\dagger(\mathbf{k}) \right]. \qquad (7.59)$$

Since we are interested in how a single mode of the field interacts with an atom, we will now expand our Hamiltonian in a basis of effective Fock states of that single mode. As in Chapter 1 we first define an effective creation operator for the mode we are considering

$$\hat{b}^\dagger = \int d\mathbf{k} \, \alpha^*(\mathbf{k}) \hat{a}^\dagger(\mathbf{k}). \qquad (7.60)$$

The annihilation operator is, as usual, the complex conjugate of the creation operator. To reduce the notational burden, we assume that the mode has well-defined polarization $\boldsymbol{\epsilon}$. Using Eq. (1.29), and so long as

$$\int d\mathbf{k} |\alpha(\mathbf{k})|^2 = 1, \qquad (7.61)$$

we satisfy the discrete boson commutation relations:

$$[\hat{b}, \hat{b}^\dagger] = 1. \qquad (7.62)$$

A set of Fock states for mode b is, according to Eq. (1.62), defined by

$$|n\rangle = \frac{(\hat{b}^\dagger)^n}{\sqrt{n!}} |0\rangle , \qquad (7.63)$$

where the normalization factor $\sqrt{n!}$ can easily be found using $\langle n | n \rangle = 1$ and repeated application of Eq. (7.62). The free Hamiltonian of the photon field-matrix elements for the Fock states on mode b are:

$$\mathcal{H}_{rs}^\alpha = \langle r | \mathcal{H}^\alpha | s \rangle = \sum_\lambda \int d\mathbf{k} \, \hbar \omega_{\mathbf{k}} \, \langle r | \hat{a}_\lambda^\dagger(\mathbf{k}) \hat{a}_\lambda(\mathbf{k}) | s \rangle . \qquad (7.64)$$

As in Chapter 1, we have removed the infinite energy of the vacuum.

Exercise 7.3: Using relation Eq. (7.60) and the commutation relations in Eq. (1.29), show that the matrix elements in Eq. (7.64) are given by

$$\mathcal{H}_{rs}^\alpha = \delta_{rs} \hbar \int d\mathbf{k} \, \alpha(\mathbf{k}) \alpha^*(\mathbf{k}) \omega_{\mathbf{k}} r \equiv \delta_{rs} \hbar \bar{\omega} r. \qquad (7.65)$$

Hint: use mathematical induction.

The matrix elements in Eq. (7.65) allow us to write the free Hamiltonian $\mathcal{H}^\alpha$ in terms of our bosonic operators in the following way

$$\mathcal{H}^\alpha = \hbar \bar{\omega} \hat{b}^\dagger \hat{b}, \qquad (7.66)$$

where $\bar{\omega}$ is defined in Eq. (7.65).

We will now consider the joint quantum state of the atom–field system. Our full state description should include both the Fock basis and the atomic state denoted by $|j\rangle \in \{|g\rangle, |e\rangle\}$. We write these joint states as $|j, n\rangle$. First, we express the interaction Hamiltonian in Eq. (7.59) in this joint basis. We already know that the dipole operator $\mathbf{r}$, which acts only on the atom and not the field, connects only different atomic states $|j\rangle$. Different Fock states are connected by the electric field. We may define the relevant matrix elements $\mathcal{H}_{d;rs}$ as follows:

$$\mathcal{H}_{d;rs} = \langle r | \mathcal{H}_d | s \rangle = \sqrt{\frac{e^2 \hbar}{2(2\pi)^3 \varepsilon_0}} \, \mathbf{r} \cdot \mathbf{M}_{rs} , \qquad (7.67)$$

where we have defined

$$\mathbf{M}_{rs} = \sum_{\lambda=1}^2 \int d\mathbf{k} \, \sqrt{\omega_{\mathbf{k}}} \, \langle r | \left[\boldsymbol{\epsilon}_\lambda(\mathbf{k}) \, \hat{a}_\lambda(\mathbf{k}) + \boldsymbol{\epsilon}_\lambda^*(\mathbf{k}) \, \hat{a}_\lambda^\dagger(\mathbf{k}) \right] | s \rangle . \qquad (7.68)$$

After some algebra, and using an induction method, one can prove that

$$\begin{aligned}
\mathbf{M}_{rs} &= \boldsymbol{\epsilon}^* \delta_{r,s+1} \sqrt{r} \int d\mathbf{k} \, \alpha^*(\mathbf{k}) \sqrt{\omega_{\mathbf{k}}} + \boldsymbol{\epsilon} \delta_{r+1,s} \sqrt{s} \int d\mathbf{k} \, \alpha(\mathbf{k}) \sqrt{\omega_{\mathbf{k}}} \\
&= \boldsymbol{\epsilon}^* \delta_{r,s+1} \sqrt{r} \mu^* + \boldsymbol{\epsilon} \delta_{r+1,s} \sqrt{s} \mu.
\end{aligned} \qquad (7.69)$$

where $\mu = \int d\mathbf{k}\,\alpha(\mathbf{k})\sqrt{\omega_{\mathbf{k}}}$. The vector $\mathbf{M}_{rs}$ has a particularly convenient form. It connects states with occupation numbers that differ by one, and has a magnitude proportional to the square root of the lower index of the two connected states. These are exactly the properties of our bosonic annihilation and creation operators and so we write the electric field operator in our Fock basis as follows:

$$\hat{\mathbf{E}} = \sqrt{\frac{\hbar}{2(2\pi)^3\varepsilon_0}}\left(\mu\boldsymbol{\epsilon}\,\hat{b} + \mu^*\boldsymbol{\epsilon}^*\,\hat{b}^\dagger\right). \tag{7.70}$$

The entire atom–field interaction can now be constructed as

$$\mathcal{H}_{\mathrm{d}} = \sqrt{\frac{e^2\hbar}{2(2\pi)^3\varepsilon_0}}\left(\boldsymbol{\epsilon}\mu\hat{b} + \mu^*\boldsymbol{\epsilon}^*\,\hat{b}^\dagger\right)\cdot\left(\mathbf{r}_{eg}^*\,|g\rangle\langle e| + \mathbf{r}_{eg}\,|e\rangle\langle g|\right). \tag{7.71}$$

Using spin language to describe atomic transitions, $|g\rangle\langle e| \equiv \sigma^-$ and $|e\rangle\langle g| \equiv \sigma^+$, we can rewrite the interaction Hamiltonian as

$$\mathcal{H}_{\mathrm{d}} = g\hat{b}\sigma_+ + g^*\hat{b}^\dagger\sigma^- + \gamma\hat{b}\sigma_- + \gamma^*\hat{b}^\dagger\sigma^+ \tag{7.72}$$

where we have defined

$$g \equiv \sqrt{\frac{e^2\hbar}{2(2\pi)^3\varepsilon_0}}\,(\mu\boldsymbol{\epsilon}\cdot\mathbf{r}_{eg}) \quad\text{and} \tag{7.73}$$

$$\gamma \equiv \sqrt{\frac{e^2\hbar}{2(2\pi)^3\varepsilon_0}}\,(\mu\boldsymbol{\epsilon}\cdot\mathbf{r}_{eg}^*). \tag{7.74}$$

It is interesting to compare these coupling constants with those of standard derivations of the atom–field interaction where the field modes are constructed inside a box of volume $V \sim L^3$. This leads to a discretization of the electromagnetic field energy levels (see e.g., Gerry and Knight, 2005). If, in our approach, we impose such a constraint, we would need to use a distribution function $\alpha(\mathbf{k})$ with an approximate width $1/L$ in all three dimensions. The normalization condition in Eq. (7.61) then forces the distribution height to be of order $\sqrt{L}$. Consequently, the factor μ above becomes proportional to $\sqrt{\tilde{\omega}/V}$, with $\tilde{\omega}$ some suitably defined angular frequency. This is exactly the same dependence as found in the more standard, but less general, approach.

Including the free energies of the atom and the quantum field, we obtain the full Hamiltonian of the atom–photon system

$$\mathcal{H} = \frac{1}{2}\hbar\omega_0\,\sigma_z + \hbar\tilde{\omega}\hat{b}^\dagger\hat{b} + g\hat{b}\sigma_+ + g^*\hat{b}^\dagger\sigma^- + \gamma\hat{b}\sigma_- + \gamma^*\hat{b}^\dagger\sigma^+. \tag{7.75}$$

If the cavity mode frequency is close to resonance with the atom, i.e., $\tilde{\omega} \approx \omega_0$, then the final two terms of the Hamiltonian couple widely off-resonant states. As we will see in more detail in the next section, this means that for small enough coupling γ can be neglected.

This is equivalent to the rotating-wave approximation for classical fields and leads, finally, to the 'Jaynes–Cummings Hamiltonian':

$$\mathcal{H} = \frac{\hbar\omega_0}{2}\sigma_z + \hbar\bar{\omega}\hat{b}^\dagger\hat{b} + g\hat{b}\sigma^+ + g^*\hat{b}^\dagger\sigma^-. \tag{7.76}$$

The most important and simplifying feature of this Hamiltonian is that it conserves excitation number. The photon annihilation operator in the interaction is coupled to the atom state-raising operator, and the photon creation operator is coupled to an atom state-lowering operator. Since the first two terms merely account for the individual energies of the atom and field, the dynamics splits into uncoupled subspaces, each characterized by a different number of excitations. For n excitations we may write a 2×2 matrix in the basis $\{|e, n-1\rangle, |g, n\rangle\}$

$$\mathcal{H}_n = \begin{pmatrix} \hbar\left(\frac{\omega_0}{2} + (n-1)\bar{\omega}\right) & \sqrt{n}g \\ \sqrt{n}g^* & \hbar\left(-\frac{\omega_0}{2} + n\bar{\omega}\right) \end{pmatrix}, \tag{7.77}$$

which describes the dynamics of a two-level atom in a quantum field. The energies of the eigenstates are shown in Fig. 7.5a. When the field is in resonance with the atomic transition, the diagonal elements are equal. The eigenstates are then entangled states of the atom–photon system $(|e, n-1\rangle \pm |g, n\rangle)/\sqrt{2}$. Under such conditions, intializing the system into a single Fock-atom state leads to oscillations, with energy transferred periodically between the atom and the field. The frequency of the oscillations depends on the number of photons in the mode, as shown in Fig. 7.5b.

Since the oscillation frequency is dependent on the number of photons in the mode, it is obvious that if we start our system in the atomic ground state and a superposition of Fock states, the subsequent evolution will consist of a combination of several different oscillation frequencies. For example, if the initial state is a coherent state, the Rabi oscillations decay

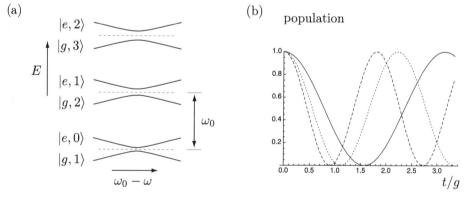

Fig. 7.5. (a) Energy level structure of the Jaynes–Cummings Hamiltonian, close to the resonance condition. The levels come in pairs, each of which corresponds to one value of the total excitation number. The levels anticross as a function of the detuning of the mode and atom. (b) Rabi oscillations at resonance for an initial atom state $|g\rangle$ for different numbers of photons in the mode (solid line: $n = 1$, dotted line: $n = 2$, dashed line: $n = 3$). The population of the atom ground state is shown as a function of normalized time.

altogether as the different component frequencies interfere with one another. Eventually, the frequencies come back into phase and there is a revival of coherent oscillations.

In contrast to the classical case, we can no longer always regard the atom here as an isolated qubit and the photon as a simple manipulation tool. The Rabi oscillations described here are now between states that involve both atom and photon quantum numbers. During the coherent evolution entanglement is generated between for example, a Fock state and an atom. Tracing out the photon component leads to loss of atomic coherence: but if both the atom and photon are regarded as qubits, the Jaynes–Cummings Hamiltonian means that simple quantum logic gates can be executed between them. Classical behaviour of the field is recovered from the quantum description in the case of a coherent state, since it is an eigenstate of the annihilation operator.

Let us briefly divert our discussion by making a comparison between the classical and quantum interactions of atoms and electromagnetic fields. Specifically, we write out some of the full quantum Hamiltonian in Eq. (7.75) in a matrix form, using the basis $\{|g, 0\rangle, |e, 0\rangle, |g, 1\rangle, |e, 1\rangle, \ldots\}$:

$$
\mathcal{H} = \hbar \begin{pmatrix}
0 & 0 & 0 & \gamma' & 0 & 0 & 0 & \cdots \\
0 & \omega_0 & g' & 0 & 0 & 0 & 0 & \cdots \\
0 & g'^* & \bar{\omega} & 0 & 0 & \gamma'\sqrt{2} & 0 & \cdots \\
\gamma'^* & 0 & 0 & \bar{\omega} + \omega_0 & g'\sqrt{2} & 0 & 0 & \cdots \\
0 & 0 & 0 & g'^*\sqrt{2} & 2\bar{\omega} & 0 & 0 & \cdots \\
0 & 0 & \gamma'^*\sqrt{2} & 0 & 0 & 2\bar{\omega} + \omega_0 & g'\sqrt{3} & \cdots \\
0 & 0 & 0 & 0 & 0 & g'^*\sqrt{3} & 3\bar{\omega} & \cdots \\
\vdots & \vdots & \vdots & \vdots & \vdots & \vdots & \vdots &
\end{pmatrix}. \tag{7.78}
$$

We have subtracted a constant energy from the diagonal and defined $g' \equiv g/\hbar$ and $\gamma' \equiv \gamma/\hbar$.

The similarity between the quantum mechanical formulation in Eq. (7.78), and the classical formulation in terms of Floquet states in Eq. (7.39) is compelling. There is a clear correspondence between our photon-number states and the Fourier series components we used to expand the Hamiltonian for a sinusoidal classical field. Even though the similarities are striking, the differences are also telling: in the fully quantized theory the off-diagonal matrix elements depend on the number of photons in the mode, which gives rise to the variable Rabi frequency we saw in Fig. 7.5. In the semiclassical theory, the off-diagonal components are independent of the Fourier component, and so a single-frequency Rabi oscillation is predicted. For very high photon numbers the two theories do converge and the correspondence principle holds, since the ratio of successive off-diagonal components in the quantum field theory tends to unity for large n.

7.3 The optical master equation and quantum jumps

We have seen how an atom can interact with a surrounding electromagnetic field. In particular, we have shown that an atom can absorb or emit radiation into modes of the electromagnetic field with a suitable energy, and that if one mode dominates the interaction then energy can flow back and forth between the atom and the mode. However, a common situation is one in which many modes interact with the atom. In this case, it takes much longer for energy to flow back to the atom once it has left, since each mode will couple differently to it. A range of characteristic Rabi frequencies results, which prevents a focused return of energy. In fact, with a continuum of modes there will no longer be a two-way exchange of energy, but rather an irreversible process that can destroy the coherence of the atomic state. To describe this process mathematically, we need to introduce the master-equation technique. This will be useful not only for studying optical decay, but also for other situations in which a system interacts with an environment with many degrees of freedom (we will return to this in Chapter 12). We will therefore introduce master equations in a general way before looking at optical processes specifically.

Consider a small system A (for example an atom) with Hamiltonian $\mathcal{H}_A$ that exists in an environment E with Hamiltonian $\mathcal{H}_E$. We assume that A and E are quite weakly coupled through an interaction Hamiltonian $\mathcal{H}_I$. We can therefore write the full Hamiltonian in the form:

$$\mathcal{H} = \mathcal{H}_A + \mathcal{H}_E + \mathcal{H}_I \equiv \mathcal{H}_0 + \mathcal{H}_I. \tag{7.79}$$

We do not expect to be able to keep track of the environment with its many degrees of freedom, but we are interested in what happens to the system. As we have seen in Chapter 3, we can describe the state of system that interacts with a larger environment by using the density matrix formalism. In this picture, the Schrödinger equation becomes the von Neumann equation:

$$\frac{d}{dt} R^{(S)}(t) = -\frac{i}{\hbar}[\mathcal{H}, R^{(S)}(t)] \tag{7.80}$$

where $R^{(S)}$ is the density matrix in the Schrödinger picture of the combined system and environment. We can make the transformation to the interaction picture using

$$R^{(S)}(t) \rightarrow \exp\left(-\frac{i}{\hbar}\mathcal{H}_0 t\right) R^{(I)}(t) \exp\left(\frac{i}{\hbar}\mathcal{H}_0 t\right), \tag{7.81}$$

$$\mathcal{H}_I \rightarrow \exp\left(-\frac{i}{\hbar}\mathcal{H}_0 t\right) \mathcal{H}_I(t) \exp\left(\frac{i}{\hbar}\mathcal{H}_0 t\right), \tag{7.82}$$

which allows us to recast Eq. (7.80) in a form that does not explicitly involve $\mathcal{H}_0$,

$$\frac{d}{dt} R^{(I)}(t) = -\frac{i}{\hbar}[\mathcal{H}_I(t), R^{(I)}(t)]. \tag{7.83}$$

We can integrate this equation to obtain

$$R^{(I)}(t) = R_0^{(I)} - \frac{i}{\hbar} \int_0^t ds [\mathcal{H}_I(s), R^{(I)}(s)], \tag{7.84}$$

where $R_0^{(I)}$ is determined by the initial conditions. Unfortunately, this is not a closed-form solution, since $R^{(I)}$ appears on both sides of the equation. We know that the coupling Hamiltonian should be relatively weak, and so a sensible approach to solving this equation is by successive substitution of this solution into the commutator on the right-hand side. The resulting exact infinite-series solution is

$$R^{(\mathrm{I})}(t) = R_0^{(\mathrm{I})} + \sum_{n=1}^{\infty} \left(-\frac{i}{\hbar}\right)^n \int \mathcal{D}_n t \, \left[\!\left[\mathcal{H}_I, R_0^{(\mathrm{I})}\right]\!\right]_n , \tag{7.85}$$

where we have defined

$$\int \mathcal{D}_n t \equiv \int_0^t dt_1 \int_0^{t_1} dt_2 \ldots \int_0^{t_{n-1}} dt_n , \tag{7.86}$$

and

$$\left[\!\left[\mathcal{H}_I, R_0^{(\mathrm{I})}\right]\!\right]_n \equiv \left[\mathcal{H}_I(t_1), \left[\mathcal{H}_I(t_2), \ldots \left[\mathcal{H}_I(t_n), R_0^{(\mathrm{I})}\right]\right]\right] . \tag{7.87}$$

If we assume an initial state of the form $R_0 = \rho(t = 0) \otimes \rho_E$, where $\rho(t)$ is the density matrix of the system at time t, and ρ_E is the initial environment state, we can write an equation for the development of the system density matrix as a function of time

$$\rho(t) = \mathcal{U}(t)\rho_0 = \sum_{n=0}^{\infty} \mathcal{U}_n(t)\rho_0 , \tag{7.88}$$

where $\mathcal{U}$ is the time-evolution superoperator for $\rho_0^{(\mathrm{I})}$, and $\mathcal{U}_n$ is the contribution to $\mathcal{U}$ that involves n terms in $\mathcal{H}_I$

$$\mathcal{U}_n(t) = \left(-\frac{i}{\hbar}\right)^n \mathrm{Tr}_E \left[\int \mathcal{D}_n t \, \left[\!\left[\mathcal{H}_I, \rho_E \otimes (\cdot)\right]\!\right]_n\right] , \tag{7.89}$$

where $n \geq 1$. For $n = 0$ we set $\mathcal{U}_n(t) = 1$.

Our final aim is to find an equation much like the von Neumann equation for the system, but where the environmental interaction is taken into account. Let us therefore recast Eq. (7.88) into the form

$$\dot{\rho}(t) = \dot{\mathcal{U}}(t)\mathcal{U}^{-1}(t)\rho(t) , \tag{7.90}$$

which is an equation for the time derivative of ρ in terms of $\rho(t)$, and therefore is of the same general form as the von Neumann equation. The first-order approximation to Eq. (7.90) can be written as

$$\dot{\rho}(t) = \dot{\mathcal{U}}_1(t)\rho(t) = -\frac{i}{\hbar}\mathrm{Tr}_E \left(\int_0^t dt_1 \, [\mathcal{H}_I(t_1), \rho_E \otimes \rho(t)]\right) . \tag{7.91}$$

We can force this term to zero by including a term $\mathrm{Tr}_E[\mathcal{H}_I, R^{(\mathrm{I})}(0)]$ in the system Hamiltonian. This term is a c-number, and it will therefore not survive in the commutator in Eq. (7.91). We must consider the second-order contribution to obtain the lowest nontrivial approximation:

$$\dot{\rho}(t) = -\frac{1}{\hbar^2} \int_0^t dt_1 \, \mathrm{Tr}_E \left([\mathcal{H}_I(t), [\mathcal{H}_I(t_1), \rho_E \otimes \rho(t)]]\right) . \tag{7.92}$$

It turns out that the third-order correction is also zero, so this second-order approximation is generally quite good. Eq. (7.92) is called the 'Redfield equation'.

Eq. (7.92) is usually obtained in an alternative way. First, Eq. (7.84) is substituted back into Eq. (7.83) to give

$$\dot{R}(t) = -\frac{1}{\hbar^2} \int_0^t ds \, \mathrm{Tr}_E \left(\left[\mathcal{H}_I(t), \left[\mathcal{H}_I(s), R^J(s) \right] \right] \right). \tag{7.93}$$

Two approximations are then made, namely the 'Born approximation' and the 'Markov approximation'. The Born approximation states that the state of the environment is unaffected by the system evolution, so an initial tensor-product state of the system and the environment remains separable throughout, i.e., $R^{(1)}(s) = \rho_E \otimes \rho(s)$. The Markov approximation states that the time evolution of the system depends only on the present state of the system, and not on any historical record of the system. This is motivated by the assumption that excitations in the environment decay over a much faster timescale than the typical timescale of the system dynamics. Thus we make the replacement $\rho(s) \to \rho(t)$ in Eq. (7.93), which leads to Eq. (7.92). Since the two derivations give an identical equation, we conclude that both the Markov and the Born approximations arise from a second-order cut-off of the density matrix evolution operator.

Next, we assume that the coupling between the system and environment is linear. The most general form of such a coupling allows several terms, labelled by the index α:

$$\mathcal{H}_I = \sum_\alpha \Upsilon_\alpha \otimes \Lambda_\alpha, \tag{7.94}$$

where $\Upsilon_\alpha = \Upsilon_\alpha^\dagger$ and operates on the system, while $\Lambda_\alpha = \Lambda_\alpha^\dagger$ and operates on the environment. We would like to transform this into the interaction picture, so it is convenient to write the system operators in the system's eigenbasis. Let us call the eigenvalues of the system Hamiltonian ϵ, each of which have a corresponding eigenvector $|\epsilon\rangle$. Now using the projectors $P_\epsilon = |\epsilon\rangle \langle \epsilon|$, we can define

$$\Upsilon_\alpha(\omega) \equiv \sum_{\epsilon,\epsilon'} \delta_{\epsilon'-\epsilon,\hbar\omega} P_\epsilon \Upsilon_\alpha P_{\epsilon'}. \tag{7.95}$$

With this definition we can rewrite Eq. (7.94) as

$$\mathcal{H}_I = \sum_{\alpha,\omega} \Upsilon_\alpha(\omega) \otimes \Lambda_\alpha. \tag{7.96}$$

The interaction Hamiltonian in the interaction picture now assumes a relatively simple structure

$$\mathcal{H}_I(t) = \sum_{\alpha,\omega} e^{-i\omega t} \Upsilon_\alpha(\omega) \otimes \Lambda(t), \tag{7.97}$$

where $\Lambda(t) = e^{i\mathcal{H}_E t/\hbar} \Lambda e^{-i\mathcal{H}_E t/\hbar}$. Eq. (7.97) is in a convenient form that allows us to factor the system and environment parts of the Redfield equation in Eq. (7.92). After some lengthy but straightforward algebra we find

$$\dot{\rho}(t) = -\sum_{\omega,\omega',\alpha,\alpha'} e^{i(\omega'-\omega)t} \gamma_{\alpha,\alpha'}(\omega) \left(\Upsilon_{\alpha'}(\omega)\rho(t)\Upsilon_\alpha^\dagger(\omega') - \Upsilon_\alpha^\dagger(\omega')\Upsilon_{\alpha'}(\omega)\rho(t) \right)$$
$$+ \text{H.c.}. \tag{7.98}$$

The reservoir correlation functions are defined by

$$\gamma_{\alpha,\alpha'}(\omega) \equiv \frac{1}{\hbar^2} \int_0^t dt_1 e^{i\omega t_1} \langle \Lambda_\alpha^\dagger(t)\Lambda_{\alpha'}(t-t_1) \rangle \tag{7.99}$$

where the angled brackets denote an average over the environment density matrix. Notice that all terms involving products $\Upsilon\Upsilon$ or $\Upsilon^\dagger\Upsilon^\dagger$ have been dropped from Eq. (7.98) since they are associated with correlation functions that must be zero.

We have already made the assumption that the environment correlations decay on a much faster timescale than any system dynamics, and we may therefore take the upper limit on the integral of Eq. (7.99) to infinity. Further, since the density matrix of the environment does not depend on time, we know that $\langle \Lambda_\alpha^\dagger(t)\Lambda_{\alpha'}(t-t_1) \rangle = \langle \Lambda_\alpha^\dagger(t_1)\Lambda_{\alpha'}(0) \rangle$, so

$$\gamma_{\alpha,\alpha'}(\omega) \equiv \frac{1}{\hbar^2} \int_0^\infty dt\, e^{i\omega t} \langle \Lambda_\alpha^\dagger(t)\Lambda_{\alpha'}(0) \rangle. \tag{7.100}$$

A further simplifying assumption that we can make in quantum optical systems is that the non-secular terms in Eq. (7.98), i.e., those with $\omega \neq \omega'$, can be dropped. We therefore find

$$\dot{\rho}(t) = -\sum_{\omega,\alpha,\alpha'} \gamma_{\alpha,\alpha'}(\omega) \left(\Upsilon_{\alpha'}(\omega)\rho(t)\Upsilon_\alpha^\dagger(\omega) - \Upsilon_\alpha^\dagger(\omega)\Upsilon_{\alpha'}(\omega)\rho(t) \right) + \text{H.c.}, \tag{7.101}$$

which is valid as long as the typical relaxation time of the system, determined by the $\gamma_{\alpha,\alpha'}^{-1}$, is small compared to the internal system dynamics. The latter occur on a timescale that is of the order $(\omega - \omega')^{-1}$.

7.3.1 The optical master equation

We have developed the theory of the master equation quite generally. We will now apply it to the specific case of the Jaynes–Cummings model. This will show us how to treat optical decay when we have no knowledge about the emission time of the photons. Our starting point is Eq. (7.76). It consists of two terms which, using the notation of Eq. (7.94), we will label $\alpha = \{+, -\}$. However, it makes no sense to consider just an atom interacting with a single mode if we are to use the technique we have just discussed. The environment must have many degrees of freedom if its correlations are to fall off much more quickly than the system dynamics. This situation is often more realistic, since optically active atoms are not usually inside perfect cavities, where the interaction is mainly between the atom and a single cavity mode. We therefore extend our model to include many different modes j, with frequencies $\bar{\omega}_j$ and coupling strengths g_j that are generated by the operators $\hat{b}_j$, $\hat{b}_j^\dagger$.

We can now immediately write down the eigenstate-projected forms of the operators that appear in Eq. (7.96), in the interaction picture, Eq. (7.97). We have

$$\Upsilon_+(-\omega_0) = \sigma^+ \quad \text{and} \quad \Lambda_+(t) = \sum_j g_j \hat{b}_j \exp(-i\bar{\omega}_j t); \qquad (7.102)$$

$$\Upsilon_-(\omega_0) = \sigma^- \quad \text{and} \quad \Lambda_-(t) = \sum_j g_j^* \hat{b}_j^\dagger \exp(i\bar{\omega}_j t). \qquad (7.103)$$

We must substitute these specific terms into Eq. (7.98). This leads to quite a complicated expression if we regard a general state of the environment. However, it is often sufficient to consider thermal states of the environment, i.e.,

$$\rho_E = \rho_{E,1} \otimes \rho_{E,2} \otimes \ldots \otimes \rho_{E,j} \otimes \ldots \qquad (7.104)$$

where

$$\rho_{E,j} = \frac{1}{1 - \exp(-\hbar\bar{\omega}_j \beta)} \sum_{n=0}^{\infty} \exp(-n\hbar\bar{\omega}_j \beta) |n\rangle_j \langle n| , \qquad (7.105)$$

with $\beta = (k_B T)^{-1}$. For this particular choice of environmental state, we can derive the relations

$$\langle \hat{b}_k \hat{b}_j \rangle = 0, \qquad (7.106)$$

$$\langle \hat{b}_k^\dagger \hat{b}_j^\dagger \rangle = 0, \qquad (7.107)$$

$$\langle \hat{b}_k \hat{b}_j^\dagger \rangle = \delta_{kj}(1 + N(\bar{\omega}_j)), \qquad (7.108)$$

$$\langle \hat{b}_k^\dagger \hat{b}_j \rangle = \delta_{kj} N(\bar{\omega}_j). \qquad (7.109)$$

Here[2], $N(\bar{\omega})$ is the Bose–Einstein occupation number $[\exp(\hbar\bar{\omega}\beta) - 1]^{-1}$. Therefore, only terms in which $\alpha = \alpha'$ survive in Eq. (7.101) and using the forms in Eq. (7.103) we can write

$$\dot{\rho}(t) = -\gamma_{+,+} \left[\sigma^+ \rho(t)\sigma^- - \sigma^-\sigma^+\rho(t) \right]$$
$$- \gamma_{-,-} \left[\sigma^-\rho(t)\sigma^+ - \sigma^+\sigma^-\rho(t) \right] + \text{H.c.}, \qquad (7.110)$$

with

$$\gamma_{+,+} = \sum_j \int_0^\infty dt \, e^{i(\bar{\omega}_j - \omega_0)t} \frac{|g_j|^2}{\hbar^2} N(\bar{\omega}_j), \qquad (7.111)$$

$$\gamma_{-,-} = \sum_j \int_0^\infty dt \, e^{i(\omega_0 - \bar{\omega}_j)t} \frac{|g_j|^2}{\hbar^2} (1 + N(\bar{\omega}_j)). \qquad (7.112)$$

---

[2] For more general environmental states such as squeezed states these relations do not hold and a more complicated master equation results.

We can go further by converting the sum over states to an integral over the density of states $D(\bar{\omega})$. We need to use

$$\int_0^\infty dt \exp(\pm i\epsilon t) = \pi\delta(\epsilon) \pm i\frac{\mathcal{P}}{\epsilon}, \tag{7.113}$$

where $\mathcal{P}$ denotes the Cauchy principal value. By defining parameters

$$\Delta_1 = \mathcal{P}\int d\bar{\omega}\,\frac{1}{\bar{\omega}-\omega_0}D(\bar{\omega})\frac{|g(\bar{\omega})|^2}{\hbar^2}N(\bar{\omega}), \tag{7.114}$$

$$\Delta_2 = \mathcal{P}\int d\bar{\omega}\,\frac{1}{\bar{\omega}-\omega_0}D(\bar{\omega})\frac{|g(\bar{\omega})|^2}{\hbar^2}, \tag{7.115}$$

and

$$\Gamma = \frac{|g(\omega_0)|^2}{\hbar^2}D(\omega_0) \tag{7.116}$$

we can write the master equation, Eq. (7.110), in the form

$$\dot{\rho} = i(2\Delta_1 + \Delta_2)[\sigma_z, \rho] + \Gamma N(\omega_0)(2\sigma^+\rho\sigma^- - \sigma^-\sigma^+\rho - \rho\sigma^-\sigma^+)$$
$$+ \Gamma[N(\omega_0)+1](2\sigma^-\rho\sigma^+ - \sigma^+\sigma^-\rho - \rho\sigma^+\sigma^-). \tag{7.117}$$

The first term is non-dissipative, and corresponds to the equation of motion for a two-level system with energy separation $\hbar(2\Delta_1 + \Delta_2)$. It is therefore an effective energy renormalization called the 'Lamb shift'. We can absorb it into the bare system Hamiltonian $\mathcal{H}_A$ and redo the interaction Hamiltonian transformation, defining the resulting density matrix as $\tilde{\rho}$. The renormalizing process has no effect on the form of the other terms in Eq. (7.117) and we can write

$$\dot{\tilde{\rho}} = \Gamma N(\omega_0)(2\sigma^+\tilde{\rho}\sigma^- - \sigma^-\sigma^+\tilde{\rho} - \tilde{\rho}\sigma^-\sigma^+)$$
$$+ \Gamma[N(\omega_0)+1](2\sigma^-\tilde{\rho}\sigma^+ - \sigma^+\sigma^-\tilde{\rho} - \tilde{\rho}\sigma^+\sigma^-). \tag{7.118}$$

This equation is our key result, and is called the 'optical master equation'.

What does the master equation mean? The first group of terms, which are proportional to $\Gamma N(\omega_0)$, came from the part of the interaction Hamiltonian whose effect was to raise the energy of the atom, taking it from state $|g\rangle$ to $|e\rangle$, and to simultaneously remove a quantum of energy from the environment. Since we are unable to track the state of the environment, which is now a continuum of modes, we are no longer able to give a state vector description of our atom's time evolution. Rather, we can use our optical master equation to encapsulate all the knowledge we can possibly have about our atom's evolution. This first group of terms describes the tendency in time of our system to increase its energy by absorbing energy from the environment. The rate of this absorption is given by $\Gamma N(\omega_0)$, which makes good sense: $N(\omega_0)$ gives us the occupation number of the environment states with the energy that matches the energy which must be gained by the atom. If there are no environment states occupied, there can be no energy gain. Similarly, the rate depends via Γ on the density of available environment states, $D(\omega_0)$, and the coupling between atom and environment,

$g(\omega_0)$. The second collection of terms, which depend on $\Gamma[N(\omega_0)+1]$, describe the tendency of an atom to emit radiation into the environment, and so move from $|e\rangle$ to $|g\rangle$. Emission can always happen if there are available states to decay into. This is the origin of the extra Γ in the rate, which is associated with spontaneous emission. The other part is proportional to $\Gamma N(\omega_0)$ and describes stimulated emission.

7.3.2 Monitored decay in the quantum-jump approach

So far, our description of system dynamics in an environment of photons has assumed that we do not have any knowledge about the environment at all. Rather, we have only been able to observe the system. However, it is often the case in an experiment that we are able to detect photons that have previously been emitted by an atom. This gives us some extra information that may affect our knowledge of the system state. In this case, we must modify our master equation and make it conditional on the observations we can make of decay photons.

Let us start by asking how the system changes when we detect a photon due to decay. The simplest assumption is that if the system has emitted a photon, it must be left in the ground state. Whatever the state of the system was before the emission, a photon detection corresponds to a projection into the state $\rho_g = |g\rangle\langle g|$. We can capture this process mathematically by defining the decay-jump operator $\mathcal{J}$ according to

$$\mathcal{J}\rho = \sigma^-\rho\sigma^+. \tag{7.119}$$

This operator leaves the system in the state $\langle e|\rho|e\rangle\,|g\rangle\langle g|$, i.e., it produces the ground state we desire, but with the prefactor $\langle e|\rho|e\rangle = \mathrm{Tr}\mathcal{J}\rho$. This is just the probability that our system was in the excited state before photon emission. During any given time interval there is a certain probability that the atom emits a photon, and that this photon is detected. Every time that such an event occurs we must apply the jump operator to the system, and renormalize the state. We call a 'trajectory' a sequence of intervals that is characterized by a string of specific detection signatures, one for each interval.

We must ensure that after averaging over many trajectories we recover the (unconditional) master equation in Eq. (7.118). To do this, let us look at an infinitesimal time interval dt. We can divide the evolution into two parts. First, if we detect a photon our system will jump to a state which we will call ρ_1; if no photon is detected we call the new state ρ_0. We can use a stochastic variable dN, which can take the value of 0 or 1, to express this in a single equation:

$$\rho(t+dt) = \rho_0(t+dt)(1-dN) + \rho_1(t+dt)dN$$
$$= \rho_0(t+dt)(1-dN) + \frac{\mathcal{J}\rho}{\mathrm{Tr}(\mathcal{J}\rho)}dN. \tag{7.120}$$

For a decay rate 2Γ and a perfect detector, the probability that a photon is detected in dt is $2\mathrm{Tr}(\mathcal{J}\rho)\Gamma$.

In practice, photodetectors have a finite-detection efficiency η. In Chapter 4 we have seen that this can be modelled as a perfect detector, preceded by a beam splitter with transmission coefficient η in the detected mode. Each photon is lost or detected independently of the other photons. We can therefore modify the probability of photon detection by multiplying by η. The expectation value $\xi(dN)$ of the stochastic variable dN then becomes

$$\xi(dN) = 2\eta\Gamma\,\mathrm{Tr}(\mathcal{J}\rho)dt, \tag{7.121}$$

and the average over many trajectories of Eq. (7.120) is given by

$$\rho(t+dt) = \rho_0(t+dt)[1 - 2\eta\Gamma\,\mathrm{Tr}(\mathcal{J}\rho)dt] + 2\eta\Gamma\,\mathcal{J}\rho dt$$
$$= \rho(t) + \Gamma(2\mathcal{J}\rho - \mathcal{A}\rho)dt, \tag{7.122}$$

where the second equation follows from Eq. (7.118), defining the superoperator $\mathcal{A}\rho = \sigma^+\sigma_-\rho + \rho\sigma^+\sigma_-$, and assuming that the temperature is sufficiently low so that $N(\omega_0) \approx 0$. By keeping only terms up to those linear in dt, we may write

$$\rho_0(t+dt) = \rho(t) + \Gamma\{2\eta\,\mathrm{Tr}[\mathcal{J}\rho(t)]\rho(t) + 2(1-\eta)\mathcal{J}\rho(t) - \mathcal{A}\rho(t)\}\,dt, \tag{7.123}$$

which is more conveniently expressed by the differential equation

$$\frac{d\rho_0}{dt} = \Gamma[2\eta\,\mathrm{Tr}(\mathcal{J}\rho_0)\rho_0 + 2(1-\eta)\mathcal{J}\rho_0 - \mathcal{A}\rho_0]. \tag{7.124}$$

This is an exact differential equation for ρ_0, but rather difficult to solve in this nonlinear form. However, it can be recast in a linear form that does not preserve the normalization of the density matrix.

Exercise 7.4: Let $\rho_0(t) = f(t)\bar{\rho}_0(t)$, where $f(t)$ is a normalization factor that depends on time. Substitute this in Eq. (7.124) and show that the resulting term in f^2 can be eliminated by choosing $f(t) = (\mathrm{Tr}[\bar{\rho}_0(t)])^{-1}$. Derive the linear equation for $\bar{\rho}_0$

$$\frac{d\bar{\rho}_0}{dt} = \Gamma[2(1-\eta)\mathcal{J}\bar{\rho}_0 - \mathcal{A}\bar{\rho}_0]. \tag{7.125}$$

We now have all we need to simulate a quantum trajectory using Eq. (7.120). For each time interval we use a random-number generator to determine whether a photon is detected or not. We make sure that the expectation value, Eq. (7.121), is satisfied by assigning an appropriate range for when the generated random number corresponds to a detection event. If a photon is indeed detected, we use the jump operator. If no photon is detected, we use our linear differential equation in Eq. (7.125) to obtain the new state, which we normalize by dividing by its trace. It is instructive to look at the form of Eq. (7.125) in the case of a perfect detector ($\eta = 1$). In that case, we can write the equation in a state vector form. It becomes a Schrödinger equation with a non-Hermitian Hamiltonian:

$$\frac{d|\bar{\psi}\rangle}{dt} = -\Gamma|e\rangle\langle e|\bar{\psi}\rangle. \tag{7.126}$$

The solution of this equation has an exponentially reducing amplitude for the excited state. In other words, as time passes with no photon detected, the observer gains confidence that the system must be in the ground state, and is incapable of producing a photon, as opposed to an excited state which has not yet decayed.

7.4 Entangling operations via path erasure

A powerful consequence of the atom–photon interactions we have discussed in this chapter is the possibility of creating entangled states of atoms by measuring the photons that they emitted. There have been several proposals for how best to do this in recent years, and we will here review some of the more promising candidates.

7.4.1 The weak driving limit

The simplest approach to entangling two atoms is to measure decay photons from a weakly driven transition. Let us consider the set-up that is depicted schematically in Fig. 7.6. We have two atoms in separate cavities that each have a Λ-configuration, i.e., two low-lying states labeled $|0\rangle$ and $|1\rangle$, and a single excited state $|e\rangle$. Both cavities are resonant with the $|e\rangle$-$|1\rangle$ transition, which must be identical for each atom. We will assume that the only significant atom–photon decay path is via photons in the resonant cavity mode. We further assume that one of the cavity mirrors is somewhat more leaky than the other, such that photons will always be emitted in the direction of that mirror.

Initialize the two atoms ($i \in \{1, 2\}$) in state $|0\rangle$, and assume that the temperature is low enough that thermal excitations are negligible. Each atom is now driven with a laser resonant with $|0\rangle \leftrightarrow |e\rangle$, but only weakly and for a short time. We therefore create only a

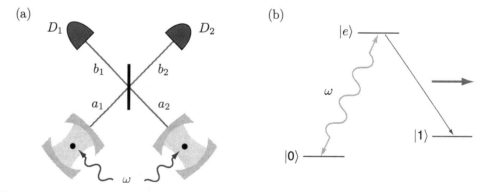

Fig. 7.6. (a) The general experimental set-up for entanglement generation through photon path erasure. Two cavities are depicted at the bottom of the figure, each of which contains an atom. In the weak driving case, each atom has the level structure shown in (b).

small amplitude for the term $|e\rangle$, and the state of each atom can be written as

$$|\psi\rangle_i = \frac{|0\rangle + \epsilon\,|e\rangle}{\sqrt{1 + \epsilon^2}}\,. \tag{7.127}$$

For simplicity, and without loss of generality, we can choose ϵ real. The state of both atoms together is then

$$|\Psi\rangle_{12} = |\psi\rangle_1 \otimes |\psi\rangle_2 = \frac{|00\rangle + \epsilon(|e0\rangle + |0e\rangle) + \epsilon^2\,|ee\rangle}{1 + \epsilon^2}\,. \tag{7.128}$$

Next, suppose that detectors D_i are set up to detect photons emitted from atom i into mode a_i. When that detector clicks we describe the system evolution with the jump operator $\mathcal{J}_i(\cdot) = \sigma_i^-(\cdot)\sigma_i^+$. Following a click in detector D_1 or D_2, we therefore project into the atom state $|1\rangle_1\,|\psi\rangle_2$ or $|\psi\rangle_1\,|1\rangle_2$, respectively. Both states are separable. Alternatively, if we place a 50:50 beam splitter in the path of the two possible photon-emission routes, in the symmetric way shown in Fig. 7.6, the photon operators after the beam splitter are written as

$$\hat{b}_1^\dagger = \frac{i\hat{a}_1^\dagger + \hat{a}_2^\dagger}{\sqrt{2}}\,, \tag{7.129}$$

$$\hat{b}_2^\dagger = \frac{\hat{a}_1^\dagger + i\hat{a}_2^\dagger}{\sqrt{2}}\,. \tag{7.130}$$

We know that a photon in mode a_i is created via the decay of atom i, and the relevant jump operator $\mathcal{J}_i$ associated with a detector click in D_i must therefore be modified as follows:

$$\mathcal{J}_1(\cdot) = \frac{1}{2}(i\sigma_1^- + \sigma_2^-)(\cdot)(-i\sigma_1^+ + \sigma_2^+)\,, \tag{7.131}$$

$$\mathcal{J}_2(\cdot) = \frac{1}{2}(\sigma_1^- + i\sigma_2^-)(\cdot)(\sigma_1^+ - i\sigma_2^+)\,. \tag{7.132}$$

Applying these operators to our initial state in Eq. (7.128) yields

$$|\Psi\rangle_1 = \frac{1}{\sqrt{2(1+\epsilon^2)}}\,[i\,|01\rangle + |10\rangle + \epsilon(i\,|1e\rangle + |e1\rangle)]\,, \tag{7.133}$$

$$|\Psi\rangle_2 = \frac{1}{\sqrt{2(1+\epsilon^2)}}\,[|01\rangle + i\,|10\rangle + \epsilon(|1e\rangle + i\,|e1\rangle)]\,, \tag{7.134}$$

where $|\Psi\rangle_i$ is the state following in a click in detector D_i. The last two terms in these states involve the excited state $|e\rangle$, and will eventually decay into $|11\rangle$. This process produces a photon in the same detector as the first decay, and this happens with a probability $\epsilon^2/(1+\epsilon^2)$. If the detector was perfect, we would be able to discount such events, since they produce an uninteresting product state. However, it is often the case that detectors are unable to count photons that are emitted in rapid succession. When this is the case, we have to describe the system by a mixed state

$$\rho_1 = \frac{1}{2(1+\epsilon^2)} \left[(i\,|01\rangle + |10\rangle)(-i\,\langle 01| + \langle 10|) + 2\epsilon^2\,|11\rangle\,\langle 11| \right], \tag{7.135}$$

$$\rho_2 = \frac{1}{2(1+\epsilon^2)} \left[(|01\rangle + i\,|10\rangle)(\langle 01| - i\,\langle 10|) + 2\epsilon^2\,|11\rangle\,\langle 11| \right]. \tag{7.136}$$

When the driving is weak, i.e., $\epsilon \ll 1$, a single detector click indicates that the atoms are almost maximally entangled, even though they never interacted with each other directly.

The crucial requirement for this procedure to work is that we detect a photon *without knowing which atom emitted it*. In this case, the beam splitter has acted to erase the 'which path' information for the detected photon. By contrast, if there is any way, practical or in principle, of telling which cavity the detected photon originated from, the state loses its entanglement, since the jump operators are no longer properly described as a superposition of atom-decay processes. For example, if the atoms emit photons of different frequency, or recoil significantly in the decay process, the path-erasure method does not work.

A drawback of the weak driving scheme is that the probability of detecting a photon on any given round of the protocol is necessarily low, since ϵ must be kept small. For any ϵ, we can calculate the fidelity of the two-atom state with respect to the maximally entangled state as

$$F = \frac{1-\epsilon^2}{1+\epsilon^2} \approx 1 - 2\epsilon^2, \tag{7.137}$$

for small ϵ, and the success probability (or efficiency) p is

$$p = 2\Gamma \left[\mathrm{Tr}(\mathcal{J}_1\rho) + \mathrm{Tr}(\mathcal{J}_2\rho) \right] = 4\Gamma\epsilon^2. \tag{7.138}$$

There is a direct trade-off between the fidelity of the entanglement and the success probability of the entanglement generation. When the procedure fails, the system ends up in a mixed state which is primarily composed of $|00\rangle\,\langle 00|$ (in the case of perfect detectors the state of the two atoms is exactly $|00\rangle\,\langle 00|$). Such a large failure rate does not really matter if we need to generate only small entangled states of perhaps two or three qubits. However, failure of the entangling procedure causes the loss of all correlations with any previously created larger entangled state, and creating large cluster states by repeated applications of this protocol becomes exponentially costly. We therefore now explore an alternative method that enjoys an increased probability of success, while keeping the fidelity close to unity.

7.4.2 The double heralding parity gate

We again consider two atoms in separate cavities, where the cavity ensures that photon emission is predominantly into a particular mode to enhance the photon collection. The beam splitter and detectors also have the same configuration as in Fig. 7.6. However, the level structure of the atoms is modified to an L-configuration, as shown in Fig. 7.7a. The states $|1\rangle$ and $|e\rangle$ are coupled by a driving laser *and* the cavity mode. The coupling between the levels $|0\rangle$ and $|e\rangle$ is highly suppressed, for example by a selection rule. The

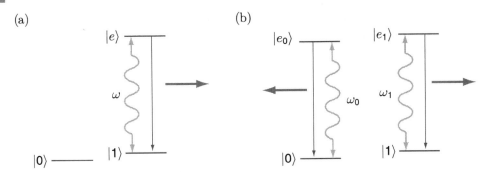

Fig. 7.7. (a) The 'L' type level structure needed for the double-heralding gate; (b) the four-level structure needed for the repeat-until-success protocol. Both $|0\rangle$ and $|1\rangle$ have a corresponding higher level, each of which decays to give distinct photons.

energy separation of $|0\rangle$ and $|1\rangle$ can be much smaller than the energy to the excited state $|e\rangle$. We assume that we have single-qubit control over both systems, e.g., using resonant electromagnetic radiation of much longer wavelength than the optics used in the rest of the protocol. Alternatively, a STIRAP or conventional Raman procedure using another excited level may be used to construct fast single-qubit gates.

We now describe the double-heralding entangling protocol. Both atoms are prepared in the state

$$|+\rangle = \frac{1}{\sqrt{2}} (|0\rangle + |1\rangle) . \tag{7.139}$$

A resonant laser pulse of frequency ω is applied to both atoms such that all population is transferred from $|1\rangle$ to $|e\rangle$. The joint state of the two atoms then becomes

$$|\Psi\rangle = \frac{1}{2} (|00\rangle + |0e\rangle + |e0\rangle + |ee\rangle) . \tag{7.140}$$

We now wait for all possible decay photons and again assume that our detectors are unable to count the number of photons. If we have perfectly efficient detectors, a detector click will occur with a probability 3/4, and depending on which detector clicks, we obtain the following mixed states after detection:

$$\rho_1 = \frac{1}{3} [(|01\rangle + i|10\rangle)(\langle 01| - i\langle 10|) + |11\rangle \langle 11|] ,$$

$$\rho_2 = \frac{1}{3} [(i|01\rangle + |10\rangle)(-i\langle 01| + \langle 10|) + |11\rangle \langle 11|] . \tag{7.141}$$

A trick is now performed to remove the right-most separable term in these mixtures. We use our single-qubit control to apply a bit-flip to each atomic qubit. The two states then become

$$\rho_1 = \frac{1}{3} [(|10\rangle + i|01\rangle)(\langle 10| - i\langle 01|) + |00\rangle \langle 00|] ,$$

$$\rho_2 = \frac{1}{3} [(i|10\rangle + |01\rangle)(-i\langle 10| + \langle 01|) + |00\rangle \langle 00|] . \tag{7.142}$$

A second round of excitation now follows. Since the state $|00\rangle$ is unable to produce any photons, a second detector click must project the two atoms onto the entangled part of the states in Eq. (7.142). Indeed, in this simplified picture, a successful operation results in a maximally entangled pure state with a probability of success on the second round of 2/3. Overall success of the protocol is marked by photon-detection events on both rounds of the protocol, hence the name 'double heralding'. In the ideal detector scenario, success happens with total probability of $3/4 \times 2/3 = 1/2$. The fidelity of this protocol is insensitive to low detector efficiency η, though of course this does adversely affect the success probability, which becomes $p = \eta^2/2$. One mechanism that can degrade the fidelity of the entangling procedure is false positives, or dark counts, in the photodetectors.

Exercise 7.5: Show that if the same detector clicks in the two successive rounds, the state of the atoms is

$$\left|\Psi^+\right\rangle = \frac{1}{\sqrt{2}}(|01\rangle + |10\rangle). \tag{7.143}$$

If different detectors click, the state is $\left|\Psi^-\right\rangle$.

The double-heralding protocol relies on the detection of two photons, one in each round. However, it is straightforward to modify the protocol to a single-round procedure. The level scheme must then be modified to accommodate the generation of two photons in a single round, as shown in Fig. 7.7b. Each qubit state $|j\rangle$ is coupled to its own excited state $|e_j\rangle$, and the photons that are produced by these two transitions must be perfectly distinguishable, for example in polarization or frequency. A probabilistic optical Bell measurement of the type described in Section 5.2 can then be used to project the atoms onto a maximally entangled state. The maximum success probability is again $\eta^2/2$, where η is the detection efficiency.

The large probability of success compared to the weak driving scheme makes it possible to use double heralding to build large cluster states efficiently. To see this, first consider what happens if we try to extend the length of a linear cluster state. We start with an $(N-1)$-qubit linear cluster state

$$|C\rangle_{2...N} = \frac{1}{\sqrt{2^{N-1}}} (|0\rangle_2 + |1\rangle_2 Z_3)(|0\rangle_3 + |1\rangle_3 Z_4)\dots(|0\rangle_N + |1\rangle_N), \tag{7.144}$$

and a separable qubit in the state $|+\rangle_1$. Applying a successful double-heralding entangling operation on qubits 1 and 2 yields the state

$$|C\rangle_{1...N} = \frac{1}{\sqrt{2}} (|0\rangle_1 |1\rangle_2 \pm |1\rangle_1 |0\rangle_2 Z_3) |C\rangle_{3...N}, \tag{7.145}$$

which can always be transformed into the N qubit cluster $|C\rangle_{1...N}$ by the local operation $H_1 X_2$. If the relative phase in Eq. (7.145) is -1, the Pauli operator Z_1 must precede this correction. When the entangling operation fails, we destroy the entanglement between qubit 2 and the remainder of the cluster, therefore shortening the total length of the cluster by one. Hence, even in the best-case scenario of 50% success probability there will be no net growth in the length of the cluster using this method. However, as was shown in Chapter 2, it is possible to build large clusters efficiently by making use of inefficient off-line creation of shorter chains, which are then attached to the main cluster. Consider a main cluster of N qubits and a short linear chain of m qubits that has been made offline. The double-heralding protocol is performed on the end qubit in each cluster. If the end qubit in the short chain is then measured in the $|\pm\rangle$ basis, the resulting state is locally equivalent to a $N + m - 1$ linear cluster. A failure leads to a shrinking of the long cluster by one qubit, i.e., to length $N - 1$. The average length of the new cluster therefore becomes

$$\langle L \rangle = p(N + m - 1) + (1 - p)(N - 1), \tag{7.146}$$

where p is the success probability of the entangling operation. Therefore as long as $p > 1/m$ the cluster will grow efficiently.

Exercise 7.6: Show that

$$|0\rangle_1 |1\rangle_2 + |1\rangle_1 |0\rangle_2 Z_3 \rightarrow (|0\rangle_1 + |1\rangle_1 Z_2)(|0\rangle_2 + |1\rangle_2 Z_3), \tag{7.147}$$

under the local transformation $H_1 X_2$.

The double-heralding procedure that we have introduced in this section is closely related to the type-II fusion gate introduced in Chapter 6. Indeed, when the double-heralding procedure is collapsed into a single-round, two-photon entangling procedure (using the level structure in Fig. 7.7b), the mathematical identification is exact. The subsequent results for optical quantum computing with single photons can therefore be translated immediately to cluster-state quantum computing with atomic qubits. The benefit of using atoms, rather than photons, is that the latter need to be stored in a quantum memory while the cluster state is prepared, whereas the atoms are themselves quantum memories.

7.4.3 The broker–client protocol

The double-heralding entanglement procedure works only some of the time. In the previous section we saw that this does not preclude the growth of cluster states, though the strategy requires significant offline resources. When the probability of successful entanglement creation becomes low, this can mean a huge overhead in the time and effort required to increase the size of a cluster, and may not be the most sensible approach, especially in the light of naturally occurring decoherence of the qubits in the cluster state.

Before we discuss a different kind of gate that enables the generation of distributed entanglement without catastrophic failure, we will present a method for using double heralding without the need for large-scale offline cluster-state generation. However, some extra resources are needed at each cluster-state node, namely one extra qubit. The two qubits at each local node are assumed to be under complete deterministic control, such that any two-qubit gate operation can be performed upon them. A system where this may be implemented is a four-level atom, where each level represents one of the two-qubit basis states $\{|00\rangle, |01\rangle, |10\rangle, |11\rangle\}$. The essential idea is that one of the qubits at each node acts as a 'broker' and the other a 'client'. The client qubits hold the main cluster state as it grows.

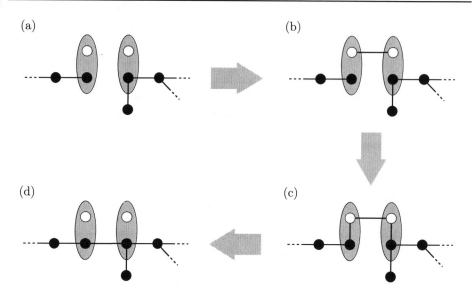

Fig. 7.8. (a) Each 'client' qubit in the main cluster (shown as black circles) has an associated 'broker' qubit (shown as white circles for two of the brokers). The client and broker are part of the same system, and entangling two-qubit operations can be performed between them in a deterministic fashion. (b) Entanglement is built up between two brokers through a heralded probabilistic scheme, such as double heralding. (c) Local two-qubit operations make cluster-state edges between the brokers and their associated clients. For example, a simple cz gate could be performed on each pair. (d) The brokers are measured in the X basis, which rearranges the cluster-state connections in the manner shown. The result is a new connection in the main cluster, which is held by the clients.

The broker qubits, on the other hand, simply build up entanglement with one or more other brokers, using a probabilistic but heralded entanglement-generation scheme like the one presented above. Once created, this entanglement can be passed into the clients deterministically without destroying any of the connections already present in the main cluster (see Fig. 7.8).

7.4.4 Repeat-until-success CX gate

We have seen that it is possible to create cluster states efficiently by erasing the which-path information for photons emitted by atoms in separate cavities. A successful entangling operation is heralded, and we have found that high-fidelity entangled states result from successful operations. The main obstacle to growing large-scale cluster states is that the success probability of the entanglement operation is bounded by 1/2, and to overcome this we need large offline resources. An alternative approach is to find a scheme where the failure of an entangling operation does not destroy any of the entanglement that already exists in the cluster. Double heralding and weak driving both fail to conserve cluster length if no photon is detected, since then one qubit is always projected out of the cluster. However, using detectors with high efficiencies ($\eta \approx 1$) and modest photon-number resolution, it is possible to devise measurements in so-called mutually unbiased bases that protect the cluster from failed entangling operations.

Instead of the L-configuration, we use the level structure in Fig. 7.7b to map qubit states directly to photonic states. Each qubit $i = 1, 2$ is again encoded in the two ground levels $|0\rangle$ and $|1\rangle$, each of which is coupled to its own excited state $|e_0\rangle$ and $|e_1\rangle$ via cavity photons with creation operators $\hat{a}_{0,i}^\dagger$ and $\hat{a}_{1,i}^\dagger$, respectively. Therefore, a π-pulse laser excitation followed by decay of a general two-qubit initial state yields the combined state of the atoms and optical modes

$$|\psi(t_0)\rangle = (c_{00}|00\rangle + c_{01}|01\rangle + c_{10}|10\rangle + c_{11}|11\rangle)|0\rangle$$

$$\rightarrow c_{00}|00\rangle\,\hat{a}_{0,1}^\dagger\hat{a}_{0,2}^\dagger|0\rangle + c_{01}|01\rangle\,\hat{a}_{0,1}^\dagger\hat{a}_{1,2}^\dagger|0\rangle$$

$$+ c_{10}|10\rangle\,\hat{a}_{1,1}^\dagger\hat{a}_{0,2}^\dagger|0\rangle + c_{11}|11\rangle\,\hat{a}_{1,1}^\dagger\hat{a}_{1,2}^\dagger|0\rangle \equiv |\psi(t_1)\rangle, \qquad (7.148)$$

where $|0\rangle$ is the vacuum state of the cavity. The kind of unbiased measurement we need should not give any information about the coefficients $\{c_{00}, c_{01}, c_{10}, c_{11}\}$. Rather, it should distinguish only between photon states that are composed of an equal superposition of all four of the two photon states that appear in each term of Eq. (7.148). Of course, we know already from Chapter 5 that we cannot project these photons deterministically onto the Bell basis, and this result ensures that we cannot construct a deterministic entangling operation. However, there is still the possibility of constructing a probabilistic entangling gate such that the failure channel leads to a known local operation on the qubits. This allows us to correct for these local operations, and leave the entanglement of the atomic qubits intact. We can then repeat the procedure until the entangling operation succeeds. This is called 'repeat-until-success' (RUS).

We implement RUS as follows. Suppose that the two different photon modes for each qubit in Fig. 7.7b are of the same frequency but have orthogonal polarization, e.g., horizontal H and vertical V. A photon from each cavity impinges on a 50:50 beam splitter, with two polarization-sensitive detectors placed in the output modes. This allows the following states to be distinguished:

$$\hat{a}_{H,1}^\dagger\hat{a}_{H,2}^\dagger|0\rangle : \text{two } H \text{ in the same detector,}$$

$$\hat{a}_{V,1}^\dagger\hat{a}_{V,2}^\dagger|0\rangle : \text{two } V \text{ in the same detector,}$$

$$\frac{1}{\sqrt{2}}(\hat{a}_{V,1}^\dagger\hat{a}_{H,2}^\dagger - \hat{a}_{H,1}^\dagger\hat{a}_{V,2}^\dagger)|0\rangle : H \text{ in one detector, } V \text{ in the other,}$$

$$\frac{1}{\sqrt{2}}(\hat{a}_{V,1}^\dagger\hat{a}_{H,2}^\dagger + \hat{a}_{H,1}^\dagger\hat{a}_{V,2}^\dagger)|0\rangle : H \text{ and } V \text{ in the same detector.} \qquad (7.149)$$

Such a measurement basis is still biased, but if we perform the following simple polarization rotation operations on the photons before they impinge on the beam splitter, we can make the required unbiased measurement:

$$U_1 = \frac{1}{\sqrt{2}}\left[\hat{a}_{H,1}^\dagger(\hat{a}_{H,1} + \hat{a}_{V,1}) + \hat{a}_{V,1}^\dagger(\hat{a}_{H,1} - \hat{a}_{V,1})\right], \qquad (7.150)$$

$$U_2 = \frac{1}{\sqrt{2}}\left[\hat{a}_{H,2}^\dagger(\hat{a}_{H,2} + \hat{a}_{V,2}) - i\hat{a}_{V,2}^\dagger(\hat{a}_{H,2} - \hat{a}_{V,2})\right], \qquad (7.151)$$

where 1 and 2 label the two beam-splitter input ports. Notice the extra factor i in Eq. (7.151).

Exercise 7.7: Using U_1 and U_2, show that making the same measurement after the beam splitter allows us to distinguish the following four photon states:

$$|\Phi_{1,2}\rangle = \frac{1}{2}(\hat{a}_{H,1}^\dagger \hat{a}_{H,2}^\dagger \mp i\hat{a}_{H,1}^\dagger \hat{a}_{V,2}^\dagger \pm i\hat{a}_{V,1}^\dagger \hat{a}_{H,2}^\dagger - \hat{a}_{V,1}^\dagger \hat{a}_{V,2}^\dagger)|0\rangle , \tag{7.152}$$

$$|\Phi_3\rangle = \frac{1}{2}(\hat{a}_{H,1}^\dagger \hat{a}_{H,2}^\dagger + \hat{a}_{H,1}^\dagger \hat{a}_{V,2}^\dagger + \hat{a}_{V,1}^\dagger \hat{a}_{H,2}^\dagger + \hat{a}_{V,1}^\dagger \hat{a}_{V,2}^\dagger)|0\rangle , \tag{7.153}$$

$$|\Phi_4\rangle = \frac{1}{2}(\hat{a}_{H,1}^\dagger \hat{a}_{H,2}^\dagger - \hat{a}_{H,1}^\dagger \hat{a}_{V,2}^\dagger - \hat{a}_{V,1}^\dagger \hat{a}_{H,2}^\dagger + \hat{a}_{V,1}^\dagger \hat{a}_{V,2}^\dagger)|0\rangle . \tag{7.154}$$

The first two states $|\Phi_1\rangle$ and $|\Phi_2\rangle$ in Exercise 7.7 are maximally entangled states, while $|\Phi_3\rangle$ and $|\Phi_4\rangle$ are separable states. We can now write the combined state $|\psi(t_1)\rangle$ of the two qubits and the two modes in Eq. (7.148) in terms of the two-photon states $|\Phi_i\rangle$, and a basis of four two-qubit states $|\Psi_i\rangle$ in the Schmidt decomposition

$$|\psi(t_1)\rangle = \frac{1}{2}\sum_{i=1}^{4}|\Phi\rangle_i |\Psi\rangle_i , \tag{7.155}$$

with

$$|\Psi\rangle_1 = e^{i\pi/4} U_{Z,1}\left(\frac{\pi}{2}\right) U_{Z,2}\left(-\frac{\pi}{2}\right) \text{CZ}\, |\psi(t_0)\rangle , \tag{7.156}$$

$$|\Psi\rangle_2 = e^{i\pi/4} U_{Z,1}\left(-\frac{\pi}{2}\right) U_{Z,2}\left(\frac{\pi}{2}\right) \text{CZ}\, |\psi(t_0)\rangle , \tag{7.157}$$

$$|\Psi\rangle_3 = |\psi(t_0)\rangle , \tag{7.158}$$

$$|\Psi\rangle_4 = -i U_{Z,1}(\pi) U_{Z,2}(\pi)\, |\psi(t_0)\rangle . \tag{7.159}$$

Measurement outcomes 1 and 2, associated with the photon state $|\Phi_{1,2}\rangle$, induce a CZ gate on the two qubits up to local rotations around the Z axis. Measurement outcome 3 does nothing to the qubit states, and outcome 4 induces two known local qubit rotations $U_{Z,1}(\pi)$ and $U_{Z,2}(\pi)$. These latter two measurement outcomes therefore do not destroy the entanglement that may already be present between qubit 1 or 2 and the rest of the cluster. The operation can therefore be repeated on the same two qubits, until success is recorded. Therefore, if the protocol is used to grow cluster states, at each step the size of the cluster either increases or remains constant.

The RUS procedure effectively allows us to create atomic cluster states for quantum computing in a deterministic fashion, based on two-photon detection signatures. However, as shown in Eq. (7.149), it requires photon detectors that can distinguish between zero, one, and two photons. Furthermore, when detection efficiencies are low, sometimes only one photon, or even zero photons will be detected. These are failure modes, and lead again to shortening of the cluster-states and low cluster-state fidelities. A combination of double-heralding techniques and RUS that produces large success probabilities ($> 1/2$) with realistic detectors was proposed by Lim *et al.* (2006).

7.5 Other entangling gates

We have seen how to use measurements of photons emitted by atoms placed in two different cavities to create entanglement between the atoms. However, these are not the only techniques that can be used to entangle two atoms for the purposes of quantum computing. In this section we discuss two alternative strategies, namely the intra-cavity entangling operation, where both atoms reside in the same cavity for the duration of the entangling operation, and a gate that takes advantage of the Zeno effect. The latter is a true two-photon gate mediated by atomic transitions.

7.5.1 Intra-cavity entangling operations

Let us first imagine a three-level atom coupled to a cavity mode, as depicted in Fig. 7.9. The cavity decay rate will be κ, but let us ignore that for the moment. We assume a maximum of one photon in the cavity at any time, so the available Fock states are $|0\rangle$ and $|1\rangle$. We will see later that this is justified by a leaky cavity in which photons quickly escape after populating the cavity mode. If the system is initially in a superposition of ground levels $|0\rangle$ and $|1\rangle$, the dynamics of the system are restricted according to the following Hamiltonian:

$$\mathcal{H} = \hbar v \, |e,0\rangle \, \langle e,0| + \left(g \, |1,1\rangle \, \langle e,0| + \frac{\hbar\Omega}{2} \, |1,0\rangle \, \langle e,0| + \text{H.c.} \right), \tag{7.160}$$

where g is the cavity–atom coupling, Ω the coupling of a classical laser and v the detuning of both cavity and laser from the $|1\rangle$-$|e\rangle$ transition. Let us further make the assumptions that

$$v \gg \kappa, \Omega, g/\hbar. \tag{7.161}$$

This set of conditions implies that the level $|e,0\rangle$ is never populated significantly and we can eliminate it using second-order degenerate perturbation theory. Defining $\Omega_{\text{eff}} = \Omega g/2v$ and $v_1 = \Omega^2/(4v)$ and $v_2 = g^2/(\hbar^2 v)$, the effective Hamiltonian becomes

$$\mathcal{H} = \hbar v_1 \, |1,0\rangle \, \langle 1,0| + \hbar v_2 \, |1,1\rangle \, \langle 1,1| + \hbar\Omega_{\text{eff}} (|1,0\rangle \, \langle 1,1| + \text{H.c.}) \,. \tag{7.162}$$

The result of running in the regime described by conditions in Eq. (7.161) is that an effective coupling is induced between two states involving the same low-lying state of the atom ($|1\rangle$), but with one or zero photons in the cavity. By avoiding significant population of the excited atomic state $|e\rangle$ the probability of emitting photons into modes other than the desired cavity mode is suppressed.

Let us now take the leakage rate of the cavity into account. Defining $\hat{a}$ and $\hat{a}^\dagger$ as the respective annihilation and creation operator for the photon in the cavity, the escape of the photon can be described using the usual optical master equation

$$\dot{\rho} = -\frac{i}{\hbar}[\mathcal{H}, \rho] + \kappa(2\hat{a}\rho\hat{a}^\dagger - \hat{a}^\dagger\hat{a}\rho - \rho\hat{a}^\dagger\hat{a}) \,, \tag{7.163}$$

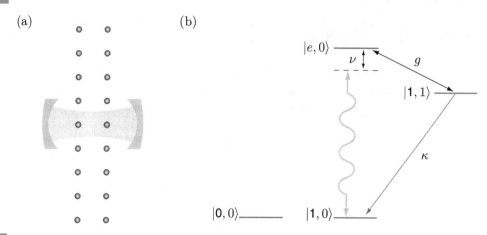

Fig. 7.9. (a) Entangling atoms via intra-cavity operations. (b) The energy-level scheme for a three-level atom coupled to a cavity photon. The three atomic levels are labelled $|0\rangle$, $|1\rangle$, and $|e\rangle$ and we allow a maximum of one photon in the cavity. The cavity decay is κ, the atom–cavity coupling g, and the optical driving strength is Ω.

where ρ is the density matrix of the system, which can be written in a basis involving the three levels $|0,0\rangle$, $|1,0\rangle$, and $|1,1\rangle$. For a sufficiently large κ it is obvious that the level $|1,1\rangle$ is hardly populated, and can be eliminated from the dynamical equations (this is called 'adiabatic elimination'). The result is a master equation for the reduced density matrix $\tilde{\rho}$, which can be expressed in the basis $\{|0,0\rangle, |1,0\rangle\}$. Since the photon quantum number is always zero we omit it from the state description, and can write the dynamical equation as

$$\dot{\tilde{\rho}} = -\frac{i}{\hbar}[\mathcal{H}_{\text{eff}}, \tilde{\rho}] + \frac{\kappa_{\text{eff}}}{2}(2P_1\tilde{\rho}P_1 - P_1\tilde{\rho} - \tilde{\rho}P_1) \tag{7.164}$$

where $\kappa_{\text{eff}} = \Omega_{\text{eff}}^2/\kappa$. We have made use of the projector $P_1 = |1\rangle\langle 1|$ and

$$\mathcal{H}_{\text{eff}} = \hbar\nu_1 P_1 . \tag{7.165}$$

This can be rewritten as

$$\dot{\tilde{\rho}} = -\frac{i}{\hbar}(\mathcal{H}_{\text{cond}}\tilde{\rho} - \mathcal{H}_{\text{cond}}^{\dagger}\tilde{\rho}) + \kappa_{\text{eff}}P_1\tilde{\rho}P_1 , \tag{7.166}$$

with $\mathcal{H}_{\text{cond}} = \mathcal{H}_{\text{eff}} - iP_1\hbar\kappa_{\text{eff}}/2$. The final term on the right-hand side describes a quantum jump, which occurs every time a photon is emitted from the cavity. The relevant jump operator is the density matrix projector $\mathcal{J}(\cdot) = P_1(\cdot)P_1$, and by detecting the decay photons using efficient detectors we know exactly when each jump occurs. The first term on the right-hand side describes the non-Hermitian evolution that occurs when a photon is not emitted and we can rewrite our unconditional master equation as a stochastic conditional equation, as we discussed in Section 7.3.2.

$$\rho(t + dt) = \rho_0(t + dt)(1 - dN) + \frac{\mathcal{J}_1\rho}{\text{Tr}(\mathcal{J}_1\rho)}dN , \tag{7.167}$$

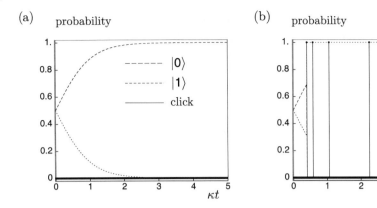

Fig. 7.10. Quantum trajectories for the measurement scheme discussed in the text. (a) The result of the measurement is state $|0\rangle$ and no detector clicks are observed. (b) The result is state $|1\rangle$ and this time a stream of detector clicks, a fluorescence signal, is observed.

and the expectation value of dN is

$$\xi(dN) = \kappa \,\text{Tr}(\mathcal{J}_1 \rho)dt. \tag{7.168}$$

Between jumps the dynamics can be written as an evolution of an (unnormalized) state vector $|\bar{\psi}\rangle$ using a Schrödinger-like equation $d\,|\bar{\psi}\rangle/dt = -i\mathcal{H}_{\text{cond}}\,|\bar{\psi}\rangle/\hbar$.

Importantly, the jump operator projects the system into the $|1\rangle$ state, which is precisely the state from which a jump is more likely to come in the first place. Moreover, the dynamics between jumps does not transfer the system between $|1\rangle$ and $|0\rangle$; it only acts to increase the population of $|0\rangle$ if no decay photons are ever emitted. Thus, we expect a stream of photons, or a macroscopic fluorescence signal, if the system is measured in state $|1\rangle$, and no light otherwise. Examples of quantum trajectories corresponding to both cases are displayed in Fig. 7.10.

The kind of projective measurement described here differs from our usual detection signatures in that it yields a macroscopic signal. However, with two atoms in the cavity we can go further and create entanglement by measuring such macroscopic signals. If the two atoms are identical, and couple with the cavity in the same way, it is straightforward to extend the analysis above.

Exercise 7.8: By following the same procedure as in the single-atom case, systematically eliminating barely populated states, show that the correct master equation for a two-atom density matrix ρ_2 is:

$$\dot{\tilde{\rho}}_2 = -\frac{i}{\hbar}(\mathcal{H}_{\text{cond},2}\tilde{\rho}_2 - \mathcal{H}^\dagger_{\text{cond},2}\tilde{\rho}_2) + \kappa_{\text{eff}}(P_{01} + P_{10})\tilde{\rho}_2(P_{01} + P_{10}) + 4\kappa_{\text{eff}}P_{11}\tilde{\rho}_2 P_{11}. \tag{7.169}$$

The two-atom projectors are defined as $P_{ij} = |ij\rangle\langle ij|$. The conditional Hamiltonian is now

$$\mathcal{H}_{\text{cond},2} = \hbar(\nu_1 - i\kappa_{\text{eff}}/2)[P_{10} + P_{01}] + \hbar(2\nu_1 - 2i\kappa_{\text{eff}})P_{11}. \tag{7.170}$$

Equations (7.169) and (7.170) show that the fluorescence rate is equal for $|10\rangle$ and $|01\rangle$, but the rate for $|11\rangle$ is four times higher than either of these. Therefore, a measurement of zero, weaker, or stronger fluorescence will determine the total number of atoms in state $|1\rangle$ as either zero, one, or two, respectively. The most interesting case is a measurement of 'weaker fluorescence', which results in a quantum jump that is a parity projection $P_{10} + P_{01}$. As we have already discussed (in Section 7.4.2), it is possible to efficiently, and probabilistically, build a cluster state from this kind of projective measurement.

7.5.2 The Zeno effect as a two-photon interaction

As a final topic in this chapter on atomic qubits, we consider an entangling gate that can be used to replace the beam splitter in path-erasure protocols for entangling atoms, or it can be used directly on single-photon qubits. It is the so-called Zeno gate. We discuss it here, because it relies critically on atom–photon interactions. A schematic of the physical implementation of the Zeno gate is shown in Fig. 7.11. It is a spliced optical fibre that is doped with atoms. The relevant level structure of the atoms is shown in the inset of Fig. 7.11.

Without the doped atoms, a spliced fibre acts as a beam splitter. An optical fibre typically consists of a 'core' region of high refractive index surrounded by 'cladding' of lower refractive index. A light beam is guided along the core because total internal reflection occurs at the fibre surfaces for an angle of incidence far enough from the normal to the interface. This happens since Maxwell's equations show that the wave solution in the cladding has an imaginary wave vector, i.e., it exponentially decays, and for a thick-enough cladding only negligible field intensity is emitted from the fibre. This is usually referred to as an 'evanescent' wave. For our beam-splitter device, we use a so-called dual-core fibre, in which the two cores are placed close enough to couple to each other via the evanescent waves. This allows the field amplitude to slowly tunnel from one core to the other. By bringing the cores of two fibres into a dual core and then splitting the cores again, we can set the length of

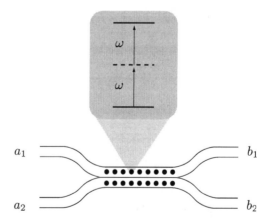

Fig. 7.11. The apparatus required for a Zeno-effect gate consists of a dual-core optical fibre filled with atoms with a strong two-photon absorption line.

the interaction region and change the amount of inter-fibre coupling. Individual photons entering the interaction region will generally be put in a superposition of the two output modes. Another way to understand the spliced fibre as a beam splitter is as follows: we have a device with two input modes and two output modes. Assuming that the mode functions are not distorted in the fibre, and using the fact that the fibre does not retain a memory of the transmitted photons, this must be a two-mode unitary transformation. As we have seen in Chapter 1, any such device can be modelled using a single beam splitter and additional single-mode phase shifters.

We now return to the Zeno gate. Let us assume that the (still empty) fibres are set such that the following unitary transformation takes place between input modes a_1 and a_2, and output modes b_1 and b_2

$$\hat{b}_1^\dagger = \cos\theta\,\hat{a}_1^\dagger + i\sin\theta\,\hat{a}_2^\dagger, \tag{7.171}$$

$$\hat{b}_2^\dagger = i\sin\theta\,\hat{a}_1^\dagger + \cos\theta\,\hat{a}_2^\dagger, \tag{7.172}$$

which is equivalent to Eq. (1.146), with $\zeta = \theta$ and $\varphi = 0$. For simplicity, we identify the photon number with the qubit states, since we can map the atomic state $|0\rangle$ to zero photons $|0\rangle$, and $|1\rangle$ to one photon $|1\rangle$ in an L-configuration. If we examine what happens to qubit states defined by zero or one photons in each input mode, we find

$$|00\rangle \to |00\rangle, \tag{7.173}$$

$$|01\rangle \to \cos\theta\,|01\rangle + i\sin\theta\,|10\rangle, \tag{7.174}$$

$$|10\rangle \to i\sin\theta\,|01\rangle + \cos\theta\,|10\rangle, \tag{7.175}$$

$$|11\rangle \to \cos 2\theta\,|11\rangle + \frac{i\sin 2\theta}{\sqrt{2}}\,(|20\rangle + |02\rangle), \tag{7.176}$$

where 0, 1, and 2 denote the photon-occupation number of the mode. Equation (7.176) is not a satisfactory two-qubit state, since the output from $|11\rangle$ involves doubly occupied modes.

In order to circumvent this problem, the two cores of the fibres in the interaction region are filled with a linear array of N atoms that have a strong two-photon absorption and a negligible single-photon absorption. The spacing between the atoms is such that over the length of the fibre between two adjacent atoms the beam-splitter evolution parameter is $\theta = \delta$.

The presence of the atoms means that every time a state with two photons encounters an atom, it absorbs the photons and spontaneously emits them into another, uninteresting, mode. The atom effectively 'measures' the number of photons in the state and distinguishes between finding 'two photons' and 'less than two photons'. The observer does not, however, know which of the two outcomes occurs. To understand how this affects the state evolution, we look at the problem using the POVM formalism, first introduced in Chapter 3, to describe the different measurement outcomes. Let us consider the ideal case where the two-photon absorption is perfect, and there is no single-photon absorption (or loss). In general, an arbitrary input state ρ will evolve according to

$$\rho \ \rightarrow \ \tilde{\rho} = \mathcal{L}(\rho) \equiv \sum_{k=1,2} A_k \rho A_k^\dagger, \qquad (7.177)$$

where the A_k are the Kraus operators (or 'effects') that define the effect of the measurement on the state. Each measurement outcome is represented by a specific A_k. Since we discard the measurement outcomes in our Zeno gate, we need to sum over all k. The two Kraus operators for the two possible measurements are

$$A_1 = |00\rangle \langle 00| + |01\rangle \langle 01| + |10\rangle \langle 10| + |11\rangle \langle 11| = \hat{\mathbb{I}}_{01}, \qquad (7.178)$$

the identity on the two-mode subspace spanned by zero and one photons in each mode, and

$$A_2 = |00\rangle \left(\langle 02| + \langle 20| \right). \qquad (7.179)$$

The latter Kraus operator induces the two-photon absorption. We apply the super-operator $\mathcal{L}(\rho)$ to both modes every time an atom is encountered. For the single-photon states, the Kraus operator is the identity operator and the evolution is a series of N rotations over angle δ. We may write this as

$$|01\rangle \rightarrow \cos N\delta \,|01\rangle + i \sin N\delta \,|10\rangle \,, \qquad (7.180)$$

$$|10\rangle \rightarrow i \sin N\delta \,|01\rangle + \cos N\delta \,|10\rangle \,. \qquad (7.181)$$

If we arrange the fibre length such that the final mixing angle is $N\delta = \pi/2$, then $|01\rangle \rightarrow i\,|10\rangle$ and $|01\rangle \rightarrow i\,|10\rangle$.

The situation for the initial $|11\rangle$ state is more complicated. Just before the encounter with the first atom, the beam-splitter evolution in Eq. (7.176) gives

$$\rho = \cos^2 2\delta \,|11\rangle \langle 11| + \frac{1}{2} \sin^2 2\delta \left(|20\rangle + |02\rangle \right) \left(\langle 20| + \langle 02| \right) \qquad (7.182)$$

$$+ \frac{i}{\sqrt{2}} \cos 2\delta \sin 2\delta \left[(|20\rangle + |02\rangle) \langle 11| - |11\rangle \left(\langle 20| + \langle 02| \right) \right]. \qquad (7.183)$$

Following the interaction with the first atom, we describe the new state using Eq. (7.179) as

$$\tilde{\rho} = \sum_k A_k \rho A_k^\dagger = \cos^2 2\delta \,|11\rangle \langle 11| + \sin^2 2\delta \,|00\rangle \langle 00| \,. \qquad (7.184)$$

The term $|00\rangle \langle 00|$ is invariant under both the beam-splitter evolution and the two-photon absorption, and so does not change during the remainder of the gate. The $|11\rangle \langle 11|$ term will again undergo the evolution in Eq. (7.184). After the full length of the joined fibre (involving all N atoms), the evolution must be

$$|11\rangle \langle 11| \ \rightarrow \ \cos^{2N} 2\delta \,|11\rangle \langle 11| + (1 - \cos^{2N} 2\delta) \,|00\rangle \langle 00| \,. \qquad (7.185)$$

Using $\delta = \pi/(2N)$, the $|11\rangle$ term becomes $\cos^{2N}(\pi/N)$. We can now suppress the two-photon absorption altogether by choosing N very large, and in the limit of $N \to \infty$ we obtain

$$|11\rangle\langle 11| \;\to\; |11\rangle\langle 11|. \tag{7.186}$$

This is an instance of the Zeno effect in quantum mechanics, where a continuous measurement effectively freezes the quantum state of the system and inhibits the normal unitary evolution of the system. Hence the name 'Zeno gate' for this device.

We may summarize the action of the Zeno gate as the following unitary matrix in the computational basis $\{|00\rangle, |01\rangle, |10\rangle, |11\rangle\}$:

$$U_{\text{Zeno}} = \begin{pmatrix} 1 & 0 & 0 & 0 \\ 0 & 0 & i & 0 \\ 0 & i & 0 & 0 \\ 0 & 0 & 0 & 1 \end{pmatrix}, \tag{7.187}$$

which is a maximally entangling two-qubit gate, closely related to the $\sqrt{\text{SWAP}}$ gate. In the present discussion we have analyzed the Zeno gate for the ideal case of perfect two-photon absorption and no single-photon absorption, using Kraus operators. When the situation is not ideal (e.g., in the case of survival of the $|20\rangle$ and $|02\rangle$ terms and photon loss), the Kraus operators need to be modified and the calculation will generally become much harder. Alternatively, the problem can be formulated in terms of a master equation, which can then be solved using standard techniques. For a detailed analysis of the lossy Zeno gate, see Leung and Ralph (2007).

7.6 References and further reading

Much effort has been devoted in this chapter to the development of quantum optical techniques describing the interaction of light with atoms. There are many excellent quantum optics books available, and here we mention Gerry and Knight (2005) as a very good introduction, and the second edition of Walls and Milburn (2008) as an excellent advanced textbook. The quantum-jump formalism was developed by Carmichael (1999). For the entangling operations based on path erasure we refer the reader to the primary literature. The concept of path erasure was introduced by Cabrillo et al. (1999), and further developed for atoms in cavities by Bose et al. (1999a). Entangling atoms in the weak driving limit was proposed by Browne et al. (2003), the double-heralding procedure is due to Barrett and Kok (2005), and the RUS scheme is due to Lim et al. (2005). The broker–client protocol was conceived by Benjamin et al. (2006). Single-round double heralding was demonstrated experimentally by Moehring et al. (2007). The intra-cavity entangling operation was proposed by Metz et al. (2006). Finally, the Zeno gate was proposed by Franson et al. (2004).

PART III

QUANTUM INFORMATION IN MANY-BODY SYSTEMS

8 Quantum communication with continuous variables

Qubits are not the only information carriers that can be used for quantum information processing. In this chapter, we will focus on quantum communication with 'continuous quantum variables', or continuous variables for short. In the context of quantum information processing we will also call continuous variables 'qunats'. We have seen in Chapter 2 that a natural representation of a continuous variable is given by the position of a particle. The conjugate continuous variable is then the momentum of the particle. Unfortunately, the eigenstates of the position and momentum operators are not physical, and we have to construct suitable approximations to these states that can be created in the laboratory. Any practical information processing device must then take into account the deviation of the actual states from the ideal position and momentum eigenstates. Rather than the position and momentum of a particle, we will consider here the two position and momentum quadratures of an electromagnetic field mode. These operators obey the same commutation relations as the canonical position and momentum operators, but they are *not* the physical position and momentum of field excitations. Approximate eigenstates of the quadrature operators can be constructed in the form of squeezed coherent states. We define a quantum mechanical phase space for quadrature operators, similar to a classical phase space for position and momentum. Probability distributions in the classical phase space then become quasi-probability distributions over the quadrature phase space. We will develop one of these distributions, namely the Wigner function, and identify certain phase-space transformations of the Wigner function with linear-optical and squeezing operations. After this brief introduction, we discuss the creation and detection of continuous-variable entanglement, which will lead us to quantum teleportation and entanglement swapping. We conclude this chapter with a discussion on entanglement distillation and purification, and quantum cryptography using continuous variables.

8.1 Phase space in quantum optics

In this section we introduce phase space and the Wigner function. We give the relation between the pure (but unphysical) continuous-variable quantum states and their squeezed-state approximations, and determine the language for later chapters. Finally, we explore how to measure single-mode optical continuous variables.

The creation and annihilation operators of a single optical mode a with sharply peaked frequency ω can be combined to define operators $\hat{q}$ and $\hat{p}$ according to

$$\hat{q} = \sqrt{\frac{\hbar}{2\omega}} \left(\hat{a} + \hat{a}^\dagger \right) \quad \text{and} \quad \hat{p} = -i\sqrt{\frac{\hbar\omega}{2}} \left(\hat{a} - \hat{a}^\dagger \right). \tag{8.1}$$

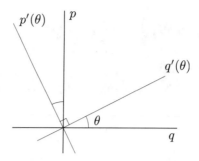

Fig. 8.1.   **Phase space of an optical mode with polarization.**

The commutation relation between $\hat{q}$ and $\hat{p}$, given $[\hat{a}, \hat{a}^\dagger] = 1$, then becomes

$$[\hat{q}, \hat{p}] = i\hbar, \tag{8.2}$$

which means that $\hat{q}$ and $\hat{p}$ form a pair of canonical position and momentum operators. We can then formally define a phase space (see Fig. 8.1) that corresponds to a single optical mode. Just as in classical physics, the state of the system (the optical mode) can be described in terms of its behaviour in phase space. The essentials of this behaviour are captured by the Wigner function. In the remainder of this chapter, we consider the rescaled position and momentum quadratures

$$\hat{q} \to \sqrt{\omega}\,\hat{q} = \sqrt{\frac{\hbar}{2}}\left(\hat{a} + \hat{a}^\dagger\right) \quad \text{and} \quad \hat{p} \to \frac{\hat{p}}{\sqrt{\omega}} = -i\sqrt{\frac{\hbar}{2}}\left(\hat{a} - \hat{a}^\dagger\right), \tag{8.3}$$

which is very convenient when all optical modes have the same (central) frequency. The commutation relations for the new $\hat{q}$ and $\hat{p}$ remain the same as in Eq. (8.2).

8.1.1 The Wigner function

Classically, the Wigner function $W(q, p)$ over phase space is a probability distribution, and the evolution of the system is given by the change of the probability distribution over time. In the quantum mechanical phase space, the Wigner function is no longer a probability distribution, but rather a *quasi*-probability distribution. The reason for this is related to the impossibility of assigning sharp values to both the position and momentum of a single system in quantum mechanics. Nevertheless, we can require that the 'marginals' of $W(q, p)$ are proper probability distributions:

$$\int_{-\infty}^{+\infty} W(q, p)\, dp = P(q) \quad \text{and} \quad \int_{-\infty}^{+\infty} W(q, p)\, dq = P(p). \tag{8.4}$$

The evolution of the system is still described by a change in the Wigner function over time. Normalization of $P(q)$ and $P(p)$ requires that

$$\int_{-\infty}^{+\infty} \int_{-\infty}^{+\infty} dp\, dp\, W(q,p) = 1. \tag{8.5}$$

Note that q and p are real numbers, while $\hat{q}$ and $\hat{p}$ are operators. Since we are dealing with quantum mechanical systems, the probability distributions over q and p must be given by the Born rule

$$P(q) = \text{Tr}(\rho \,|q\rangle \langle q|) \quad \text{and} \quad P(p) = \text{Tr}(\rho \,|p\rangle \langle p|), \tag{8.6}$$

where the state of the optical mode is denoted by the density operator ρ, and $|q\rangle$, $|p\rangle$ are position and momentum eigenstates, respectively.[1]

Next, we will derive the explicit form of the Wigner function associated with an arbitrary density operator. To this end, we exploit certain transformation properties of the marginals $P(q)$ and $P(p)$, and $W(q,p)$. We also make use of the Fourier transforms of these distributions. First, consider a pure rotation in phase space over an angle θ (see Fig. 8.1):

$$q'(\theta) = \cos\theta\, q + \sin\theta\, p \quad \text{and} \quad p'(\theta) = -\sin\theta\, q + \cos\theta\, p. \tag{8.7}$$

We can think of this either as a rotation of the coordinate system in phase space, or as an active rotation of ρ. Let us define

$$\rho_\theta = R(\theta)\, \rho\, R^\dagger(\theta), \tag{8.8}$$

where the unitary transformation R implements the rotation of Eq. (8.7). The expectation value of the rotated state must again be a probability distribution

$$P_\theta(q) \equiv \text{Tr}\left[R(\theta)\, \rho\, R^\dagger(\theta)\, |q\rangle \langle q| \right] = \text{Tr}\left[\rho R^\dagger(\theta)\, |q\rangle \langle q|\, R(\theta) \right]$$
$$= \text{Tr}\left[\rho\, |q'(-\theta)\rangle \langle q'(-\theta)| \right], \tag{8.9}$$

where we have used the cyclic property of the trace. Note that the state $|q'(\theta)\rangle$ is an eigenstate of neither $\hat{q}$, nor $\hat{p}$, but of the rotated quadrature $\hat{q}'$. In terms of the Wigner function, this rotation corresponds to the rotation of the variables q and p according to Eq. (8.7), and we therefore require

$$P_\theta(q) = \int_{-\infty}^{+\infty} W(\cos\theta\, q - \sin\theta\, p, \sin\theta\, q + \cos\theta\, p)\, dp. \tag{8.10}$$

Second, we take the Fourier transform of both $W(q,p)$ and $P_\theta(q)$:

$$\mathfrak{W}(u,v) = \int_{-\infty}^{+\infty} \int_{-\infty}^{+\infty} W(q,p) e^{-iuq - ivp}\, dq\, dp, \tag{8.11}$$

---

[1] We will label $|q\rangle$ and $|p\rangle$ with indices q and p when ambiguity would otherwise arise in the notation.

and

$$\mathfrak{P}_\theta(\zeta) = \int_{-\infty}^{+\infty} P_\theta(q) e^{-\zeta q}\, dq$$

$$= \int_{-\infty}^{+\infty} \int_{-\infty}^{+\infty} W(q',p') e^{-i\zeta q}\, dq dp$$

$$= \mathfrak{W}(\zeta \cos\theta, \zeta \sin\theta), \tag{8.12}$$

where the last equality follows from a coordinate change $(q,p) \rightarrow (q',p')$. The function $\mathfrak{W}(u,v)$ is called the 'characteristic function' of the Wigner function. Note that since $\mathfrak{P}_\theta(\zeta) = \mathfrak{W}(\zeta \cos\theta, \zeta \sin\theta)$, we can use $\mathfrak{P}_\theta(\zeta)$ for all θ to reconstruct the full two-dimensional characteristic function $\mathfrak{W}$, and hence the Wigner function itself.

Third, we calculate

$$\mathfrak{P}_\theta(\zeta) = \int \langle q| U(\theta)\rho\, U^\dagger(\theta)\, |q\rangle\, e^{-i\zeta q}\, dq = \mathrm{Tr}\left[\rho\, U^\dagger(\theta) e^{-i\zeta \hat{q}} U(\theta)\right]$$

$$= \mathrm{Tr}\left[\rho\, e^{-i\hat{q}\zeta \cos\theta - i\hat{p}\zeta \sin\theta}\right]. \tag{8.13}$$

Identifying $u = \zeta \cos\theta$ and $v = \zeta \sin\theta$, we can write the characteristic function as

$$\mathfrak{W}(u,v) = \mathrm{Tr}\left[\rho\, e^{-i\hat{q}u - i\hat{p}v}\right] = \int \langle q| \rho\, e^{-i\hat{q}u - i\hat{p}v}\, |q\rangle\, dq$$

$$= e^{-iuv/2} \int \langle q| \rho\, e^{-iu\hat{q}}\, |q + v\rangle\, dq$$

$$= \int e^{-iux} \langle x - v/2| \rho\, |x + v/2\rangle\, dx, \tag{8.14}$$

where we have made the substitution $q = x - v/2$. Finally, taking the Fourier transform back to $W(q,p)$ then yields the standard form for the Wigner function

$$W(q,p) = \frac{1}{2\pi\hbar} \int_{-\infty}^{\infty} e^{ipx/\hbar} \langle q - x/2| \rho\, |q + x/2\rangle\, dx. \tag{8.15}$$

The manipulations we have performed have given us the Wigner function for ρ, which is an operator associated with the state of a system. However, we can define the Wigner function for an arbitrary operator that has its support in the same Hilbert space as ρ. In particular, we can construct Wigner functions $W_A(q,p)$ for observables $\hat{A}$ by making the substitution $\rho \rightarrow \hat{A}$. We will find later in this chapter that this has important benefits.

From Eq. (8.15) it is clear that any density operator ρ gives rise to a Wigner function. However, there is a special class of states that is important for optical quantum information processing with continuous variables. These are the so-called Gaussian states, which are uniquely defined by the means $\langle q\rangle$ and $\langle p\rangle$, and their variances Δq and Δp. Examples of Gaussian states are the vacuum, coherent states, and squeezed coherent states. Fock states are *not* Gaussian states.

Exercise 8.1: Show that taking the Fourier transform of $\mathfrak{W}(u, v)$ gives the required result of Eq. (8.15).

Wigner functions are particularly useful for calculating expectation values of symmetrically ordered functions of $\hat{q}$ and $\hat{p}$. Consider symmetric functions $S(\hat{q}^n \hat{p}^m)$ of $\hat{q}$ and $\hat{p}$, where $S(\cdot)$ indicates the symmetrized form of the argument. For example, $S(\hat{q}\hat{p}) = (\hat{q}\hat{p} + \hat{p}\hat{q})/2$ and $S(\hat{q}^2 \hat{p}) = (\hat{q}^2 \hat{p} + \hat{q}\hat{p}\hat{q} + \hat{p}\hat{q}^2)/3$. It can be shown that the expectation value of $S(\hat{q}^n \hat{p}^m)$ is given by

$$\text{Tr}\left[\rho\, S(\hat{q}^n \hat{p}^m)\right] = \frac{1}{2\pi\hbar} \int dq\, dp\, W(q, p) q^n p^m. \tag{8.16}$$

Alternatively, one may wish to calculate normal-ordered or anti-normal-ordered products of the creation and annihilation operators that constitute $\hat{q}$ and $\hat{p}$. In these cases, one should use the P-representation or the Q-representation, respectively, instead of the Wigner function. However, the discussion of these alternative quasi-probability distributions is beyond the scope of this book, and the reader is referred to Leonhardt (1997) or Barnett and Radmore (1997) for comprehensive introductions to quasi-probability distributions and their characteristic functions.

8.1.2 Transformations in phase space

There exists a one-to-one correspondence between the evolution of the quantum state of an optical mode and the transformations of the corresponding quasi-probability distribution in phase space. In this section, we consider three basic transformations in phase space, namely translations, rotations, and squeezing operations. We will show that these correspond to the evolution of quantum states due to coherent displacements $D(\alpha)$, phase shifts $e^{-i\theta\hat{n}}$, and single-mode squeezers $S(\xi)$. The operators $D(\alpha)$ and $S(\xi)$ are in fact named after their effect on the Wigner function in phase space.

Translations in phase space

Translations in phase space correspond to the displacement operator that was introduced in Chapter 1:

$$D^\dagger(\alpha)\, \hat{a}\, D(\alpha) = \hat{a} + \alpha, \tag{8.17}$$

where α is a complex number. We can apply this operator to the quadrature operators $\hat{q}$ and $\hat{p}$, and obtain the transformed position quadrature

$$D^\dagger(\alpha)\, \hat{q}\, D(\alpha) = \hat{q} + \sqrt{2\hbar}\,\text{Re}(\alpha), \tag{8.18}$$

and the transformed momentum quadrature

$$D^\dagger(\alpha)\, \hat{p}\, D(\alpha) = \hat{p} + \sqrt{2\hbar}\,\text{Im}(\alpha). \tag{8.19}$$

The expectation values of $\hat{q}$ and $\hat{p}$ are displaced by an amount dependent upon the argument of the displacement operator. The variances Δq and Δp are unaffected by these displacements. In fact, any higher-order moments of $\hat{q}$ and $\hat{p}$ are unaffected.

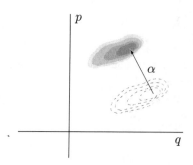

Fig. 8.2.   Translations of the Wigner function correspond to the displacement operator $D(\alpha)$.

Exercise 8.2: Show that the higher-order moments of $\hat{q}$ and $\hat{p}$ are unaffected by the displacement operator $D(\alpha)$.

It is now convenient to change to so-called 'photon-number units', in which $\hbar = 1/2$. In this way, the value of α corresponds directly to the translation in phase space. The effect of the displacement operator on a position eigenstate becomes

$$D(\alpha)\,|q\rangle = e^{2iq\mathrm{Im}(\alpha)}\,|q + \mathrm{Re}(\alpha)\rangle\,. \tag{8.20}$$

We will use this convention in the remainder of this chapter and the next. The Wigner function can then be written as

$$W(q,p) = \frac{1}{\pi}\int_{-\infty}^{\infty} dx\, e^{2ipx}\,\langle q - x/2|\,\rho\,|q + x/2\rangle\,, \tag{8.21}$$

and the transformed Wigner function $W_\alpha(q,p)$ is obtained by substituting the transformed density operator $D(\alpha)\rho D^\dagger(\alpha)$ into ρ:

$$
\begin{aligned}
W_\alpha(q,p) &= \frac{1}{\pi}\int_{-\infty}^{\infty} dx\, e^{2ipx}\,\langle q - x/2|\,D(\alpha)\rho D^\dagger(\alpha)\,|q + x/2\rangle \\
&= \frac{1}{\pi}\int_{-\infty}^{\infty} dx\, e^{2ipx} e^{-2ix\mathrm{Im}(\alpha)}\,\langle q - \mathrm{Re}(\alpha) - x/2|\,\rho\,|q - \mathrm{Re}(\alpha) + x/2\rangle \\
&= W(q - \mathrm{Re}(\alpha), p - \mathrm{Im}(\alpha))\,. \tag{8.22}
\end{aligned}
$$

In other words, the displacement operator $D(\alpha)$ of an optical mode corresponds to a translation of the Wigner function in the quadrature phase space (see Fig. 8.2), which shows up as the opposite translation $(-\alpha)$ in the coordinate system (q,p).

The translations form a group of transformations in phase space, since the combination of any two translations α_1 and α_2 forms a third translation $\alpha_3 = \alpha_1 + \alpha_2$, and every translation α has an inverse $-\alpha$. The group identity corresponds to $\alpha = 0$. These group properties follow directly from the definition of $D(\alpha)$, and are useful in many calculations.

Rotations in phase space

We derived the explicit form of the Wigner function in Eq. (8.15) by requiring that a rotation in phase space corresponds to a counter-rotation of the quadratures according to Eq. (8.8):

$$\rho_\theta = R(\theta)\,\rho\,R^\dagger(\theta). \tag{8.23}$$

This means that R transforms the quadrature operators according to

$$R^\dagger(\theta)\,\hat{q}\,R(\theta) = \cos\theta\,\hat{q} + \sin\theta\,\hat{p}$$
$$R^\dagger(\theta)\,\hat{p}\,R(\theta) = -\sin\theta\,\hat{q} + \cos\theta\,\hat{p}. \tag{8.24}$$

It is straightforward to prove that the number operator is the generator of these rotations

$$R(\theta) = e^{-i\theta\hat{n}}. \tag{8.25}$$

Rotations of the Wigner function therefore correspond to phase shifts of the optical mode (see Fig. 8.3). These phase-space transformations form a simple two-dimensional rotation group, which can be proved directly from Eq. (8.25). The transformation of the Wigner function then becomes

$$W_\theta(q,p) = W(\cos\theta\,q - \sin\theta\,p, \sin\theta\,q + \cos\theta\,p), \tag{8.26}$$

which is the counter-rotation corresponding to Eq. (8.24).

Exercise 8.3: Prove Eq. (8.25), and show that the number state $|n\rangle$ is represented by a rotationally invariant phase-space distribution.

Squeezing in phase space

We can formally write the phase-space squeezing operation shown in Fig. 8.4 as a transformation

$$W(q,p) \to W_\Delta(q',p') = W\left(q\Delta, \frac{p}{\Delta}\right), \tag{8.27}$$

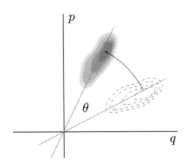

Fig. 8.3.   Rotations of the Wigner function correspond to the phase shift $e^{-i\theta\hat{n}}$.

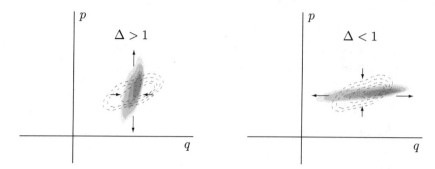

Squeezing of the Wigner function corresponds to the single-mode squeezing operator $S(\xi)$. We have defined $\Delta = e^r$.

where Δ is real and positive. It is clear that such a transformation leaves volumes in phase space invariant, since Δ is a simple scale factor of position and momentum, and $q'p' = qp$. This is a necessary requirement for keeping the marginal probability distributions normalized. When $\Delta > 1$ the distribution is squeezed in the q direction (i.e., the distribution becomes narrower in the q direction), and when $\Delta < 1$ the squeezing occurs in the p direction. Squeezing in other directions can be implemented by rotating $W(q,p)$ over the desired angle θ, squeezing in the q (or p) direction, and rotating W again over an angle $-\theta$.

The quadrature operators $\hat{q}$ and $\hat{p}$ transform under the single-mode squeezing operator

$$S(\xi) = \exp\left(\xi^* \frac{\hat{a}^2}{2} - \xi \frac{\hat{a}^{\dagger 2}}{2}\right) \tag{8.28}$$

with $\xi = re^{i\phi}$. If we assume $\xi = r$ is real, and either positive or negative (meaning $\phi = 0$ or $\phi = \pi$), we obtain

$$S(r)\,\hat{q}\,S^\dagger(r) = \hat{q}\,e^{+r} \equiv \hat{q}\Delta \quad \text{and} \quad S(r)\,\hat{p}\,S^\dagger(r) = \hat{p}\,e^{-r} \equiv \frac{\hat{p}}{\Delta}. \tag{8.29}$$

From this, we can establish the effect of a squeezing operation on a position eigenstate:

$$S(r)\hat{q}\,|q\rangle = S(r)q\,|q\rangle = q\,[S(r)\,|q\rangle]$$

$$= S(r)\hat{q}S^\dagger(r)S(r)\,|q\rangle = e^{+r}\hat{q}\,[S(r)\,|q\rangle]. \tag{8.30}$$

Defining $|\psi\rangle = S(r)\,|q\rangle$, we deduce from this that $|\psi\rangle$ is proportional to the rescaled position eigenstate $|qe^{-r}\rangle$. When we normalize $|\psi\rangle$, we find that

$$S(r)\,|q\rangle = e^{-r/2}\,|qe^{-r}\rangle = \frac{1}{\sqrt{\Delta}}\,|q/\Delta\rangle. \tag{8.31}$$

The squeezing of a state $\rho \to S(r)\rho S^\dagger(r)$ therefore leads to the Wigner function

$$W_r(q,p) = \frac{1}{\pi} \int dx\, e^{2ipx} \langle q - x/2| S(r)\rho S^\dagger(r) |q + x/2\rangle$$

$$= \frac{1}{\pi} \int dx\, e^{2ipx} e^r \langle e^r(q - x/2)| \rho |e^r(q + x/2)\rangle$$

$$= \frac{1}{\pi} \int dy\, e^{2ipye^{-r}} \langle qe^r - y/2| \rho |qe^r + y/2\rangle$$

$$= W\left(qe^r, pe^{-r}\right)$$

$$= W(q\Delta, p/\Delta), \tag{8.32}$$

where we have used $S^\dagger(r) = S(-r)$, and we have changed the integration variable from x to $y = xe^r$ in the third line. This transformation has the same form as Eq. (8.27), and the rescaling of q and p by Δ is therefore equivalent to single-mode squeezing in the q and p directions. Squeezing in arbitrary directions ξ can be derived by using $S(\xi) = R^\dagger(\phi)S(r)R(\phi)$.

Exercise 8.4: Prove the normalization of the state $S(r)|q\rangle$ in Eq. (8.31).

We have shown that the translation, rotation, and squeezing operations in phase space correspond to the state evolution that is generated by linear and quadratic Hamiltonians of the single optical mode. These are the 'linear' mode transformations, because the mode operators are transformed into a linear function of themselves:

$$\hat{a} \to u\hat{a} + v\hat{a}^\dagger + w \quad \text{and} \quad \hat{q} \to r\hat{q} + s\hat{p} + t, \tag{8.33}$$

and so on. They are known from Chapter 1 as the Bogoliubov transformations. This raises the question: what geometric transformations correspond to nonlinear mode transformations? We will not discuss this in detail, but merely state that each nonlinear optical transformation has a well-defined operation on the Wigner function. This must be true, since the Wigner function is a complete representation of a quantum state. Next, we consider the Wigner function of a number of quantum states that play an important role in continuous-variable quantum information processing.

8.1.3 Vacuum, coherent, and squeezed coherent states

From the correspondence between the simple phase-space operations with optical mode transformations we now find a straightforward recipe for finding the Wigner function of coherent states, squeezed vacuum, and squeezed coherent states. We first determine the Wigner function of the vacuum state, and create the other states by displacements, rotations, and squeezing. The wave function of the vacuum state in the one-dimensional position quadrature space can be defined as

$$\psi_0(q) = \langle q|0\rangle. \tag{8.34}$$

Since $\hat{a}\,|0\rangle = 0$, we can write

$$\langle q|\hat{a}\,|0\rangle = \langle q|\,(\hat{q} + i\hat{p})\,|0\rangle = \left(q + \frac{\partial}{\partial q}\right)\psi_0(q) = 0, \tag{8.35}$$

where we have again chosen $\hbar = 1/2$. Furthermore, we have used the commutation relation between $\hat{q}$ and $\hat{p}$, which allows us to write $p = -i\partial/\partial q$. The normalized solution to this differential equation is given by

$$\psi_0(q) = \sqrt[4]{\frac{2}{\pi}}\,\exp\left(-q^2\right). \tag{8.36}$$

Substituting this into Eq. (8.15) yields

$$W_0(q,p) = \frac{2}{\pi}\,\exp\left(-2q^2 - 2p^2\right), \tag{8.37}$$

which is the Wigner function for the vacuum state of a single optical mode.

In order to find the Wigner function of a coherent state, we translate the Wigner function of the vacuum in Eq. (8.37) along $\alpha \equiv q_\alpha + ip_\alpha$, where q_α and p_α are the real and imaginary parts of α, respectively. We then obtain

$$W_\alpha(q,p) = W_0(q - q_\alpha, p - p_\alpha)$$
$$= \frac{2}{\pi}\,\exp\left[-2(q - q_\alpha)^2 - 2(p - p_\alpha)^2\right]. \tag{8.38}$$

Similarly, the Wigner function of the squeezed vacuum state can be found immediately using Eq. (8.27):

$$W_\Delta(q,p) = W_0(\Delta\,q, p/\Delta)$$
$$= \frac{2}{\pi}\,\exp\left(-2\Delta^2 q^2 - \frac{2p^2}{\Delta^2}\right). \tag{8.39}$$

Plots of the Wigner functions for the vacuum and the squeezed vacuum with $\Delta = 2$ are given in Fig. 8.5. Combining the squeezing and translation operations, we obtain the Wigner function for the squeezed coherent state

$$W_{\alpha,\Delta}(q,p) = \frac{2}{\pi}\,\exp\left[-2\Delta^2(q - q_\alpha)^2 - \frac{2(p - p_\alpha)^2}{\Delta^2}\right]. \tag{8.40}$$

Note that we have applied the squeezing operator first, followed by the phase-space translation. This agrees with the definition of the squeezed coherent states in Chapter 1. It should also be clear that changing the order of squeezing and displacement gives rise to a different Wigner function.

Exercise 8.5: Verify that the integral of $W_{\alpha,\Delta}(q,p)$ over phase space is 1.

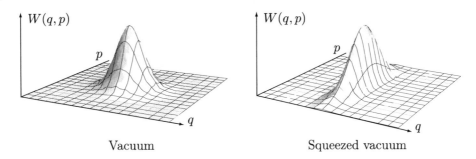

Vacuum   Squeezed vacuum

Fig. 8.5. Wigner functions of the vacuum and the squeezed vacuum ($\Delta = 2$). The peaks are centred around the origin; we have shifted the vertical axis for clarity.

8.1.4 Multi-mode Wigner functions

The extension of the Wigner function to describe multi-mode fields is immediate. For two modes in (separable) states ρ_1 and ρ_2, the Wigner function of the combined system is the product of the individual systems

$$W(q_1,p_1;q_2,p_2) = W_1(q_1,p_1)W_2(q_2,p_2), \tag{8.41}$$

where we have defined

$$W_j(q_j,p_j) = \frac{1}{\pi} \int dx\, e^{2ip_j x} \langle q_j - x/2|\, \rho_j\, |q_j + x/2\rangle \tag{8.42}$$

for $j = 1, 2$. We can now define transformations that mix the quadratures of mode 1 with the quadratures of mode 2.

The most widely considered optical transformation that couples two modes is the beam splitter. The general beam-splitter rule of Eq. (1.147) in Chapter 1 leads to the following transformations for the quadratures of mode 1

$$\hat{q}_1 \to \cos\theta\, \hat{q}_1 + \sin\theta\,(\sin\varphi\, \hat{q}_2 + \cos\varphi\, \hat{p}_2)$$
$$\hat{p}_1 \to \cos\theta\, \hat{p}_1 - \sin\theta\,(\cos\varphi\, \hat{q}_2 - \sin\varphi\, \hat{p}_2), \tag{8.43}$$

and to the transformations of mode 2

$$\hat{q}_2 \to \cos\theta\, \hat{q}_2 + \sin\theta\,(-\sin\varphi\, \hat{q}_1 + \cos\varphi\, \hat{p}_1)$$
$$\hat{p}_2 \to \cos\theta\, \hat{p}_2 - \sin\theta\,(\cos\varphi\, \hat{q}_1 + \sin\varphi\, \hat{p}_1). \tag{8.44}$$

The phase φ can be used to choose which quadratures are mixed in the beam splitter. We usually take $\varphi = \pi/2$, such that position and momentum quadratures do not mix. To find the transformed Wigner function, we invert the operator transformation, remove the hats from the operators, and use the resulting relations as substitution rules.

In practice, squeezing can be obtained by pumping a medium that has a $\chi^{(2)}$ nonlinearity, which can be written as a Hamiltonian that is quadratic in the annihilation (creation) operators. The quadratic operators correspond to linear-optical elements such as beam splitters, polarization rotations, and phase shifters. Consequently, we have a complete (linear) optical description for creating and manipulating continuous quantum variables. Next, we show how continuous variables can be measured with homodyne detection.

8.1.5 Homodyne detection

The Wigner function determines, via its marginals, the probability of finding outcome x_ζ when measuring the $\hat{x}_\zeta$ quadrature of an optical mode in a state ρ. The next natural question is how to implement such measurements. The answer is that quadratures of optical fields are measured by 'homodyne detection'. Consider a 50:50 beam splitter with input modes a_1 and a_2, and output modes b_1 and b_2 (see Fig. 8.6). The intensity difference (measured with photocurrent detectors) in the output modes can be written in terms of the input modes:

$$\Delta I = \langle \hat{b}_1^\dagger \hat{b}_1 - \hat{b}_2^\dagger \hat{b}_2 \rangle = \langle \hat{a}_1^\dagger \hat{a}_2 - \hat{a}_2^\dagger \hat{a}_1 \rangle. \tag{8.45}$$

Next, we assume that mode a_1 is the signal mode (whose quadrature we wish to measure), and the field in mode a_2 is a strong local oscillator, i.e., a bright coherent state. We can then replace the operators $\hat{a}_2$ and $\hat{a}_2^\dagger$ with the complex amplitude $\alpha e^{\pm i\zeta}$ of the (now classical) field:

$$\Delta I = \langle \hat{a}^\dagger \alpha e^{i\zeta} + \alpha e^{-i\zeta} \hat{a} \rangle = \sqrt{2}\alpha \left\langle \frac{e^{-i\zeta}\hat{a} + e^{i\zeta}\hat{a}^\dagger}{\sqrt{2}} \right\rangle. \tag{8.46}$$

Therefore, up to a known scaling factor $\sqrt{2}\alpha$, the intensity difference ΔI measures $(e^{-i\zeta}\hat{a}_1 + e^{i\zeta}\hat{a}_1^\dagger)/\sqrt{2}$, which is the quadrature $\hat{x}_\zeta$. We can choose which quadrature we measure by adjusting the phase of the local oscillator. Clearly, the local oscillators of the homodyne detectors in a set-up must have a well-defined phase relationship with the signal modes. In practice, this often means that both the local oscillators and the signals must derive their phase relationship from a master laser that serves as a clock.

How do we describe homodyne detection in terms of the Wigner function? To establish this, recall that we can write down the Wigner function for an operator. We first give the

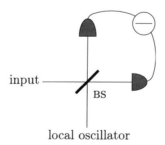

input

BS

local oscillator

Fig. 8.6. Homodyne detection: the detectors record the difference of the intensities.

overlap formula for two Wigner functions W_A and W_B of operators A and B, respectively:

$$\text{Tr}(AB) = \frac{\pi}{8} \int_{-\infty}^{\infty} \int_{-\infty}^{\infty} dq\, dp\, W_A(q,p)\, W_B(q,p), \qquad (8.47)$$

where A and B do not have to be Hermitian. This formula thus allows us to calculate expectation values of operators. The expectation value of the position quadrature is then

$$\text{Tr}(\rho \hat{q}) = \frac{\pi}{8} \int_{-\infty}^{\infty} \int_{-\infty}^{\infty} dq\, dp\, q\, W(q,p), \qquad (8.48)$$

which agrees with our result in Eq. (8.16) above. A particular outcome q' of the homodyne detection can then be implemented by using the Wigner function for the position eigenstate (or its regularized squeezed coherent-state approximation) instead of the operator $\hat{q}$. The probability $\text{Tr}(\rho\,|q'\rangle\langle q'|)$ is of course zero for ideal position eigenstates, but this technique can still be used to determine the evolution of a system based on specific homodyne-detection outcomes of some subsystems. The trace is then interpreted as the partial trace.

Exercise 8.6: Prove the overlap formula in Eq. (8.47), and calculate the Wigner function for the single-mode displacement, rotation, and squeezing operators.

The finite-detection efficiency η of a detector can be described using the beam-splitter model introduced in Chapter 4 (see Fig. 4.1). The quadrature operators transform according to

$$\hat{q} \to \hat{q}' = \sqrt{\eta}\,\hat{q} + \sqrt{1-\eta}\,\hat{q}_v$$
$$\hat{p} \to \hat{p}' = \sqrt{\eta}\,\hat{p} + \sqrt{1-\eta}\,\hat{p}_v, \qquad (8.49)$$

where $\hat{q}$ and $\hat{p}$ are the quadratures of the measured mode, and $\hat{q}_v$ and $\hat{p}_v$ are the quadratures of the (vacuum) input mode of the beam splitter.

In the next section, we study how to produce entanglement in optical continuous variables, and how to establish quantum communication.

8.2 Continuous-variable entanglement

One of the main operations in quantum information processing is the creation of entanglement, and by extension, the detection of entanglement. In this section, we describe how to perform these tasks with optical continuous variables. We first consider the ideal case of entangling position and momentum eigenstates, and then we show that in terms of optical operations, this corresponds to a two-mode squeezing operator. In addition, we show how the same entangling procedure can be implemented using only linear-optical transformations, such as a 50:50 beam splitter. Reversibility of optical elements then leads directly to a method for entanglement detection.

For discrete d-level systems, the generalized Bell states are given by

$$|\Psi_{nm}\rangle = \frac{1}{\sqrt{d}} \sum_{k=0}^{d-1} e^{2\pi i n k/(d-1)} |k, k \oplus m\rangle. \tag{8.50}$$

It is easily verified that for $d = 2$ these states reduce to the familiar Bell states for qubits. We can take the continuum limit, where k, n, and m become the continuous variables q, u, and v, and the sum is replaced with an integral:

$$|\Psi(u, v)\rangle = \frac{1}{\mathcal{N}} \int dq\, e^{-iuq} |q, q + v\rangle_q, \tag{8.51}$$

with $\mathcal{N}$ a normalization constant. For these states, the following expectation values hold:

$$\langle \hat{q}_1 - \hat{q}_2 \rangle = -v \quad \text{and} \quad \langle \hat{p}_1 + \hat{p}_2 \rangle = u, \tag{8.52}$$

and the variances are

$$\text{Var}(\hat{q}_1 - \hat{q}_2) = 0 \quad \text{and} \quad \text{Var}(\hat{p}_1 + \hat{p}_2) = 0, \tag{8.53}$$

which means that both the positions and momenta of the two systems are perfectly correlated. Therefore, they form an ideal 'Einstein–Podolski–Rosen' (EPR) pair. The states $|\Psi(u, v)\rangle$ are the continuous-variable versions of the Bell states. Using the cx gate for continuous variables in Eq. (2.98), we can write $|\Psi(u, v)\rangle$ in terms of position and momentum eigenstates $|v\rangle_q$ and $|u\rangle_p$:

$$
\begin{aligned}
|\Psi(u, v)\rangle &= \frac{1}{\mathcal{N}} \int dq\, e^{-iuq} \exp\left(-2i\hat{q}_1\hat{p}_2\right) |q, v\rangle_q \\
&= \text{cx} \left(\frac{1}{\mathcal{N}} \int dq\, e^{-iuq} |q\rangle_q \right) |v\rangle_q = \text{cx}\, |u\rangle_p |v\rangle_q.
\end{aligned} \tag{8.54}
$$

We now have to find a realistic (meaning approximate) optical implementation for this state.

First of all, we note that we can again use squeezed coherent states as approximate eigenstates of the position and momentum operators. The cx gate in Eq. (8.54) can be written in terms of creation and annihilation operators, which reveals that it is a two-mode squeezing operator:

$$\text{cx}_{ij} = \exp\left(-2i\hat{q}_i\hat{p}_j\right) = \exp\left[-\frac{1}{2} \left(\hat{a}_i\hat{a}_j - \hat{a}_i^\dagger\hat{a}_j^\dagger + \hat{a}_i^\dagger\hat{a}_j - \hat{a}_i\hat{a}_j^\dagger \right) \right]. \tag{8.55}$$

This is completely analogous to the creation of two-qubit entanglement, where first a Hadamard gate is applied to the control qubit, and then a cx is applied. However, the large amounts of squeezing needed for the continuous-variable cx gate are hard to implement in practice. We will therefore investigate whether we can entangle two squeezed coherent states with passive linear-optical elements, such as phase shifters and beam splitters.

It turns out that this is indeed possible: consider two separate Wigner functions $W_1(q_1, p_1)$ and $W_2(q_2, p_2)$, corresponding to two squeezed coherent states. Suppose that these are position and momentum squeezed coherent states, respectively:

$$W_1(q_1, p_1) = \frac{2}{\pi} \exp\left[-2\Delta^2 (q_1 - q_\alpha)^2 - \frac{2(p_1 - p_\alpha)^2}{\Delta^2} \right], \tag{8.56}$$

$$W_2(q_2, p_2) = \frac{2}{\pi} \exp\left[-\frac{(q_2 - q_\alpha)^2}{2\Delta^2} - 8\Delta^2 (p_2 - p_\alpha)^2 \right]. \tag{8.57}$$

We write the Wigner function of the combined system as the product of W_1 and W_2, $W(q_1, p_1; q_2, p_2)$, and apply a rotation in phase space. This yields

$$W(q_1 \cos\theta + q_2 \sin\theta, p_1 \cos\theta + p_2 \sin\theta; q_2 \cos\theta - q_1 \sin\theta, p_2 \cos\theta - p_1 \sin\theta). \tag{8.58}$$

If we set $\theta = \pi/2$, the rotated Wigner function $W = W(q_1', p_1'; q_2', p_2')$ becomes

$$W = \frac{4}{\pi^2} \exp\left[-\frac{((q_1' - q_2')/\sqrt{2} - q_\alpha)^2}{2\Delta^2} - \frac{2((p_1' + p_2')/\sqrt{2} - p_\alpha)^2}{\Delta^2} \right]$$

$$\times \exp\left[-2\Delta^2 \left(\frac{q_1' + q_2'}{\sqrt{2}} - q_\alpha \right)^2 - 8\Delta^2 \left(\frac{p_1' - p_2'}{\sqrt{2}} - p_\alpha \right)^2 \right], \tag{8.59}$$

where

$$q_1' = \frac{q_1 + q_2}{\sqrt{2}}, \quad q_2' = \frac{q_2 - q_1}{\sqrt{2}}, \quad p_1' = \frac{p_1 + p_2}{\sqrt{2}}, \quad p_2' = \frac{p_2 - p_1}{\sqrt{2}} \tag{8.60}$$

are the rotated coordinates. In the limit of infinite squeezing $\Delta^2 \to \infty$, the Wigner function reduces to a product of two delta functions $\delta(q_1' - q_2' - q_\alpha)\, \delta(p_1' + p_2' - p_\alpha)$, which means that the difference of positions is sharply peaked around q_α, and the total momentum is sharply peaked around p_α. These approach the EPR correlations in Eqs. (8.52) and (8.53). Since the entangled squeezed coherent state has a sharp position difference as well as a sharp total momentum, we should be able to perform a simultaneous measurement of $\hat{q}_1 - \hat{q}_2$ and $\hat{p}_1 + \hat{p}_2$. Indeed, it is straightforward to show that these two operators do commute.

The remaining question is how to induce the rotation of the Wigner function. From Eq. (8.58) we find that

$$\begin{pmatrix} q_1 \\ q_2 \end{pmatrix} \to \begin{pmatrix} q_1' \\ q_2' \end{pmatrix} = \begin{pmatrix} \cos\theta & \sin\theta \\ -\sin\theta & \cos\theta \end{pmatrix} \begin{pmatrix} q_1 \\ q_2 \end{pmatrix}, \tag{8.61}$$

and similarly

$$\begin{pmatrix} p_1 \\ p_2 \end{pmatrix} \to \begin{pmatrix} p_1' \\ p_2' \end{pmatrix} = \begin{pmatrix} \cos\theta & \sin\theta \\ -\sin\theta & \cos\theta \end{pmatrix} \begin{pmatrix} p_1 \\ p_2 \end{pmatrix}. \tag{8.62}$$

This is the same as a rotation of the corresponding operators $\hat{q}_i$ and $\hat{p}_i$, which can be translated immediately to a rotation of the creation and annihilation operators:

$$\begin{pmatrix} \hat{a}_1 \\ \hat{a}_2 \end{pmatrix} \to \begin{pmatrix} \hat{a}_1' \\ \hat{a}_2' \end{pmatrix} = \begin{pmatrix} \cos\theta & \sin\theta \\ -\sin\theta & \cos\theta \end{pmatrix} \begin{pmatrix} \hat{a}_1 \\ \hat{a}_2 \end{pmatrix}. \tag{8.63}$$

However, a rotation acting on two annihilation operators is generated by the quadratic Hamiltonian $\hat{a}_1^\dagger \hat{a}_2 - \hat{a}_1 \hat{a}_2^\dagger$, which we identified in Chapter 1 with the interaction Hamiltonian of a beam splitter. Therefore, if we are given two states that are squeezed in orthogonal directions, we can use a 50:50 beam splitter to create two entangled squeezed modes.

In order to perform a complete Bell measurement, we mix modes a_1 and a_2 on a 50:50 beam splitter

$$\hat{a}_1 \rightarrow \frac{\hat{b}_1 + \hat{b}_2}{\sqrt{2}} \quad \text{and} \quad \hat{a}_2 \rightarrow \frac{\hat{b}_1 - \hat{b}_2}{\sqrt{2}}. \qquad (8.64)$$

The position difference is then transformed into

$$\hat{q}_1 - \hat{q}_2 = \tfrac{1}{2}\left(\hat{a}_1 + \hat{a}_1^\dagger - \hat{a}_2 - \hat{a}_2^\dagger\right) \;\rightarrow\; \frac{1}{\sqrt{2}}\left(\hat{b}_2 + \hat{b}_2^\dagger\right) = \sqrt{2}\hat{q}_2', \qquad (8.65)$$

that is, the position observable in mode b_2. We can measure this using homodyne detection. Similarly, the total momentum is transformed to

$$\hat{p}_1 + \hat{p}_2 = -\frac{i}{2}\left(\hat{a}_1 - \hat{a}_1^\dagger + \hat{a}_2 - \hat{a}_2^\dagger\right) \;\rightarrow\; -\frac{i}{\sqrt{2}}\left(\hat{b}_2 - \hat{b}_2^\dagger\right) = \sqrt{2}\hat{p}_1', \qquad (8.66)$$

the momentum observable of mode b_1. The quadratures $\hat{p}_1'$ and $\hat{q}_2'$ can be measured simultaneously, since they pertain to different modes. It should not come as a surprise that the detection of continuous-variable Bell states involves a 50:50 beam splitter. After all, we used the same beam splitter to create the entanglement from single-mode squeezed coherent states in the first place. The second beam splitter merely reverts the entangling operation.

At this point, the reader may notice a striking difference between the creation and detection of entanglement for single-photon qubits and for continuous variables. Indeed, we found in Chapter 6 that it is impossible to deterministically create and detect two-qubit entanglement when those qubits are represented by single-photon states. Here, we have seen that entanglement between the squeezed quadratures of two field modes is created deterministically by a single 50:50 beam splitter. This is one of the attractive features of continuous-variable quantum information processing. However, we will see later in this Chapter that the use of continuous variables introduces other difficulties.

Finally in this section, we ask: how much entanglement is contained in a two-mode squeezed vacuum state, such as the one prepared above? One measure of entanglement that can be used here is the von Neumann entropy of the state of one of the parties. To calculate this, we need to express the state in the Schmidt basis. The Wigner function is not so useful for this, and we therefore recall the state of the two-mode squeezed vacuum expressed in the number basis in Eq. (1.235)

$$|\Psi(r,\phi)\rangle_{12} = \frac{1}{\cosh r} \sum_{n=0}^{\infty} e^{in\phi} \tanh^n r \, |n\rangle_1 \, |n\rangle_2, \qquad (8.67)$$

where r is the squeezing parameter and ϕ determines the squeezing orientation via $\xi = re^{i\phi}$. This description of the state is manifestly in the Schmidt basis, in this case the photon-number basis. Tracing out one subsystem, say system 2, yields the mixed state

$$\rho_1 = \mathrm{Tr}_2\left[\lvert\Psi(r,\phi)\rangle_{12}\,\langle\Psi(r,\phi)\rvert\right] = \frac{1}{\cosh^2 r}\sum_{n=0}^{\infty}\tanh^{2n} r\,\lvert n\rangle_1\,\langle n\rvert. \qquad (8.68)$$

The von Neumann entropy of the subsystem, in other words the entanglement of the combined system, is then given by

$$E(r) = -\mathrm{Tr}\left(\rho_1\log_2\rho_1\right) = -\sum_{n=0}^{\infty}\lambda_n\log_2\lambda_n, \qquad (8.69)$$

where λ_n are the eigenvalues of ρ_1. Using Eq. (8.68), the entanglement of the two-mode squeezed vacuum is

$$E(r) = \cosh^2 r\log_2\cosh^2 r - \sinh^2 r\log_2\sinh^2 r. \qquad (8.70)$$

The entanglement increases asymptotically linearly with the squeezing parameter r, shown in Fig. 8.7. Calculating this measure of entanglement for other Gaussian states is also possible, but in general this can be done only numerically. A general procedure was developed by Parker *et al.* (2000).

Exercise 8.7: Show that $E(r)$ is given by Eq. (8.70).

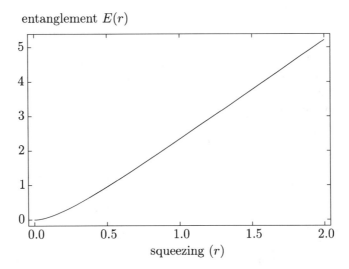

Fig. 8.7. The von Neumann entropy for two-mode entangled squeezed vacuum, with squeezing parameter r.

8.3 Teleportation and entanglement swapping

In this section we will describe how to use continuous variables in teleportation and entanglement swapping, and how to build quantum repeaters for long-distance communication.

8.3.1 Quantum teleportation

In Chapter 2 we saw that teleportation allows the transfer of the quantum state of a qubit from one system to another, using entanglement. In this section we describe quantum teleportation with optical continuous variables, which was introduced by Braunstein and Kimble (1998). We first give the protocol in terms of ideal position and momentum states, and then we describe the approximate, but physically realistic, protocol.

The arbitrary pure state of system 1, which we wish to teleport, is up to a normalization

$$|\psi\rangle_1 \propto \int dq \, \psi(q) \, |q\rangle_q \,, \tag{8.71}$$

where we have used the notation $\psi(q) = \langle q|\psi\rangle$. Second, the ideal EPR states $|\Psi(u,v)\rangle$ on two systems are

$$|\Psi(u,v)\rangle = \frac{1}{\mathcal{N}} \int dq \, e^{2\pi i v q} \, |q, q+u\rangle_q \,. \tag{8.72}$$

We use the EPR state $|\Psi(0,0)\rangle_{23}$ in the teleportation protocol, but in principle any u and v can be used. Next, we follow the usual teleportation procedure: we project systems 1 and 2 onto the Bell basis, which will produce a pair of continuous variables u and v. Hence systems 1 and 2 are now in the state $|\Psi(u,v)\rangle$, for some u and v. The output state is then evaluated as

$$_{12}\langle\Psi(u,v)|\,\psi\rangle_1\,|\Psi(0,0)\rangle_{23} = Z(v)X(u)\,|\psi\rangle_3 \,. \tag{8.73}$$

The operators $X(u)$ and $Z(v)$ were defined in Eqs. (2.79) and (2.82) as the displacement and phase operators, respectively. They depend on the measurement outcome u and v. Owing to the linearity of quantum mechanics, this protocol still works when the input state is a mixed state.

Exercise 8.8: Prove Eq. (8.73).

With squeezed coherent states as approximations of the position and momentum eigenstates, the teleportation protocol also becomes approximate. Using the techniques of the previous section, we can calculate the Wigner function of the output state, given an arbitrary input state. However, rather than calculating the Wigner function at each stage of the teleportation protocol, we can determine the transformation rules of the q_i and p_i, and substitute their inverse into $W(q_i, p_i)$. We have seen in the previous section that this is a valid approach. This will lead us to evaluate first the transformation of the quadrature operators in the Heisenberg picture. We will recall the operator transformations of the single-mode

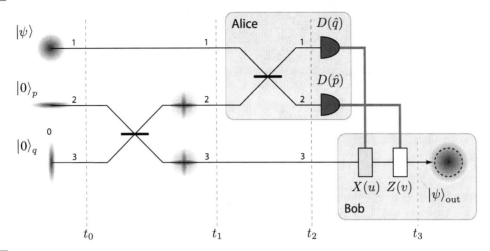

Fig. 8.8. Quantum teleportation with continuous variables. The labels on the horizontal lines correspond to the indices of the quadratures, which have time dependence $t = t_0, t_1, t_2, t_3$. Before the initial squeezing operation we set $t = 0$.

squeezer and the beam splitter, and then we will proceed with the step-wise teleportation protocol shown in Fig. 8.8. We explicitly state the time dependence of the quadrature operators in order to avoid confusion.

First, recall from Eq. (8.29) that the squeezing $S(\xi)$ of the quadrature operator with real $\xi = r$ leads to the transformation of the quadrature operators at time $t = 0$ to operators at time t:

$$\hat{q}(t) = S(r)\hat{q}(0)\, S^\dagger(r) = \hat{q}(0)\, e^{+r} = \hat{q}(0)\Delta,$$

$$\hat{p}(t) = S(r)\hat{p}(0)\, S^\dagger(r) = \hat{p}(0)\, e^{-r} = \frac{\hat{p}(0)}{\Delta}, \tag{8.74}$$

where, again, we have set $\Delta = e^{+r}$.

Second, the beam-splitter transformations derived in Eqs. (8.43) and (8.44) for the special choice of $\varphi = \pi/2$ and $\theta = \pi/4$ can be written as

$$\hat{q}_j(t) = \frac{\hat{q}_j(0) + \hat{q}_k(0)}{\sqrt{2}} \quad \text{and} \quad \hat{q}_k(t) = \frac{\hat{q}_j(0) - \hat{q}_k(0)}{\sqrt{2}} \tag{8.75}$$

for the position quadratures of modes j and k. Similarly, the transformation rule for the momentum quadratures is given by

$$\hat{p}_j(t) = \frac{\hat{p}_j(0) + \hat{p}_k(0)}{\sqrt{2}} \quad \text{and} \quad \hat{p}_k(t) = \frac{\hat{p}_j(0) - \hat{p}_k(0)}{\sqrt{2}}. \tag{8.76}$$

Note that these beam splitters are asymmetric (there is a phase factor of -1 associated with transmission of mode k).

We will now use these transformations to implement the squeezers and beam splitters in the teleportation protocol in Fig. 8.8. We can straight away make the following identifications: $\hat{q}_1(t_1) = \hat{q}_1(t_0) = \hat{q}_1(0)$ and $\hat{p}_1(t_1) = \hat{p}_1(t_0) = \hat{p}_1(0)$, and $\hat{q}_3(t_1) = \hat{q}_3(t_2)$ and $\hat{p}_3(t_1) = \hat{p}_3(t_2)$. These will become useful in the derivation below.

Step 1: We squeeze the vacuum state in modes 2 and 3 in momentum and position, respectively. This will induce the transformation

$$\hat{q}_2(t_0) = \hat{q}_2(0)\,\Delta \quad \text{and} \quad \hat{q}_3(t_0) = \frac{\hat{q}_2(0)}{\Delta},$$
$$\hat{p}_2(t_0) = \frac{\hat{q}_2(0)}{\Delta} \quad \text{and} \quad \hat{p}_3(t_0) = \hat{q}_2(0)\,\Delta. \tag{8.77}$$

Step 2: The 50:50 beam splitter in modes 2 and 3 creates the approximate EPR pair necessary for teleportation, and this induces the quadrature transformations

$$\hat{q}_2(t_1) = \frac{\hat{q}_2(t_0) + \hat{q}_3(t_0)}{\sqrt{2}} \quad \text{and} \quad \hat{q}_3(t_1) = \frac{\hat{q}_2(t_0) - \hat{q}_3(t_0)}{\sqrt{2}}. \tag{8.78}$$

There are similar transformations for the momentum quadratures. We will concentrate on the position quadratures in the remaining steps, and leave the momentum quadratures as an exercise for the reader.

Step 3: The beam splitter for the continuous-variable Bell measurement induces the transformations

$$\hat{q}_1(t_2) = \frac{\hat{q}_1(t_1) - \hat{q}_2(t_1)}{\sqrt{2}} = \frac{\hat{q}_1(0)}{\sqrt{2}} - \frac{\hat{q}_2(t_0) + \hat{q}_3(t_0)}{2}$$
$$\hat{q}_2(t_2) = \frac{\hat{q}_1(t_1) + \hat{q}_2(t_1)}{\sqrt{2}} = \frac{\hat{q}_1(0)}{\sqrt{2}} + \frac{\hat{q}_2(t_0) + \hat{q}_3(t_0)}{2}, \tag{8.79}$$

where we have used the identification $\hat{q}_1(t_1) = \hat{q}_1(0)$, and have substituted Eq. (8.78). Notice the orientation of the beam splitter. Using the squeezing transformations from Step 1, this leads to the relation

$$\hat{q}_2(t_0) = -2\hat{q}_1(t_2) + \sqrt{2}\hat{q}_1(0) - \frac{\hat{q}_3(0)}{\Delta}. \tag{8.80}$$

We can now write the position quadrature $\hat{q}_3(t_2) = \hat{q}_3(t_1)$ as

$$\hat{q}_3(t_2) = \hat{q}_1(0) - \sqrt{2}\hat{q}_1(t_2) - \frac{\sqrt{2}\,\hat{q}_3(0)}{\Delta}. \tag{8.81}$$

Similarly, the momentum quadrature $\hat{p}_3(t_2)$ becomes

$$\hat{p}_3(t_2) = \hat{p}_1(0) - \sqrt{2}\hat{p}_2(t_2) + \frac{\sqrt{2}\,\hat{p}_2(0)}{\Delta}. \tag{8.82}$$

Step 4: Alice measures the position quadrature $\hat{q}_1(t_2)$ in mode 1, yielding the outcome $u/\sqrt{2}$, and she measures the momentum quadrature $\hat{p}_2(t_2)$ in mode 2, yielding the outcome $v/\sqrt{2}$. These outcomes allow Bob (who holds mode 3) to apply the displacements

$$X(u) = \exp\left(-2iu\,\hat{p}_3\right) \quad \text{and} \quad Z(v) = \exp\left(2iv\,\hat{p}_3\right). \tag{8.83}$$

These displacements lead to the output quadratures

$$\hat{q}_3(t_3) = \hat{q}_1(0) - \frac{\sqrt{2}\,\hat{q}_3(0)}{\Delta} \quad \text{and} \quad \hat{p}_3(t_3) = \hat{p}_1(0) + \frac{\sqrt{2}\,\hat{p}_3(0)}{\Delta}. \tag{8.84}$$

In this protocol, the quadrature operators of mode 3 are transformed into the quadrature modes of mode 1 at time $t = 0$ plus an extra part that is proportional to the squeezing of mode 3. In other words, the input quadratures are teleported to the output quadratures up to a factor. In the limit of perfect squeezing $\Delta \to \infty$, the teleportation protocol is perfect.

Exercise 8.9: Verify Eq. (8.82).

We can now construct the Wigner function of the teleported state, given the Wigner functions of the unknown input state and the (approximate) EPR pair. Suppose that the input state is in an arbitrary state $W_{\text{in}}[q_1(0), p_1(0)]$, and the form of W_{in} is unknown to Alice and Bob. The input is then described as a maximally mixed state

$$\int dq_1\, dp_1\, W_{\text{in}}(q_1, p_1). \tag{8.85}$$

The Wigner function of the EPR pair is written as

$$W_{\text{EPR}}\left[q_2(t_1), p_2(t_1); q_3(t_1), p_3(t_1)\right].$$

Using the beam-splitter transformation in Eq. (8.79) of Alice's continuous-variable Bell measurement, this becomes

$$W_{\text{EPR}} \to W_{\text{EPR}}\left[\frac{q_2(t_2) - q_1(t_2)}{\sqrt{2}}, \frac{p_2(t_2) - p_1(t_2)}{\sqrt{2}}; q_3(t_2), p_3(t_2)\right]. \tag{8.86}$$

In addition, Alice's measurement can be incorporated in the Wigner function by including two delta functions with the measurement outcomes $q_1(t_2)$ and $p_2(t_2)$. Given the beam-splitter transformations of Eq. (8.79), these become

$$\delta\left[\frac{q_1(t_2) + q_2(t_2)}{\sqrt{2}} - q_1(0)\right] \quad \text{and} \quad \delta\left[\frac{p_1(t_2) + p_2(t_2)}{\sqrt{2}} - p_1(0)\right]. \tag{8.87}$$

The Wigner function $W = W[q_1(t_2), p_1(t_2); q_2(t_2), p_2(t_2); q_3(t_2), p_3(t_2)]$ of the entire three-mode system then becomes

$$W = \int dq_1(0)\, dp_1(0)\, W_{\text{in}}[q_1(0), p_1(0)]$$
$$\times W_{\text{EPR}}\left[\frac{q_2(t_2) - q_1(t_2)}{\sqrt{2}}, \frac{p_2(t_2) - p_1(t_2)}{\sqrt{2}}; q_3(t_2), p_3(t_2)\right]$$
$$\times \delta\left[\frac{q_1(t_2) + q_2(t_2)}{\sqrt{2}} - q_1(0)\right] \delta\left[\frac{p_1(t_2) + p_2(t_2)}{\sqrt{2}} - p_1(0)\right]. \qquad (8.88)$$

Since Alice detects $q_1(t_2)$ and $p_2(t_2)$, we can integrate over the conjugate variables $p_1(t_2)$ and $q_2(t_2)$, and evaluation of the delta functions then gives

$$W' = \int dq_2(t_2)\, dp_1(t_2)\, W[q_2(t_2), p_2(t_2); q_3(t_2), p_3(t_2)]$$
$$= \int dq_1(0)\, dp_1(0)\, W_{\text{in}}[q_1(0), p_1(0)]$$
$$\times W_{\text{EPR}}\left[q_1(0) - \sqrt{2}q_1(t_2), \sqrt{2}p_2(t_2) - p_1(0); q_3(t_2), p_3(t_2)\right]. \qquad (8.89)$$

Bob's displacement after Alice's classical communication (u, v) induces the transformation

$$q_3(t_2) \to q_3(t_3) - u \quad \text{and} \quad p_3(t_2) \to p_3(t_3) - v, \qquad (8.90)$$

which leads to

$$W'' = \int dq_1(0)\, dp_1(0)\, W_{\text{in}}[q_1(0), p_1(0)]$$
$$\times W_{\text{EPR}}\left[q_1(0) - u, v - p_1(0); q_3(t_3) - u, p_3(t_3) - v\right]. \qquad (8.91)$$

The Wigner function of the teleported state in mode 3 is then found by integrating over u and v, since the teleportation procedure should be independent of the actual values of u and v. The integration can be performed explicitly using Eq. (8.59), resulting in the Wigner function of the teleported state $W_{\text{tel}}(q_3, p_3)$

$$W_{\text{tel}}(q_3, p_3) = \frac{\Delta^2}{\pi} \int dq\, dp\, W_{\text{in}}(q, p) \exp\left[-\Delta^2(q - q_3)^2 - \Delta^2(p - p_3)^2\right], \qquad (8.92)$$

where we have suppressed the argument (t_3) of q_3 and p_3. We can write the teleported Wigner function more compactly as a convolution of the input state with a Gaussian noise function

$$W_{\text{tel}}(q_3, p_3) = \int dq\, dp\, G_\Delta(q_3 - q, p_3 - p)\, W_{\text{in}}(q, p)$$
$$= G_\Delta \circ W_{\text{in}}, \qquad (8.93)$$

where

$$G_\Delta(q, p) = \frac{\Delta^2}{\pi} \exp\left[-\Delta^2(q^2 + p^2)\right]. \qquad (8.94)$$

The teleportation procedure therefore works for *any* input state, not just Gaussian states. This is clear from the derivation of Eq. (8.93), where we made no assumptions about W_{in}, other than that it is a reasonably well-behaved function.

Exercise 8.10: Show that in the limit of $\Delta \to \infty$, teleportation is perfect.

8.3.2 Entanglement swapping

In continuous-variable teleportation, as in teleportation of discrete systems, the teleported state may have been part of an entangled system. The teleportation operation then results in entanglement swapping. We can repeat the arguments of the previous section to find the quadrature operators in the Heisenberg picture and the Wigner function of the entangled system. Here, we introduce two generalizations to the teleportation protocol: first we include a 'gain factor' $g \in [0, 1]$ in the phase-space displacements. Second, we introduce the efficiency of the photodetectors in the continuous-variable Bell measurement.

A schematic of entanglement swapping with continuous variables is shown in Fig. 8.9. Alice and Bob initially each create an approximate EPR pair locally, and then each feed one half of their pair into a swapping station. A Bell measurement is performed at the swapping station and the outcome is communicated with Bob, but it could just as well have been communicated with Alice, or indeed partly with Alice and partly with Bob. The result is shared entanglement between Alice and Bob, the nature of which we will now calculate.

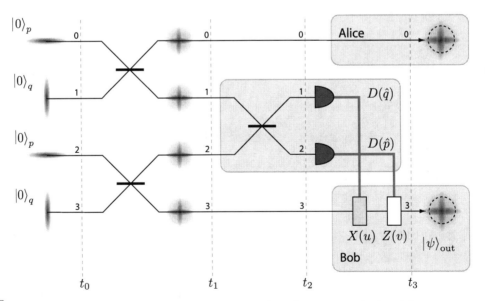

Fig. 8.9. Entanglement swapping. Modes 1 and 2 are measured in a 'swapping station', and the measurement outcomes are sent to Bob. After the Bell measurement, Alice and Bob share entanglement.

We assume that modes 0 and 1 are squeezed by an amount Δ_1 in conjugate quadratures, and similarly that modes 2 and 3 are squeezed by an amount Δ_2 in conjugate variables. The position quadratures of the four input modes can then be written as

$$\hat{q}_0(t_1) = \frac{1}{\sqrt{2}} \left[\Delta_1 \hat{q}_0(0) + \frac{\hat{q}_1(0)}{\Delta_1} \right],$$

$$\hat{q}_1(t_1) = \frac{1}{\sqrt{2}} \left[\Delta_1 \hat{q}_0(0) - \frac{\hat{q}_1(0)}{\Delta_1} \right],$$

$$\hat{q}_2(t_1) = \frac{1}{\sqrt{2}} \left[\Delta_2 \hat{q}_2(0) + \frac{\hat{q}_3(0)}{\Delta_2} \right],$$

$$\hat{q}_3(t_1) = \frac{1}{\sqrt{2}} \left[\Delta_2 \hat{q}_2(0) - \frac{\hat{q}_3(0)}{\Delta_2} \right], \tag{8.95}$$

with similar expressions for the momentum quadratures, keeping in mind that $\Delta_j \to \Delta_j^{-1}$ for the momentum. The Bell measurement in the swapping station induces the transformation

$$\hat{q}_1(t_2) = \frac{\hat{q}_1(t_1) - \hat{q}_2(t_1)}{\sqrt{2}} \quad \text{and} \quad \hat{q}_2(t_2) = \frac{\hat{q}_1(t_1) + \hat{q}_2(t_1)}{\sqrt{2}}, \tag{8.96}$$

just as before in Eq (8.79). Using the identifications $\hat{q}_0(t_1) = \hat{q}_0(t_2)$ and $\hat{q}_3(t_1) = \hat{q}_3(t_2)$, we write the quadratures of modes 0 and 3 as

$$\hat{q}_0(t_2) = \hat{q}_2(t_1) + \left[\hat{q}_0(t_1) - \hat{q}_1(t_1) \right] + \sqrt{2}\hat{q}_1(t_2),$$

$$\hat{p}_0(t_2) = \hat{p}_2(t_1) + \left[\hat{p}_0(t_1) + \hat{p}_1(t_1) \right] - \sqrt{2}\hat{p}_2(t_2),$$

$$\hat{q}_3(t_2) = \hat{q}_1(t_1) - \left[\hat{q}_2(t_1) - \hat{q}_3(t_1) \right] - \sqrt{2}\hat{q}_1(t_2),$$

$$\hat{p}_3(t_2) = \hat{p}_1(t_1) + \left[\hat{p}_2(t_1) + \hat{p}_3(t_1) \right] - \sqrt{2}\hat{p}_2(t_2). \tag{8.97}$$

Using the relations in Eq. (8.95), we obtain

$$\hat{q}_0(t_2) = \hat{q}_2(t_1) + \frac{\sqrt{2}\,\hat{q}_1(0)}{\Delta_2} + \sqrt{2}\hat{q}_1(t_2),$$

$$\hat{p}_0(t_2) = \hat{p}_2(t_1) + \frac{\sqrt{2}\,\hat{p}_1(0)}{\Delta_2} - \sqrt{2}\hat{p}_2(t_2),$$

$$\hat{q}_3(t_2) = \hat{q}_1(t_1) - \frac{\sqrt{2}\,\hat{q}_3(0)}{\Delta_2} - \sqrt{2}\hat{q}_1(t_2),$$

$$\hat{p}_3(t_2) = \hat{p}_1(t_1) + \frac{\sqrt{2}\,\hat{p}_3(0)}{\Delta_2} - \sqrt{2}\hat{p}_2(t_2). \tag{8.98}$$

We can now apply Bob's displacements based on the measurement outcomes of the swapping station, namely

$$\hat{q}_3(t_3) = \hat{q}_3(t_2) + g\sqrt{2}\,\hat{q}_1(t_2) \quad \text{and} \quad \hat{p}_3(t_3) = \hat{q}_3(t_2) + g\sqrt{2}\,\hat{p}_2(t_2), \tag{8.99}$$

where we have used the gain factor g. After some algebra, we find that the quadrature operators in modes 0 and 3 become

$$\hat{q}_0(t_2) = \frac{1}{\sqrt{2}}\left[\Delta_1 \hat{q}_0(0) + \frac{\hat{q}_1(0)}{\Delta_1}\right],$$

$$\hat{p}_0(t_2) = \frac{1}{\sqrt{2}}\left[\Delta_2 \hat{p}_0(0) + \frac{\hat{p}_1(0)}{\Delta_2}\right],$$

$$\hat{q}_3(t_2) = \frac{1}{\sqrt{2}}\left[g\Delta_1 \hat{q}_0(0) - g\frac{\hat{q}_1(0)}{\Delta_1} - (g-1)\Delta_2 \hat{q}_2(0) - (g+1)\frac{\hat{q}_3(0)}{\Delta_2}\right],$$

$$\hat{p}_3(t_2) = \frac{1}{\sqrt{2}}\left[g\frac{\hat{p}_0(0)}{\Delta_1} - g\Delta_1 \hat{p}_1(0) - (g-1)\frac{\hat{p}_2(0)}{\Delta_2} - (g+1)\Delta_2 \hat{p}_3(0)\right]. \quad (8.100)$$

We can now compare the swapped entanglement with the initial entanglement. Recall that a perfect EPR pair on modes 1 and 2 has correlations $q_1 - q_2 = 0$ and $p_1 + p_2 = 0$. For the squeezed entanglement in modes 0 and 1 considered above, the corresponding observables are

$$\hat{q}_0(t_1) - \hat{q}_1(t_1) = \frac{\sqrt{2}\,\hat{q}_1(0)}{\Delta_1} \quad \text{and} \quad \hat{p}_0(t_1) + \hat{p}_1(t_1) = \frac{\sqrt{2}\,\hat{p}_0(0)}{\Delta_1}. \quad (8.101)$$

Similar relations hold for the entanglement in modes 2 and 3, with $\Delta_1 \to \Delta_2$. In order to compare the swapped entanglement with the initial entanglement, we calculate $Q \equiv \hat{q}_0(t_3) - \hat{q}_3(t_3)$ and $P \equiv \hat{p}_0(t_3) + \hat{p}_3(t_3)$:

$$Q = \frac{g-1}{\sqrt{2}}\left[\Delta_1 \hat{q}_0(0) - \Delta_2 \hat{q}_2(0)\right] - \frac{g+1}{\sqrt{2}}\left[\frac{\hat{q}_1(0)}{\Delta_1} + \frac{\hat{q}_3(0)}{\Delta_2}\right],$$

$$P = \frac{g-1}{\sqrt{2}}\left[\Delta_2 \hat{p}_3(0) - \Delta_1 \hat{p}_1(0)\right] + \frac{g+1}{\sqrt{2}}\left[\frac{\hat{p}_0(0)}{\Delta_1} + \frac{\hat{p}_2(0)}{\Delta_2}\right]. \quad (8.102)$$

It is clear from these relations that perfect entanglement swapping is possible only if $g = 1$ and either Δ_1 or Δ_2 (or both) tend to infinity. When both Δ_1 and Δ_2 are finite, there are situations where it may be advantageous to use a reduced gain g. For example, Van Loock and Braunstein (2000) show, using the swapped entanglement for the teleportation of a coherent state, that the fidelity attains its optimum for

$$g = \tanh(r_1 + r_2), \quad (8.103)$$

which is smaller than one for finite r_1 and r_2, and $\Delta_i = e^{r_i}$. The optimal fidelity then becomes

$$F = \left[1 + \frac{\cosh(2r_1 - 2r_2) + 1}{\cosh 2r_1 + \cosh 2r_2}\right]^{-1}. \quad (8.104)$$

The classical boundary occurs when either one of the pairs has vanishing squeezing ($r_1 = 0$ or $r_2 = 0$). This leads to a fidelity threshold of $F = 1/2$ that truly quantum entanglement swapping must surpass.

8.4 Entanglement distillation

In discrete entanglement swapping, the measurement outcomes can be used to purify or distil entanglement, such that the reduced density matrices held by Alice and Bob have larger von Neumann entropy. This is not the case for continuous-variable entanglement swapping. To see why, recall that in the discrete case described in Chapter 5, entanglement distillation is possible due to the rejection of swapped entanglement based on certain outcomes in the Bell measurement. However, the teleportation protocol for continuous-variable quantum states is deterministic, and each measurement outcome is completely equivalent to any other in the swapping protocol. This means that there is no rule according to which we can reject swapped pairs based on measurement outcomes. Combined with the fact that perfect entanglement (and consequently perfect teleportation) is impossible for continuous variables, we must conclude that entanglement swapping with continuous variables always decreases the entanglement.

In general, it is possible to modify the entanglement swapping protocol to construct quantum distillation and purification protocols for Gaussian states. However, an important theorem states that it is impossible to construct a distillation protocol using arbitrary local Gaussian unitary operations, homodyne detection, and classical feed-forward. We therefore have to include some non-Gaussian operations in the distillation protocol. This was derived independently by Eisert *et al.* (2002), Fiurášek (2002), and Giedke and Cirac (2002).

There are several protocols of this nature, and most include a form of photon detection since it is the most accessible non-Gaussian operation. Here we briefly outline the distillation protocol of Browne *et al.* (2003), shown in Fig. 8.10. The protocol consists of four steps:

Step 1: Two sources (1 and 2) of weakly entangled pure Gaussian states each send one mode to station A, and the other to station B. In each station, the two input modes are combined on a 50:50 beam splitter.

Step 2: In each station, one of the output modes is sent into a bucket detector, which will determine whether or not there were photons in that output mode.

Step 3: If either detector registers photons, the undetected output modes are discarded; when both detectors do *not* detect photons (i.e., they are projections onto the vacuum), the output modes are more entangled than the states produced by source 1 and source 2.

Step 4: Repeat steps 1 to 3 for distilled output modes to increase the entanglement even further.

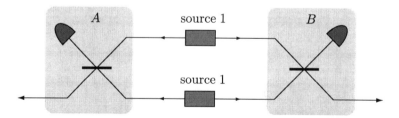

Fig. 8.10.   A distillation protocol for Gaussian states. The detectors are bucket detectors, without number resolution.

Since there is a probabilistic element into the protocol, namely the detection outcomes, we have a mechanism for rejecting unwanted outcomes. This also means that the protocol is not always successful, and it is perhaps somewhat counterintuitive that the success probability can be quite high. However, the success probability decreases with each iteration, and there will be an optimal number of iterations given an ensemble of sources.

The protocol still works when the detectors have finite efficiency, but now the decision mechanism allows false positives. If the losses are too big, the protocol will actually degrade the entanglement with respect to the sources. The minimum allowable detector efficiency generally depends on the amount of entanglement in the source: the more entanglement is present in the sources, the easier it is to degrade it with imperfect detectors. Given a fixed detector efficiency, there will be an optimal number of iterations, beyond which no more entanglement is distilled. For more details, see Eisert *et al.* (2004).

Finally, an entanglement distillation protocol can be used in the construction of a quantum repeater. It is fairly straightforward to find a parameter regime in which the distillation outperforms the degradation due to entanglement swapping, and a combination of the two protocols can then be used to distribute entangled qunats over increasingly large distances. However, the distillation protocol relies on a comparison between two detector outputs that are typically far away from each other. We therefore need a storage mechanism for the qunats, or a quantum memory. We will return to this in Chapter 10.

8.5 Quantum cryptography

In quantum communication with discrete systems, the possibility of cryptography follows directly from the no-cloning theorem. At first glance, the reader may suspect that cryptography with continuous variables is therefore impossible, since continuous-variable quantum states typically have a large average photon number, and also a large variation in the photon number. This may allow an eavesdropper to gain information about the state via an ordinary beam-splitter attack. Surprisingly, this turns out not to be the case. Due to the inevitable addition of noise in an eavesdropper attack, secure key distribution is possible with continuous variables. Indeed, this is an example of imperfect copying of continuous-variable quantum states, again due to the no-cloning theorem. Below, we describe the extension of the BB84 protocol to continuous variables due to Hillery (2000). This will provide us with some physical insight why the protocol is likely to be secure. Subsequently, we will give the outline of a security proof of this protocol following Gottesman and Preskill (2001).

8.5.1 Transmission of squeezed coherent states

Cryptographic schemes based on continuous variables have two important advantages over quantum cryptography with single photons: first, there is no need for single-photon sources, and only a relatively small amount of squeezing suffices to attain security of the protocol (approximately 2.51 dB of squeezing is required, which is well below the current achievable level of 10 dB). Second, cryptography with continuous variables requires homodyne

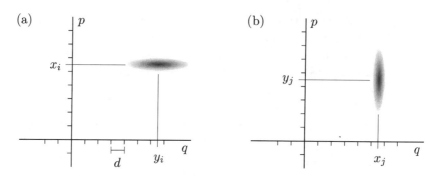

Fig. 8.11. Quantum cryptography with continuous variables. Alice sends a sequence of squeezed coherent states to Bob.

detection, which is generally easier to perform with high efficiency than the detection of single photons.

As in the discrete version of quantum cryptography, the main aim of quantum cryptography with continuous variables is for Alice and Bob to establish a shared random key with absolute security. The key may subsequently be used to encode a message. Alice sends a string of N squeezed coherent states to Bob, with each state squeezed either in the q or the p direction. Furthermore, the choice of q or p is made randomly for each transmission of a single squeezed coherent state. The expectation values of the squeezed quadrature (the narrow direction in Fig. 8.11) are denoted by x_i, with i running from 1 to N. The expectation value of the other quadrature is y_i. The preparation procedure for the ith squeezed coherent state is therefore

1. Alice randomly chooses a quadrature q or p to squeeze, and applies the corresponding squeezing operator to the vacuum:

$$W_0(q,p) \;\rightarrow\; W_q(\Delta q, p/\Delta) \quad \text{or} \quad W_p(q/\Delta, \Delta p), \tag{8.105}$$

where $\Delta = e^r$ is the strength of the squeezing.

2. She displaces the squeezed vacuum state along x_i in the squeezing direction, and along y_i in the anti-squeezed direction

$$W_q(\Delta q, p/\Delta) \rightarrow W_q\left[\Delta(q+x_i), \frac{p+y_i}{\Delta}\right]$$

$$W_p(q/\Delta, \Delta p) \rightarrow W_q\left[\frac{q+y_i}{\Delta}, \Delta(p+x_i)\right]. \tag{8.106}$$

Bob receives these squeezed coherent states and makes a homodyne measurement on each received state. He and Alice have previously established a shared reference frame for q and p (e.g., by means of a shared local oscillator), and he chooses his quadrature measurement randomly in the q or p direction. About fifty per cent of the time, Alice and Bob respectively prepare and measure the squeezed coherent state in the same quadrature. The number of matching measurements is M.

When Alice and Bob prepare and measure in the same quadrature, and Alice's displacement value was x_i, Bob will generally *not* find the measurement result x_i, but rather something close to x_i. This is due to the finite squeezing of the state. The consequence of this is that Alice cannot send exact continuous variables x_i to Bob using this protocol. Instead, we can divide the position and momentum axes in phase space in units of size d. When Alice sends a squeezed coherent state with expectation value x_i, the probability that Bob measures a value in an interval d_i of size d around x_i can be made arbitrarily large. Rather than having Alice communicate a string $\vec{x} = (x_1, \ldots, x_M)$ to Bob, Alice communicates a string of intervals $\vec{d} = (d_1, \ldots, d_M)$. Each interval d_i is then associated with a code word, or a letter in a specified alphabet. The simplest example of such an alphabet is a binary system, i.e., the logical value is 0 when the bin is even, and 1 when the bin is odd.

There are now two considerations we need to address: first, we want a good transmission fidelity, and second, we want to protect the message from the eavesdropper Eve. Good transmission means that, in the squeezed direction, the probability of measuring a value for the quadrature outside the bin is very small. Roughly, this means that the bin size d must be larger than the variance of the squeezed coherent state. More accurately, the probability that the quadrature measurement yields a value in the interval d_i is calculated using the Error function. The marginal distribution for Bob's measurement outcome is given by the integration of the Wigner function over the quadrature coordinate that is the canonical conjugate with respect to the measured quadrature:

$$
P(x, x_i) = \frac{2}{\pi} \int_{-\infty}^{\infty} dy \, \exp\left[-2\Delta^2(x - x_i)^2 - \frac{2(y - y_i)^2}{\Delta^2} \right]
$$

$$
= \sqrt{\frac{2}{\pi}} \Delta \exp\left[-2\Delta^2(x - x_i)^2 \right]. \tag{8.107}
$$

The probability that the measurement outcome lies in the interval $d_i = [x_i - d/2, x_i + d/2)$ is then given by

$$
P(d_i) = \int_{x_i - \frac{d_i}{2}}^{x_i + \frac{d_i}{2}} dx \, P(x, x_i) = \frac{2}{\sqrt{\pi}} \int_0^{\frac{d\Delta}{\sqrt{2}}} du \, \exp[-u^2] = \mathrm{Erf}\left(\frac{d\Delta}{\sqrt{2}} \right), \tag{8.108}
$$

where we have made the change of variable $u = \sqrt{2}\Delta(x - x_i)$, and $\mathrm{Erf}(x)$ is the Error function. It can be determined numerically, and if we require a transmission fidelity of $1 - P(d_i) < 10^{-3}$, we find the relation between the size of the bin d and the amount of squeezing $\Delta = e^r$ as

$$
d \gtrsim 5\Delta^{-1}, \tag{8.109}
$$

which must be satisfied if transmission errors are to be less than one in a thousand.

Any cryptographic protocol must be suitably protected from eavesdroppers. There are several aspects to this: first, measuring the squeezed coherent state in the wrong quadrature must not reveal any information about d_i. This requirement is satisfied automatically, because the expectation value in the wrong quadrature is y_i, which is not correlated to x_i. A measured value close to y_i may lead Eve to guess that the code word Alice sent corresponded to the bin $d_i' = [y_i - d/2, y_i + d/2)$. Furthermore, for appreciable amounts of squeezing

$\Delta \gg d$, which means that the wrong quadrature measurement gives a large spread over bins. When the number of code words is much less than the number of available bins, the transmitted signal is essentially random. This randomness can be increased by Alice when she chooses the displacements y_i randomly. This removes all information about the correctness of the choice of Eve's quadrature measurement.

Second, *whatever* Eve does between Alice and Bob (as long as it is allowed by the laws of quantum physics), she must not be able to gain any information about the code words d_i that Alice tries to communicate to Bob, without being detected. We consider two possible strategies for Eve:

1. Measure and resend;
2. partial measurement, or beam-splitter attack.

It is clear that the first strategy will reveal the presence of Eve: when she measures the wrong quadrature she will also send the wrong squeezed coherent state to Bob, who will now see a much larger transmission error.

Exercise 8.11: Show that $d < 1/2$ must be satisfied in order to foil the 'measure and resend' strategy.

The beam-splitter attack is a more subtle strategy, but it will also reveal the presence of Eve. In this attack, Eve places a beam splitter in the transmission mode between Alice and Bob, and 'siphons off' a small amount of the field amplitude. She can then perform any operation she desires on the reflected mode. If the reflectivity is low enough, Alice and Bob may interpret the reduced amplitude as signal loss in the transmission line. Nevertheless, this attack by Eve also turns out not to work.

As we will show shortly, placing a beam splitter in the transmission line between Alice and Bob will alter the Wigner function of the state sent by Alice to Bob. Generally, the Wigner function will become broader, and this is detectable statistically by Alice and Bob (this has the same effect as losses). Therefore, Eve must keep the reflectivity of the beam splitter as small as possible in order to escape detection. On the other hand, she cannot choose the reflectivity arbitrarily small because that would not give her enough field amplitude to extract information about d_i, which is her ultimate goal. To make this precise, we now

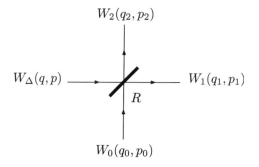

Fig. 8.12.   Transformation of the Wigner functions due to a beam splitter with reflectivity coefficient R.

calculate the Wigner function of a squeezed coherent state that is sent through a beam splitter with arbitrary reflectivity, and with a vacuum input on the secondary port.

The Wigner function of the vacuum and the squeezed coherent state in, say, the q quadrature is given by

$$W(q,p;q_0,p_0) = W_\Delta(q,p)\, W_0(q_0,p_0), \tag{8.110}$$

with

$$W_0(q_0,p_0) = \frac{2}{\pi} \exp\left(-2q_0^2 - 2p_0^2\right) \tag{8.111}$$

the Wigner function of the vacuum state, and

$$W_\Delta(q,p) = \frac{2}{\pi} \exp\left[-2\Delta^2(q-x_i)^2 - \frac{2(p-y_i)^2}{\Delta^2}\right] \tag{8.112}$$

the Wigner function of the squeezed coherent state. We apply the beam-splitter transformation between $\hat{q}$ and $\hat{q}_0$ (and $\hat{p}$ and $\hat{p}_0$) according to the substitution rule

$$
\begin{aligned}
q &\to \sqrt{T}\,q_1 + \sqrt{R}\,q_2 & \text{and} && q_0 &\to \sqrt{R}\,q_1 - \sqrt{T}\,q_2 \\
p &\to \sqrt{T}\,p_1 + \sqrt{R}\,p_2 & \text{and} && p_0 &\to \sqrt{R}\,p_1 - \sqrt{T}\,p_2,
\end{aligned}
\tag{8.113}
$$

where R is the reflection coefficient and T the transmission coefficient, and $R + T = 1$. This leads to a transmitted Wigner function, which is received by Bob:

$$W_1(q_1,p_1) = \mathcal{N} \exp\left[-\frac{2\Delta^2(q_1 - \sqrt{T}\,x_i)^2}{T + R\Delta^2} - \frac{2(p_1 - \sqrt{T}\,y_i)^2}{T\Delta^2 + R}\right], \tag{8.114}$$

where $\mathcal{N}$ is the normalization constant

$$\mathcal{N}^{-1} = \frac{\pi}{2} \frac{1}{\sqrt{2}\Delta} \sqrt{(\Delta^4 + 1) - (\Delta^2 - 1)^2(R^2 + T^2)}. \tag{8.115}$$

In the case of perfect transmission ($R \to 0$, $T \to 1$), this Wigner function reduces to the Wigner function in Eq. (8.112). The reflected state, collected by Eve, has the Wigner function

$$W_2(q_2,p_2) = \mathcal{N} \exp\left[-\frac{2\Delta^2(q_2 - \sqrt{R}\,x_i)^2}{T\Delta^2 + R} - \frac{2(p_2 - \sqrt{R}\,y_i)^2}{T + \Delta^2 R}\right]. \tag{8.116}$$

Again, the Wigner function reduces to that of the vacuum state when $R \to 0$ and $T \to 1$. The marginal probability distributions over q_1 and q_2 for Bob and Eve, respectively, are

$$
\begin{aligned}
P_{\text{Bob}}(q_1) &= \int_{-\infty}^{\infty} dp_1\, W_1(q_1,p_1) \\
&= \frac{\Delta}{\sqrt{T + R\Delta^2}} \sqrt{\frac{2}{\pi}} \exp\left[-\frac{2\Delta^2(q_1 - \sqrt{T}\,x_i)^2}{T + R\Delta^2}\right]
\end{aligned}
\tag{8.117}
$$

and

$$P_{\text{Eve}}(q_2) = \int_{-\infty}^{\infty} dp_2\, W_2(q_2, p_2)$$

$$= \frac{\Delta}{\sqrt{T\Delta^2 + R}} \sqrt{\frac{2}{\pi}} \exp\left[-\frac{2\Delta^2(q_2 - \sqrt{R}\,x_i)^2}{T\Delta^2 + R}\right]. \tag{8.118}$$

It is clear from Eq. (8.117) that the beam splitter rescales the expectation value x_i by a factor $\sqrt{T}$, and as a consequence Bob's marginal probability distribution over q_1 is shifted by an amount $(1 - \sqrt{T})x_i$, and broadens by a factor $T + R\Delta^2 > 1$. This will lead to transmission errors between Alice and Bob. Similarly, when Eve decides to measure the same quadrature as Bob (q_2), her distribution has shifted by an amount $\sqrt{R}\,x_i$, and it has narrowed by a factor $T + R/\Delta^2 < 1$. In order for Eve to extract some information about x_i from her state, the minimal shift $\sqrt{R}\,d$ of her distribution must be comparable to the reduced width of the distribution:

$$\frac{T\Delta^2 + R}{2\Delta^2} \approx Rd^2 \quad \text{or} \quad R \approx \frac{1}{1 + 2d^2 - \Delta^{-2}}. \tag{8.119}$$

At the same time, the shift in Bob's marginal distribution must be smaller than the width of the distribution:

$$d^2\left(1 - \sqrt{T}\right)^2 < \frac{T + R\Delta^2}{2\Delta^2}, \tag{8.120}$$

As shown in Fig. 8.13, this is consistent with the requirement that

$$R < \frac{d}{\Delta}. \tag{8.121}$$

From Eqs. (8.119) and (8.121) we see that the competing requirements of Eve's sufficient resolution and Bob's inability to detect her are incompatible: when squeezing is large

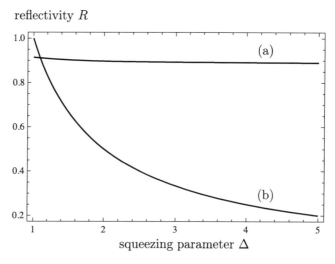

reflectivity R

squeezing parameter Δ

Fig. 8.13. The reflectivity must be smaller than a specific value if Bob is not to detect the presence of Eve in his measurement statistics. Curve (a) is the requirement in Eq. (8.121), while curve (b) is given by Eq. (8.120). Both curves are in units where $d = 1$.

$(\Delta \gg 1)$, and the code-word bins are small $d \ll 1$, Eq. (8.119) shows that R must be close to 1, while according to Eq. (8.121) R must be smaller than a very small number. The stronger the squeezing (i.e. the larger Δ is), the smaller the bin size d can be, and therefore the larger the discrepancy between the two requirements.

The physical reason why Eve cannot extract information about d_i without being detected (for large squeezing), is that the beam splitter adds *vacuum noise* both to Eve's state, and to the state that is transmitted to Bob. Eve can try to reduce noise by squeezing the vacuum in the unused input port of her beam splitter. However, she must choose a quadrature for this squeezing, and when she chooses incorrectly, she adds even more noise to the system. Alice and Bob can detect this statistically. Of course, since we have only shown that Eve cannot gain information about the key while remaining undetected for this specific strategy, this is not a security proof. Such a proof will be given later in this section.

8.5.2 Sharing approximate EPR pairs

Next, we describe quantum cryptography using shared EPR pairs, or approximate EPR pairs in the practical limit. Similar to our approach in Section 5.4, we construct an ideal protocol (using the non-physical exact quadrature eigenstates) that is provably secure, and then we modify this protocol by allowing for approximate EPR pairs, and finally reducing the entanglement-based protocol to a version that closely resembles the above protocol by Hillery. To this end, we use the quantum code for encoding a qubit in a continuous variable that was introduced towards the end of Chapter 2. For the moment, we consider general continuous variables. In particular, we do not restrict ourselves to optical states. Later in this section we will specifically return to the optical implementation.

In order to encode a qubit in a continuous variable we need a stabilizer for the continuous-variable state that leaves room for a qubit degree of freedom. Following Gottesman and Preskill (2001), consider the following stabilizer generators:

$$S(q) = \exp\left(i2\sqrt{\pi}\hat{q}\right) \quad \text{and} \quad S(p) = \exp\left(-i2\sqrt{\pi}\hat{p}\right). \tag{8.122}$$

These operators commute, due to the special value of $\alpha = \sqrt{\pi}$ in the exponent, where α was defined in Eq. (2.113) as the spacing in the GKP code.

Exercise 8.12: Show that $[S(q), S(p)] = 0$.

The eigenvalues of $S(q)$ and $S(p)$ are $+1$ if the values of q and p are $s\sqrt{\pi}$, with $s \in \mathbb{Z}$. The corresponding eigenstates are then the GKP code states introduced in Chapter 2:

$$|\bar{0}\rangle \propto \sum_{s=-\infty}^{+\infty} |2s\sqrt{\pi}\rangle_q \propto \sum_{s=-\infty}^{+\infty} |s\sqrt{\pi}\rangle_p$$

$$|\bar{1}\rangle \propto \sum_{s=-\infty}^{+\infty} |(2s+1)\sqrt{\pi}\rangle_q \propto \sum_{s=-\infty}^{+\infty} (-1)^s |s\sqrt{\pi}\rangle_p, \tag{8.123}$$

where the subscripts q and p indicate whether the states are position or momentum eigenstates, respectively. We can formally construct single-qubit Pauli operators in this code space according to

$$\bar{X} : |\bar{0}\rangle \to |\bar{1}\rangle, \ |\bar{1}\rangle \to |\bar{0}\rangle \quad \text{and} \quad \bar{Z} : |\bar{0}\rangle \to |\bar{0}\rangle, \ |\bar{1}\rangle \to -|\bar{1}\rangle. \tag{8.124}$$

We can write these operators explicitly as

$$\bar{X}(0) = \exp\left(-i\sqrt{\pi}\,\hat{p}\right) \quad \text{and} \quad \bar{Z}(0) = \exp\left(i\sqrt{\pi}\,\hat{q}\right), \tag{8.125}$$

where we have added the argument (0) because we will generalize these operators in a moment. The eigenstates of the operator $\bar{Z}(0)$ are $|\bar{0}\rangle$ and $|\bar{1}\rangle$, and the eigenstates of $\hat{\bar{X}}(0)$ are

$$|\bar{+}\rangle = \frac{|\bar{0}\rangle + |\bar{1}\rangle}{\sqrt{2}} \quad \text{and} \quad |\bar{-}\rangle = \frac{|\bar{0}\rangle - |\bar{1}\rangle}{\sqrt{2}}. \tag{8.126}$$

Given that the states $|\bar{0}\rangle$ and $|\bar{1}\rangle$ were constructed as eigenstates of $S(q)$ and $S(p)$ with eigenvalues $+1$, a measurement of position or momentum must always give an outcome that is a multiple of $\sqrt{\pi}$. If an actual outcome (after some noisy transmission, for example) differs slightly from a multiple of $\sqrt{\pi}$, we can choose the nearest value, and recover the encoded value. The code therefore protects against small translations ($< \frac{1}{2}\sqrt{\pi}$) in q and p.

This code can be generalized to states with different stabilizer-generator eigenvalues, for example

$$S(q)|\psi\rangle = \exp(2\pi i\sigma_q)|\psi\rangle \quad \text{and} \quad S(p)|\psi\rangle = \exp(-2\pi i\sigma_p)|\psi\rangle, \tag{8.127}$$

where σ_q and σ_p take arbitrary values in the interval $[-1, 1]$. The eigenstates of these stabilizer generators are translated with respect to the eigenstates of the $+1$ eigenvalue stabilizers. The state defines a different code, which we shall denote by (σ_q, σ_p). The code $(0, 0)$ is then our original $+1$ eigenvalue code. This means that we also need to generalize the Pauli operators, which now become

$$\bar{X}(\sigma_p) = \exp\left[-i\sqrt{\pi}\left(\hat{p} - \sigma_p\sqrt{\pi}\right)\right] \tag{8.128}$$

and

$$\bar{Z}(\sigma_q) = \exp\left[i\sqrt{\pi}\left(\hat{q} - \sigma_q\sqrt{\pi}\right)\right]. \tag{8.129}$$

The need for an argument for $\bar{X}$ and $\bar{Z}$ is now clear, and for the (0,0) code the Pauli operators reduce to those in Eq. (8.125).

Exercise 8.13: Show that $\bar{X}(\sigma_p)$ and $\bar{Z}(\sigma_q)$ are the Pauli operators for the (σ_q, σ_p) code.

Establishing encoded Bell pairs

Next, we construct a secure quantum key distribution protocol using entanglement. Assume that Alice and Bob share a number of EPR pairs. In particular, the pairs are stabilized by

$$S_A(q) \otimes S_B^{-1}(q) = \exp\left[2i\sqrt{\pi}\left(\hat{q}_A - \hat{q}_B\right)\right] \tag{8.130}$$

and

$$S_A(p) \otimes S_B(p) = \exp\left[-2i\sqrt{\pi}\left(\hat{p}_A + \hat{p}_B\right)\right]. \tag{8.131}$$

The states with $+1$ eigenvalues of these stabilizers uniquely define the EPR pair with

$$q_A - q_B = p_A + p_B = 0, \tag{8.132}$$

that is, perfect position correlation, and perfect momentum anti-correlation. Different eigenvalues of the stabilizer generators define EPR states where $q_A - q_B$ and $p_A + p_B$ differ from zero. The stabilizer generators for the ideal EPR pair are constructed from the stabilizer generators in Eq. (8.122), which defined the continuous-variable code for a qubit in Eq. (8.123). Alice can therefore perform a measurement of $S_A(q)$ and maintain a potentially perfect correlation with Bob's measurement statistics (should Bob measure $S_B(q)$). Her measurement outcome will be an eigenvalue $\exp(2\pi i \sigma_q)$, for some value of σ_q. Note that this is *not* the same as measuring the stabilizer generator in Eq. (8.130), since Bob is still free to choose whatever observable he wants to measure. Since we are not confining our discussion to optical states, we assume that the measurement does not destroy the quantum system. Alice continues by measuring the second stabilizer generator $S_A(p)$, obtaining the outcome σ_p. She is allowed to do this, since the stabilizers $S_A(q)$ and $S_A(p)$ commute. She thus obtained two values σ_q and σ_p that characterize the state shared by Alice and Bob.

We now make an important observation: the initial (ideal) EPR pair is also an eigenstate of the operators

$$\bar{X}_A(\sigma_p) \otimes \bar{X}_B(\sigma_p) \quad \text{and} \quad \bar{Z}_A(\sigma_q) \otimes \bar{Z}_B^{-1}(\sigma_q). \tag{8.133}$$

This means that Alice's measurement has projected the state of the initial EPR pair onto the maximally entangled encoded qubit state

$$|\text{EPR}\rangle \to |\bar{\Phi}^+\rangle \equiv \frac{|\bar{0}_\sigma\rangle_A |\bar{0}_\sigma\rangle_B + |\bar{1}_\sigma\rangle_A |\bar{1}_\sigma\rangle_B}{\sqrt{2}} \tag{8.134}$$

in the (σ_q, σ_p) encoding, where

$$|\bar{0}_\sigma\rangle \propto \sum_{s=-\infty}^{+\infty} \left|(2s + \sigma_q)\sqrt{\pi}\right\rangle_q \propto \sum_{s=-\infty}^{+\infty} \left|(s + \sigma_p)\sqrt{\pi}\right\rangle_p$$

$$|\bar{1}_\sigma\rangle \propto \sum_{s=-\infty}^{+\infty} \left|(2s + 1 + \sigma_q)\sqrt{\pi}\right\rangle_q \propto \sum_{s=-\infty}^{+\infty} (-1)^s \left|(s + \sigma_p)\sqrt{\pi}\right\rangle_p. \tag{8.135}$$

The eigenstates of $\bar{Z}(\sigma_p)$ are $|\bar{0}_\sigma\rangle$ and $|\bar{1}_\sigma\rangle$, and the eigenstates of $\bar{X}(\sigma_p)$ are $|\bar{+}_\sigma\rangle$ and $|\bar{-}_\sigma\rangle$. Alice's two measurements are a form of 'distillation' that takes an EPR pair with arbitrary q_A and $q_B = q_A$, and p_A and $p_B = -p_A$ to an entangled pair in which position and momentum are multiples of $\sqrt{\pi}$ (up to a shift σ_q and σ_p). The crucial point is that Alice's measurement does not project onto a single state (in which case any entanglement between her and Bob would be destroyed), but rather onto an infinite dimensional subspace.

At this point, Alice may (publicly) announce to Bob her measurement outcomes σ_q and σ_p, which allows Bob to measure his stabilizer generators $S_B(q)$ and $S_B(p)$ with measurement outcomes τ_q and τ_p, and determine the quantities

$$\exp[i(\sigma_q - \tau_q)] \quad \text{and} \quad \exp[-i(\sigma_p + \tau_p)].$$

These form the syndrome of the correlated measurements, and they will differ from 1 if the transmission of the state from Alice to Bob induced translations in q and p. Based on the syndrome, Bob can adjust the position and momentum of his system (that is, perform error correction) such that he and Alice share a near-perfect state $|\bar{\Phi}^+\rangle$ in the (σ_q, σ_p) code space.

Alice and Bob have now established a shared entangled qubit pair, encoded in a continuous-variable quantum system. They can use this in a QKD protocol that is provably secure (see Section 5.4). Furthermore, the entire discussion in this section can be generalized for d-level (qudit) systems.

Reduction to practical QKD

The above discussion was framed in terms of ideal EPR pairs, which are of course non-physical states, and can never be created by Alice in her lab. We therefore need to investigate how this protocol is affected when Alice creates approximate EPR pairs. When she chooses a Gaussian approximation to an EPR pair, we can describe the entangled state with the Wigner function

$$W(q_A, p_A; q_B, p_B) = \left(\frac{2}{\pi}\right)^2 \exp\left[-2\Delta^2(q_A - q_B)^2 - 2\Delta^2(p_A + p_B)^2\right]$$
$$\times \exp\left[-\frac{2(q_A + q_B)^2}{\Delta^2} - \frac{2(p_A - p_B)^2}{\Delta^2}\right]. \tag{8.136}$$

Note that we are still considering general continuous-variable systems. In particular, we assume that we can still perform 'quantum non-demolition' (QND) measurements. The correlation between the position and momentum found by Alice and Bob will be imperfect. However, we will assume that the approximation is sufficiently good that the position and momentum correlations are perfect to an accuracy of m bits. The protocol then proceeds as follows:

1. Alice prepares her approximate EPR state, and sends one part to Bob;
2. each measures the stabilizer generators $S_i(q)$ and $S_i(p)$ to an accuracy of m bits ($i = A, B$);
3. Alice broadcasts her measurement outcomes for σ_q and σ_p, which requires $2m$ bits;
4. Bob uses Alice's broadcast to correct their state for transmission errors;
5. the state has been distilled to a two-qubit Bell state based on a continuous variable with m-bit accuracy;
6. the entanglement-based QKD protocol commences.

Note that this protocol still requires QND measurements. In addition, Alice and Bob need to store their systems in perfect quantum memories. In the remainder of this section, we will eliminate these requirements, which leads to Hillery's protocol of the previous section. We will establish the minimum amount of squeezing required for its optical implementation.

The entanglement-based QKD protocol is essentially the EKERT91 protocol, in which Alice and Bob share n Bell pairs in the state $|\Phi^+\rangle$. Alice measures randomly in the $\bar{X}$ or $\bar{Z}$ basis, and sends information about her choice of bases to Bob in the form of a bit string. Bob then measures his qubits in the bases that were indicated by Alice, and they establish a random shared bit string as a result.

We first observe that Alice can measure her qubits at any time, since at no point does she have to operate on her qubits based on information she receives from Bob. In particular, this means that she can measure her qubits before she sends the entangled partners to Bob. However, now the 'measure and send' procedure is indistinguishable from sending Bob unentangled qubits that are randomly chosen from a set of four states. To anyone but Alice, these states are highly mixed (and therefore constitute a 'proper mixture', as discussed in Chapter 3). The QKD protocol is now modified as follows:

1. Alice chooses a random m-bit value for the eigenvalues $e^{2\pi i\sigma_q}$ and $e^{-2\pi i\sigma_p}$, a random basis $\bar{X}$ or $\bar{Z}$, a random bit to send to Bob, and she prepares the corresponding (approximate) encoded state $|\bar{0}_\sigma\rangle$, $|\bar{1}_\sigma\rangle$, $|\bar{+}_\sigma\rangle$, or $|\bar{-}_\sigma\rangle$ for Bob;
2. after transmission of the system to Bob, Alice broadcasts the m-bit stabilizer eigenvalues, and her choice of basis $\bar{X}$ or $\bar{Z}$;
3. Bob measures $S(q)$ or $S(p)$ to m-bit accuracy, depending on Alice's choice of basis $\bar{Z}$ or $\bar{X}$, respectively;
4. he then subtracts Alice's m-bit value, and corrects to the nearest multiple of $\sqrt{\pi}$ to correct for any translation errors in the transmission of the quantum system;
5. the shared bit is determined whether the multiple of $\sqrt{\pi}$ is even or odd.

In this protocol, there is no longer any need for QND measurements, since the systems held by Alice and Bob are measured only once. Previously, the systems were measured twice: once for entanglement distillation, and a second time for the QKD protocol. However, Bob still needs a near-perfect quantum memory, because he needs to wait for Alice's broadcast before he knows in which basis he needs to measure his system. Another difficulty with this protocol is that the state Alice sends to Bob is very difficult to make. We will therefore simplify the protocol further.

If Alice tells Bob to measure the stabilizer generator $S(q)$, she does not need to tell him the value of σ_p. Likewise, if she tells him to measure the stabilizer generator $S(p)$, she does not need to broadcast the value of σ_q. Therefore, Alice does not send the unnecessary information. This does not compromise the security of the protocol, because Alice broadcasts *less* information than before. Moreover, if Alice decided to send $|\bar{0}_\sigma\rangle$ without information about σ_p, Bob and any present eavesdropper must average over σ_p and will describe the state of the system as

$$\rho_0(\sigma_q) \propto \sum_{s=-\infty}^{+\infty} \left|(2s+\sigma_q)\sqrt{\pi}\right\rangle_q \left\langle(2s+\sigma_q)\sqrt{\pi}\right|. \qquad (8.137)$$

Similarly, if Alice decides to send the state $|\bar{1}_\sigma\rangle$ without information about σ_p, the state of the system as described by Bob (and Eve) would become

$$\rho_1(\sigma_q) \propto \sum_{s=-\infty}^{+\infty} \left|(2s+1+\sigma_q)\sqrt{\pi}\right\rangle_q \left\langle(2s+1+\sigma_q)\sqrt{\pi}\right|. \qquad (8.138)$$

If in addition we also average over σ_q (which would be the correct procedure for Bob and Eve to establish the state without knowledge of σ_q), the state becomes a random-position eigenstate. Likewise, if Alice chose the eigenbasis of $\bar{X}$ to send her bit, the state averaged over σ_q and σ_p is a random-momentum eigenstate. Therefore, the protocol may be modified such that Alice sends a random position or momentum eigenstate. Bob's measurement reduces to a measurement in the position or momentum basis.

In addition, Bob may circumvent the need for a quantum memory by making his measurement in either the position or momentum basis, and hoping that his choice of basis coincides with that of Alice. If it doesn't, that particular run will be discarded, and they will try again. Since *not* using a transmitted state for the QKD protocol cannot yield any information about the key, no eavesdropper can benefit from this modification, and the protocol is still secure to within the *m*-bit accuracy.

With these modifications, we can state the protocol in terms of an optical implementation. The position and momentum operators become the corresponding quadratures, and their approximate eigenstates become squeezed coherent states. The protocol is then

1. Alice randomly chooses a quadrature $\hat{q}$ or $\hat{p}$ and again randomly picks a corresponding continuous variable q or p over a broad range;
2. she prepares a squeezed coherent state centered around q or p and sends it to Bob;
3. Bob randomly chooses a quadrature measurement $\hat{q}$ or $\hat{p}$ and finds the measurement outcome q_A or q_B;
4. Alice and Bob publicly compare their choice of quadratures, and keep the measurement data for coinciding quadrature measurements;
5. if their quadratures coincide, Alice broadcasts the *m*-bit value q or p modulo $\sqrt{\pi}$ (thus not revealing whether it is an even or odd multiple, which constitutes her random bit), and Bob subtracts this from his measurement outcome q_A or q_B;
6. Bob corrects the difference to the nearest multiple of $\sqrt{\pi}$, and the shared bit has value 0 or 1, depending on whether the multiple is even or odd.

This protocol is not quite identical to the protocol by Hillery (when the code-word alphabet is binary), but the differences are merely cosmetic. The security of this protocol therefore implies the security of that of Hillery.

How good must the approximation to quadrature eigenstates be for this protocol to be secure? In Section 5.4 we mentioned that the security proof of BB84 employs CSS codes to demonstrate that the maximum error rate for which the protocol is still secure is about 11%. The same proof applies here since we use continuous variables to define a qubit code space, and we therefore also limit the error rate due to finite squeezing and transmission losses to 11%. This corresponds to a squeezing strength that exceeds 2.51 dB. Current experimental levels of squeezing are about 10 dB, which means that practical QKD with squeezed states is possible.

Finally, we must stress again that these QKD protocols are not authentication protocols. In other words, we assume that Alice and Bob are already confident that they are communicating with each other, and not with some impostor. Authentication is a different problem from key distribution.

8.6 References and further reading

For an introduction to Wigner functions in quantum optics we refer to the excellent texts by Leonhardt (1997), and Barnett and Radmore (1997). Dedicated books on general quantum information with continuous variables were edited by Braunstein and Pati (2003) and Cerf *et al.* (2007). A review article by Braunstein and van Loock (2005) gives a complete overview of the literature. Quantum teleportation of qunats was demonstrated experimentally by Furusawa *et al.* (1998). For more information on entanglement distillation and purification protocols for Gaussian states we refer to Duan *et al.* (2000a, 2000b), Browne *et al.* (2003), Fiurášek *et al.* (2003), Heersink *et al.* (2006), and Fiurášek *et al.* (2007). Finally, quantum cryptography for continuous variables was introduced by Hillery (2000), Ralph (2000), and Cerf *et al.* (2002), and investigated experimentally by Lodewyck *et al.* (2007) and Zhang *et al.* (2008).

9 Quantum computation with continuous variables

In the previous chapter we have discussed some important quantum information processing techniques based on continuous variables, or 'qunats'. In this chapter, we describe methods for extending these techniques to quantum computation with optical qunats. We will first show in Section 9.1 how the optical qunat states are initialized using single-mode squeezing, and we give optical implementations of some of the most important single-qunat gates, such as the Heisenberg–Weyl operators and the Fourier transform. In Section 9.2, we study the theory of two-mode Gaussian operations, and we construct a general optical circuit that can implement any two-mode Gaussian operation. In Section 9.3, we introduce the stabilizer formalism for qunats, and use it to derive a Gottesman–Knill theorem. This leads to the general requirement of nonlinear gates for qunats, and their optical implementation will be discussed in Section 9.4. Having thus established the optical implementation of universal quantum computing with continuous variables, we extend this to the one-way model in Section 9.5, where we introduce cluster states for continuous variables. Finally, in Section 9.6, we discuss various quantum error-correction techniques for qunat quantum computing.

9.1 Single-mode optical qunat gates

We have already seen in Chapter 2 that the single qunat gates are fundamentally different from qubit gates. In the context of optical implementations, this difference arises due to the complementary nature of qubits and qunats: qubits that are represented by single photons can be considered particles, whereas qunats are encoded in the wave properties of the optical field. In this section, we will show how to create the optical qunat by squeezing the vacuum, and give optical implementations of the phase-space displacements (the Heisenberg–Weyl operators), the Fourier transform, and the phase gate.

9.1.1 Initializing the qunat states

One of the most important operations in qunat quantum information processing is the creation of the squeezed states that are to become the approximations to the (unphysical) position and momentum eigenstates. This is accomplished with the single-mode squeezing operation. We have developed the mathematical description of this operation in Chapter 1 and in the previous chapter, and we saw in Chapter 4 how it describes photon-pair production. Here we will describe how quadrature squeezing can be implemented in the laboratory.

Using Eq. (1.200), the single-mode squeezing of a mode a can be written in terms of the mode operators as

$$S(\xi) = \exp\left(\xi^* \frac{\hat{a}^2}{2} - \xi \frac{\hat{a}^{\dagger 2}}{2}\right). \tag{9.1}$$

We will now show that this operation can be implemented using a material with a so-called $\chi^{(2)}$ nonlinearity. Such a material can be described by the Hamiltonian in the rotating-wave approximation

$$\mathcal{H} = \hbar\omega\,\hat{a}^\dagger\hat{a} + \hbar\omega_P\,\hat{b}^\dagger\hat{b} + i\hbar\chi^{(2)}\left(\hat{a}^2\hat{b}^\dagger - \hat{a}^{\dagger 2}\hat{b}\right), \tag{9.2}$$

where $\chi^{(2)}$ is the strength of the nonlinear interaction of the light with the medium. In the parametric approximation, we set $\hat{b} \to \beta e^{-i\omega_P t}$ and up to an irrelevant constant term the Hamiltonian then becomes

$$\mathcal{H}_p = \hbar\omega\,\hat{a}^\dagger\hat{a} + i\hbar\chi^{(2)}\left(\hat{a}^2\beta^* e^{i\omega_P t} - \hat{a}^{\dagger 2}\beta e^{-i\omega_P t}\right). \tag{9.3}$$

We can remove the first term by going to the interaction picture, and find

$$\mathcal{H}_I = i\hbar\chi^{(2)}\left(\hat{a}^2\beta^* e^{i(\omega_P - 2\omega)t} - \hat{a}^{\dagger 2}\beta e^{-i(\omega_P - 2\omega)t}\right). \tag{9.4}$$

When we select the modes where $\omega_P = 2\omega$, and we exponentiate the Hamiltonian to obtain the corresponding unitary operation, we find the squeezing operator

$$S = \exp\left(-\frac{i}{\hbar}\mathcal{H}_I t\right) = \exp\left(\chi^{(2)}t\beta^* \hat{a}^2 - \chi^{(2)}t\beta\,\hat{a}^{\dagger 2}\right). \tag{9.5}$$

Since the strength of the nonlinearity $\chi^{(2)}$ is a real number, as is the interaction time t, we can define $\xi = 2\chi^{(2)}t\beta$, and the pumped second-order nonlinear medium acts as the squeezing operator in Eq. (9.1). A schematic of the set-up is shown in Fig. 9.1. For an in-depth treatment of nonlinear optical media, we refer the reader to Boyd (2003).

Experimentalists typically measure squeezing in decibels. If we write $\xi = re^{i\phi}$ and s is the amount of squeezing in dB, we define

$$s = 10\log_{10}\left(e^{2r}\right) \approx 8.69\,r. \tag{9.6}$$

Achievable levels of optical squeezing currently stand at about $10\,\text{dB}$, or $r \approx 1.15$. Even though higher values of ξ can in principle be obtained by choosing β sufficiently large, *any* source of imperfection in the system will cause s to deteriorate, and increasing the amount of squeezing further is a challenging task.

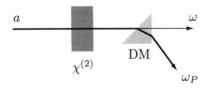

Fig. 9.1.   A bright coherent state with frequency ω_P pumps a nonlinear medium $\chi^{(2)}$. The mode a with frequency $\omega = \omega_P/2$ is squeezed, and the two modes are separated with a dichroic mirror (DM).

9.1.2 The Heisenberg–Weyl operators

Next, we consider two basic logic gates for qunat quantum information processing, namely the Heisenberg–Weyl operators. They are defined as

$$X(q) = \exp\left(-\frac{i}{\hbar}q\hat{p}\right) \quad \text{and} \quad Z(p) = \exp\left(\frac{i}{\hbar}p\hat{q}\right), \tag{9.7}$$

which using

$$\hat{q} = \sqrt{\frac{\hbar}{2\omega}}\left(\hat{a} + \hat{a}^\dagger\right) \quad \text{and} \quad \hat{p} = -i\sqrt{\frac{\hbar\omega}{2}}\left(\hat{a} - \hat{a}^\dagger\right) \tag{9.8}$$

can be written in terms of the creation and annihilation operators, leading to the expressions

$$X(q) = \exp\left[\sqrt{\frac{\omega}{2\hbar}}q\left(\hat{a}^\dagger - \hat{a}\right)\right] \quad \text{and} \quad Z(p) = \exp\left[\frac{ip}{\sqrt{2\hbar\omega}}\left(\hat{a}^\dagger + \hat{a}\right)\right]. \tag{9.9}$$

These can be interpreted as displacement operators

$$D(\alpha) = \exp\left(\alpha\hat{a}^\dagger - \alpha^*\hat{a}\right), \tag{9.10}$$

such that

$$X(q) = D\left(\sqrt{\frac{\omega}{2\hbar}}q\right) \quad \text{and} \quad Z(p) = D\left(\frac{ip}{\sqrt{2\hbar\omega}}\right). \tag{9.11}$$

In other words, $X(q)$ is a displacement along the real axis in phase space, and $Z(p)$ is a displacement along the imaginary axis. The question is then how we implement the displacement operator with optical elements.

Consider the optical set-up shown in Fig. 9.2. According to Eq. (1.146), the Hamiltonian for a beam splitter on modes a and b with $\varphi = 3\pi/2$ can be written as

$$\mathcal{H} = -i\hbar\zeta\left(\hat{a}^\dagger\hat{b} - \hat{a}\hat{b}^\dagger\right). \tag{9.12}$$

The beam-splitter transformation of mode a is then given by

$$\hat{a} \rightarrow \cos\zeta\,\hat{a} - \sin\zeta\,\hat{b}. \tag{9.13}$$

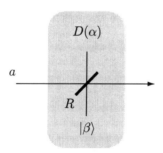

Fig. 9.2. Optical implementation of a displacement operator $D(\alpha)$. The displacement is ideal in the limit where the reflection coefficient R vanishes and $\beta = -\alpha/\sqrt{R}$.

We can now use the same trick that we used in our description of parametric downconversion. When mode b is in a bright coherent state β, produced by a laser operating far above threshold, the transformation becomes

$$\hat{a} \rightarrow \cos \zeta \, \hat{a} - \sin \zeta \, \beta \,. \tag{9.14}$$

Next, we assume that $\zeta \ll 1$, so we can approximate $\cos \zeta \simeq 1$, and $\sin \zeta \simeq \zeta$. Furthermore, if we define $\alpha \equiv -\zeta\beta$, we find

$$\hat{a} \rightarrow \hat{a} + \alpha \,, \tag{9.15}$$

which, according to Eq. (1.186), is the transformation of the annihilation operator due to a displacement α. The reflection coefficient of the beam splitter must therefore be small ($R = \zeta^2$) and the coherent input state must be $|\beta\rangle$ with large β. We can understand this behaviour physically by noting that a beam splitter generally entangles the modes it operates on, except in special cases (such as the vacuum or coherent states). Therefore, if the state of the input mode is e.g., a squeezed coherent state, and the reflected mode will be traced over, the entanglement induced by the beam splitter will cause the reduced density matrix of the (displaced) output mode to become mixed. The lower the reflectivity, the smaller this mixing effect, and the better this system approximates a displacement operator. In the teleportation experiment by Furusawa *et al.* (see Section 8.3), the reflectivity was $R = 0.01$.

9.1.3 The Fourier transform

The next logic gate for qunats that we consider here is the Fourier transform $\mathcal{F}$ of mode a. This transformation can be written as

$$\mathcal{F} = \exp\left[\frac{i\pi}{4\hbar}\left(\omega \hat{q}^2 + \frac{\hat{p}^2}{\omega}\right)\right]. \tag{9.16}$$

Using Eq. (9.8) we can write this as

$$\mathcal{F} = \exp\left[\frac{i\pi}{4}\left(\hat{a}^\dagger\hat{a} + \hat{a}\hat{a}^\dagger\right)\right] = \exp\left(\frac{i\pi}{2}\hat{a}^\dagger\hat{a}\right)\exp\left(\frac{i\pi}{4}\right). \tag{9.17}$$

This is a simple phase shift in the optical mode (up to an overall phase, which can be ignored). The Fourier transform can therefore be implemented either via a delay line or a material with a linear index of refraction and low losses.

We could have guessed that the Fourier transform is related to a phase shift, since it transforms position eigenstates into momentum eigenstates, and vice versa:

$$\mathcal{F}\hat{q}\,\mathcal{F}^\dagger = \frac{\hat{p}}{\omega} \quad \text{and} \quad \mathcal{F}\hat{p}\,\mathcal{F}^\dagger = -\omega\hat{q}\,. \tag{9.18}$$

In phase space, these eigenstates are rotated over $\pi/2$ with respect to each other, and we have already seen in the previous chapter that rotations in phase space are equivalent to phase shifts.

The Fourier transform $\mathcal{F}$ transforms the Heisenberg–Weyl operators into each other, but it also introduces factors of ω

$$\mathcal{F}X(q)\mathcal{F}^\dagger = Z(\omega q) \quad \text{and} \quad \mathcal{F}Z(p)\mathcal{F}^\dagger = X\left(-\frac{p}{\omega}\right) = X^{-1}\left(\frac{p}{\omega}\right). \tag{9.19}$$

Since the Heisenberg–Weyl operators are implemented by the displacement operator, we calculate the effect of $\mathcal{F}$ on $D(\alpha)$:

$$\mathcal{F}D(\alpha)\mathcal{F}^\dagger = D(i\alpha). \tag{9.20}$$

It therefore turns real displacements into imaginary displacements and vice versa. This can be understood as a rotation over $\pi/2$ in phase space.

Finally, the Fourier transform of the single-mode squeezer gives the inverse squeezing operator

$$\mathcal{F}S(\xi)\mathcal{F}^\dagger = S(-\xi) = S(\xi)^{-1}. \tag{9.21}$$

Again, this is easily understood as a rotation in phase space.

Exercise 9.1: Verify Eq. (9.21).

9.1.4 The phase gate

The final single-qunat gate we consider here is the phase gate $\Phi(\theta)$. It can be written as

$$\Phi(\theta) = \exp\left(\frac{i}{\hbar}\frac{\theta}{2}\omega\hat{q}^2\right). \tag{9.22}$$

In Fig. 9.3 we show the effect of this gate on the Wigner function of the vacuum state for three values of θ. The phase gate is a combination of single-mode squeezing and a rotation of the quadratures. It can therefore be implemented using a pumped second-order nonlinear material and a phase shift.

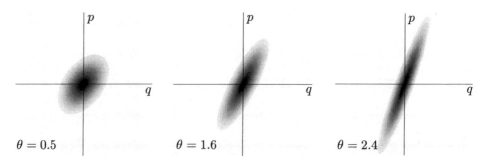

$\theta = 0.5$ $\theta = 1.6$ $\theta = 2.4$

Fig. 9.3. The effect of the phase gate $\Phi(\theta)$ on the Wigner function of the vacuum state. Notice how the squeezing angle increases with increasing θ.

This gate immediately suggests another phase gate $\Phi_P(\theta)$ that is generated by the momentum quadrature

$$\Phi_P(\theta) = \exp\left(\frac{i}{\hbar}\frac{\theta}{2}\frac{\hat{p}^2}{\omega}\right).$$

(9.23)

It is related to $\Phi(\theta)$ via a Fourier transform

$$\Phi_P(\theta) = \mathcal{F}\,\Phi(\theta)\,\mathcal{F}^\dagger,$$

(9.24)

and is therefore implemented optically with a second-order nonlinear material and phase shifts as well.

Exercise 9.2: Show that the phase gates can be decomposed into single-mode squeezing and a quadrature rotation.

9.2 Two-mode Gaussian qunat operations

The quadratic operators in $\hat{q}$ and $\hat{p}$ of the previous section form a group of Gaussian operators on a single optical mode. However, any quantum computation requires entanglement, and we can obtain this only with multi-mode operations, as shown in Chapter 8. In this section, we consider a special case of multi-mode operations, namely two-mode Gaussian operations. We ask which sets of basic one- and two-mode Gaussian operators are sufficient to efficiently simulate any other Gaussian operator, and we will describe an optical implementation that can be used to implement any two-mode Gaussian operator.

Rather than the unitary transformations U, we will study the Hamiltonians $\mathcal{H}$ that give rise to $U = \exp(-i\mathcal{H}t/\hbar)$. We can do this without loss of generality. Following Kraus *et al.* (2003), and setting $\omega = 1$, we define the local Hamiltonian $\mathcal{H}_j^\ell$

$$\mathcal{H}_j^\ell(\phi) = \phi\left(\hat{q}_j^2 + \hat{p}_j^2\right),$$

(9.25)

that is, a phase shift ϕ on mode a_j. We assume that we can apply gates like these instantaneously. While this is obviously not true in general, we can adjust the optical path length from one gate to the next such that these local phase shifts are implemented without additional cost in time. In addition, we define the two-mode 'interaction' Hamiltonian on modes a_1 and a_2

$$\mathcal{H} = c_{qq}\,\hat{q}_1\hat{q}_2 + c_{pp}\,\hat{p}_1\hat{p}_2 + c_{qp}\,\hat{q}_1\hat{p}_2 + c_{pq}\,\hat{p}_1\hat{q}_2.$$

(9.26)

Alternatively, this can be written as

$$\mathcal{H} = (\hat{q}_1, \hat{p}_1)\,M\begin{pmatrix}\hat{q}_2 \\ \hat{p}_2\end{pmatrix} \quad \text{with} \quad M = \begin{pmatrix} c_{qq} & c_{qp} \\ c_{pq} & c_{pp} \end{pmatrix}.$$

(9.27)

The matrix M has singular values λ_1 and λ_2, and we define

$$s_1 = \lambda_1 \quad \text{and} \quad s_2 = \text{sign}(\det M)\,\lambda_2,$$

(9.28)

as the *restricted* singular values of M.

Next, consider the Hamiltonian evolution given by

$$e^{-\frac{i}{\hbar}\mathcal{H}'t'} = U_1^{(N)} U_2^{(N)} e^{-\frac{i}{\hbar}\mathcal{H}t_N} \cdots e^{-\frac{i}{\hbar}\mathcal{H}t_2} U_1^{(1)} U_2^{(1)} e^{-\frac{i}{\hbar}\mathcal{H}t_1} U_1^{(0)} U_2^{(0)}, \qquad (9.29)$$

where the local unitary transformations

$$U_j^{(k)} = \exp\left[-\frac{i}{\hbar}\mathcal{H}_j^\ell(\phi_k)\right] \qquad (9.30)$$

are the single-mode phase shifts. We say that $\mathcal{H}$ simulates $\mathcal{H}'$ *efficiently* if Eq. (9.29) holds true for some choice of phase shifts $U_j^{(k)}$ and

$$t' = t \equiv \sum_k t_k, \qquad (9.31)$$

where we assumed that the phase shifts are implemented instantaneously. It turns out that $\mathcal{H}'$ can be simulated efficiently by $\mathcal{H}$ if and only if

$$s_1 + s_2 \geq s_1' + s_2' \quad \text{and} \quad s_1 - s_2 \geq s_1' - s_2', \qquad (9.32)$$

where s_j and s_j' are the restricted singular values of $\mathcal{H}$ and $\mathcal{H}'$, respectively. The proof is given by Kraus *et al.* (2003).

We now give two examples of this theorem: consider the cz gate described by the Hamiltonian $\mathcal{H}_{\text{cz}} = \hat{q}_1 \hat{q}_2$. The restricted singular values are $s_1 = 1$ and $s_2 = 0$. Can this Hamiltonian simulate the Hamiltonians of the beam splitter

$$\mathcal{H}_{BS} = \hat{q}_1 \hat{p}_2 - \hat{p}_1 \hat{q}_2, \qquad (9.33)$$

and the two-mode squeezer

$$\mathcal{H}_{S_2} = \hat{q}_1 \hat{q}_2 - \hat{p}_1 \hat{p}_2 ? \qquad (9.34)$$

For the beam-splitter Hamiltonian, the restricted singular values are $s_1' = 1$ and $s_2' = 1$. We then find immediately that $s_1 + s_2 = 1$ and $s_1' + s_2' = 2$. Therefore, the requirement $s_1 + s_2 \geq s_1' + s_2'$ is violated and the cz Hamiltonian cannot efficiently simulate the beam-splitter interaction Hamiltonian. Similarly, the restricted singular values of the two-mode squeezing Hamiltonian are $s_1' = 1$ and $s_2' = -1$. These values violate the requirement $s_1 - s_2 \geq s_1' - s_2'$, and the $\mathcal{H}_{\text{cz}}$ cannot efficiently simulate the two-mode squeezing Hamiltonian.

In our definition of efficient simulation, we required that the total time of the two unitary transformations t' and t is equal. If $\mathcal{H}'$ cannot be simulated by $\mathcal{H}$ efficiently, we may still be able to simulate $\mathcal{H}'$ with $\mathcal{H}$ in time $t > t'$. The minimum time $t_{\min}$ needed to simulate $\mathcal{H}'$ is given by

$$t_{\min} = \min_t \left\{ t : (s_1 + s_2)t \geq (s_1' + s_2')t', (s_1 - s_2)t \geq (s_1' - s_2')t' \right\}. \qquad (9.35)$$

Since each Hamiltonian is characterized by two real parameters s_1 and s_2, this behaviour can be represented graphically as shown in Fig. 9.4. Each point (s_1, s_2) in the (s_1', s_2') plane corresponds to a class of Hamiltonians $\mathcal{H}(s_1, s_2)$. We can restrict our discussion to the region $s_1' \geq |s_2'|$, because we can always apply single-mode phase shifts such that $\mathcal{H}$ lies in this

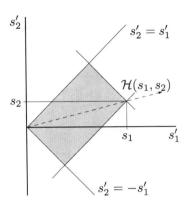

Fig. 9.4. The set of Hamiltonians $\mathcal{H}'$ that can be simulated efficiently with $\mathcal{H}$ is represented by the shaded rectangle. As the interaction time of $\mathcal{H}$ increases, the point (s_1, s_2) moves out along the dashed line (see Kraus *et al.*, 2003).

quadrant. From Fig. 9.4 it is easily seen that, except in the special cases of Hamiltonians on the lines $s_2' = |s_1'|$, any $\mathcal{H}$ of the form in Eq. (9.26) can simulate any other, given local phase shifts and sufficient time. The beam splitter and the two-mode squeezer both lie on the lines $s_2' = |s_1'|$, which can be understood as follows: the beam splitter has a bilinear Hamiltonian in which each term is a product of a creation and an annihilation operator. No matter how this is combined with single-mode phase shifts, this can never produce squeezing, where the Hamiltonian contains products of two creation (annihilation) operators. Similarly, a pure squeezing Hamiltonian cannot simulate the action of a beam splitter.

Exercise 9.3: Show that every Hamiltonian given by Eq. (9.26) can be described by a point (s_1, s_2) in the quadrant $s_1' \geq |s_2'|$.

The second question we address in this section is how we can construct *any* Gaussian operation with a restricted set of gates. We explicitly include single-mode quadratic terms in this construction. These were excluded from our previous discussion.

First, we recall from Chapter 1 that, according to the Bloch–Messiah reduction, we can write any linear Bogoliubov transformation of a set of modes as a passive interferometer (beam splitters and phase shifters), followed by a collection of single-mode squeezers, followed by another passive interferometer. The Bogoliubov transformation is generated by a bilinear Hamiltonian; see Eq. (1.172). We will now translate the general two-mode interferometric set-up, including the two single-mode squeezers, into a sequence of passive two-mode interferometers and two-mode squeezers.

In Chapter 1 we wrote the Bloch–Messiah reduction in terms of creation and annihilation operators $\hat{a}^\dagger$ and $\hat{a}$, whereas here we consider the quadrature operators $\hat{q}$ and $\hat{p}$. To fix our terminology, consider a unitary transformation U on N modes generated by a bilinear Hamiltonian $\mathcal{H}$:

$$\mathcal{H} = \frac{1}{2} \sum_{j,k=1}^{N} \left(\hat{q}_j A_{jk} \hat{q}_k + \hat{p}_j B_{jk} \hat{p}_k + \hat{q}_j C_{jk} \hat{p}_k + \hat{p}_j D_{jk} \hat{q}_k \right), \tag{9.36}$$

with Hermitian matrices A, B, C, and D, such that

$$U(t) = \exp\left(-\frac{i}{\hbar}\mathcal{H}t\right).$$

(9.37)

When the unitary operator is used to transform the quadrature operators $U(t)\hat{q}_j U^\dagger(t)$, the mode transformations can be written as

$$\begin{pmatrix} \vec{q}_{\text{out}} \\ \vec{p}_{\text{out}} \end{pmatrix} = K \begin{pmatrix} \vec{q}_{\text{in}} \\ \vec{p}_{\text{in}} \end{pmatrix},$$

(9.38)

where $\vec{q} = (\hat{q}_1, \ldots, \hat{q}_N)$, etc., and the matrix K obeys

$$K^T \Omega K = \Omega \quad \text{with} \quad \Omega = \begin{pmatrix} 0 & \hat{\mathbb{I}}_N \\ -\hat{\mathbb{I}}_N & 0 \end{pmatrix}.$$

(9.39)

Matrices with this property are 'symplectic', and their form ensures that the output quadratures again obey the canonical commutation relation $[\hat{q}_j, \hat{p}_k] = i\hbar\delta_{jk}$. The matrix K can be diagonalized using two orthogonal symplectic matrices O and R, and the resulting singular value decomposition for two modes yields

$$K = ODR^T \quad \text{with} \quad D = \begin{pmatrix} e^{r+s} & 0 & 0 & 0 \\ 0 & e^{-(r+s)} & 0 & 0 \\ 0 & 0 & e^{r-s} & 0 \\ 0 & 0 & 0 & e^{-(r-s)} \end{pmatrix},$$

(9.40)

where r and s are real. We wrote D in this way because it allows us to further decompose it into a product of two matrices D_r and D_s, obeying

$$D_r = \begin{pmatrix} e^r & 0 & 0 & 0 \\ 0 & e^{-r} & 0 & 0 \\ 0 & 0 & e^r & 0 \\ 0 & 0 & 0 & e^{-r} \end{pmatrix} \quad \text{and} \quad D_s = \begin{pmatrix} e^s & 0 & 0 & 0 \\ 0 & e^{-s} & 0 & 0 \\ 0 & 0 & e^{-s} & 0 \\ 0 & 0 & 0 & e^s \end{pmatrix}.$$

(9.41)

Next, we show that these matrices can be generated with unitary transformations such as those generated by $\mathcal{H}_{BS}$ and $\mathcal{H}_{S_2}$. We define the following four Hamiltonians

$$\begin{aligned} \mathcal{H}_{BS} &= \hat{q}_1\hat{p}_2 - \hat{p}_1\hat{q}_2 \quad \text{and} \quad \tilde{\mathcal{H}}_{BS} = -(\hat{q}_1\hat{q}_2 + \hat{p}_1\hat{p}_2) \\ \mathcal{H}_{S_2} &= \hat{q}_1\hat{q}_2 - \hat{p}_1\hat{p}_2 \quad \text{and} \quad \tilde{\mathcal{H}}_{S_2} = \hat{q}_1\hat{p}_2 + \hat{p}_1\hat{q}_2, \end{aligned}$$

(9.42)

which generate the unitary transformations $U_{BS}(t)$, $\tilde{U}_{BS}(t)$, $U_{S_2}(t)$, and $\tilde{U}_{S_2}(t)$, respectively. The two-mode squeezers D_r and D_s are then produced by

$$\begin{aligned} D_r &: \tilde{U}_{BS}^\dagger(\pi/4)\, U_{S_2}(r)\, \tilde{U}_{BS}(\pi/4) \\ D_s &: U_{BS}^\dagger(\pi/4)\, \tilde{U}_{S_2}(s)\, U_{BS}(\pi/4). \end{aligned}$$

(9.43)

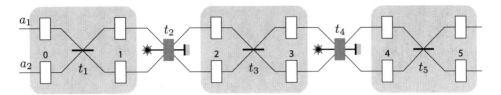

Fig. 9.5. Any two-mode Gaussian operator can be constructed from five elementary operations. The shaded boxes are arbitrary two-mode operations, consisting of four single-mode phase gates and a beam splitter. The elements indicated by t_2 and t_4 are two-mode squeezers. The interaction times t_i determine the strength of the gates.

Let's choose $r = t_2$ and $s = t_4$. If we put all the components in the right order, the most general two-mode Gaussian operation G can then be written as

$$G = U_1^{(5)} U_2^{(5)} \, U_{BS}(t_5) \, U_1^{(4)} U_2^{(4)} \, U_{S_2}(t_4) \, U_1^{(3)} U_2^{(3)} \, U_{BS}(t_3)$$
$$\times \, U_1^{(2)} U_2^{(2)} \, U_{S_2}(t_2) \, U_1^{(1)} U_2^{(1)} \, U_{BS}(t_1) \, U_1^{(0)} U_2^{(0)} \, , \tag{9.44}$$

which is shown in Fig. 9.5.

The two entangling gates that we are particularly interested in are the CX gate and the CZ gate. From Chapter 2, Eqs. (2.97) and (2.100), we infer that

$$\text{CX}_{ij} = \exp\left(-\frac{i}{\hbar} \hat{q}_i \hat{p}_j\right) \quad \text{and} \quad \text{CZ}_{ij} = \exp\left(\frac{i}{\hbar} \hat{q}_i \hat{q}_j\right). \tag{9.45}$$

The CZ gate is related to the CX gate via two Fourier transforms on the target system, indicated by j, since $\mathcal{F} \hat{p}_j \mathcal{F}^\dagger = -\hat{q}_j$ (recall that we set $\omega = 1$).

Exercise 9.4: Construct the optical implementation of the CZ gate. How much squeezing does it need?

When the object is to create (approximate) EPR pairs, however, we have seen in the previous chapter that we can circumvent the need for two-mode squeezing and send two single-mode squeezed vacuum states into a 50:50 beam splitter. The two vacuum states must be squeezed in orthogonal directions. This is therefore *not* an actual implementation of a CZ gate. Rather, it is a way to create entanglement. Later in this chapter we will see how this can be used to construct a measurement-based model of computation on qunat cluster states.

9.3 The Gottesman–Knill theorem for qunats

So far, we have considered only Gaussian operations. We explicitly discussed two-mode Gaussian operations, which can be extended to multi-mode Gaussian operations in a straightforward manner, according to Eq. (9.36). In this section we will show that if we restrict our qunat quantum computation to Gaussian operations on squeezed coherent states, and include feed-forward based on (perfect or lossy) homodyne detection, then we can efficiently simulate the computation on a classical computer. This is the Gottesman–Knill theorem for continuous variables. We will again use the stabilizer formalism in order to explicitly construct such an efficient classical description.

First, recall that the cx gate, the cz gate, the Fourier transform $\mathcal{F}$, and the phase gate $\Phi(\theta)$ transform the operators $X(q)$ and $Z(p)$ into each other. The Fourier transform acts as

$$\mathcal{F} X(q)\, \mathcal{F}^\dagger = Z(q) \quad \text{and} \quad \mathcal{F} Z(p)\, \mathcal{F}^\dagger = X^{-1}(p)\,, \tag{9.46}$$

and the phase gate yields

$$\Phi(\theta)\, X(q)\, \Phi^{-1}(\theta) = X(q) Z(\theta q) \exp\left(\frac{i}{2\hbar}\theta q^2\right)$$

$$\Phi(\theta)\, Z(p)\, \Phi^{-1}(\theta) = Z(p)\,. \tag{9.47}$$

Under the cx operation the Heisenberg–Weyl operators transform as

$$\begin{aligned} X_1(q) &\to X_1(q) X_2(q) \quad &\text{and} \quad Z_1(p) &\to Z_1(p) \\ X_2(q) &\to X_2(q) \quad &\text{and} \quad Z_2(p) &\to Z_1^{-1}(p) Z_2(p)\,. \end{aligned} \tag{9.48}$$

Finally, the cz gate produces the transformation

$$X_1(q) \to X_1(q) Z_2(q)\,, \tag{9.49}$$

which is analogous to the transformation of Pauli operators for qubits. These four operators are exponents of quadratic functions in the quadrature operators and, together with $X(q)$ and $Z(p)$, they efficiently generate all possible quadratic operations on N modes.

Exercise 9.5: Calculate the remaining transformations of $X(q)$ and $Z(p)$ under cz, not shown in Eq. (9.49).

Next, we note that the above operators can be combined to construct single-mode squeezers. When applied to the vacuum, these operators create the required initial states for the computation. If we want to construct an efficient description using the stabilizer formalism of a quantum computation based on Gaussian operations, we need to construct the stabilizer V_j of the vacuum state of mode a_j. This can be achieved by exponentiating the annihilation operator $\hat{a}_j$

$$V_j\, |0\rangle = \exp(c\,\hat{a}_j)\, |0\rangle = \exp\left[c\left(\sqrt{\frac{\omega}{2\hbar}}\hat{q}_j + \frac{i}{\sqrt{2\hbar\omega}}\hat{p}_j\right)\right] |0\rangle = |0\rangle\,. \tag{9.50}$$

The inclusion of $\hbar$ and ω makes this notation rather cumbersome, and their presence is not essential to the stabilizer formalism. We assume that all modes have the same discrete frequency, and we adopt photon-number units in which $\hbar = 1/2$. This yields the stabilizer generator for the vacuum

$$V_j\, |0\rangle = \exp\left[c\left(\hat{q}_j + i\hat{p}_j\right)\right] |0\rangle = |0\rangle\,. \tag{9.51}$$

Clearly, the vacuum is the $+1$ eigenstate of the operator V_j for arbitrary $c \in \mathbb{C}$. When we restrict ourselves to pure states, this stabilizer generator determines the vacuum uniquely. Even though $\hat{a}$ is not a Hermitian operator, and V_j is therefore not unitary, the stabilizer generator V_j is linear in $\hat{q}$ and $\hat{p}$, and is therefore a Gaussian operator.

The input states of an optical qunat quantum computer are squeezed coherent states. We will now construct the stabilizer generators for these states. First, the vacuum is squeezed via a single-mode squeezing operator, for example in the $\hat{q}$ quadrature when $r > 0$

$$S_q(r) = \exp\left[ir\left(\hat{q}\hat{p} + \hat{p}\hat{q}\right)\right].$$ (9.52)

Under the influence of this operator, the stabilizer generator evolves according to $(\Delta = e^r)$

$$V_j \rightarrow V_j(\Delta) \equiv S_q(r)\, V_j\, S_q^\dagger(r) = \exp\left(c\Delta\, \hat{q} + \frac{ic\hat{p}}{\Delta}\right).$$ (9.53)

When the vacuum is squeezed in the $\hat{p}$ quadrature with $S_p(r) = S_q(-r)$, the stabilizer generator becomes

$$V_j \rightarrow V_j(\Delta^{-1}) \equiv S_p(r)\, V_j\, S_p^\dagger(r) = \exp\left(\frac{c\hat{q}}{\Delta} + ic\Delta\hat{p}\right).$$ (9.54)

These stabilizer generators are again Gaussian functions.

Exercise 9.6: Verify the evolution of V_j under $S_q(r)$ and $S_p(r)$.

The stabilizer generator V_j transforms under the Heisenberg–Weyl operators $X(q)$ and $Z(p)$ according to

$$X(q)\, V_j\, X^{-1}(q) = \exp\left[c(\hat{q} + q) + ic\hat{p}\right],$$ (9.55)

and

$$Z(p)\, V_j\, Z^{-1}(p) = \exp\left[c\hat{q} + ic(\hat{p} + p)\right],$$ (9.56)

which are also Gaussian operators. When we have N qunats, the N-mode vacuum input state is stabilized by the set of N stabilizer generators $\{V_j\}$ with $j = 1, \ldots, N$. To prepare the input states for a quantum computation, each mode is squeezed in the $\hat{q}$ quadrature, and displaced using the operators $X_j(q_j)Z_j(p_j)$. The stabilizer generator for mode a_j then becomes

$$
\begin{aligned}
S_j &= X_j(q_j)Z_j(p_j)\, V_j(\Delta)\, Z_j^{-1}(p_j)X_j^{-1}(q_j) \\
&= \exp\left[c\Delta(\hat{q}_j + q_j) + \frac{ic(\hat{p}_j + p_j)}{\Delta}\right] \\
&= Z_j\left(-\frac{ic\Delta}{2}\right) X_j\left(-\frac{c}{2\Delta}\right) \exp\left(\frac{c^2}{4} + c\Delta q_j + \frac{icp_j}{\Delta}\right).
\end{aligned}
$$ (9.57)

Therefore, up to a complex exponential factor related to the displacement (q_j, p_j) and the squeezing strength $\Delta = e^r$, the stabilizer generator S_j is a product of Heisenberg–Weyl operators. This leads to the following stabilizer generators for the squeezed vacuum:

$$S_j = X_j\left(-\frac{c}{2\Delta}\right) Z_j\left(\frac{c\Delta}{2i}\right). \tag{9.58}$$

When $r \to +\infty$ the operator X tends towards the identity operator, and a choice of $c = 2i$ means that the stabilizer generator for the squeezed vacuum in the quadrature $\hat{q}$ is the operator $Z_j(\Delta)$. Similarly, when $r \to -\infty$ and $c = -2$, the squeezing occurs in the quadrature $\hat{p}$, and the state is stabilized by the operator $X_j(\Delta^{-1})$.

Exercise 9.7: Derive the stabilizer generator for the squeezed coherent state.

A quantum computation based solely on Gaussian operations U_G, generated by the Hamiltonian in Eq. (9.36)

$$U_G = \exp\left[-\frac{i}{2}\sum_{j,k=1}^{N}\left(\hat{q}_j A_{jk}\hat{q}_k + \hat{p}_j B_{jk}\hat{p}_k + \hat{q}_j C_{jk}\hat{p}_k + \hat{p}_j D_{jk}\hat{q}_k\right)\right], \tag{9.59}$$

formally transforms the Heisenberg–Weyl operators as

$$U_G X_j(q_j) U_G^\dagger \propto \prod_{k=1}^{N} X_k\left(q_j F_{jk}^{(X)}\right) Z_k\left(q_j G_{jk}^{(X)}\right)$$

$$U_G Z_j(p_j) U_G^\dagger \propto \prod_{k=1}^{N} X_k\left(p_j F_{jk}^{(Z)}\right) Z_k\left(p_j G_{jk}^{(Z)}\right), \tag{9.60}$$

up to exponential factors that also depend on $F_{jk}^{(l)}$ and $G_{jk}^{(l)}$, which in turn depend on the matrices A, B, C, and D. In order to keep track of the computation, we need to keep track of the numbers $F_{jk}^{(l)}$ and $G_{jk}^{(l)}$. Each one of them is a matrix with N^2 elements, and there are four of them. The total number of real parameters we need to track in order to describe the computation is therefore $4N^2$. This is a polynomial in the size of the input, and if the computer takes a polynomial number of steps (which is necessary for the computation to be efficient), then the classical description is also efficient.

Implicit in the above discussion is our assumption that the final readout is done in a quadrature basis for all the qunats. Without loss of generality, we can assume that the readout takes place in the position quadrature for all variables. This is implemented by homodyne detection. In practice, homodyne detection is performed with lossy photodetectors, and it is *a priori* not clear that this can be described as a Gaussian operation as well. However, we have seen in the previous chapter that loss can be modelled with a beam splitter of transmittance η. The beam splitter is a Gaussian operation, and therefore lossy homodyne detection is included in the Gottesman–Knill theorem for qunat quantum computing.

Similarly, we can include a large class of feed-forward operations in the theorem. Feed-forward is defined as a measurement of one or more qunats, the outcome of which is used to choose a new operation. It is well known that any feed-forward operation can formally be written as a controlled operation with the measurement postponed to the end of the computation. If this controlled operation is again a controlled Gaussian, the Gottesman–Knill theorem holds. Thus we have arrived at a very large class of operations that can be simulated efficiently on a classical computer by keeping track of the stabilizer. Consequently, for a quantum computer based on qunat computation to offer a fundamental speed-up, we need to include non-Gaussian operations.

9.4 Nonlinear optical qunat gates

An intuitive reason that Gaussian operators lead to a Gottesman–Knill theorem for qunats is that they are at most quadratic in their interaction Hamiltonians, and the commutator of two quadratic functions of the quadrature operators is again a quadratic function of the quadrature operators:

$$[\hat{q}^2, \hat{p}^2] = i(\hat{q}\hat{p} + \hat{p}\hat{q}), \quad [\hat{q}^2, \hat{p}\hat{q}] = i\hat{q}^2, \qquad (9.61)$$

and so on. When we determine the operator evolution in the Heisenberg picture, we evaluate repeated commutators according to the Baker–Campbell–Hausdorff relation

$$e^{\mu A} B e^{-\mu A} = B + \mu[B, A] + \frac{\mu^2}{2!}[B, [B, A]] + \cdots \qquad (9.62)$$

When repeated commutators of A and B do not increase the order of the polynomial in the quadrature operators, we cannot create arbitrary polynomials with A and B. We therefore need operators that are higher-order polynomials in the quadrature operators.

To make this more concrete, we consider the so-called single-mode Kerr nonlinearity introduced in Chapter 2:

$$\mathcal{H}_K = (\hat{q}^2 + \hat{p}^2)^2, \qquad (9.63)$$

which leads to the unitary transformation

$$K(\kappa) \equiv \exp(-2i\mathcal{H}_K\kappa) = \exp\left[-\frac{i\kappa}{2}\left(\hat{a}^\dagger\hat{a} + \frac{1}{2}\right)^2\right]. \qquad (9.64)$$

When we take the commutator between $\mathcal{H}_K$ and linear or quadratic Hamiltonians, we obtain, for example,

$$[\mathcal{H}_K, \hat{q}] = -i(\hat{q}^2\hat{p} + \hat{p}\hat{q}^2 + 2\hat{p}^3)$$
$$[\mathcal{H}_K, \hat{p}] = i(\hat{p}^2\hat{q} + \hat{q}\hat{p}^2 + 2\hat{q}^3)$$
$$[\mathcal{H}_K, \hat{q}\hat{p} + \hat{p}\hat{q}] = 4i(\hat{q}^4 - \hat{p}^4). \qquad (9.65)$$

These commutators are polynomials of order 3 and 4. We can now prove by induction that a polynomial of any degree can be constructed with linear and quadratic Hamiltonians and the Kerr Hamiltonian.

Exercise 9.8: Show that any polynomial term $\hat{q}^n \hat{p}^m$ can be constructed with the Kerr Hamiltonian and linear plus quadratic Hamiltonians.

In order to find out how we can create optical elements that are properly described by the Kerr Hamiltonian, we first investigate the effect of $\mathcal{H}_K$ on a single optical mode. It is convenient to perform this calculation for the mode operators $\hat{a}$ and $\hat{a}^\dagger$, rather than the quadrature operators $\hat{q}$ and $\hat{p}$. The mode transformation becomes

$$K(\kappa)\,\hat{a}\,K^\dagger(\kappa) = \hat{a}\,e^{-2i\kappa\hat{a}^\dagger\hat{a}} = e^{-2i\kappa\hat{a}\hat{a}^\dagger}\,\hat{a}\,. \tag{9.66}$$

This is equivalent to an intensity-dependent phase shift in mode a (called 'self-phase modulation'), and can be implemented with a lossless and dispersionless nonlinear medium. Classically, the optical response of such a medium can be expressed as a power series in the electric field of the polarization field in the medium:

$$P(t) = \chi^{(1)}E(t) + \chi^{(2)}E(t)^2 + \chi^{(3)}E(t)^3\,, \tag{9.67}$$

where $P(t) = \sum_k P(\omega_k)e^{-i\omega_k t} + \text{c.c.}$ and $E(t) = \sum_n E(\omega_n)e^{-i\omega_n t} + \text{c.c.}$ are multi-mode expansions of the polarization and electric field. Here, we are particularly interested in the third-order term $P^{(3)}(t) = \chi^{(3)}E(t)^3$. The polarizability and the electric fields are of course vector fields with three spatial components, and in general we should write

$$P_i^{(3)}(t) = \sum_{jkl} \chi_{ijkl}^{(3)} E_j(t) E_k(t) E_l(t)\,. \tag{9.68}$$

The nonlinear susceptibility $\chi_{ijkl}^{(3)}$ is then a fourth-rank tensor with 81 independent components. However, most materials are highly symmetric, and the actual number of independent components is far less. Here, we take $\chi^{(3)}$ to be independent of the orientation of the fields, and treat the electric field and the polarization as scalar quantities.

Suppose that the incoming wave is a monochromatic field $E(t) = E_0 e^{-i\omega t} + E_0^* e^{i\omega t}$. It is straightforward to show that the polarization field then becomes

$$\begin{aligned} P(t) &= \chi^{(1)}E(t) + \chi^{(3)}E(t)^3 \\ &= \left[\chi^{(1)} + 3\chi^{(3)}|E_0|^2\right]E(t) + \chi^{(3)}\left(E_0 e^{-3i\omega t} + E_0^* e^{3i\omega t}\right), \end{aligned} \tag{9.69}$$

where $|E_0|^2$ is the intensity of the incoming beam. The nonlinear polarizing ($\chi^{(3)}$) medium produces a second wave with frequency 3ω. The amplitude of the incoming wave is modified by a factor $\chi^{(1)} + 3\chi^{(3)}|E_0|^2$. The index of refraction for the wave with frequency ω in this material then becomes

$$n^2 = 1 + 4\pi\left[\chi^{(1)} + 3\chi^{(3)}|E_0|^2\right]. \tag{9.70}$$

Since phase shifts can be implemented using materials with a refractive index that is different from the refractive index of the propagating medium, the intensity-dependent phase shift can be implemented using an optical medium with a $\chi^{(3)}$ nonlinearity. The unwanted wave at frequency 3ω must be filtered out (Fig. 9.6).

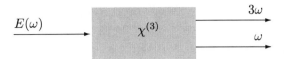

Fig. 9.6. An optical medium with a $\chi^{(3)}$ nonlinearity. The incoming wave with frequency ω experiences an intensity-dependent phase shift, and a secondary wave of frequency 3ω is created.

From Eq. (9.66) we can calculate the transformation of the quadrature operators due to a Kerr Hamiltonian:

$$\hat{q} \to \hat{q} \cosh \phi - i\hat{p} \sinh \phi$$
$$\hat{p} \to \hat{p} \cosh \phi + i\hat{q} \sinh \phi, \tag{9.71}$$

where

$$\phi(\hat{q}, \hat{p}) \equiv K(\hat{q}\hat{p} - \hat{p}\hat{q} - i\hat{q}^2 - i\hat{p}^2). \tag{9.72}$$

This is a highly nonlinear transformation.

Exercise 9.9: Verify the quadrature transformations in Eq. (9.71).

In principle, we now have all the ingredients for optical quantum computing with qunats: we have single-mode Gaussian gates, such as the Heisenberg–Weyl gates, the Fourier transform, and the phase gates; we also have two-mode entangling gates such as the cx and the cz gates, and we have a non-Gaussian gate based on a Kerr nonlinearity that makes the set of gates universal for quantum computing. Nevertheless, putting all these gates together in a practical quantum computer is extremely challenging, with many in-line squeezing elements and potentially noisy Kerr nonlinearities. Once again, we would like to invoke the Bloch–Messiah reduction and execute all squeezing in advance followed by passive linear interferometry, but this time the Kerr nonlinearities get in the way, and the reduction is generally not possible. However, we may be able to circumvent this problem if we can defer the use of Kerr nonlinearities to the final stage, just before the measurement of the optical modes. This approach leads to the one-way model for qunats.

9.5 The one-way model for qunats

Now that we have constructed a universal set of gates for quantum computing with qunats, we consider the one-way model of quantum computation. We first give a description of generalized cluster states for qunats, and show how to perform a universal set of measurements.

9.5.1 Qunat cluster states

Recall that cluster states (or graph states) are stabilizer states that act as the entanglement resource for a quantum computation. Here we consider the optical implementation of qunat cluster states, including finite squeezing. We have already seen in Chapter 2 that we can extend the qubit teleportation circuit to qunats, so we expect that we can also create cluster states for quantum computing with qunats.

We saw in the previous section that the stabilizer generator for finite squeezing in the $\hat{q}$ quadrature was given by Eq. (9.58). When we squeeze in the $\hat{p}$ quadrature and choose $c = -2$, the stabilizer for the momentum-squeezed state becomes

$$S_j = X_j(\Delta)Z_j(i\Delta^{-1}) \tag{9.73}$$

with $\Delta = e^r$. The states defined by these stabilizer generators are approximate momentum quadrature eigenstates with eigenvalue $p = 0$, in the limit of large squeezing $\Delta \to \infty$. These states are the building blocks for qunat cluster states, and we apply the CZ gates to construct the desired cluster state. The effect of the gate CZ_{jk} on the stabilizer S_j is

$$\begin{aligned}
S_j' &= \text{CZ}_{jk}S_j\text{CZ}_{jk}^\dagger = \text{CZ}_{jk}X_j(\Delta)\text{CZ}_{jk}^\dagger\text{CZ}_{jk}Z_j(i\Delta^{-1})\text{CZ}_{jk}^\dagger \\
&= X_j(\Delta)Z_j(\Delta)Z_k(i\Delta^{-1})\,. \tag{9.74}
\end{aligned}$$

In other words, the evolution of the stabilizer generator is identical to that of the ideal qunat states and qubit states, up to an operator $Z_j(i\Delta^{-1}) = \exp(-2\hat{q}/\Delta)$ that becomes the identity operator in the limit of infinite squeezing $\Delta \to \infty$. Cluster states can then be constructed in the same way as for qubits, except that each stabilizer has the extra factor $Z_j(i\Delta^{-1})$. The cluster is then completely determined by the stabilizer generators

$$S_j = X_j(\Delta_j) \prod_{k \in n(j)} Z_k(\Delta_j)Z_j(i\Delta_j^{-1})\,, \tag{9.75}$$

with $j = 1, \ldots, N$, and each mode j may be squeezed by a different amount Δ_j. The symbol $n(j)$ again denotes the set of all neighbours of mode j.

Unlike the Pauli operators in the qubit case, the Heisenberg–Weyl operators for qunat systems are exponentials of linear functions in $\hat{q}_k$ and $\hat{p}_k$. It is therefore more economical to consider only the argument of the exponential. When we write the stabilizer generator S_j in Eq. (9.75) as a single exponential, we have

$$S_j = \exp\left[-2i\Delta_j\left(\hat{p}_j - \sum_{k\in n(j)}\hat{q}_k\right) - \frac{2\hat{q}_j}{\Delta_j}\right]. \tag{9.76}$$

Given that for a cluster state $|C\rangle$ we have $S_j|C\rangle = |C\rangle$, the argument of the exponential must be zero. Consequently, in the limit of large squeezing the term $\hat{q}/\Delta$ vanishes, and the correlation between modes j and k can be compactly written as

$$\hat{p}_j - \sum_{k \in n(j)} \hat{q}_k \to 0. \tag{9.77}$$

There are N such relations, which completely specify the correlations in the qunat cluster state. In vector notation, this can be summarized as

$$\hat{\mathbf{p}} - A\hat{\mathbf{q}} \to 0, \tag{9.78}$$

where $\mathbf{q} = (\hat{q}_1, \ldots, \hat{q}_N)$, $\mathbf{p} = (\hat{p}_1, \ldots, \hat{p}_N)$, and A is the adjacency matrix that determines which qunats are neighbours. Any multi-mode Gaussian state that obeys Eq. (9.78) in the limit of large squeezing is a cluster state. The variance of the correlation in Eq. (9.78), denoted by $\text{Var}(\hat{\mathbf{p}} - A\hat{\mathbf{q}})$, is called the 'excess noise' in the cluster. The excess noise of a single qunat j in the cluster is given by $\text{Var}(\hat{p}_j - \sum_k \hat{q}_k)$, where k runs over all the neighbours of j, or $k \in n(j)$.

Exercise 9.10: Using the stabilizer generators S_j, show that in the absence of squeezing no entanglement is generated.

In practice, passive linear-optical elements for which the photon number is invariant are easier to implement than squeezing operations. It is therefore worthwhile investigating to what extent we can create squeezed states first, and create entanglement via passive linear optics such as beam splitters and phase shifters. It is clear that this is always possible by using the Bloch–Messiah reduction presented in Chapter 1: any multi-mode Gaussian operation can be decomposed into a passive linear interferometer, followed by single-mode squeezers, followed by another passive interferometer. The construction of the cluster state we described above fits in this category. We can therefore apply the single-mode squeezing first, followed by passive linear-optical elements (the passive interferometer acting on the vacuum can be omitted). An explicit construction was given by Van Loock *et al.* (2007). A note of caution: the standard construction of cluster states based on momentum-squeezed vacuum and cz gates allows us to grow the cluster locally. In particular, the construction procedure is independent of the graph we wish to create. This is no longer the case when all squeezing is performed in advance, since the amount of squeezing for a single vertex in the cluster state depends on the number of edges of that vertex. This is in itself not a show-stopper, although the state-preparation procedure may become more complex. However, in the next section we will encounter a more serious problem when we wish to use the Bloch–Messiah reduction.

9.5.2 Information propagation and processing

How does the quantum information propagate through a linear qunat cluster state? As in the qubit case, propagation of information through the cluster state is driven by local teleportation. In the previous chapter we saw that the teleportation of qunats is imperfect, and that the teleported Wigner function is a convolution of the Wigner function of the input state with a Gaussian distribution. In the case of infinite squeezing, this Gaussian distribution becomes a Dirac delta function, and teleportation becomes perfect.

The fidelity of information propagation in qunat cluster states is also imperfect when squeezing is finite. Suppose that we wish to propagate the state

$$|\psi\rangle = \int dq\, \psi(q)\, |q\rangle_q \tag{9.79}$$

through the cluster. The local teleportation circuit for qunats is then

$$|\psi\rangle = X(s)\mathcal{F}\,|\psi\rangle\,.$$

To demonstrate that this corresponds to information propagation, we outline the teleportation procedure for an ideal (and unphysical) ancilla state $|0\rangle_p$. After the cz gate, the state can be written as

$$\text{cz}\,|\psi\rangle\,|0\rangle_p = \frac{1}{\sqrt{\pi}} \int dq\, dq'\, \psi(q)e^{2iqq'}\,|q, q'\rangle_q \,. \tag{9.80}$$

Applying the Fourier transform $\mathcal{F}_1^\dagger$ on qunat 1, and projecting it onto the position eigenstate $|s\rangle\,\langle s|$ is equivalent to projecting straight onto the momentum eigenstate $|s\rangle_p\,\langle s|$, which then yields the teleported state

$$|\psi'\rangle = \frac{1}{\pi} \int dq\, dq'\, \psi(q)e^{2iq(q'-s)}\,|q'\rangle_q \,. \tag{9.81}$$

We will now prove that this state is identical to $X(s)\mathcal{F}\,|\psi\rangle$ by showing that we retrieve the input state $|\psi\rangle$ when we first apply $X^\dagger(s) = X(-s)$ followed by $\mathcal{F}^\dagger$. The $X(-s)$ operator yields

$$X(-s)\,|\psi'\rangle = \frac{1}{\pi} \int dq\, dq'\, \psi(q)e^{2iq(q'-s)}\,|q'-s\rangle_q \,, \tag{9.82}$$

and the inverse Fourier transform turns this state into

$$\mathcal{F}^\dagger X(-s)\,|\psi'\rangle = \frac{1}{\pi} \int dq\, dq'\, dq''\, \psi(q)e^{2i(q'-s)(q-q'')}\,|q''\rangle_q$$

$$= \int dq\, dq''\, \psi(q)\delta(q-q'')\,|q''\rangle = |\psi\rangle \,. \tag{9.83}$$

In the derivation of the last line we made a change of variables of integration $y = q' - s$, and integration over y yielded the delta function. Since we used the unnormalizable position and momentum eigenstates, the teleportation protocol has perfect fidelity.

What happens when we use normalizable, imperfectly squeezed states? The input state is still represented by $|\psi\rangle$, but the auxiliary state $|0\rangle_p$ is no longer a perfect zero-momentum quadrature eigenstate. Instead, we can write this state as a squeezed coherent state peaked around $p = 0$ with squeezing Δ. A natural distribution is the Gaussian distribution with width Δ^{-1}, which gives

$$
\begin{aligned}
|0_\Delta\rangle_p &= \sqrt[4]{\frac{4\Delta^2}{\pi}} \int_{-\infty}^{+\infty} dp\, e^{-2p^2\Delta^2} |p\rangle_p \\
&= \sqrt[4]{\frac{4\Delta^2}{\pi^3}} \int_{-\infty}^{+\infty} dp\, dq\, e^{-2p^2\Delta^2 + 2ipq} |q\rangle_q \\
&= \frac{1}{\sqrt[4]{\pi\Delta^2}} \int_{-\infty}^{+\infty} dq\, \exp\left(-\frac{q^2}{2\Delta^2}\right) |q\rangle_q .
\end{aligned} \tag{9.84}
$$

In the second line we wrote the momentum eigenstate as a Fourier transform of position eigenstates. Using this normalizable ancilla state, instead of the infinitely squeezed state $|0\rangle_p$, and following the same protocol as in the case of perfect squeezing, results in a convolution of the input state with a Gaussian distribution

$$
|\psi_{\text{tel}}\rangle = \sqrt[4]{\frac{4\Delta^2}{\pi}} \int_{-\infty}^{+\infty} du\, dq\, e^{-2\Delta^2 u^2 - 2ius} \psi(q+u) |q\rangle_q \equiv G_\Delta(s) |\psi\rangle , \tag{9.85}
$$

where $G_\Delta(s)$ is a Gaussian convolution operator

$$
G_\Delta(s) = \int_{-\infty}^{+\infty} dq\, du\, \exp\left(-2\Delta^2 u^2 - 2ius\right) |q\rangle \langle q+u| . \tag{9.86}
$$

In the limit of infinite squeezing ($\Delta \to \infty$) this becomes the identity operator. The phase in the convolution operator depends on the measurement outcome s, and can therefore not be predicted beforehand. This means that we cannot cancel this type of noise at the preparation stage of the cluster state, and other noise-reduction methods have to be found. We will return to this issue in Section 9.6.

Exercise 9.11: Verify that measurement-based propagation of qunat quantum information through a cluster state is described by the convolution in Eq. (9.85).

Now we consider the implementation of gates in a qunat cluster state. The circuit for a two-qunat cluster state that implements a gate U is given by

$$
\begin{array}{c}
|\psi\rangle \;\; \boxed{U} \;\; \boxed{\mathcal{F}^\dagger} \;\; \measuredangle \; \boxed{s} \\
|0\rangle_p \;\;\rule{4cm}{0.4pt}\;\; |\psi'\rangle = X(s)\mathcal{F}U |\psi\rangle ,
\end{array}
$$

where U is a single-qunat gate that commutes with the cz gate, meaning $U = \exp[if(\hat{q})]$ for an arbitrary polynomial function f of $\hat{q}$. The measurement basis can then be interpreted as the eigenbasis of the operator $U^\dagger \hat{p} U$. Measuring in this basis has the effect of teleporting U into the propagation of $|\psi\rangle$. This is completely analogous to the qubit case.

This analogy also means that U is not a general single-qunat gate: it is generated by a polynomial $f(\hat{q})$ that does not depend on $\hat{p}$. We therefore need to adopt a technique similar to the qubit case, where we apply several gates U_j in a row in order to implement a general gate. The Fourier transform in the output state can then be used to introduce gates generated by the momentum quadrature $\hat{p}$. Consider the three-qunat linear cluster state

that gives rise to the output state

$$|\psi_{\text{out}}\rangle = X(s_2)\mathcal{F}\,U_2 X(s_1)\mathcal{F}\,U_1\,|\psi_{\text{in}}\rangle\,. \tag{9.87}$$

Just as in the qubit case, we would like to move the operator $X(s_1)$ to the left, since we can then postpone the corrective Heisenberg–Weyl operations and keep track of them in a classical register. However, $X(s_1)$ and U_2 do not commute (except when s_1 is zero, and the probability for this outcome is vanishingly small). For a general $U = \exp[if(\hat{q})]$ and s, we can write

$$UX(s) = X(s)U'(s)\,, \tag{9.88}$$

where U' is a modified gate dependent on s. Using the relation

$$X(-s)\,UX(s) = X(-s)\,\exp\left[if(\hat{q})\right]X(s)$$
$$= \exp\left[if(X(-s)\hat{q}X(s))\right] = \exp\left[if(\hat{q}-s)\right]\,, \tag{9.89}$$

we find that

$$U(\hat{q})\,X(s) = X(s)\,U(\hat{q}-s)\,, \tag{9.90}$$

where we have included the dependence of U on $\hat{q}$, since it allows us to show how U changes. Remember, this is valid only for gates that are generated by polynomials of the position quadrature. The output state in Eq. (9.87) then becomes

$$|\psi_{\text{out}}\rangle = X(s_2)Z(s_1)\mathcal{F}\,U_2'(s_1)\mathcal{F}\,U_1\,|\psi_{\text{in}}\rangle\,, \tag{9.91}$$

where we have also used $\mathcal{F}X(q) = Z(q)\mathcal{F}$. As an example, consider $U = \exp(iu\hat{q}^3)$: we have

$$U(\hat{q})X(s) = X(s)\exp\left[iu(\hat{q}-s)^3\right] = X(s)e^{3iu\hat{q}s(s-\hat{q})}U(\hat{q})\,, \tag{9.92}$$

up to an overall phase factor that can be ignored.

What types of gate can we perform with qunat cluster states and measurements? To answer this question, it is convenient to define three different classes of U, namely the linear, quadratic, and order $n > 2$ polynomials in $\hat{q}$ and $\hat{p}$, denoted by $U^{(1)}$, $U^{(2)}$, and $U^{(n)}$, respectively. The linear class consists of the Heisenberg–Weyl operators and phase shifts, and the operators in the quadratic class include the Fourier transform, the phase gate, etc. It is straightforward to show that for any Heisenberg–Weyl operator $V = \exp(2ip\hat{q} - 2iq\hat{p})$, a quadratic operator $U^{(2)}$ obeys

$$U^{(2)}V = V'U^{(2)}\,, \tag{9.93}$$

where V' is again a Heisenberg–Weyl operator. This is a special case of the relation in Eq. (9.88) where the operator $U^{(2)}$ remains unchanged under the conjugation operation. Mathematically, both the quadratic operators and the linear Heisenberg–Weyl operators generate a (continuous) group. The latter plays the role of the Pauli group for qunats, and the group of quadratic operators can therefore be considered an extended Clifford group. With these definitions, we can turn our attention to the construction of arbitrary single-qunat gates.

We will now discuss what types of measurement must be performed in order to execute the different gates in measurement-based quantum computing. First we will consider the Heisenberg–Weyl operators. An arbitrary linear gate $U^{(1)} = \exp(2ip\hat{q} - 2iq\hat{p})$ can be decomposed into a product of $X(q)$ and $Z(p)$ up to a global phase. Since $Z(p) = \exp(2ip\hat{q})$, this gate can be performed with a single measurement of the observable $Z(-p)\hat{p}Z(p) = \hat{p} + p$. In other words, a homodyne detection of the quadrature $\hat{p}$ with the addition of p to the measurement outcome. The operator $X(q)$ must be applied with the second measurement. The Fourier transforms in Eq. (9.91) allow us to implement a gate $\tilde{U} = \exp[-2iq\hat{p}]$:

$$X(s_2)Z(s_1)\mathcal{F}\, U_2'\mathcal{F}\, U_1 = X(s_2)\mathcal{F}\, \mathcal{F}^\dagger Z(s_1)\mathcal{F}\, \mathcal{F}\, \mathcal{F}^\dagger U_2'\mathcal{F}\, U_1$$

$$= X(s_2)Z(s_1)\mathcal{F}\, \tilde{U}_2' U_1 . \tag{9.94}$$

The gate $\tilde{U}_2' = \mathcal{F}^\dagger U_2'\mathcal{F}$ must now be equal to $X(q)$. When $U_2' = Z(-q)$, this condition is fulfilled, and the gate can again be implemented with a homodyne detection.

Next, we discuss how quadratic gates can be implemented with measurements. A quadratic gate from the extended Clifford group $U^{(2)}$ can be written as

$$U^{(2)} = \exp\left[2iu\hat{q}^2 + 2iv\hat{p}^2 + 2iw\left(\hat{q}\hat{p} + \hat{p}\hat{q}\right)\right], \tag{9.95}$$

where we ignored the linear contribution for simplicity. Using a Baker–Campbell–Hausdorff relation (see Appendix 1 on how to derive such relations) this can be written as

$$U^{(2)} = \exp\left(2iu'\hat{q}^2\right)\exp\left[2iw'\left(\hat{q}\hat{p} + \hat{p}\hat{q}\right)\right]\exp\left(2iv'\hat{p}^2\right), \tag{9.96}$$

for some choice of u', v', and w'. First, we show what measurement we need to make to implement the phase gate $\Phi(u') = \exp(2iu'\hat{q}^2)$. From this, the gate $\exp(2iv'\hat{p}^2)$ follows immediately by sandwiching between Fourier transforms. Finally, we construct the gate $\exp[2iw'(\hat{q}\hat{p} + \hat{p}\hat{q})]$, which is a squeezing operation. The quadratic gate in $\hat{q}$ commutes with the cz operator, and can therefore be implemented immediately. The observable to be measured is

$$e^{-2iu'\hat{q}^2}\hat{p}\, e^{2iu'\hat{q}^2} = \hat{p} + u'\hat{q}, \tag{9.97}$$

a linear combination of quadrature operators. Introducing the parameterization $u' = -\tan\theta$, this can be written as a rotation over an angle θ with an overall scaling factor $\cos\theta$. Similarly, the phase gate in the momentum quadrature $\exp(2iv'\hat{p}^2)$ can be implemented on the second measurement round due to the appearance of the Fourier transforms

$$\mathcal{F}\exp\left(2iv\hat{q}^2\right)\mathcal{F}\exp\left(2iu\hat{q}^2\right) = \mathcal{F}^2\exp\left(2iv\hat{p}^2\right)\exp\left(2iu\hat{q}^2\right). \tag{9.98}$$

When $u = v$, we can write this as

$$\mathcal{F} e^{2iu\hat{q}^2} \mathcal{F} e^{2iu\hat{q}^2} = \mathcal{F}^2 e^{2iu\hat{p}^2} e^{2iu\hat{q}^2}$$

$$= \mathcal{F}^2 e^{2iu(\hat{q}^2+\hat{p}^2)} e^{2iu^2(\hat{q}\hat{p}+\hat{p}\hat{q})} + O(u^2), \tag{9.99}$$

which, to lowest order and up to a rotation, is a single-mode squeezing operation with $r = 4u^2$. We can remove the rotation by repeating this operation for the value $-u$. This yields

$$\mathcal{F}^2 e^{-2iu\hat{p}^2} e^{-2iu\hat{q}^2} \mathcal{F}^2 e^{2iu\hat{p}^2} e^{2iu\hat{q}^2} = e^{4iu^2(\hat{q}\hat{p}+\hat{p}\hat{q})} + O(u^3). \tag{9.100}$$

This corresponds to a single-mode squeezing operator for reasonably low values of u (see Van Loock, 2007). We can therefore implement Gaussian, or extended Clifford operations such as rotations and squeezing using homodyne measurements. Notice that the Gaussian operations do not rely on feed-forward of the measurement result. This is a general property of extended Clifford operators.

Exercise 9.12: Show that vertical connections in a cluster, together with momentum quadrature measurements, can be used to implement a CZ gate.

The Gottesman–Knill theorem for qunat quantum computing tells us that homodyne measurements and Gaussian operations are not sufficient for implementing efficient quantum computing. More precisely, if the operation of the quantum computer is restricted to the vacuum, extended Clifford operations, and homodyne detection, we can always simulate the quantum computation efficiently on a classical computer. Therefore, we need another type of measurement, for example measuring an observable that leads to the cubic gate

$$U^{(3)} = \exp\left(i\theta_3 \hat{q}^3\right), \tag{9.101}$$

where θ_3 is the strength of the induced gate (see Gu *et al.*, 2009). In particular, we wish to implement the cluster-state computation that is symbolically written as

In standard circuit-model language, this can be written as

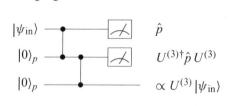

where $|\psi_{\text{in}}\rangle$ is the input state to this part of the computation (most likely mixed), $\hat{p}$ denotes homodyne detection of the momentum quadrature, and $U^{(3)\dagger}\hat{p}\,U^{(3)}$ indicates the measurement that leads to the implementation of $U^{(3)}$, up to some corrective operations. We are done if we have an optical implementation for this particular measurement. However, it is not straightforward to find an optical implementation for this measurement, and we have to be a bit more resourceful.

Since $U^{(3)}$ commutes with the CZ operation, we can bring $U^{(3)}$ to the left-hand side, and write the circuit as

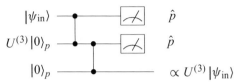

where both measurements are now implemented with regular homodyne detection. The question is then how to create the state $U^{(3)}|0\rangle_p$. In cluster-state language, this state can itself be created via the teleportation of the gate $U^{(3)}$, and the cluster we need to create becomes

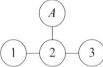

where we have labelled the qunats such that a measurement of the (as of yet unspecified) observable A creates a cubic gate $U^{(3)}|0\rangle_p$ acting on qunat 2. In general, A yields unpredictable measurement outcomes n, and will therefore not create the gate $U^{(3)}$ with the required strength θ_3 directly. Instead, it will create a gate $U^{(3)}$ with some strength $\gamma(n)$. In order to correct this gate, we need to implement the transformation

$$\gamma(n)\,\hat{q}^3 \to \theta_3\,\hat{q}^3 \quad \text{or} \quad \hat{q} \to \sqrt[3]{\frac{\theta_3}{\gamma(n)}}\,\hat{q} \equiv t_n\hat{q}\,. \qquad (9.102)$$

This is a rescaling of $\hat{q}$, which can be implemented optically with a squeezing operation

$$\hat{q} \to S^{\dagger}(t_n)\,\hat{q}\,S(t_n)\,. \qquad (9.103)$$

In other words, we need to apply two squeezing operators, one before the measurement of qunat 2, and one after. Therefore, we must measure A before we measure qunats 1, 2, and 3, which will then allow us to implement the two corrective squeezing operations t_n and $-t_n$. The corresponding subgraph then becomes

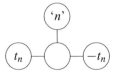

These squeezing operations may require multiple qunat systems, as shown in Eq. (9.100). This necessary time ordering of the non-Clifford part determines the temporal direction of the computation.

Finally, how do we implement a measurement of the observable A with optics? It was shown by Gottesman *et al.* (2001) that a good approximation to the cubic gate $U^{(3)}$ can be implemented using photon-number measurements. Adapted to cluster-state quantum computing, this leads to the measurement of the observable

$$A = X^{\dagger}(r)\,\hat{n}\,X(r)\,, \tag{9.104}$$

where the displacement r is much larger than the typical squeezing parameter in the qunats, $r \gg \ln \Delta$. The measurement outcome n is related to the cubic gate strength via

$$\gamma(n) = \frac{1}{6\sqrt{2n+1}}\,. \tag{9.105}$$

This is an approximate cubic gate, and the errors introduced by this gate are analyzed by Gottesman *et al.* (2001). In particular, this gate requires photon-number counting with a precision Δn that is much smaller than $\langle \hat{n} \rangle^{\frac{1}{3}}$. For large squeezing, this is extremely challenging to achieve with realistic photodetectors.

9.6 Quantum error correction for qunats

In the final section of this chapter we consider the optical implementation of several error-correction protocols for qunat quantum computing. We have already seen in Chapters 2 and 8 that we can encode qubits (and more generally qudits) in a qunat via the Gottesman–Kitaev–Preskill (GKP) code. This code protects against small phase-space displacements. We will present a protocol for the state preparation of such codes based on a cross-Kerr nonlinearity. Next, we present another error-correction code that can be implemented entirely with passive linear-optical elements such as beam splitters and phase shifters. Finally, we will present a cluster-state code that allows us to correct for the accumulated noise in the information propagation through qunat clusters that are constructed with finite squeezing.

9.6.1 State preparation of GKP codes

How can we create the encoded states of the Gottesman–Kitaev–Preskill (GKP) code? It is not possible to prepare these states from squeezed coherent states with linear optics alone. Following Pirandola *et al.* (2004), we briefly sketch how to create these states with coherent input states, a cross-Kerr nonlinearity, and homodyne detection. These optical operations are similar to the ones we introduced in the construction of qunat quantum computing, except that here we require a cross-Kerr nonlinearity, rather than a single-mode Kerr nonlinearity. The physical implementation of cross-Kerr nonlinearities is discussed in Chapter 10 and Appendix 3.

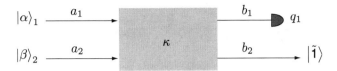

Fig. 9.7.
Schematic for creating approximate GKP code states. The input modes to the cross-Kerr region κ are prepared in coherent states, and output mode b_1 is measured with homodyne detection, yielding outcome q_1. The state in mode b_2 is an approximate GKP code state $|\tilde{1}\rangle$.

Consider two modes, a_1 and a_2, that are coupled via a cross-Kerr nonlinearity described by the interaction Hamiltonian

$$\mathcal{H}_K = \hbar\kappa\,\hat{a}_1^\dagger\hat{a}_1\,\hat{a}_2^\dagger\hat{a}_2\,, \tag{9.106}$$

shown in Fig. 9.7. When the two modes are in coherent states $|\alpha\rangle_1$ and $|\beta\rangle_2$, the interaction produces an output state that can be written as

$$|\psi_{\text{out}}\rangle_{12} = e^{-|\alpha|^2/2} \sum_{n=0}^{\infty} \frac{\alpha^2}{\sqrt{n!}}\,|n\rangle_1\,\big|\beta\,e^{-in\kappa t}\big\rangle_2\,. \tag{9.107}$$

We can make the following approximation: since κt is very small in practice, we assume that $|\alpha|^2\kappa t \ll 1$, which allows us to write $\beta\,e^{-in\kappa t} = \beta(1 - in\kappa t)$. If we also displace the phase-space coordinate for mode a_2 such that $\hat{a}_2 \to \hat{a}_2 + \beta$, we can write this state as

$$|\psi_{\text{out}}\rangle_{12} = e^{-|\alpha|^2/2} \sum_{n=0}^{\infty} \frac{\alpha^2}{\sqrt{n!}}\,|n\rangle_1\,|-in\beta\kappa t\rangle_2\,. \tag{9.108}$$

This state can be regarded as the output state of an interaction due to the Hamiltonian

$$\mathcal{H}' = \hbar\beta\kappa\,\hat{a}_1^\dagger\hat{a}_1\left(\hat{a}_2 + \hat{a}_2^\dagger\right) = \sqrt{2}\hbar\beta\kappa\,\hat{a}_1^\dagger\hat{a}_1\,\hat{q}_2\,, \tag{9.109}$$

which is analogous to the Hamiltonian of a ponderomotive force. The approximate code words are now created in mode a_2 by performing a homodyne measurement of the quadrature $\hat{q}_1$ in mode a_1, and recording the outcome q_1. The state in mode a_2 then becomes

$$|\psi_{\text{out}}\rangle_2 = e^{-|\alpha|^2/2} \sum_{n=0}^{\infty} \frac{\alpha^2}{\sqrt{n!}}\langle q_1|n\rangle\,|-in\beta\kappa t\rangle_2 \equiv |\tilde{1}\rangle\,, \tag{9.110}$$

where $|\tilde{1}\rangle$ is an approximation to the GKP code state $|\bar{1}\rangle$. The output state $|\tilde{0}\rangle$ corresponding to the GKP code state $|\bar{0}\rangle$ is obtained by a regular displacement $X(\frac{1}{4})$.

(a) $|\langle q|\tilde{1}\rangle|^2 = |\psi(q)|^2$

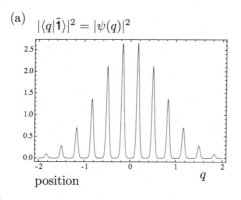

position $\qquad q$

(b) $|\langle p|\tilde{1}\rangle|^2 = |\Psi(p)|^2$

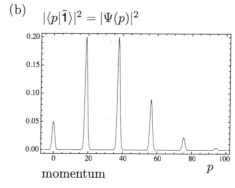

momentum $\qquad p$

Fig. 9.8. The probability distribution of finding measurement outcome q and p for the wavefunctions $\psi(q)$ and $\Psi(p)$, respectively. The amplitude of the coherent state is $\alpha = 2$, and $\sqrt{2}\beta\kappa t = 3$.

We can calculate the wavefunction $\psi(q) = \langle q|\tilde{1}\rangle$ and its Fourier transform $\Psi(p) = \langle p|\tilde{0}\rangle = \langle p|\tilde{1}\rangle$ as

$$\psi(q) = \mathcal{N} \sum_{n=0}^{\infty} \frac{\alpha^n H_n(q_1)}{\sqrt{2^n}\,n!} \exp\left[-\frac{\alpha^2 + q_1^2}{2} - \frac{q^2}{2} + in\sqrt{2}\beta\kappa tq\right]$$

$$\Psi(p) = \mathcal{N} \sum_{n=0}^{\infty} \frac{\alpha^n H_n(q_1)}{\sqrt{2^n}\,n!} \exp\left[-\frac{\alpha^2 + q_1^2}{2} - \frac{1}{2}\left(p - n\sqrt{2}\beta\kappa t\right)^2\right], \qquad (9.111)$$

where $\mathcal{N}$ is a normalization constant

$$\mathcal{N} = \sum_{n,m=0}^{\infty} \frac{\alpha^{n+m} H_n(q_1)H_m(q_1)}{\sqrt{2^{n+m}}\,n!m!} \exp\left[-(\alpha^2 + q_1^2) - \frac{\beta^2\kappa^2 t^2}{2}(n-m)^2\right], \qquad (9.112)$$

and the $H_n(q_1)$ are Hermite polynomials. The corresponding probability distributions for position $|\langle q|\tilde{1}\rangle|^2 = |\psi(q)|^2$ and momentum $|\langle p|\tilde{1}\rangle|^2 = |\Psi(p)|^2$ are shown in Fig. 9.8. The output states are also well behaved when we consider the wavefunctions for $|\langle q|\tilde{\pm}\rangle|^2$ and $|\langle p|\tilde{\pm}\rangle|^2$. For more details, see Pirandola *et al.* (2004).

Exercise 9.13: Show that in the limit of $\kappa \to 0$ the wave functions $\psi(q)$ and $\Psi(p)$ reduce to coherent states.

9.6.2 Qunat error correction with linear optics and squeezing

So far we have considered GKP codes, which encode a qubit (or, more generally, a qudit) in a continuous variable such that its value is protected from small displacements in the position and momentum quadratures. Now we will describe an error-correction code that protects a qunat from (arbitrary) displacements in phase space. Remarkably, this code uses only linear-optical elements, including squeezing.

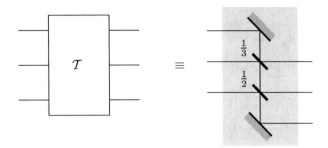

Fig. 9.9. A tritter $\mathcal{T}$: the top input mode is evenly distributed over the three output modes. The fractions indicate the reflection coefficients.

We consider the nine-qunat code introduced by Braunstein (1998a, 1998b), which is the continuous-variable extension of the nine-qubit Shor code. It is a concatenation of two majority codes, one for position errors, and one for momentum errors. These codes rely critically on the fact that we can distribute a single non-zero qunat q over three modes according to

$$|q, 0, 0\rangle \rightarrow \left| \frac{q}{\sqrt{3}}, \frac{q}{\sqrt{3}}, \frac{q}{\sqrt{3}} \right\rangle . \tag{9.113}$$

This is not possible with discrete systems, since qudit values cannot be divided by a factor $\sqrt{3}$. Here, $|0\rangle$ is the position eigenstate with eigenvalue zero, not the vacuum state. Note that this distribution is not the same as cloning, since we can define this procedure in one basis only (the position basis, in this case). The distribution can be implemented with a three-mode beam splitter, called a 'tritter' $\mathcal{T}$, shown in Fig. 9.9 as a combination of two beam splitters. In other words, the tritter is a balanced 3-port that acts on the three quadrature operators according to

$$\mathcal{T} = \frac{1}{\sqrt{3}} \begin{pmatrix} 1 & 1 & 1 \\ 1 & \varpi^2 & \varpi^4 \\ 1 & \varpi^4 & \varpi^8 \end{pmatrix} , \tag{9.114}$$

with $\varpi \equiv \exp(i\pi/3)$.

To gain an intuition about the nine-qunat code, we now show in some detail how this three-qunat code can correct displacement errors. Consider an encoded state $\left| q/\sqrt{3}, q/\sqrt{3}, q/\sqrt{3} \right\rangle$. Since

$$1 + \varpi^2 + \varpi^4 = 1 + \varpi^4 + \varpi^8 = 0 , \tag{9.115}$$

applying the inverse tritter $\mathcal{T}^\dagger$ to the encoded state retrieves the input state $|q, 0, 0\rangle$. Now consider that one of the transmitted modes experiences a displacement $\delta/\sqrt{3}$ in the position quadrature. For definiteness, suppose the displacement occurs in mode a_2. The state after decoding then becomes

$$\mathcal{T} \left| \frac{q}{\sqrt{3}}, \frac{q+\delta}{\sqrt{3}}, \frac{q}{\sqrt{3}} \right\rangle = \frac{1}{3} \begin{pmatrix} 1 & 1 & 1 \\ 1 & \varpi^2 & \varpi^4 \\ 1 & \varpi^4 & \varpi^8 \end{pmatrix} \begin{pmatrix} q \\ q+\delta \\ q \end{pmatrix} = \begin{pmatrix} q+\delta/3 \\ \varpi^2\delta/3 \\ \varpi^4\delta/3 \end{pmatrix} . \tag{9.116}$$

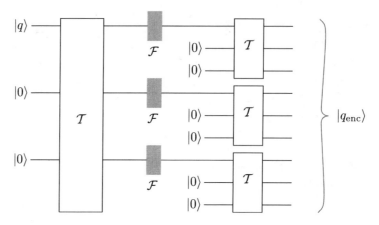

Fig. 9.10.   The optical implementation of a nine-qunat encoder.

The ideal measurement outcomes in modes a_2 and a_3 are then $\delta/3$, since ϖ^2 and ϖ^4 are pure phases (i.e., displacements in the momentum quadrature). These outcomes tell us to displace the position quadrature of the undetected mode by an amount $-\delta/\sqrt{3}$.

This code does not protect against displacements in the momentum quadrature, and we therefore concatenate this code with a similar code for the momentum. We apply the Fourier transform

$$\mathcal{F}|q\rangle = \frac{1}{\sqrt{2\pi\hbar}} \int_{-\infty}^{\infty} dx\, e^{\frac{i}{\hbar}xq} |x\rangle \,, \tag{9.117}$$

on the three output modes of the tritter, and couple each mode to two new modes in the position eigenstates with eigenvalue zero, again using the tritters defined above. The total operator for the encoding of a single qunat in nine qunats is

$$\hat{E}_q = \hat{T}_{789}\hat{T}_{456}\hat{T}_{123}\mathcal{F}_7\mathcal{F}_4\mathcal{F}_1\hat{T}_{147}\,, \tag{9.118}$$

and the decoding device is described by the operator $\hat{D}_q = \hat{E}_q^\dagger$. The state after the encoding procedure then becomes

$$|q_{\mathrm{enc}}\rangle = \frac{1}{\sqrt{(2\pi\hbar)^3}} \int dx\, dy\, dz\, e^{\frac{i}{\hbar}(x+y+z)q} \left|\frac{x}{3}, \frac{x}{3}, \frac{x}{3}, \frac{y}{3}, \frac{y}{3}, \frac{y}{3}, \frac{z}{3}, \frac{z}{3}, \frac{z}{3}\right\rangle. \tag{9.119}$$

The input modes are denoted by a_j, and the output modes by b_j. In the full nine-qunat code, we perform eight position-quadrature measurements on modes b_2 to b_9. The measurement outcomes are used to determine the total displacement in the position and momentum quadratures we need to apply to the undetected output mode b_1.

Exercise 9.14: Show that the nine-qunat code can correct for arbitrary phase-space displacements.

As we have seen in Section 9.1, the Fourier transform can be implemented with a simple phase shift, and therefore the entire encoder can be implemented using passive linear optics acting on highly squeezed states in the position quadrature. Consequently, the decoder can be

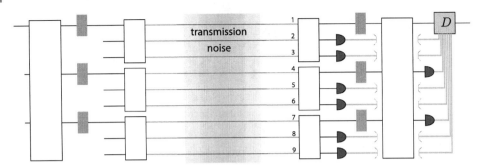

Fig. 9.11. The nine-qunat code in action. The corrective displacement D depends on the syndrome provided by the measurement outcomes of the eight homodyne detections.

implemented with passive linear optics and homodyne detection. The entire quantum error-correction code for an optical qunat is shown in Fig. 9.11. The measurement outcomes of the eight auxiliary modes form a list of eight real numbers $\vec{q} = (q_2, \ldots, q_9)$, which constitute the syndrome.

We now introduce three types of noise. First, we have to take into account the fact that any optical implementation can achieve only finite squeezing in the auxiliary modes. This will add a small random number to the outcomes of the homodyne measurements. Second, the homodyne detection will make use of photodetectors with finite efficiency η, and this will result in a vacuum noise component proportional to $\sqrt{\eta^{-1} - 1}$ in the measurement outcomes. Finally, there is the transmission noise of the channel that we want to correct. The first two sources of noise are strictly due to the physical limitations of the error-correction procedure itself, and have nothing to do with the transmission noise. Clearly, the amount of squeezing in the auxiliary modes and the efficiency of the homodyne detection must be such that the error in the undetected output mode is smaller than the error that would have been accumulated without error correction.

Suppose that a small transmission error occurs in mode a_k. The measurement outcome of a detector in mode b_j can be decomposed into contributions from the different types of noise. The ideal syndrome is denoted by $q_j^{(\text{id})}$, which is the measurement outcome for perfect squeezing and ideal homodyne detection. Added to this are the fluctuating components due to finite squeezing, denoted by $q_l^{(\text{sq})}$, and the vacuum fluctuation $q_j^{(\text{vac})}$

$$q_j^{(\text{syn})} = q_j^{(\text{id})} + \sum_{l=2}^{9} c_{jl}^k q_l^{(\text{sq})} + \sqrt{\eta^{-1} - 1}\, q_j^{(\text{vac})}. \qquad (9.120)$$

The coefficients c_{jl}^k take into account how the squeezing parameter Δ_l of the auxiliary input mode a_l contributes to mode a_j. Each c_{jl}^k is proportional to Δ_l^{-1}. In the simplest case, where all auxiliary input modes are squeezed by the same amount Δ, the Wigner function of the output will be a convolution of the input Wigner function and the noise function G

$$W_{\text{out}}(q, p) = G \circ W_{\text{in}} \equiv \int dq' dp'\, G(q - q', p - p') W_{\text{in}}(q', p'), \qquad (9.121)$$

where the noise function G is given by

$$G(q,p) = \frac{1}{2\pi\hbar^2(\Delta^2 + \eta^{-1} - 1)} \exp\left[-\frac{q^2 + p^2}{2\hbar^2(\Delta^2 + \eta^{-1} - 1)}\right]. \qquad (9.122)$$

In the limit of infinite squeezing *and* perfect homodyne detection, the noise function becomes a delta function, and error correction becomes ideal.

In practice, this code will cause small amounts of noise in the qunat, and multiple invocations of this error-correction protocol will have a cumulative effect on the noise. In this sense, the GKP codes are more robust against small errors, since a measurement of the optical mode yields a real number that is used to decide which qubit state was detected. The discreteness of qubits (or qudits) ensures that the GKP codes do not accumulate errors. However, currently it seems that high-fidelity GKP are also much harder to create experimentally.

9.6.3 Propagation errors in cluster states

Finally, we discuss an error-correction protocol for qunat cluster states due to Van Loock *et al.* (2007). In Section 9.5 we have seen that we can construct optical cluster states with qunats, but that the finite squeezing induces errors in the local teleportation protocol that is used for information propagation. In fact, as will be clear from the previous section as well, detector efficiencies in homodyne measurements will also contribute to the degradation of information propagation.

The error-correction protocol works as follows. The quadratures of mode a_4 associated with qunat 4 in Fig. 9.12a after the momentum measurements on modes a_1, a_2, and a_3 with outcomes s_1, s_2, and s_3 must be corrected with the operator

$$\mathcal{F}^\dagger X(-s_1)\,\mathcal{F}^\dagger X(-s_2)\,\mathcal{F}^\dagger X(-s_3), \qquad (9.123)$$

acting on mode a_4. The first operator $X(-s_3)$ transforms the quadratures of mode a_4 according to

$$\hat{q}_4(t_1) = \hat{q}_4(t_0) - s_3 \quad \text{and} \quad \hat{p}_4(t_1) = \hat{p}_4(t_0), \qquad (9.124)$$

and the subsequent Fourier transform produces

$$\hat{q}_4(t_2) = \hat{p}_4(t_1) = \hat{p}_4(t_0) \quad \text{and} \quad \hat{q}_4(t_2) = -\hat{q}_4(t_1) = -\hat{q}_4(t_0) + s_3. \qquad (9.125)$$

(a)   (b)   (c)

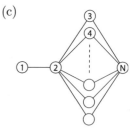

Fig. 9.12. A redundant encoding protecting information propagation.

Repeating this procedure for the remaining corrective operators yields

$$\hat{q}_4(t_6) = -\hat{p}_4(t_0) + s_2 \quad \text{and} \quad \hat{p}_4(t_6) = \hat{q}_4(t_0) - s_3 + s_1 \,. \tag{9.126}$$

Next, we replace s_1, s_2, and s_3 by the corresponding correlations

$$s_1 = \hat{p}_1(t_0) + \hat{q}_2(t_0) \,, \quad s_2 = \hat{p}_2(t_0) + \hat{q}_1(t_0) \quad \text{and} \quad s_3 = \hat{p}_3(t_0) \,, \tag{9.127}$$

which leads to the quadratures of the output mode:

$$\hat{q}_4(t_6) = \hat{q}_1(t_0) + \left[\hat{p}_2(t_0) - \hat{q}_3(t_0)\right] - \left[\hat{p}_4(t_0) - \hat{q}_3(t_0)\right]$$
$$\hat{p}_4(t_6) = \hat{p}_1(t_0) - \left[\hat{p}_2(t_0) - \hat{q}_2(t_0) - \hat{q}_4(t_0)\right] . \tag{9.128}$$

When the squeezing and homodyne detection is perfect, the cluster correlations $\hat{p}_3 - \hat{q}_2 - \hat{q}_4$, $\hat{p}_2 - \hat{q}_3$, and $\hat{p}_3 - \hat{q}_2$ tend to zero, and we recover ideal propagation

$$\hat{q}_4(t_6) = \hat{q}_1(t_0) \quad \text{and} \quad \hat{p}_4(t_6) = \hat{p}_1(t_0) \,. \tag{9.129}$$

When the squeezing is imperfect, the cluster correlations are imperfect and contribute a non-zero variance to the excess noise. For the linear cluster, this added noise can be calculated as

$$\text{Var}\left[(\hat{p}_2 - \hat{q}_3) - (\hat{p}_3 - \hat{q}_2)\right] = 2\Delta^{-2}$$
$$\text{Var}\left[\hat{p}_3 - \hat{q}_2 - \hat{q}_4\right] = \Delta^{-2} \,, \tag{9.130}$$

where $\Delta = e^r$ is the squeezing parameter, and we assume that all modes have equal squeezing.

Next, consider the diamond-shaped cluster in Fig. 9.12b. The quadratures of the fifth qunat in mode a_5 must be corrected by the operator

$$\mathcal{F}^\dagger X(-s_1)\, \mathcal{F}^\dagger X(-s_2)\, \mathcal{F}^\dagger X(-\frac{1}{2}(s_3 + s_4)) \,, \tag{9.131}$$

which leads to quadrature operators

$$\hat{q}_5(t) = \hat{q}_1 + \left(\hat{p}_2 - \hat{q}_3 - \hat{q}_4\right) - \left(\hat{p}_5 - \hat{q}_3 - \hat{q}_4\right)$$
$$\hat{p}_5(t) = \hat{p}_1 - \frac{1}{2}\left(\hat{p}_3 - \hat{q}_2 - \hat{q}_5\right) - \frac{1}{2}\left(\hat{p}_4 - \hat{q}_2 - \hat{q}_5\right) , \tag{9.132}$$

where we suppressed the initial time dependence (t_0) and wrote $t_6 = t$ for clarity. These expressions lead to the excess noise

$$\text{Var}\left[(\hat{p}_2 - \hat{q}_3 - \hat{q}_4) - (\hat{p}_5 - \hat{q}_3 - \hat{q}_4)\right] = 2\Delta^{-2}$$
$$\text{Var}\left[\frac{1}{2}\left(\hat{p}_3 - \hat{q}_2 - \hat{q}_5\right) - \frac{1}{2}\left(\hat{p}_4 - \hat{q}_2 - \hat{q}_5\right)\right] = \frac{\Delta^{-2}}{2} \,. \tag{9.133}$$

The variance of the position quadrature remains unchanged, but the variance of the momentum quadrature is reduced by a factor two. We can generalize this result to the situation in Fig. 9.12c, where we find that the variance of the position quadrature remains unchanged

$$\text{Var}\left[(\hat{p}_2 - \hat{q}_3 - \hat{q}_4) - (\hat{p}_N - \hat{q}_3 - \cdots - \hat{q}_{N-1})\right] = 2\Delta^{-2}, \qquad (9.134)$$

but the variance in the momentum quadrature

$$\text{Var}\left[\frac{\hat{p}_3 - \hat{q}_2 - \hat{q}_N}{N-2} - \frac{\hat{p}_4 - \hat{q}_2 - \hat{q}_N}{N-2} - \cdots - \frac{\hat{p}_{N-1} - \hat{q}_2 - \hat{q}_N}{N-2}\right] = \frac{\Delta^{-2}}{N-2} \qquad (9.135)$$

is reduced by a factor $N-2$. This is not a general error-correction code, because it does not affect the accumulation of errors in the position quadrature. Nevertheless, this procedure may be beneficial when the quantum information is encoded such that errors in the position quadrature affect the computation much less than errors in the momentum quadrature. Since the variance of the position quadrature does not deteriorate, it seems likely that this procedure can be repeated with the position and momentum quadratures exchanged. This would completely counteract the propagation of errors through the cluster state.

9.7 References and further reading

The concept of quantum computing with continuous variables was first introduced by Lloyd and Braunstein (1999). For a recent review of quantum information with continuous variables, see Braunstein and van Loock (2005), and the book by Braunstein and Pati (2003). Cluster-state quantum computing with qunats was introduced by Menicucci *at al.* (2006), and Zhang and Braunstein (2006). Other important results for qunat cluster states were derived by van Loock *at al.* (2007), Zhang (2008a, 2008b), and Gu *at al.* (2009). An implementation of cluster-state quantum computing in an optical frequency comb was proposed by Flammia *et al.* (2009). Quantum error correction for qunats was introduced by Braunstein (1998a), Lloyd and Slotine (1998), and Gottesman *et al.* (2001). The nine-qubit error-correction code by Braunstein was recently demonstrated experimentally by Aoki *et al.* (2008). Finally, it was proved by Niset *et al.* (2009) that it is impossible to design a protocol based on Gaussian (Clifford) operations, homodyne detection, and classical feed-forward that corrects arbitrary Gaussian errors.

Atomic ensembles in quantum information processing

In Chapter 7 we discussed the interaction between atoms and photons. We found that the coupling between a single atom and a single photon is rather weak, unless the atom and the photon are contained within a small cavity. The only way to significantly affect a free atom with light is to use a state of light with many photons, and in particular we showed how to use classical laser fields to prepare particular atomic states. However, there is another way of enhancing the interaction in a useful way. We can use many atoms rather than a single one. In this chapter, we will show that the interaction between a single photon and an ensemble of atoms can be greatly increased through the phenomenon of collective enhancement. We will show that this gives an atomic ensemble a relatively large susceptibility, and that multilevel atoms can be used to generate large optical nonlinearities. These offer the possibility of slowing, and even stopping, a beam of light; a phenomenon that goes hand in hand with electromagnetically induced transparency. We will discuss how a quantum state of light can be stored in and retrieved from an ensemble of atoms, and how the states of different atomic ensembles can be entangled with one another. We will show that collective states of atoms can even be used as a single qubit or several qubits, and finally, we will show that atomic ensembles can be use to mediate a photon–photon interaction.

10.1 An ensemble of identical two-level atoms

In this section, we extend the description of a two-level atom interacting with an optical field to an 'ensemble' of identical two-level atoms. Much of the argument in this section and the next follows Beausoleil *et al.* (2004). We treat the electromagnetic field first as a classical field, and then we consider the quantized field. The main aim of this section is to give a fully quantum mechanical derivation of the susceptibility of the ensemble, including its frequency dependence and the induced losses. In the next section, we will use these results to discuss electromagnetically induced transparency.

Consider an atom with the level structure shown in Fig. 10.1. The density matrix of this atom is $\rho(t)$, the ground state is $|g\rangle$, and the excited state is $|e\rangle$. The frequency of the transition between the ground and excited state is ω_0. The system dynamics are as usual described by a master equation (see Chapter 7 for the derivation):

$$\dot{\rho}(t) = -\frac{i}{\hbar}[\mathcal{H}, \rho(t)] - \Gamma[\rho(t)]. \tag{10.1}$$

The Hamiltonian $\mathcal{H}$ describes our two-level atom, which has a transition dipole $\mathbf{r}_{eg}$ driven by an electric field $\mathbf{E} = E_0 \boldsymbol{\epsilon} \cos(\omega_l t)$. In the frame rotating at frequency ω_l, and using the

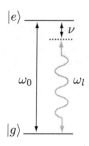

Fig. 10.1. **Level structure of a two-level atom.**

'rotating-wave approximation' (RWA, see Chapter 7), the Hamiltonian takes the form of Eq. (7.22), i.e.,

$$\mathcal{H} = \frac{\hbar}{2} \begin{pmatrix} -\nu & \Omega^* \\ \Omega & \nu \end{pmatrix}, \tag{10.2}$$

where we define the Rabi frequency $\Omega \equiv 2e\mathbf{r}_{eg} \cdot \epsilon E_0 / \hbar$, and $\nu \equiv \omega_0 - \omega_l$ is the detuning between the frequency of the transition and that of the field.

We include dissipation in the master equation via the term $\Gamma[\rho(t)]$, which can be written in the form first introduced in Eq. (7.101):

$$\Gamma[\rho(t)] = \frac{1}{2} \sum_m \gamma_m \left([\rho L_m^\dagger, L_m] + [L_m^\dagger, L_m \rho] \right). \tag{10.3}$$

This gives rise to a master equation in so-called Lindblad form (after Lindblad, 1976), and the operators L_m are Lindblad operators. We will include two dissipative processes in a phenomenological way, as opposed to looking at a specific microscopic Hamiltonian for the system–environment interaction. This allows the theory to be applied generally to a range of environmental processes. The first process to be included is spontaneous emission, which from the arguments of Chapter 7 has an associated operator $L_m = \sigma^-$. The second is pure dephasing ($L_m = \sigma_z$), which can only change the coherence between $|e\rangle$ and $|g\rangle$, not their populations. The two processes have rates γ_s and γ_d, respectively. With these Lindblad operators we derive three coupled differential equations:

$$\dot{\rho}_{gg}(t) = \text{Im}[\rho_{eg}(t)\Omega^*] + \gamma_s \rho_{ee}(t), \tag{10.4}$$

$$\dot{\rho}_{ee}(t) = -\text{Im}[\rho_{eg}(t)\Omega^*] - \gamma_s \rho_{ee}(t), \tag{10.5}$$

$$\dot{\rho}_{eg}(t) = -(\gamma_T + i\nu)\rho_{eg}(t) + i\frac{\Omega}{2}[\rho_{ee}(t) - \rho_{gg}(t)], \tag{10.6}$$

where γ_T is the combined dissipation rate $\gamma_T = \gamma_d + \gamma_s/2$. The first two differential equations are not really independent, since $\rho_{gg}(t) + \rho_{ee}(t) = 1$ by virtue of the trace property of density operators.

To proceed, we assume that our system is initially in the ground state. Equations (10.5) to (10.6) can be solved using a 'bootstrapping' approach in which we use the fact that, after some fast initial changes, the value of $\dot{\rho}_{eg}(t)$ becomes rather small in comparison with the magnitude of terms on the right-hand side of Eq. (10.6). The element $\rho_{eg}(t)$ then adiabatically follows the population difference $[\rho_{gg}(t) - \rho_{ee}(t)]$:

$$\rho_{eg}(t) = \tilde{\rho}_{eg}[\rho_{gg}(t) - \rho_{ee}(t)] = \tilde{\rho}_{eg}[2\rho_{gg}(t) - 1]. \tag{10.7}$$

with

$$\tilde{\rho}_{eg} = -\frac{\Omega}{2(\nu - i\gamma_T)}. \tag{10.8}$$

We can now rewrite Eq. (10.4) as

$$\dot{\rho}_{gg}(t) = -\Gamma_1 \rho_{gg} + \Gamma_2, \tag{10.9}$$

with

$$\Gamma_1 = \gamma_s - 2\text{Im}[\tilde{\rho}_{eg}\Omega^*], \tag{10.10}$$

$$\Gamma_2 = \gamma_s - \text{Im}[\tilde{\rho}_{eg}\Omega^*]. \tag{10.11}$$

Therefore

$$\rho_{gg}(t) = \frac{\Gamma_2}{\Gamma_1} + \left(1 - \frac{\Gamma_2}{\Gamma_1}\right) \exp(-\Gamma_1 t). \tag{10.12}$$

Substituting this closed form into Eq. (10.6) yields

$$\rho_{eg}(t) = [1 - \exp(-(\gamma_T + i\nu)t]\tilde{\rho}_{eg}\left[\frac{2\Gamma_2}{\Gamma_1} - 1 + \left(2 - \frac{2\Gamma_2}{\Gamma_1}\right)\exp(-\Gamma_1 t)\right]. \tag{10.13}$$

If $|\text{Im}[\tilde{\rho}_{eg}\Omega^*]| \ll \gamma_s$, then the excited state is only rarely populated, and we can assume that $\rho_{gg} \approx 1$ and $\rho_{ee} \approx 0$. In this case the coherence soon stabilizes to a steady-state value of $\tilde{\rho}_{eg}$.

Next, we calculate the absorption coefficient of the atom via its linear susceptibility. With our original definition of the atom–photon coupling in Chapter 7, we can define a linear atomic susceptibility by the relation

$$\mathbf{p} \equiv -\chi_a^{(1)}(-\omega_l, \omega_l)\frac{\epsilon_0 E_0}{2}\boldsymbol{\epsilon} + \text{c.c.}, \tag{10.14}$$

where $\boldsymbol{\epsilon}$ is the polarization of the driving field, and $\mathbf{p}$ is the steady-state induced dipole

$$\mathbf{p} = e\,\text{Tr}\left[\rho(\mathbf{r}_{eg}\,|e\rangle\,\langle g| + \text{H. c.})\right] = e(\tilde{\rho}_{ge}\mathbf{r}_{eg} + \tilde{\rho}_{eg}\mathbf{r}_{ge}). \tag{10.15}$$

Here, $\mathbf{r}_{eg}$ is the transition dipole matrix element between states $|e\rangle$ and $|g\rangle$. From this, we obtain

$$\chi_a^{(1)}(-\omega_l, \omega_l) = -\frac{2\boldsymbol{\epsilon}^* \cdot \mathbf{r}_{ge}}{\epsilon_0 E_0}\tilde{\rho}_{eg} \tag{10.16}$$

and so

$$\chi_a^{(1)}(-\omega_l, \omega_l) = \frac{\hbar|\Omega|^2}{2E_0^2\epsilon_0(\nu - i\gamma_T)}. \tag{10.17}$$

Both the field and polarization oscillate with the frequency ω_l back in the original (non-rotating) reference frame, and hence we include this frequency as an argument in our susceptibility function. Since the Rabi frequency Ω is proportional to the field amplitude E_0, this expression does not depend on E_0. Nonetheless, we choose to leave our expression in this form since then it only has a dependence on parameters that are directly accessible in experiments.

Exercise 10.1: Derive Eq. (10.17).

What we have described is an exceedingly weak response that is difficult to measure directly. However, two things can enhance the interaction and make the response readily observable. First, in Chapter 7 we exploited the possibility of the atom being placed in a cavity, which greatly increases the atom–light interaction since each photon effectively makes several cavity 'round trips' before being emitted. Second, we can increase the number of atoms, and make them all interact with the same field. This is the approach we describe in this chapter.

First, we simplify the description of light interacting with an ensemble of identical two-level atoms and assume that a gas of atoms is confined such that the spatial variation of the light is negligible in that region. We label the state with atom i in its excited state $|e\rangle_i$, and all other atoms in the state $|g\rangle_{j\neq i}$ as $|w_g^{e,i}\rangle$. The state of all atoms in the ground level will be called $|\mathbf{g}\rangle$. The Hamiltonian may now be written in the basis $\{|\mathbf{g}\rangle, |w_g^{e,1}\rangle, |w_g^{e,2}\rangle, \ldots, |w_g^{e,N}\rangle\}$ as

$$\mathcal{H} = \frac{\hbar}{2} \begin{pmatrix} -\nu & \Omega^* & \Omega^* & \ldots & \Omega^* \\ \Omega & \nu & 0 & \ldots & 0 \\ \Omega & 0 & \nu & \ldots & 0 \\ \vdots & \vdots & \vdots & \ddots & \vdots \\ \Omega & 0 & 0 & \ldots & \nu \end{pmatrix}. \tag{10.18}$$

This matrix has a particularly symmetrical form, and it is straightforward to show that it has $(N-1)$ degenerate eigenvalues ν. The eigenvectors $|\lambda_j\rangle$, with $j = 1$ to $N-1$, corresponding to these eigenvalues ν, are all linear combinations of states involving one atomic excitation, and have no $|\mathbf{g}\rangle$ component. Furthermore, all the $|\lambda_j\rangle$ must be orthogonal to the symmetric linear combination (the W state)

$$\left|W_g^e\right\rangle \equiv \frac{1}{\sqrt{N}} \sum_{i=1}^{N} \left|w_g^{e,i}\right\rangle. \tag{10.19}$$

We may therefore simplify our calculation significantly by rewriting the Hamiltonian in the basis $\{|\mathbf{g}\rangle, |W_g^e\rangle, |\lambda_1\rangle, \ldots, |\lambda_{N-1}\rangle\}$, such that

$$\mathcal{H} = \frac{\hbar}{2} \begin{pmatrix} -\nu & \Omega^*\sqrt{N} & 0 & \ldots & 0 \\ \Omega\sqrt{N} & \nu & 0 & \ldots & 0 \\ 0 & 0 & \nu & \ldots & 0 \\ \vdots & \vdots & \vdots & \ddots & \vdots \\ 0 & 0 & 0 & \ldots & \nu \end{pmatrix}. \tag{10.20}$$

The closed Hamiltonian dynamics of a system intialized in the state $|\mathbf{g}\rangle$ are therefore completely decoupled from all of the states $|\lambda_j\rangle$, and the only state that couples to $|\mathbf{g}\rangle$ is $\left|W_g^e\right\rangle$. Even if we allow each atom i to decay (with rate γ_d) and dephase (with rate γ_s) as our single atom did in the preceding discussion, we find that there is no coupling from $|\mathbf{g}\rangle$ and $\left|W_g^e\right\rangle$ to the $|\lambda_j\rangle$ states, and we can therefore ignore them completely in a calculation of the susceptibility. In fact, we can derive equations completely analogous to Eqs. (10.4)–(10.6) but with modified decay constants:

$$\dot{\rho}_{\mathbf{gg}}(t) = \text{Im}[\rho_{\text{wg}}(t)\Omega\sqrt{N}] + \gamma_s'\rho_{\text{ww}}(t) \tag{10.21}$$

$$\dot{\rho}_{\text{ww}}(t) = -\text{Im}[\rho_{\text{wg}}(t)\Omega\sqrt{N}] - \gamma_s'\rho_{\text{ww}}(t) \tag{10.22}$$

$$\dot{\rho}_{\text{wg}}(t) = -(\gamma_T' + i\nu)\rho_{\text{wg}}(t) + i\frac{\Omega\sqrt{N}}{2}[\rho_{\text{ww}}(t) - \rho_{\mathbf{gg}}(t)], \tag{10.23}$$

where we have defined

$$\gamma_s' = \gamma_s, \tag{10.24}$$

$$\gamma_d' = \frac{N+3}{4}\gamma_d, \tag{10.25}$$

$$\gamma_T' = \frac{\gamma_s'}{2} + \gamma_d'. \tag{10.26}$$

In order to calculate the susceptibility of the ensemble, we may again make the weak driving approximation, so that the population is primarily in the ground state, with very little excitation to state $\left|W_g^e\right\rangle$. We obtain an expression analogous to Eq. (10.8) for steady-state coherence:

$$\tilde{\rho}_{\text{wg}} = -\frac{\Omega\sqrt{N}}{2(\nu - i\gamma_T')}. \tag{10.27}$$

It is straightforward to find the polarization of the entire ensemble:

$$\mathbf{p} = e\sum_i \text{Tr}\left(\rho\mathbf{r}_{e_i g}|e_i\rangle\langle g|\right) + \text{H.c.} \tag{10.28}$$

$$= e\sum_i \tilde{\rho}_{eg}\mathbf{r}_{ge} + \text{c.c.} \tag{10.29}$$

$$= e\sqrt{N}\tilde{\rho}_{\text{wg}}\mathbf{r}_{ge} + \text{c.c.} \tag{10.30}$$

$$= -\frac{eN\Omega\mathbf{r}_{eg}^*}{(\nu - i\gamma_T')} + \text{c.c..} \tag{10.31}$$

We then find that that the susceptibility per unit volume is

$$\chi_V^{(n_a)}(-\omega_l, \omega_l) = \frac{n_a\hbar|\Omega|^2}{2E_0^2\epsilon_0(\nu - i\gamma_T')}, \tag{10.32}$$

where $n_a = N/V_a$ is the number density of atoms in the ensemble which is assumed to occupy a volume V. Notice that $\chi_V^{(n_a)}$ approaches n_a times $\chi_V^{(1)}$ only if the pure dephasing rate γ_d' of the atomic ensemble is much smaller than the spontaneous decay rate γ_s'.

From classical electromagnetism, it is well known that the real part of the volume susceptibility gives a polarization in phase with the applied field, and so affects the refractive index. On the other hand, the imaginary part of $\chi_V^{(1)}$ corresponds to an out-of-phase polarization which leads to a dissipation of energy, i.e., absorption. The mathematical relations are:

$$\eta^2(\omega_l) = 1 + \mathrm{Re}\left[\chi_V^{(1)}(-\omega_l, \omega_l)\right], \tag{10.33}$$

$$\kappa(\omega_1) = \frac{\omega_l}{\eta(\omega_l)c}\mathrm{Im}\left[\chi_V^{(1)}(-\omega_l, \omega_l)\right], \tag{10.34}$$

where η is the refractive index and κ the absorption coefficient.

In Fig. 10.2 we display the typical dependence of both η and κ on the detuning parameter ν. A refractive index less than unity implies a phase velocity larger than the speed of light in vacuum c. The group velocity is a measure of the speed of a wave packet and is given by

$$v_g = \frac{d\omega}{dk} = \frac{c}{\eta + \dfrac{d\eta}{d\omega}\omega}. \tag{10.35}$$

Close inspection reveals that v_g can also be greater than c. This happens when the gradient of the curve in Fig. 10.2 is steeply positive (since $d\eta/d\omega = -d\eta/d\nu$). How can this result be reconciled with the theory of special relativity? This question has been debated for many years, and was thoroughly addressed by Brillouin (1960). Importantly, the effect here occurs in a region of high dispersion, where the phase velocities for the Fourier components of a wave packet are all different. In addition, there is strong absorption in the regions where the group velocity becomes superluminal. These factors mean that the shape and amplitude of a wave packet changes as it passes through the medium and so it is no longer straightforward to

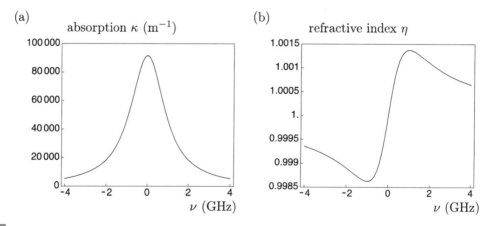

Fig. 10.2. (a) Absorption coefficient and (b) refractive index for a gas of atoms in a weak optical field. The following parameters were used: $n = 10^{19}$ m^{-3}, $r_{eg} \cdot \epsilon = 0.1$ nm, $\gamma_T' = 1$ GHz, $\omega_0 = 5 \times 10^{15}$ Hz.

define what is meant by its velocity. The group velocity might most accurately be regarded as the velocity of the maximum amplitude of the packet, which can of course exceed c without violating causality when the shape of the packet changes (i.e., if the peak moves forward relative to the rest of the packet). One definition of the signal velocity, proposed by Brillouin, is based on the speed of the front of the wave, i.e., the first wavelike disturbance associated with the light pulse. This is a good definition for 'analytic' pulses whose shape is described by a well-behaved mathematical function. In this case, all the information about the packet is transferred once any part of it is detected. This velocity will never exceed the speed of light in vacuum, and causality is assured.

10.1.1 The interaction of atomic ensembles with quantized fields

The analysis presented so far has been for classical fields. However, later in the chapter we will be concerned not only with classical concepts such as the susceptibility, but also with photon interactions with the atomic ensembles. We therefore show next how the results obtained thus far transfer to an atomic ensemble in a quantized field.

We found in Chapter 7 that the Jaynes–Cummings Hamiltonian in the rotating-wave approximation couples combined states of a single atom and a single mode with the same total excitation number. An atom in state $|g\rangle$ with a quantized field of n photons, which we will label $|g, n\rangle$, will have an interaction Hamiltonian matrix element with state $|e, n - 1\rangle$, as shown in Eq. (7.77). In the absence of decoherence, it is not necessary to consider any dynamics outside this two-dimensional Hilbert space, and the Hamiltonian can be written as

$$\mathcal{H} = \begin{pmatrix} \hbar\omega_l(n - \frac{1}{2}) - \frac{1}{2}\hbar v & g^*\sqrt{n} \\ g\sqrt{n} & \hbar\omega_l(n - \frac{1}{2}) + \frac{1}{2}\hbar v \end{pmatrix}. \tag{10.36}$$

Here, v is again the detuning of the field and the atomic transition $v \equiv \omega_0 - \omega_l$, where ω_l is now the frequency of the quantized field mode. The coupling constant g was defined in Eq. (7.73). Apart from an unimportant constant energy on the diagonal, this quantized Hamiltonian for the subspace characterized by n field excitations is identical to the earlier classical Hamiltonian in Eq. (10.2) if we choose $\Omega_n \equiv 2g\sqrt{n}/\hbar$. The unitary evolution is therefore identical in the two cases.

However, decoherence does complicate this direct mapping. Spontaneous-emission processes from $|e\rangle$ cause photons to populate environmental modes that are different from the single mode we have so far considered. It is therefore necessary to append an extra, third, quantum number to our state description to take account of the number of photons emitted into the environment. Following spontaneous emission, the state $|e, n - 1, 0\rangle$ therefore becomes $|g, n - 1, 1\rangle$, and the total excitation number of our single mode–atom system is reduced by one. Further decay events could of course take place until eventually there would be no photons remaining in the mode. At first glance, this seems to complicate things enormously, since the dynamics can now traverse vastly more levels. However, we can make an important simplification: since we are interested only in high-fidelity dynamics for quantum computing applications, even a single photon-emission process should be unlikely. It therefore suffices to restrict ourselves to only the three coupled levels

$\{|g, n, 0\rangle, |e, n - 1, 0\rangle, |g, n - 1, 1\rangle\}$ with the Hamiltonian

$$\mathcal{H} = \begin{pmatrix} \hbar\omega_l(n - \frac{1}{2}) - \frac{1}{2}\hbar\nu & g^*\sqrt{n} & 0 \\ g\sqrt{n} & \hbar\omega_l(n - \frac{1}{2}) + \frac{1}{2}\hbar\nu & 0 \\ 0 & 0 & \hbar\omega_l(n - \frac{3}{2}) - \frac{1}{2}\hbar\nu \end{pmatrix}. \tag{10.37}$$

We now include decoherence processes analogous to the single-atom case. First, spontaneous emission with the rate γ_s is described by the operator $|g, n - 1, 1\rangle \langle e, n - 1, 0|$. Second, pure dephasing with rate γ_d corresponds to the same σ_z operator as before, acting in the subspace spanned by $|g, n, 0\rangle$ and $|e, n - 1, 0\rangle$. The dynamics is then described by equations similar to those for the classical field:

$$\dot{\rho}_{11}(t) = 2\text{Im}\left[\frac{\rho_{21}(t)g^*\sqrt{n}}{\hbar}\right],$$

$$\dot{\rho}_{22}(t) = -2\text{Im}\left[\frac{\rho_{21}(t)g^*\sqrt{n}}{\hbar}\right] - \gamma_s\rho_{22}(t),$$

$$\dot{\rho}_{21}(t) = -(\gamma_T + i\nu)\rho_{21}(t) + i\frac{g\sqrt{n}}{\hbar}[\rho_{22}(t) - \rho_{11}(t)],$$

$$\dot{\rho}_{33}(t) = \gamma_s\rho_{22}(t), \tag{10.38}$$

where the indices run over the basis $\{|g, n, 0\rangle, |e, n - 1, 0\rangle, |g, n - 1, 1\rangle\}$. We do not explicitly write the equation for ρ_{31} since it is not needed in our subsequent calculations. A similar bootstrapping approach to the one we considered for the classical case may also be used here. Again assuming the weak driving limit, $\rho_{11}(t) \gg \rho_{22}(t)$ for all t we find:

$$\rho_{11}(t) = \exp(-\Gamma t),$$

$$\rho_{21}(t) = [1 - \exp(-(i\nu + \gamma_T)t)]\tilde{\rho}_{21}(n)\exp(-\Gamma t),$$

$$\rho_{22}(t) = \frac{\Gamma}{\gamma_s - \Gamma}[e^{-\Gamma t} - e^{-\gamma_s t}], \tag{10.39}$$

with

$$\tilde{\rho}_{21}(n) = -\frac{g\sqrt{n}}{\hbar(\nu - i\gamma_T)}, \tag{10.40}$$

and

$$\Gamma = -\frac{2\text{Im}[\tilde{\rho}_{21}g^*\sqrt{n}]}{\hbar}. \tag{10.41}$$

The expressions in Eq. (10.39) have much in common with the semiclassical density matrix elements, derived at the beginning of this Section. The essential difference is that all three of these elements now eventually decay to zero. This is a consequence of the fact that we explicitly include the field in our description of the quantum state, and it is possible for a photon to be scattered from the coherent mode with which we excite the system. With an ensemble of N atoms, the rate of absorption of the field is

$$R = N\Gamma = \frac{2N|g|^2 n}{\hbar^2}\text{Im}\left[\frac{1}{\nu - i\gamma_T}\right] = N|\Omega_n|^2\text{Im}\left[\frac{1}{2(\nu - i\gamma_T)}\right]. \tag{10.42}$$

Similarly, the mean energy of the system is altered according to:

$$\langle E(t) \rangle = N\text{Tr}[\mathcal{H}, \rho] = -N\text{Re}[\rho_{21}(t)g^*\sqrt{n}], \tag{10.43}$$

where we have used a minus sign in the definition of the energy shift for consistency with the absorption-rate definition. Following the decay of initial transients and before the slow absorption of the field becomes dominant, this is given in frequency units by

$$\frac{\langle E \rangle}{\hbar} = -N\text{Re}[\tilde{\rho}_{21}g^*\sqrt{n}] = N|\Omega_n|^2\text{Re}\left[\frac{1}{2(\nu - i\gamma_T)}\right]. \tag{10.44}$$

The right-hand sides of Eqs. (10.42) and (10.44) are the real and imaginary parts of the same quantity, a sort of complex energy shift whose imaginary part corresponds to field absorption. This is important for the subsequent discussion of the susceptibility.

In order to define a susceptibility in this fully quantized model, we need to relate the coupling constant g in Ω_n to an electric field strength. To make the connection between the phase behaviour of a classical field and a quantum field it is most convenient to consider a coherent state. It is defined in terms of Fock states $|n\rangle$ by

$$|\alpha(t)\rangle = \exp\left(\frac{1}{2}|\alpha(t)|^2\right)\sum_{n=0}^{\infty}\frac{\alpha^n(t)}{\sqrt{n!}}|n\rangle, \tag{10.45}$$

where $\alpha(t) = \sqrt{\langle n \rangle}\exp(i\omega_l t)$. In Chapter 7, Eq. (7.70), we used the following operator for a single-mode electric field

$$\hat{\mathbf{E}} = \xi_a(\epsilon\hat{a} + \epsilon^*\hat{a}^\dagger), \tag{10.46}$$

where we have defined a new constant

$$\xi_a = \sqrt{\frac{\hbar}{2(2\pi)^3\varepsilon_0}}\mu. \tag{10.47}$$

It is also useful to relate ξ_a to an effective mode volume for the field, which would arise from a simple box-normalization picture. We define

$$\mathcal{V} \equiv \frac{8\pi^3\omega_l}{\mu^2}, \tag{10.48}$$

such that

$$\xi_a = \sqrt{\frac{\hbar\omega_l}{2\varepsilon_0\mathcal{V}}}. \tag{10.49}$$

The amplitude of the electric field is given by the expectation value of $\hat{\mathbf{E}}$ for the coherent state

$$\langle\alpha(t)|\hat{\mathbf{E}}|\alpha(t)\rangle = \xi_a(\epsilon\alpha(t) + \text{c. c.}). \tag{10.50}$$

Referring back to our original definition of g in Eq. (7.73), and using Eq. (10.47) we find that

$$g = e\xi_a(\boldsymbol{\epsilon}\cdot\mathbf{r}). \tag{10.51}$$

To find the susceptibility, we need to determine the polarization of the sample under illumination by a coherent state. After initial transients, but before the long, slow decay of polarization, we know that the coherence between states $|g, n\rangle$ and $|e, n-1\rangle$ is given by $\tilde{\rho}_{21}(n)\rho_{11}(t=0)$. Therefore, the polarization of the atom following coherent-state excitation is

$$\mathbf{p} = -e\mathbf{r}_{12}\exp(-|\alpha|^2)\sum_n \frac{g\sqrt{n+1}}{\hbar(\nu - i\gamma_T)}\frac{\alpha^n\alpha^{n+1}}{\sqrt{n!(n+1)!}} = -e\mathbf{r}_{12}\frac{g\sqrt{\langle n\rangle}}{\hbar(\nu - i\gamma_T)}. \tag{10.52}$$

The single-atom susceptibility is therefore given by:

$$\chi^{(1)} = 2e\frac{(\boldsymbol{\epsilon}\cdot\mathbf{r})^*}{\epsilon_0\xi_a\sqrt{\langle n\rangle}}\frac{g\sqrt{\langle n\rangle}}{\hbar(\nu - i\gamma_T)} = \frac{\hbar\Omega_{coh}^2}{2\epsilon_0\langle n\rangle\xi_a^2(\nu - i\gamma_T)}, \tag{10.53}$$

where we have used the definition $\Omega_{coh} = 2g\sqrt{\langle n\rangle}/\hbar$ for the effective Rabi frequency of a coherent state. This equation is entirely analogous to the one we found for a classical field, in Eq. (10.17). The extension to n_a atoms follows as before, from which we can find the volume susceptibility:

$$\chi^{(n_a)} = \frac{n_a\hbar|\Omega_{coh}|^2}{2\epsilon_0\langle n\rangle\xi_a^2(\nu - i\gamma_T')} = \frac{n_a|\Omega_{coh}|^2V}{\omega_l\langle n\rangle(\nu - i\gamma_T')}. \tag{10.54}$$

Expressions for refractive index and absorption coefficient follow in the same way as before.

10.1.2 Phase shift of a single mode

In the previous discussion, we saw that the phase of the electric field follows the phase of the coefficient $\alpha(t)$. If the field mode volume is V and the ensemble occupies a small relative volume V, we expect a modification in the frequency of the mode, since the wave vector of the mode is fixed by the physical constraints of the cavity

$$\omega' = \left(1 - \frac{V}{V}\right)\omega_l + \frac{V}{V}\frac{1}{\sqrt{1 + \chi^{(n)}}}\omega_l = \omega_l - W. \tag{10.55}$$

The susceptibility $\chi^{(n_a)}$ has real and imaginary parts, which correspond to a phase shift of the mode, and a decay in amplitude, respectively. The shift W is then

$$W = \frac{V}{2V}\omega_l\chi^{(n)} = \frac{N|\Omega_{coh}|^2}{\langle n\rangle}\frac{1}{2(\nu - i\gamma_T')}. \tag{10.56}$$

The fundamental connection between energy shift and absorption is now clearly seen, and this leads to a fuller appreciation of the very similar pair of expressions we obtained for Fock states in Eqs. (10.42) and (10.44).

10.2 Electromagnetically induced transparency

In this section, we will discuss the optical properties of ensembles of three- and four-level atoms. In particular, we will show how an ensemble of identical atoms becomes opaque for certain frequencies of light, but that a strong 'control' laser of a different frequency can open up a transparent window in this absorption line. This is the phenomenon of 'electromagnetically induced transparency', or EIT.

10.2.1 Three-level systems

Let us now consider the three-level Λ system that we first encountered in Chapter 7, which is the simplest system that can exhibit EIT. The level structure and parameter definitions are displayed in Fig. 10.3. The idea is to apply a strong field that is close to resonance to one arm of a Λ system and then to probe the system with a weak field applied to the other arm. The Hamiltonian in this situation, with two applied laser fields, was introduced in Eq. (7.47)

$$\mathcal{H}_\Lambda(t) = \hbar \begin{pmatrix} 0 & 0 & \Omega_1^* \cos \omega_1 t \\ 0 & \delta & \Omega_2^* \cos \omega_2 t \\ \Omega_1 \cos \omega_1 t & \Omega_2 \cos \omega_2 t & \omega \end{pmatrix}. \tag{10.57}$$

In Chapter 7, we went on to use a rotating frame and made the rotating-wave approximation to remove the time dependence. We found in Eq. (7.50) that

$$\mathcal{H}_\Lambda = \frac{\hbar}{2} \begin{pmatrix} 0 & 0 & \Omega_1 \\ 0 & 2(\nu_1 - \nu_2) & \Omega_2 \\ \Omega_1 & \Omega_2 & 2\nu_1 \end{pmatrix}. \tag{10.58}$$

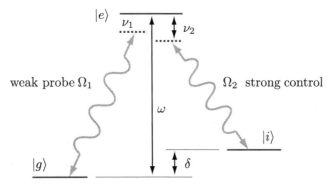

Fig. 10.3. The atomic energy-level structure required for observation of electromagnetically induced transparency. Assuming that all of the population is initially in state $|g\rangle$, the control pulse opens a window in the absorption spectrum of the probe.

In addition to the Hamitonian dynamics, let us also take into account some forms of dissipation. Specifically, we assume that the highest-lying level $|e\rangle$ can decay to both lower levels $|g\rangle$ and $|i\rangle$. To do this we introduce Lindblad operators σ_{eg}^- and σ_{ei}^-, which are the usual Pauli σ^- operators acting on the subspace of $|e\rangle$ and $|g\rangle$ or $|i\rangle$. The decay rates are γ_s^{eg} and γ_s^{ei}, respectively. Without complicating matters, we may also allow pure dephasing terms on all three pairs of levels, $\{\sigma_z^{gi}, \sigma_z^{eg}, \sigma_z^{ei}\}$, with rates $\{\gamma_d^{gi}, \gamma_d^{eg}, \gamma_d^{ei}\}$.

The master equation becomes a set of nine differential equations, some of which are coupled. However, we are not interested in evaluating all density matrix elements; we are concerned only with those leading to expressions for the susceptibility. We can also make some approximations by assuming that the atom is initially in its ground state $|g\rangle$ and the probe field is sufficiently weak, and the decay rate from $|e\rangle$ sufficiently strong that population in the other two levels is always negligible (i.e. $\Omega_1 \ll \gamma_s^{eg}$). Then $\rho_{gg} \approx 1$, $\rho_{ii} \approx 0$, and $\rho_{ee} \approx 0$. It follows that $\rho_{ei} \approx 0$. We may then find relatively simple differential equations for the two remaining density matrix elements

$$\dot{\rho}_{ig}(t) = -[G_{ig} + i(\nu_1 - \nu_2)]\rho_{ig}(t) - \frac{i\Omega_2^*}{2}\rho_{eg}(t), \tag{10.59}$$

$$\dot{\rho}_{eg}(t) = -[G_{eg} + i\nu_1]\rho_{eg}(t) - \frac{i\Omega_2}{2}\rho_{ig}(t) - \frac{i\Omega_1}{2}, \tag{10.60}$$

with

$$G_{ig} = \gamma_d^{ig} + \tfrac{1}{4}(\gamma_d^{eg} + \gamma_d^{ei}) \tag{10.61}$$

$$G_{eg} = \gamma_d^{eg} + \tfrac{1}{4}(\gamma_d^{ig} + \gamma_d^{ei}) + \tfrac{1}{2}(\gamma_s^{eg} + \gamma_s^{ei}). \tag{10.62}$$

Following the decay of all the transients ($\dot{\rho}_{ig}(t) = 0$ and $\dot{\rho}_{eg}(t) = 0$), we settle to a steady-state solution ($\rho_{ig}(t) = \tilde{\rho}_{ig}, \rho_{eg}(t) = \tilde{\rho}_{eg}$). Since the probe field addresses the transition between $|g\rangle$ and $|e\rangle$, the susceptibility is proportional to ρ_{eg}. We therefore obtain

$$\chi_a^{(1)}(-\omega_l, \omega_l) = \frac{\hbar\Omega_1^2}{2E_0^2\epsilon_0} \frac{2[(\nu_1 - \nu_2) - iG_{ig}]}{4[(\nu_1 - \nu_2) - iG_{ig}][\nu_1 - iG_{eg}] - |\Omega_2|^2}, \tag{10.63}$$

for the susceptibility of a single three-level atom in the Λ-configuration.

The extension of our calculation to the case of N atoms follows along similar lines to that of a two-level atomic ensemble in the previous section. We first notice that only the symmetric combination of excited states $\left|W_g^e\right\rangle = N^{-\frac{1}{2}} \sum_j |w_g^{e,j}\rangle$ couples to the state $|g\rangle$ in which all the atoms are in the ground state. Each state $|w_g^{e,j}\rangle$ couples via Ω_2 to the state with atom j in state $|i\rangle$ and all the other atoms in state $|g\rangle$. We label this state $|w_g^{i,j}\rangle$. The state $\left|W_g^e\right\rangle$ can couple only to the symmetric combination of states $|w_g^{i,j}\rangle$

$$\left|W_g^i\right\rangle \equiv \frac{1}{\sqrt{N}} \sum_{j=1}^{N} \left|w_g^{i,j}\right\rangle. \tag{10.64}$$

The $(N-1)$ linear combinations of $\left|w_i^{e,j}\right\rangle$ that are orthogonal to $\left|W_g^i\right\rangle$ have no coupling to $\left|W_g^e\right\rangle$, and cannot be accessed. We may therefore write the Hamiltonian in the reduced basis $\{|\mathbf{g}\rangle, \left|W_g^i\right\rangle, \left|W_g^e\right\rangle\}$ as

$$\mathcal{H}_\Lambda = \frac{\hbar}{2}\begin{pmatrix} 0 & 0 & \Omega_1\sqrt{N} \\ 0 & 2(\nu_1 - \nu_2) & \Omega_2 \\ \Omega_1\sqrt{N} & \Omega_2 & 2\nu_1 \end{pmatrix}. \tag{10.65}$$

It is reasonably straightforward to show that the decay constants become

$$G'_{ig} = \frac{N+3}{4}\gamma_d^{ig} + \frac{1}{4}(N\gamma_d^{eg} + \gamma_d^{ei}) \tag{10.66}$$

$$G'_{eg} = \frac{N+3}{4}\gamma_d^{eg} + \frac{1}{4}(N\gamma_d^{ig} + \gamma_d^{ei}) + \frac{1}{2}(\gamma_s^{eg} + \gamma_s^{ei}), \tag{10.67}$$

and finally we obtain

$$\chi_a^{(na)}(-\omega_l, \omega_l) = \frac{\hbar\Omega_1^2}{2E_0^2\epsilon_0} \frac{2[(\nu_1 - \nu_2) - iG'_{ig}]}{4[(\nu_1 - \nu_2) - iG'_{ig}][\nu_1 - iG'_{eg}] - |\Omega_2|^2}. \tag{10.68}$$

Again, the relation $\chi_a^{(na)} = n_a\chi_a^{(1)}$ is satisfied only if the decoherence of each atom is dominated by spontaneous-emission processes, and in the following discussion we will assume that this is indeed the case. This will allow us to use a one-atom model to explain the phenomena we will encounter.

We can use Eqs. (10.34) to relate the refractive index and absorption coefficients to the atomic susceptibility, and an example is shown in Fig. 10.4. There are two important qualitative differences with ensembles of two-level atoms, as we shall now discuss.

The first effect is that a window opens up in the absorption spectrum that becomes larger for a larger Ω_2. This is the famous EIT effect. In order to get a better insight into why this phenomenon occurs, let us first consider the effect of the strong field without a probe field present. Setting $\Omega_1 = 0$ and rewriting the single-atom Hamiltonian in Eq. (10.57) in terms of a new time-dependent state vector $\left|i'\right\rangle = e^{i\omega_2 t}|i\rangle$, we obtain in the basis $\{|\mathbf{g}\rangle, \left|i'\right\rangle, |e\rangle\}$

$$\mathcal{H}_\Lambda = \frac{\hbar}{2}\begin{pmatrix} 0 & 0 & 0 \\ 0 & 2(\omega - \nu_2) & \Omega_2 \\ 0 & \Omega_2 & 2\omega \end{pmatrix}, \tag{10.69}$$

where we have again made use of a rotating-wave approximation. We see that two of the eigenstates are now linear combinations of $\left|i'\right\rangle$ and $|e\rangle$, and that these have energies

$$\epsilon_\pm = \hbar\omega - \frac{\hbar}{2}\left(\nu_2 \pm \sqrt{\Omega^2 + \nu_2^2}\right), \tag{10.70}$$

above the energy of state $|g\rangle$. We therefore expect the absorption spectrum to be a doublet of lines, equally spaced around the energy $\omega - \nu/2$. A probe field with this energy is therefore

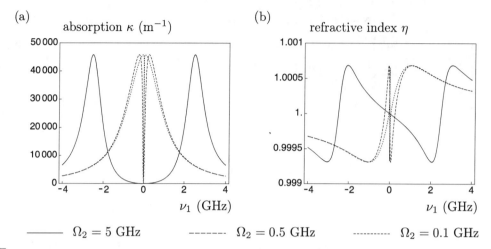

Fig. 10.4. (a) Absorption coefficient and (b) refractive index for a gas of atoms with a Λ structure. One arm of the Λ is addressed by a relatively strong field characterized by Rabi coupling Ω_2. A weak optical field applied to the other arm is then used to probe the structure and its absorption and refractive index are displayed here for different strengths of the strong field. The following parameters were used: $n = 10^{19}$ m^{-3}, $r_{eg} \cdot \epsilon = 0.1$ nm, $G_{eg} = 1$ GHz, $G_{ig} = 0$, $\omega_0 = 5 \times 10^{15}$ Hz, $\nu_2 = 0$.

not absorbed, so long as the two absorption lines are not too broad, i.e., for a sufficiently coherent system. For the case $\nu_2 = 0$ the expected transparency window lies at exactly the energy of the original transition between $|g\rangle$ and $|e\rangle$. This is shown in Fig. 10.4.

Second, there is a strong modulation of the refractive index near $\nu_1 = 0$ in Fig. 10.4. Indeed, the derivative of the refractive index with respect to the frequency becomes large and positive, and therefore the group velocity, as defined by Eq. (10.35), can become very small. A pulse of light is therefore expected to slow down in an electromagnetically transparent medium, and this 'slow light' effect has been observed experimentally in several systems.

We can also interpret EIT in terms of the STIRAP protocol discussed in Chapter 7. From the discussion in Section 7.1, we know that for equal detuning of the control and probe fields there is always a dark eigenstate of the form

$$|\psi(\theta)\rangle = \cos\theta \, |g\rangle - \sin\theta \, |i\rangle, \tag{10.71}$$

with the mixing angle related to the coupling strengths by $\tan\theta = (\Omega_1/\Omega_2)$. With the strong control field applied, and before application of the weak probe field, $|g\rangle$ is an eigenstate. Now imagine applying a finite pulse of the weak field. When the forward tail of the pulse enters the ensemble, the amplitude of the field increases slowly. This does not move the system from the dark state; it merely introduces a very small admixture of the $|i\rangle$ level. The pulse then passes through the gas, and the amplitude slowly reduces towards the tail of the pulse. This causes all the atoms to return to the state $|g\rangle$. The speed of the probe beam is reduced while it is interacting with the gas, because a portion of the information contained in the pulse is transferred to the atoms in the gas, whose average velocity is zero. We will

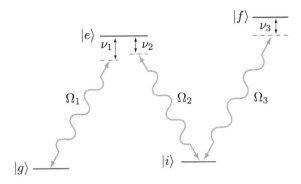

Fig. 10.5. The four-level atomic system described by the Hamiltonian in Eq. (10.72).

see later in the chapter that it is possible to transfer all of the information contained in a weak light pulse to the atoms in the ensemble.

10.2.2 Four-level systems

Next, we explore the possibilities offered by four-level atomic systems, of the type displayed in Fig. 10.5. Let us again first look at the single-atom dynamics; by simple extension of the analysis presented in the previous section we find the Hamiltonian in the basis $\{|g\rangle, |i\rangle, |e\rangle, |f\rangle\}$:

$$\mathcal{H}_\Lambda = \frac{\hbar}{2}\begin{pmatrix} 0 & 0 & \Omega_1 & 0 \\ 0 & 2\delta_1 & \Omega_2 & \Omega_3 \\ \Omega_1 & \Omega_2 & 2\nu_1 & 0 \\ 0 & \Omega_3 & 0 & 2\delta_2 \end{pmatrix}, \tag{10.72}$$

where $\delta_1 \equiv \nu_1 - \nu_2$ and $\delta_2 \equiv \nu_1 - \nu_2 + \nu_3$. We can again allow for pure dephasing processes between any pair of levels k and l at rate γ_d^{kl}. In addition we admit decay processes between excited levels $|e\rangle$ and $|f\rangle$ and low-lying levels $|g\rangle$ and $|i\rangle$ with rates γ_s^{ab}, where $a = e, f$ and $b = g, i$. Following the methods presented earlier, and using the assumption that $|g\rangle$ is the only level which is significantly populated we may, once again, find the steady-state value $\tilde{\rho}_{eg}$ of ρ_{eg}

$$\tilde{\rho}_{eg} = \frac{4\Omega_1(iG_{ig} - \delta_1)(\delta_2 - iG_{fg}) + \Omega_1\Omega_3^2}{2(\delta_2 - iG_{fg})[4(\nu_1 - iG_{eg})(\delta_1 - iG_{ig}) - \Omega_2^2] - 2(\nu_1 - iG_{eg})\Omega_3^2}, \tag{10.73}$$

where, as before, we have combined the total decay rates for each pair of levels k and l into parameters G_{kl}.

We may also straightforwardly extend our analysis to N atoms in a completely analogous way to the three-level system, by defining the symmetric excited state for the fourth level $|f\rangle$, i.e.

$$\left|W_g^f\right\rangle \equiv \frac{1}{\sqrt{N}}\sum_{j=1}^{N}\left|w_g^{fj}\right\rangle. \tag{10.74}$$

Recalling that the states $\left|W_g^e\right\rangle$ and $\left|W_g^i\right\rangle$ were similarly defined symmetric excitation states involving $|e\rangle$ and $|i\rangle$ respectively, we find that the coherence between the collective $|\mathbf{g}\rangle$ and $\left|W_g^e\right\rangle$ in this four-level system is

$$\tilde{\rho}_{eg}^{(N)} = \frac{4N\Omega_1(iG_{ig}' - \delta_1)(\delta_2 - iG_{fg}') + \Omega_1\Omega_3^2}{2(\delta_2 - iG_{fg}')[4(\nu_1 - iG_{eg}')(\delta_1 - iG_{ig}') - \Omega_2^2] - 2(\nu_1 - iG_{eg}')\Omega_3^2}, \qquad (10.75)$$

where the new collective rates are equal to the corresponding individual decay rates only if spontaneous decay dominates. Assuming this to be true, we find that the volume susceptibility depends linearly on the number density of atoms n_a via $\tilde{\rho}_{eg}^{(N)}$

$$\chi_a^{(n_a)}(-\omega_l, \omega_l) = -\frac{\hbar\Omega_1}{2E_0^2\epsilon_0}\,\tilde{\rho}_{eg}^{(N)}. \qquad (10.76)$$

In the strictest sense, this is no longer a *linear* susceptibility, since $\chi_a^{(n_a)}$ has a strong dependence on the quantities Ω_2^2 and Ω_3^2, which are themselves proportional to the intensities of two different laser sources. We can therefore expect strongly *nonlinear* effects. The refractive index and absorption for the laser coupling levels $|g\rangle$ and $|i\rangle$ will depend on the intensities of the other lasers. An example of the profound effect of the laser in the third arm is shown in Fig. 10.6. The relevant parameters in this plot are exactly as they were for the three-level case discussed above, for $\Omega_2 = 0.1$ GHz. When the third laser is turned off, a transparency window is evident in the frequency dependence of the absorption spectrum. However, increasing the value of Ω_3 quickly leads to closure of this window, an effect that

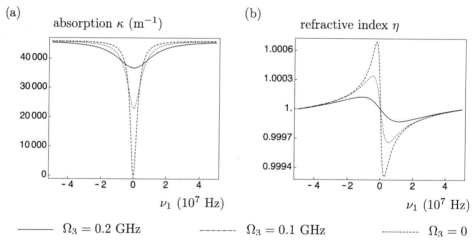

(a) absorption κ (m^{-1}) (b) refractive index η

ν_1 (10^7 Hz) ν_1 (10^7 Hz)

——— $\Omega_3 = 0.2$ GHz   - - - - - - $\Omega_3 = 0.1$ GHz   ·········· $\Omega_3 = 0$

Fig. 10.6. (a) Absorption coefficient and (b) refractive index for a gas of atoms with the four-level structure displayed in Fig. 10.5. The relevant parameters are the same as for the three-level figure with $\Omega_2 = 0.1$ GHz, i.e. $n = 10^{19}$ m^{-3}, $r_{eg} \cdot \epsilon = 0.1$ nm, $G_{eg}' = 1$ GHz, $G_{ig}' = 0$, $\omega_0 = 5 \times 10^{15}$ Hz, $\nu_2 = 0$. The effect of increasing the coupling parameter Ω_3 of a resonant ($\nu_3 = 0$) laser in the third arm is displayed. The new decay rate G_{fg}' is set equal to G_{eg}'.

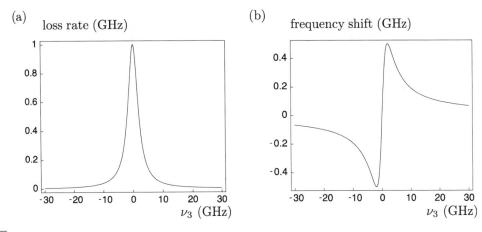

Fig. 10.7. (a) Amplitude loss rate and (b) energy shift for a photon interacting with a gas of atoms with the four-level structure displayed in Fig. 10.5. The photon is resonant with the transition connecting states $|0\rangle$ and $|e\rangle$ and has effective Rabi coupling $\tilde{\Omega}_{n_1=1} = 0.2$ GHz. A classical laser, with coupling strength $\Omega_2 = 0.2$ GHz resonantly couples states $|e\rangle$ and $|1\rangle$. A second single photon, with $\Omega_{n_3=1} = \Omega_{n_1=1}$ is also present and couples $|1\rangle$ and $|f\rangle$. All of the transitions are weakly driven, so that $G_{e0} = G_{f0} = 1$ GHz. Only 100 atoms are assumed to be present in the ensemble. The plots display the dependence on ν_3, the detuning of the second single photon from resonance.

can be exploited in optical switching. A recent experimental demonstration of the closure of the window is described in Bason *et al.* (2009).[1]

More significant is the effect of taking the third laser Ω_3 off resonance. The loss rate and the frequency shift are displayed in Fig. 10.7. The transparency is quickly restored by increasing ν_3, but the refractive index does not return immediately to unity. One would therefore expect the third laser to induce a significant change in the velocity of the probe beam, without any substantial absorption. A change in velocity is translated into a change in the phase of the probe beam.

Can we describe the interaction of the modes (call them a_1 and a_3) that support the classical fields Ω_1 and Ω_3 as a quantum *cross-Kerr* nonlinearity for two photons? In order to answer this question, we must again switch to a quantum description of fields 1 and 3, while assuming that the field in mode a_2 remains in a classical (or coherent) state. The number of quanta in field mode a_1 will be n_1, while a_3 contains n_3 photons. For simplicity, let us take $\nu_1 = \nu_2 = 0$, and $G_{ig} = 0$. Using the results of the discussion of Section 10.1, we can write the steady-state value of the coherence $\tilde{\rho}_{eg}$ in this quantized field case as

$$\tilde{\rho}_{eg} = -\frac{N}{2} \frac{i\tilde{\Omega}_{n_1}|\tilde{\Omega}_{n_3}|^2}{|\Omega_2|^2(G_{fg} + i\nu_3) + G_{eg}|\tilde{\Omega}_{n_3}^2|}, \tag{10.77}$$

[1] In this study, the authors in fact observe a very narrow absorption line as a function of probe detuning, much narrower than is predicted in the theory we present here. The reason for this is the Doppler effect, which means that atoms travelling with different velocities with respect to the light beam have a modified detuning.

where $\tilde{\Omega}_{n_i} = g_i \sqrt{n_i}/\hbar$. Equations (10.42) and (10.44) allow us to relate $\tilde{\rho}_{eg}$ to a quantity κ, whose real and imaginary parts correspond to the absorption rate and frequency shift of the field, respectively.

$$\kappa = -2\tilde{\rho}_{eg}\tilde{\Omega}_{n_1} = \frac{N}{2} \frac{i|\tilde{\Omega}_{n_1}|^2 |\tilde{\Omega}_{n_3}|^2}{|\Omega_2|^2(G_{fg} + i\nu_3) + G_{eg}|\tilde{\Omega}_{n_3}^2|}. \tag{10.78}$$

The values of κ are shown in Fig 10.7. Importantly, we find that significant energy shifts can be generated, even when the rate of field absorption is rather small. Moreover, the form in Eq. (10.78) shows that the interaction between photons in modes a_1 and a_3 is of a cross-Kerr form, since the energy shift is directly proportional to the product of n_1 and n_3. The reduced Hamiltonian for modes a_1 and a_3 then assumes the familiar form

$$\mathcal{H}_K = \hbar\kappa \, \hat{a}_1^\dagger \hat{a}_1 \, \hat{a}_3^\dagger \hat{a}_3, \tag{10.79}$$

that is, the four-level atomic ensemble can be used to construct a cross-Kerr nonlinearity. Moreover, the self-phase modulation due to terms $(\hat{a}_1^\dagger \hat{a}_1)^2$ and $(\hat{a}_3^\dagger \hat{a}_3)^2$ in the Hamiltonian is strongly suppressed.

One might think that the cross-Kerr nonlinearity discussed here could be used to entangle pairs of photons. Indeed, the size of the parameter κ would certainly not preclude this possibility. However, we have not taken into account the time dependence of any photon wave packet. This turns out to be critical: as we show in Appendix 3, it is *impossible* to produce a *strong* cross-Kerr nonlinearity under any circumstances. Nonetheless, the nonlinearity here can be used to build up small phases between photons and this can be used in a practical way, as we discuss in Section 10.5.

10.3 Quantum memories and quantum repeaters

We have seen that atomic ensembles can be used to store a pulse of light. In this section, we explore the possibilities this behaviour offers for quantum information processing. In particular, we will consider atomic ensembles as single-photon memories, and we discuss the possibility of creating quantum repeaters for photonic quantum communication.

10.3.1 Quantum memories

We return to an ensemble of N three-level atoms, shown in Fig. 10.8. Instead of the two classical fields discussed in Section 10.2.1, let us imagine that the probe field is in fact a single-mode quantum field, initially populated by a single photon. We follow the methods presented in Section 10.2 for deriving the Hamiltonian of an ensemble of N atoms, and for converting classical fields into quantum fields. Recall that the symmetric collective excitation states involving levels $|i\rangle$ and $|e\rangle$ are $\left|W_g^i\right\rangle$ and $\left|W_g^e\right\rangle$, respectively. In the extended

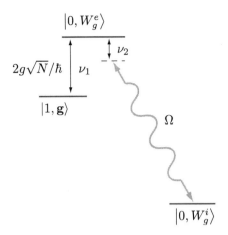

Fig. 10.8. Energy levels and couplings for the quantum-memory protocol. One arm of the three-level system of each atom is now coupled to a quantized field mode, while the other is addressed by a classical laser of coupling Ω. We depicted the effective level scheme before a transformation is carried out to move the system to a frame rotating with the classical laser frequency. Following this and a rotating-wave approximation, the Hamiltonian is given by Eq. (10.80).

basis $\{|1, \mathbf{g}\rangle, \left|0, W_g^i\right\rangle, \left|0, W_g^e\right\rangle\}$, where the first entry in the ket refers to the number of photons in the mode, we can write down the Hamiltonian

$$\mathcal{H} = \mathcal{H}_\Lambda = \frac{\hbar}{2} \begin{pmatrix} 0 & 0 & 2g\sqrt{N}/\hbar \\ 0 & 2(\nu_1 - \nu_2) & \Omega \\ 2g\sqrt{N}/\hbar & \Omega & 2\nu_1 \end{pmatrix}. \tag{10.80}$$

In Chapter 7, we found an essentially identical STIRAP Hamiltonian for a single atom, and described how it always has an eigenstate of the form

$$|\psi(\theta, 1)\rangle \equiv \cos\theta\, |1, \mathbf{g}\rangle - \sin\theta\, \left|0, W_g^i\right\rangle \tag{10.81}$$

when $\nu_1 = \nu_2$. We have defined

$$\tan\theta = \frac{2g\sqrt{N}}{\hbar\Omega}. \tag{10.82}$$

The state $|\psi(\theta, 1)\rangle$ does not have any $\left|0, W_g^e\right\rangle$ component and is therefore immune to spontaneous emission from the level $|e\rangle$. Furthermore, $|\psi(\theta, 1)\rangle$ is a coherent superposition of an excitation in the quantum field and an excitation in the atomic ensemble. It is often called a 'dark state polariton'. These can be interpreted as quasi-particles, and are of utmost importance for single-photon quantum memories.

Imagine that initially we have $\Omega \gg g\sqrt{N}/\hbar$, i.e., we arrange our system such that a strong control laser is present when our single photon enters the ensemble. The system is in the eigenstate $|1, \mathbf{g}\rangle$. The intensity of the control laser is adiabatically reduced, and the system follows state $|\psi(\theta, 1)\rangle$ until it eventually becomes $\left|0, W_g^i\right\rangle$. We have now transferred a single photon into a single excitation of the atomic ensemble! Moreover, this excitation is quite stable, since it only involves a collective mode of the metastable atomic state $|i\rangle$ with a suppressed decay rate (see Fig. 10.3). Second, the ensemble is in a W state $\left|W_g^i\right\rangle$, which has the desirable property that decoherence of a single atom hardly affects the fidelity of the state of the ensemble. Notice also that the energy of the photon no longer resides in the atomic ensemble. It has been carried away coherently by the classical laser field Ω.

The stored photon can be recovered at any time by increasing the control laser intensity once again. Adiabatic switching means that state $|\psi(\theta, 1)\rangle$ is followed back to $|1, \mathbf{g}\rangle$. In principle, the spatiotemporal mode shape of the outgoing photon can be made identical to that of the incoming photon, since both processes couple to the same optical mode. This is essential if the photons are subsequently used in second-order interference experiments, such as the quantum information processing described in Chapter 6. If the photon carries a qubit degree of freedom, two quantum memories must be used to store the qubit value: one for the mode associated with qubit state $|0\rangle$, and one for the mode associated with $|1\rangle$.

The quantum-memory protocol can be extended to Fock states with more than one photon. We must first define a series of multiply excited collective states. For example, a two-excitation symmetric state involving atomic state $|e\rangle$ is

$$\left|W_g^{2e}\right\rangle = \sqrt{\frac{2}{N(N-1)}} \sum_{j=1}^{N} \sum_{k>j}^{N} |g_1, g_2, \ldots, e_j, \ldots, e_k, \ldots, g_N\rangle. \qquad (10.83)$$

There are $\frac{1}{2}N(N-1)$ different states with the form in the summation, hence the normalization factor. A similar state $\left|W_g^{2i}\right\rangle$ can of course be defined for two collective excitations of the atom state $|i\rangle$. A combination involving one excitation of $|i\rangle$ and one of $|e\rangle$ will also be accessible, and the distinguishability of these two states leads to a slightly modified definition:

$$\left|W_g^{e+i}\right\rangle = \sqrt{\frac{1}{N(N-1)}} \sum_{j=1}^{N} \sum_{k \neq j}^{N} |g_1, g_2, \ldots, e_j, \ldots, i_k, \ldots, g_N\rangle. \qquad (10.84)$$

Starting with a two-photon Fock state, and all atoms in their ground states, the levels and coupling scheme are as shown in Fig. 10.9. The coupling matrix elements are found using the definition of the states given in the above equations, together with the Jaynes–Cummings Hamiltonian and properties of bosonic operators. The Jaynes–Cummings coupling Hamiltonian for N atoms interacting with an extended field mode a is

$$\mathcal{H}_{JC} = \sum_{j} [g \hat{a} \sigma_j^+ + g^* \hat{a}^\dagger \sigma_j^-], \qquad (10.85)$$

where σ_j^+ and σ_j^- are the raising and lowering operators connecting the two states $|g\rangle$ and $|e\rangle$ on atom j.

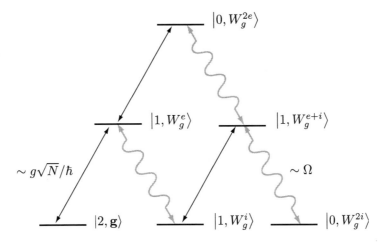

Fig. 10.9.   Energy levels and couplings for the quantum-memory protocol.

Exercise 10.2: Show that the matrix element of $\mathcal{H}_{JC}$ between states $|2, \mathbf{g}\rangle$ and $\left|1, W_g^e\right\rangle$ is $\langle 2, \mathbf{g}|\, \mathcal{H}_{JC}\, |1, W_g^e\rangle = g\sqrt{2N}$. Similarly, show that

$$\langle 1, W_g^e|\, \mathcal{H}_{JC}\, |0, W_g^{2e}\rangle = g\sqrt{2(N-1)}. \tag{10.86}$$

The full Hamiltonian for the six levels displayed in Fig. 10.9 consists of six coupling terms, which are most clearly written in the basis

$$\{|2, \mathbf{g}\rangle,\, \left|1, W_g^i\right\rangle,\, \left|0, W_g^{2i}\right\rangle,\, \left|1, W_g^e\right\rangle,\, \left|0, W_g^{e+i}\right\rangle,\, \left|0, W_g^{2e}\right\rangle\},$$

yielding the Hamiltonian

$$\mathcal{H} = \begin{pmatrix} 0 & 0 & 0 & g\sqrt{2N} & 0 & 0 \\ 0 & 0 & 0 & \Omega/2 & g\sqrt{N-1} & 0 \\ 0 & 0 & 0 & 0 & \Omega/\sqrt{2} & 0 \\ g\sqrt{2N} & \Omega/2 & 0 & \hbar\nu & 0 & g\sqrt{2(N-1)} \\ 0 & g\sqrt{N-1} & \Omega/\sqrt{2} & 0 & \hbar\nu & \Omega/2 \\ 0 & 0 & 0 & g\sqrt{2(N-1)} & \Omega/2 & \hbar\nu \end{pmatrix}, \tag{10.87}$$

where we set $\nu_1 = \nu_2 \equiv \nu$. By inspection, such a Hamiltonian must have one eigenvector, with eigenvalue zero, that is a linear combination of the first three levels in our basis set, i.e., of the form $c_1 |2, \mathbf{g}\rangle + c_2 \left|1, W_g^i\right\rangle + c_3 \left|0, W_g^{2i}\right\rangle$. The coefficients can be found by multiplying the matrix above with the general form of the expected eigenvector, leading to two simultaneous equations:

$$\begin{pmatrix} g\sqrt{2N} & \Omega/2 & 0 \\ 0 & g\sqrt{N-1} & \Omega/\sqrt{2} \end{pmatrix} \begin{pmatrix} c_1 \\ c_2 \\ c_3 \end{pmatrix} = 0. \tag{10.88}$$

This yields the dark state

$$|\psi(\theta, 2)\rangle = \cos^2\theta\, |2, \mathbf{g}\rangle - \sqrt{2}\sin\theta\cos\theta\, \left|1, W_g^i\right\rangle + \sin^2\theta\, \left|0, W_g^{2i}\right\rangle, \qquad (10.89)$$

with θ given by Eq. (10.82). We see that by varying θ in the same way as before it is possible to move a state of two photons into a state of two excitations of the atomic ensemble, and vice versa.

Indeed, such an argument can be extended to any number of photons, as long as it is less than the total number of atoms in the ensemble. There are a family of dark states corresponding to different numbers of excitations n, which take the form:

$$|\psi(\theta, n)\rangle = \sum_{k=0}^{n} \sqrt{\frac{n!}{k!(n-k)!}}(-1)^k \sin^n\theta\cos^{n-k}\theta\, \left|n-k, W_g^{ki}\right\rangle, \qquad (10.90)$$

where $|W_g^{ki}\rangle$ is the symmetric superposition of k excited atoms in the state $|i\rangle$, with the remaining atoms in the ground state $|g\rangle$. Superpositions of different numbers of Fock states can also be converted into superpositions of different numbers of atomic excitations, since the dynamics of each component exists in an independent Hilbert subspace, and θ does not depend on n. Even a travelling wavepacket can be stopped and stored in a gas of atoms, as long as the amplitude of the packet does not change so quickly that adiabaticity conditions are violated. This means that the atomic ensemble of identical three-level atoms can also be used as a memory for the optical qunats discussed in Chapters 8 and 9.

10.3.2 Quantum repeaters

In practical quantum-communication protocols using optical fibres, the losses put a limit on the attainable distance for quantum communication. This is due to the fact that quantum information cannot be cloned, and we therefore cannot amplify the single-photon signal without losing the quantum mechanical behaviour of the information carriers. One way to circumvent this is to construct a 'quantum repeater' using the quantum memories discussed above. Our aim is to first create high-fidelity maximal entanglement between two atomic ensembles. In particular, we distribute a single excitation coherently over two ensembles A_1 and A_2. We repeat this procedure with two other ensembles B_1 and B_2, and then use entanglement swapping between A_2 and B_1 to create entanglement in A_1 and B_2. Assuming that A_1 and A_2 are a distance L apart, and so are B_1 and B_2, we can in principle create entanglement spanning a distance $2L$. We will find, however, that the entangling procedure is probabilistic, and some care must be taken in characterizing the scaling of the quantum repeater.

We return once more to the three-level system shown in Fig. 10.8, except that here the quantum and classical fields are exchanged. In Chapter 7, we discussed Raman transitions and described how effective Rabi oscillations can be induced between two uncoupled levels through virtual optical transitions with an intermediate level. The effective Rabi frequency is proportional to the product of coupling matrix elements between the intermediate state and each of the other two states, and inversely proportional to the detuning of both coupling lasers. We have exactly this situation in our atomic-ensemble system. When the ensemble is initially in the ground state with zero photons, we expect oscillations between levels $|0, \mathbf{g}\rangle$ and $|1, W_g^i\rangle$ with the effective Rabi frequency given by

$$\Omega_R = \frac{g\sqrt{N}\Omega}{\hbar \nu}. \tag{10.91}$$

The effective Hamiltonian for the system becomes

$$\mathcal{H} = \Omega_R |0, \mathbf{g}\rangle \left\langle 1, W_g^i \right| + \text{H.c.}. \tag{10.92}$$

If a short pulse of the classical coupling laser with a duration τ is applied to the ensemble, the state becomes

$$|\phi\rangle = |0, \mathbf{g}\rangle + i\sqrt{p_c} \left| 1, W_g^i \right\rangle + O(p_c), \tag{10.93}$$

where $p_c \ll 1$ is the excitation probability

$$\sqrt{p_c} = \frac{g\sqrt{N}\Omega\tau}{\hbar \nu}. \tag{10.94}$$

The terms of order p_c in $|\phi\rangle$ are components with more than one excitation, but we will assume that these are negligible. Two ensembles A_1 and A_2 that are similarly excited then have the joint state $|\phi_1\rangle \otimes |\phi_2\rangle$.

The creation of entanglement between A_1 and A_2 is closely related to the distributed schemes for entanglement by measurement discussed in Chapter 7. Here, we follow the weak driving protocol, but any other entangling protocol can also be adapted for our purposes. Again, the operation depends on measuring the photons emitted from the ensembles. We can assume that the ensembles are contained in optical cavities[2] with leakage rates κ_1 and κ_2. The paths of emitted photons are arranged such that they meet on a beam splitter, and are detected by a pair of detectors D_1 and D_2 (see Fig. 10.10). The photons that trigger these detectors are related to the two input modes by (see Section 7.4)

$$\hat{b}_1^\dagger = \frac{i\hat{a}_1^\dagger + \hat{a}_2^\dagger}{\sqrt{2}} \quad \text{and} \quad \hat{b}_2^\dagger = \frac{\hat{a}_1^\dagger + i\hat{a}_2^\dagger}{\sqrt{2}}. \tag{10.95}$$

We can use the quantum-jump formalism developed in Section 7.4 to calculate which atomic-ensemble states exist following a single click in either detector. The relevant jump operators are $\mathcal{J}_1(\cdot) = \hat{b}_1(\cdot)\hat{b}_1^\dagger$ and $\mathcal{J}_2(\cdot) = \hat{b}_2(\cdot)\hat{b}_2^\dagger$. Following a jump (i.e., a detection event) we may therefore describe the system in the pure state

$$|\Psi^\pm\rangle_{A_1 A_2} = \frac{1}{\sqrt{2}} \left(|\mathbf{g}\rangle_{A_1} \left| W_g^i \right\rangle_{A_2} \pm i \left| W_g^i \right\rangle_{A_1} |\mathbf{g}\rangle_{A_2} \right) + O(\sqrt{p_c}), \tag{10.96}$$

where the relative phase is determined by which detector clicks. The probability of obtaining a click in any given round is, of course, only p_c. Nonetheless, with a small p_c a high-fidelity

---

[2] This is not a necessary condition, but it increases the photon-collection efficiency.

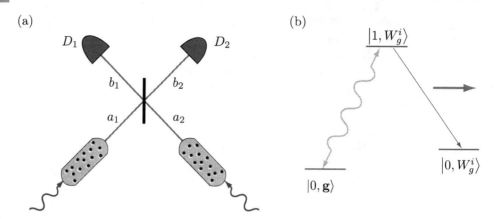

(a) Schematic diagram for the experimental set-up required to entangle two atomic ensembles. (b) Diagram of the relevant energy levels in each ensemble.

entangled state of the two ensembles results from a successful operation. Moreover, the entangled state is again quite stable since it involves the metastable states $|i\rangle$, rather than the excited state $|e\rangle$.

Using an atomic ensemble, as opposed to a single atom in a cavity, for this kind of entanglement-generation scheme makes sense when one calculates the proportion of photons emitted into the monitored cavity mode, as opposed to spontaneously emitted photons into other environmental modes. We can easily work out this figure of merit using methods which we developed earlier in this chapter. We append the state of a third system, namely the environmental mode, to our basis of the ensemble and the optical mode. We now have an effective three-level system. Cavity decay may now be modelled using the master equation given in Eq. (10.1), with Lindblad operator $L_m = |0, W_g^i, 1\rangle \langle 1, W_g^i, 0|$ and corresponding rate κ. We analyzed a system with exactly this form before, when we discussed the interaction of a quantized field with a two-level system. We may therefore carry over the results we derived in Eqs. (10.37) to (10.41), making sure that the relevant parameters take on their new definitions. If we take the 'bad cavity limit', in which $\kappa \gg \Omega_1$, we find that the population P ending up in state $|0, W_g^i, 1\rangle$ is

$$P = 1 - \exp(-\Gamma \tau) \approx \Gamma \tau, \tag{10.97}$$

with

$$\Gamma = \frac{8|\Omega_R|^2}{\kappa} = \frac{8g^2 N |\Omega|^2}{\hbar^2 \nu^2 \kappa}. \tag{10.98}$$

On the other hand, the number of spontaneously emitted photons is given by the probability Q that the collective state $|W_g^i\rangle$ is populated during the short pulse, multiplied by the optical decay rate γ, that is,

$$Q = \frac{|\Omega|^2 \gamma \tau}{\nu^2}. \tag{10.99}$$

The figure of merit is then

$$\frac{P}{Q} = \frac{8g^2 N}{\hbar^2 \kappa \gamma}.$$ (10.100)

The improved figure of merit for larger N is a consequence of 'collective enhancement' and achieving this is the essential purpose of using atomic ensembles for quantum processing, rather than single atoms in a cavity.

So far, we have shown how to create entanglement between two ensembles A_1 and A_2. The entanglement procedure involved a path-erasure procedure for photons, and so the limiting distance between ensembles over which the procedure works is determined by photon loss. Let L be this maximum distance between A_1 and A_2. There is a way of increasing this maximum distance over which entanglement can be generated, by first creating a second pair of entangled ensembles B_1 and B_2, and then using the principle of entanglement swapping to create entanglement between A_1 and B_2. The joint state of the four ensembles before swapping is $\left|\Psi^\pm\right\rangle_{A_1 A_2} \otimes \left|\Psi^\pm\right\rangle_{B_1 B_2}$. The swap can be performed by again exploiting the principle of photon-path erasure.

First the atomic excitation states $|W_g^i\rangle$ in A_2 and B_1 are excited by a classical laser, slightly detuned from the $|W_g^e\rangle$ level. The quantum field then couples back to $|g\rangle$. In effect, by reversing the quantum and classical fields, any atomic-ensemble excitation is converted back into a photon that can be measured. Emitted photons from A_2 and B_1 are again passed through a beam splitter, thus removing the 'which path?' information that they contain. We leave it as an exercise below to show that an entangled state between A_1 and B_2 results from measuring a single photon in either detector placed downstream of the beam splitter. Since a successful entanglement swap depends on detecting a single photon, this is also a probabilistic procedure, and successful generation of long-range entanglement also requires the techniques developed in Chapters 5, 6, and 7.

Exercise 10.3: Show that the above protocol can indeed swap the entanglement from pairs $\{A_1, A_2\}$ and $\{B_1, B_2\}$ to entanglement in the pair $\{A_1, B_2\}$.

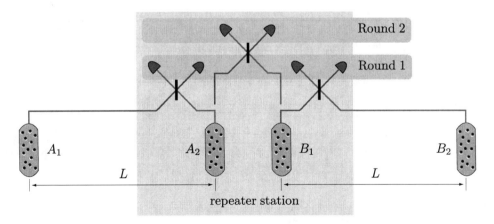

Fig. 10.11. Two rounds of path erasure probabilistically entangles ensemble A_1 with B_2 over an extended distance $2L$.

Once long-range entanglement is established between distant ensembles, it is possible to perform a variety of quantum-communication tasks. In particular, both quantum cryptography and quantum teleportation, which were introduced in Chapters 5 and 8, can be performed efficiently. Notice that this quantum repeater works for the single-rail encoding, i.e., when the degree of freedom is in the photon number. We can in fact double the size of the repeater to accommodate the dual-rail encoding. In the next section, we move on to consider how an atomic ensemble can be used as a single qubit.

10.4 The atomic ensemble as a single qubit

In the previous section, we demonstrated how to entangle a pair of atomic ensembles. We used the method of weak excitation to prevent the creation of more than one collective excitation, and so restrict the accessible Hilbert space of the system to that of a two-level system. We could therefore regard the states $|\mathbf{g}\rangle$ and $|W_g^i\rangle$ as a single qubit. This representation of a qubit has rather attractive properties, including fast manipulation through collective enhancement and long coherence times (since the state $|W_g^i\rangle$ is a collective mode of metastable states). However, the weak excitation method for preventing departures from the restricted Hilbert space greatly constrains the family of single-qubit operations that can be performed in one step. In this section, we will discuss a particular type of excited atom, the Rydberg state, and show how this difficulty can be overcome.

Rydberg states are highly excited electronic states of an atom. Such states resemble atomic hydrogen, since one electron occupies a state with very high principal quantum number and the other electrons screen the nucleus, giving it an effective charge of one. The excited electron also has a very large orbit, and the transition dipole matrix element p_{ab} between two different Rydberg states $|a\rangle$ and $|b\rangle$ is therefore much larger than that of transitions between lower-lying levels. Such large transition dipoles can lead to significant atom–atom interactions. To see this, consider the situation shown in Fig. 10.12, which shows the energy level structure of two atoms j and k, each with Rydberg states $|a\rangle$, $|r\rangle$, and $|b\rangle$. The wave functions of each Rydberg state are such that of the three states, only $|r\rangle$ has allowed dipole transitions to the low-lying storage levels $|g\rangle$ and $|i\rangle$. In addition, $|a\rangle$ and $|r\rangle$ are coupled, with a large transition dipole element, as are $|r\rangle$ and $|b\rangle$. Moreover, the levels $|a\rangle$ and $|b\rangle$ are symmetrically disposed about $|r\rangle$ so that the joint state $|r\rangle_j |r\rangle_k$ is resonant with $|a\rangle_j |b\rangle_k$ and $|b\rangle_j |a\rangle_k$. Such a scenario leads to resonant energy transfer between the atoms, mediated by the dynamic dipole–dipole interaction that leads to microwave photon exchange. It has much in common with the Förster interaction, which we will introduce in Chapter 11. The coupling strength depends on the distance R_{jk} between each atom pair and is given by $\kappa_{jk} = p_{ar}p_{rb}R_{jk}^{-3}$. The relevant Hamiltonian for the whole ensemble is

$$\mathcal{H} = \hbar \sum_{k=1}^{N} \sum_{j>k} \kappa_{jk} |r\rangle_j |r\rangle_k \left(\langle a|_j \langle b|_k + \langle b|_j \langle a|_k \right) + \text{H. c.}. \tag{10.101}$$

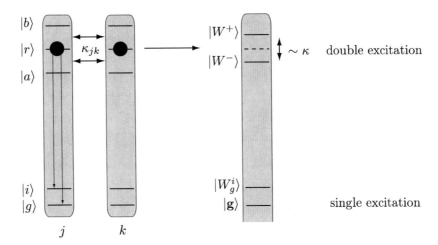

Fig. 10.12. The dipole blockade mechanism. Left: if two atoms j and k are excited into a Rydberg level $|r\rangle$ then the large spatial wavefunction leads to strong interactions κ. Right: such interactions shift the two-level eigenstates such that the optically active states, here shown as solid lines, are shifted.

Consider now the doubly excited states $|W_g^{2r}\rangle$, which are defined in an analogous way to those introduced in Eq. (10.83)

$$\left|W_g^{2r}\right\rangle = \sqrt{\frac{2}{N(N-1)}} \sum_{j=1}^{N} \sum_{k>j}^{N} \left|g_1, g_2, \ldots, r_j, \ldots, r_k, \ldots, g_N\right\rangle. \tag{10.102}$$

In the presence of the interaction Hamiltonian of Eq. (10.101), the terms in the superposition $|W_g^{2r}\rangle$ are no longer valid eigenstates. Therefore $|W_g^{2r}\rangle$ may no longer be regarded as an addressable energy level. Rather, the correct collective states must become a superposition of eigenstates of the Hamiltonian in Eq. (10.101), namely

$$\left|W_g^{a+b}\right\rangle \propto \sum_{j=1}^{N} \sum_{k>j} \left(\left|g_1, \ldots, a_j, b_k, \ldots, g_N\right\rangle + \left|g_1, \ldots, b_j, a_k, \ldots, g_N\right\rangle\right),$$

$$\left|W_g^{a-b}\right\rangle \propto \sum_{j=1}^{N} \sum_{k>j} \left(\left|g_1, \ldots, a_j, b_k, \ldots, g_N\right\rangle - \left|g_1, \ldots, b_j, a_k, \ldots, g_N\right\rangle\right),$$

$$\left|W^{\pm}\right\rangle \equiv \frac{1}{\sqrt{2}} \left(\left|W_g^{2r}\right\rangle \pm \left|W_g^{a+b}\right\rangle\right), \tag{10.103}$$

where the normalization of the first two states is $[2/N(N-1)]^{\frac{1}{2}}$. The state $|W_g^{a-b}\rangle$ has no coupling to the storage levels of the ensemble, so we need not consider it further. On the other hand, the levels $|W^{\pm}\rangle$ will couple to the ground state $|\mathbf{g}\rangle$ and the state $|W_g^i\rangle$. However, the dipole–dipole interaction shifts their energies by an amount $\pm\hbar\kappa\sqrt{2}$. Therefore, the collective states involving two excitations must also necessarily have an energy shift of order $\pm p_{ar}p_{rb}/V$, where V is the volume of the ensemble. As a consequence, a laser tuned to the transition between the ground state $|\mathbf{g}\rangle$ and the single-excitation Rydberg state $|W_g^r\rangle$ cannot create more than one excitation, since the first excitation will move the collective state off resonance. This is called the 'dipole blockade effect', and it requires that the collectively enhanced driving strength $\hbar\Omega\sqrt{N}$ is smaller than the energy shift $\hbar\kappa\sqrt{2}$.

The dipole blockade allows for the preparation and storage of an arbitrary single-qubit state in an atomic ensemble. The qubit states are defined by

$$|0\rangle \equiv |\mathbf{g}\rangle \quad \text{and} \quad |1\rangle \equiv \left|W_g^i\right\rangle. \tag{10.104}$$

If a laser is tuned to the transition between $|\mathbf{g}\rangle$ and $|W_g^r\rangle$, the dipole blockade means that the quantum dynamics is restricted to this two-level system, and varying the length of the laser pulse allows the preparation of any superposition of the two levels. Furthermore, since Rydberg states are highly excited and fragile states, once a qubit state has been prepared, we drive the system from the state $|W_g^r\rangle$ to the qubit state $|W_g^i\rangle$ in such a way that $|\mathbf{g}\rangle$ remains unaffected. This can be accomplished with a π pulse on the transition between $|W_g^r\rangle$ and $|W_g^i\rangle$.

Any single-qubit operation is also easily implemented using this method. The qubit is first transferred from the qubit state $|1\rangle$ to the Rydberg level $|W_g^r\rangle$, while leaving the state $|0\rangle$ untouched. Single-qubit manipulations are then carrried out by varying the length and phase of the laser that is resonant with the transition between $|\mathbf{g}\rangle$ and $|W_g^r\rangle$.

Quantum computing with atomic ensembles

A number of proposals exist for scaling up from a single qubit to the many qubits needed to perform quantum algorithms. This is very much an active area of research and we will only briefly touch on some of the main ideas.

First, an ensemble of atoms may have a reasonably large number p of stable levels, which can be used as a small-scale quantum information processor. A single system with p levels can generally represent only $\log_2 p$ qubits, but in an ensemble the encoding can be much more efficient. To see this, let us label the collective state associated with one atom in level $|u\rangle$ with $u = \{1, \ldots, p\}$, and all others in the ground level $|g\rangle$, as $|W_g^u\rangle$. It is also possible to define symmetric states involving more excitations. For example, one atom in the state $|u\rangle$, another in $|v\rangle$ (with $v = \{1, \ldots, p\}$), and all others in the ground state $|g\rangle$ can be written as the collective state

$$\left|W_g^{u+v}\right\rangle = \sqrt{\frac{1}{N(N-1)}} \sum_{j=1}^{N} \sum_{k\neq j}^{N} |g_1, \ldots, u_j, v_k, \ldots, g_N\rangle. \tag{10.105}$$

Naturally, so long as the number of atoms exceeds the number of levels in one atom, this argument can be extended to anything up to $p-1$ different excitations. We might therefore express a general state as $|W_g^{\vec{n}}\rangle$, with $\vec{n}$ the $(p-1)$-tuple $(n_1, \ldots, n_u, \ldots, n_v, \ldots, n_{p-1})$. For each n_u we have $n_u \in 0, 1$, which means that the level u is occupied ($n_u = 1$) or not ($n_u = 0$). The state in Eq. (10.105), for example, would have $n_u = n_v = 1$, and all other n_i zero. With this encoding, the number of available qubits becomes $p - 1$. Using the dipole

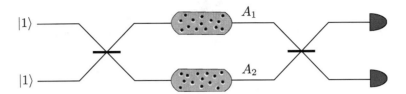

Fig. 10.13. A pair of atomic ensembles can be entangled using a Mach–Zehnder interferometer with single-photon input states.

blockade effect, both single- and two-qubit operations are possible in this system. For more details, see Brion *et al.* (2007).

Second, atomic ensembles can be used to generate cluster states efficiently. To this end, we need to construct a robust entangling procedure. The essence of the idea is summarized in Fig. 10.13. Each atomic ensemble has a pair of storage levels $|\mathbf{g}\rangle = |0\rangle$ and $|W_g^i\rangle = |1\rangle$. Furthermore, $|\mathbf{g}\rangle$ is coupled to a Rydberg state $|W_g^r\rangle$ via a quantized field mode. The entangling procedure now works as follows: the ensembles may be prepared in their ground states $|\mathbf{g}\rangle$. Two identical single photons $|1\rangle$ are incident on a 50:50 beam splitter. The Hong–Ou–Mandel effect ensures that the output state is $i(|20\rangle + |02\rangle)/\sqrt{2}$. Each output mode of the beam splitter interacts with an atomic ensemble. When there are two photons in the mode, one photon will create a Rydberg excitation, and the dipole blockade mechanism renders the ensemble transparent to the second photon. The joint state of the ensemble–photon system therefore becomes

$$\frac{i}{\sqrt{2}} \left(\left| \mathbf{g}, W_g^r \right\rangle |0, 1\rangle + \left| W_g^r, \mathbf{g} \right\rangle |1, 0\rangle \right). \tag{10.106}$$

The second beam splitter induces path erasure of the transmitted photons, and the joint state of the ensemble–photon system becomes

$$\frac{1}{2} \left(\left| \mathbf{g}, W_g^r \right\rangle + \left| W_g^r, \mathbf{g} \right\rangle \right) |1, 0\rangle + \frac{1}{2} \left(\left| \mathbf{g}, W_g^r \right\rangle - \left| W_g^r, \mathbf{g} \right\rangle \right) |0, 1\rangle. \tag{10.107}$$

After detection of a single photon, and driving the Rydberg atom in $|r\rangle$ to the storage state $|i\rangle$, the ensembles are in the Bell state

$$\left| \Psi^\pm \right\rangle = \frac{1}{\sqrt{2}} \left(|01\rangle \pm |10\rangle \right). \tag{10.108}$$

This scheme can be scaled up to generate larger cluster states by appropriate measurements, described by Zwierz and Kok (2009). Moreover, it can be made robust against detector inefficiencies and photon losses using the techniques presented in Chapters 6 and 7. An alternative scheme is presented in Barrett *et al.* (2008).

10.5 Photon–photon interactions via atomic ensembles

In Section 10.2, we saw that the atomic ensemble can be used to implement a cross-Kerr nonlinearity between two modes a_1 and a_2, governed by the interaction Hamiltonian $\mathcal{H}_K = \hbar \kappa \, \hat{a}_1 \hat{a}_1^\dagger \hat{a}_2 \hat{a}_2^\dagger$. Furthermore, we have seen in Section 6.2 that such an interaction, if sufficiently strong, can be used in a coherent photon switch. In turn, this switch can be reformulated as a CZ gate for photonic qubits. Unfortunately, due to the finite spatiotemporal structure of the photonic wavepackets, a large Kerr nonlinearity unavoidably induces noise, which reduces the fidelity of the CZ gate (see Appendix 3). A Kerr-mediated photon–photon interaction therefore cannot be used for coherent quantum information processing (see again Appendix 3). However, instead of a direct interaction between photons, we may use weaker cross-Kerr nonlinearities to couple single photons to a macroscopic field mode. In this section we will study such mediated photon interactions.

10.5.1 Weak Kerr nonlinearities

We investigate how we can create a deterministic two-photon entangling operation by coupling each photon to an auxiliary mode in a coherent state. Before we construct the full entangling gate, we will describe the core interaction between a single photon and a coherent state, by means of a weak interaction. Using $|\alpha\rangle_0$ as the auxiliary bright coherent state in mode a_0, we write

$$
|1\rangle_1 |\alpha\rangle_0 = e^{-|\alpha|^2/2} \sum_{k=0}^{\infty} \frac{\alpha^k \hat{a}_0^{\dagger k}}{k!} \hat{a}_1^\dagger |0,0\rangle_{10}
$$

$$
\rightarrow e^{-|\alpha|^2/2} \sum_{k=0}^{\infty} \frac{\alpha^k (\hat{a}_0^\dagger e^{i\theta \hat{n}_1})^k}{k!} \hat{a}_1^\dagger e^{i\theta \hat{n}_0} |0,0\rangle_{10}
$$

$$
= e^{-|\alpha|^2/2} \sum_{k=0}^{\infty} \frac{(\alpha e^{i\theta})^k \hat{a}_0^{\dagger k}}{k!} |1,0\rangle_{10} = |1\rangle_1 |\alpha e^{i\theta}\rangle_0 , \qquad (10.109)
$$

where $\theta \ll \pi$ is the interaction strength of the cross-Kerr nonlinearity. Note that the interaction appears as a phase shift of the coherent state, rather than a global phase shift on the product state.

Now that we have established the core interaction, we will use this to create a deterministic parity projection for two photonic qubits. Consider two single-photon polarization qubits, in modes 1 and 2, each of which are coupled to the auxiliary coherent mode $|\alpha\rangle_0$ with coupling strengths θ_1 and θ_2, respectively. The input state is given by

$$
(c_{00} |00\rangle + c_{01} |01\rangle + c_{10} |10\rangle + c_{11} |11\rangle) |\alpha\rangle .
$$

The interaction of both qubits with the coherent state turns this into

$$
c_{00} |00\rangle |\alpha\rangle + c_{01} |01\rangle |\alpha e^{i\theta_2}\rangle + c_{10} |10\rangle |\alpha e^{i\theta_1}\rangle + c_{11} |11\rangle |\alpha e^{i(\theta_1+\theta_2)}\rangle . \qquad (10.110)
$$

Next, we can choose $\theta_1 = -\theta_2 \equiv \theta$, which means that the output state can be written as

$$
(c_{00} |00\rangle + c_{11} |11\rangle) |\alpha\rangle + c_{01} |01\rangle |\alpha e^{i\theta}\rangle + c_{10} |10\rangle |\alpha e^{-i\theta}\rangle . \qquad (10.111)
$$

The coherent state is clearly entangled with the photonic qubits, and we can disentangle it by performing a measurement of the coherent auxiliary mode a_0. We want to project the qubits

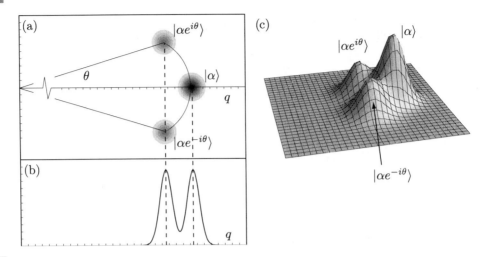

Fig. 10.14.   Parity projection using weak nonlinearities. (a) The distribution in phase space; (b) the projection on the quadrature q; and (c) the Wigner function of the coherent mode.

onto the even or odd parity subspace (given by $|00\rangle \langle 00| + |11\rangle \langle 11|$ and $|01\rangle \langle 01| + |10\rangle \langle 10|$, respectively). This means that the measurement outcomes must distinguish between input states $|\alpha\rangle$ and $|\alpha e^{\pm i\theta}\rangle$, but must *not* distinguish $|\alpha e^{+i\theta}\rangle$ and $|\alpha e^{-i\theta}\rangle$.

We assume without loss of generality that α is real, and the quasi-probability distribution in phase space (the Wigner function) for the three possibilities is shown in Fig. 10.14a. If we perform a measurement of the quadrature $\hat{q}$, then we can obey the measurement requirements above. If the separation between the peaks is sufficiently large, a high-fidelity deterministic parity projection can be implemented.

We treat the measurement of $\hat{q}$ as a von Neumann measurement with outcome q. The two-qubit output state is therefore projected onto the state

$$\langle q|\alpha\rangle (c_{00} |00\rangle + c_{11} |11\rangle) + \langle q|\alpha e^{-i\theta}\rangle c_{01} |01\rangle + \langle q|\alpha e^{-i\theta}\rangle c_{10} |10\rangle . \qquad (10.112)$$

We have to find $\langle q|\alpha\rangle$ and $\phi(q)$. From Chapter 1, we find that for arbitrary α

$$\langle q|\alpha\rangle = \frac{1}{\sqrt[4]{\pi}} \exp\left[-\frac{1}{2}\left(q - \sqrt{2}\alpha \right)^2 + \frac{1}{2}\alpha(\alpha - \alpha^*) \right]. \qquad (10.113)$$

When α is real, we have

$$\langle q|\alpha\rangle = \frac{1}{\sqrt[4]{\pi}} \exp\left[-\frac{1}{2}\left(q - \sqrt{2}\alpha \right)^2 \right]$$

$$\langle q|\alpha e^{i\theta}\rangle = \frac{1}{\sqrt[4]{\pi}} \exp\left[-\frac{1}{2}\left(q - \sqrt{2}\alpha \cos\theta \right)^2 + i\sin\theta \left(\sqrt{2}q - \alpha\cos\theta \right) \right]$$

$$\langle q|\alpha e^{-i\theta}\rangle = \frac{1}{\sqrt[4]{\pi}} \exp\left[-\frac{1}{2}\left(q - \sqrt{2}\alpha \cos\theta \right)^2 - i\sin\theta \left(\sqrt{2}q - \alpha\cos\theta \right) \right]. \qquad (10.114)$$

We can write this in terms of the real functions $f_0(q)$ and $f_\theta(q)$

$$\langle q|\alpha \rangle \equiv f_0(q) \quad \text{and} \quad \langle q|\alpha e^{\pm i\theta} \rangle = |\langle q|\alpha e^{i\theta} \rangle| e^{\pm i\phi(q)} \equiv f_\theta(q) e^{\pm i\phi(q)}, \quad (10.115)$$

with $\phi(q) = \sin\theta(\sqrt{2}q - \alpha\cos\theta)$. The two-qubit output state can therefore be written as

$$f_0(q) (c_{00}|00\rangle + c_{11}|11\rangle) + f_\theta(q) \left(e^{-i\phi(q)}c_{01}|01\rangle + e^{i\phi(q)}c_{10}|10\rangle \right), \quad (10.116)$$

that is, the even parity state has amplitude $f_0(q)$ and the odd parity state has amplitude $f_\theta(q)$. In addition, the odd parity state requires a corrective phase shift of $2\phi(q)$ on the photonic qubit in mode a_2.

The functions $f_0(q)$ and $f_\theta(q)$ are Gaussians, and we obtain good discrimination in the measurement only when the overlap of the two functions with respect to the measurement outcome q is very small. In other words, we must choose θ and α such that f_0 and f_θ form clearly separate peaks (see Fig. 10.14b). The width of (the real part of) these distributions is of order one. We can distinguish the two peaks (and correspondingly obtain a high fidelity) when the distance between the peaks is larger than 2:

$$\sqrt{2}\alpha - \sqrt{2}\alpha\cos\theta > 2. \quad (10.117)$$

Solving this for α and θ, we obtain the distinguishability criterion

$$\alpha\theta^2 > 2\sqrt{2}. \quad (10.118)$$

The parity projection thus obtained is practically deterministic if the peak separation is big enough. In turn, the parity projection can be used to deterministically create cluster states for quantum computing.

Exercise 10.4: Calculate the fidelity of the weak Kerr parity projection.

Fig. 10.14 shows that the parity gate based on weak Kerr nonlinearities is based on the definite separation of the coherent-state distributions in phase space: the two peaks must be clearly distinguishable in order to obtain a high-fidelity parity projection. However, the clear separation along the quadrature q (in Fig. 10.14b) occurs for much larger θ than the clear separation of the peaks in the complete phase space (that is, in Fig. 10.14a). An obvious question is therefore whether we can improve the distinguishability criterion in Eq. (10.118).

First, we note that the smallest angle θ for which separation of the phase-space distributions occurs is in the radial direction from the distribution on the horizontal axis. Second, we can displace the entire distribution by means of the operator $D(-\alpha)$, which amounts to a translation along the coordinate q. After such a displacement, the largest of the 'blobs' in Fig. 10.14a is centered on the origin. The two satellite blobs are at equal distance from the origin. A measurement of the photon number, or distance from the origin, then distinguishes between the distribution at the origin, but not between the satellite distributions. This is exactly what we need for the parity gate to work. The displaced state after the weak nonlinear interaction becomes

$$(c_{00}\,|00\rangle\,c_{11}\,|11\rangle)\,|0\rangle + c_{01}\,|01\rangle\,\big|\alpha(1 - e^{i\theta_2})\big\rangle + c_{10}\,|10\rangle\,\big|\alpha(1 - e^{i\theta_1})\big\rangle. \qquad (10.119)$$

Using the general result $\langle\alpha|\,\hat{n}\,|\alpha\rangle = |\alpha|^2$ we evaluate the separation of the expectation values as

$$\big\langle\alpha(1 - e^{\pm i\theta})\big|\,\hat{n}\,\big|\alpha(1 - e^{\pm i\theta})\big\rangle - \langle 0|\,\hat{n}\,|0\rangle = 2\alpha^2(1 - \cos\theta). \qquad (10.120)$$

This must again be larger than the added variances:

$$2\alpha^2(1 - \cos\theta) > 1 \quad \text{or for } \theta \ll 1: \quad \alpha\theta > 1. \qquad (10.121)$$

Compared to Eq. (10.118), this is a much improved distinguishability criterion.

The second improvement we can make is via squeezing. If the squeezing occurs in the quadrature q, the variance is multiplied by a factor e^{-r}. Consequently, the original distinguishability criterion in Eq. (10.118) becomes

$$\alpha\theta^2 e^r > 2\sqrt{2}. \qquad (10.122)$$

Similarly, we can apply number-phase squeezing in the displaced protocol, which yields

$$\alpha\theta e^r > 1. \qquad (10.123)$$

A note of caution however: one must make sure that the squeezing is perfectly aligned with the quadrature, otherwise the increased uncertainty will show up in the measurements. This may be very challenging in practice.

10.5.2 Intensity coupling to matter qubits

We have described the implementation of a deterministic parity projection by coupling qubits to a coherent state. Other than the phase shift the qubits impart on the coherent state, the physical nature of the qubits is immaterial. This means that the above protocol works not only for photonic qubits, but *any* physical system that can be used to generate a conditional phase shift in a coherent state.

In particular, the phase shift can be generated by any Hamiltonian of the type:

$$\mathcal{H}_I = \hbar\nu\,\sigma_z\,\hat{a}_0^\dagger\hat{a}_0. \qquad (10.124)$$

Here, σ_z is the diagonal Pauli operator of the qubit, and $\hat{a}_0^\dagger\hat{a}_0$ is the number operator for the object, called a 'bus', we use to couple the qubits. In the specific case of the cross-Kerr nonlinearity, the bus is the coherent state. If the fragile quantum state of the bus can be maintained in transportation, we can use this technique for distributed quantum computing. For more details about bus-based quantum computing, see Munro *et al.* (2005).

10.6 References and further reading

Multi-level atoms and electromagnetically induced transparency is reviewed in detail in Fleischhauer *et al.* (2005). Applications of coherent population transfer to quantum information processing are discussed in Beausoleil *et al.* (2004). Specific experimental demonstrations of slow light were described in an atomic vapour in Kasapi *et al.* (1995) and in a solid in Bigelow *et al.* (2003). Dark state polaritons and their applications to quantum memories were introduced in Fleischhauer and Lukin (2000) and expanded upon in Fleischhauer and Lukin (2002) and Lukin (2003). Dipole blockade for quantum information processing was first discussed in Brennen *et al.* (1999), and Jaksch *et al.* (1999). Lukin *et al.* (2001) show how to use the blockade in atomic ensembles to perform quantum logic gates. For specific details of how to implement repeater protocols in atomic ensembles, as well as how to overcome potential errors, we refer to Duan *et al.* (2001).

11 Solid-state quantum information carriers

In this chapter we will discuss solid-state quantum computing, concentrating on systems where qubit manipulation, initialization or readout is performed optically. We will begin with a discussion of crystals with a periodic lattice and derive Bloch's theorem, which sets constraints on the form of electronic wave functions in crystals. We will then introduce semiconductor heterostructures and show that these have a discrete energy-level structure with transitions corresponding to the optical region of the electromagnetic spectrum. The discrete levels can be used as several different kinds of qubit, and we will discuss two that can be manipulated optically, namely an electron spin and an exciton. We will touch upon crystal defects and their importance in optical quantum computing. The emphasis will be on the NV^- centre in diamond, which has produced some of the most important experimental results in recent years. Towards the end of the chapter, we will discuss specific implementations of single- and two-qubit gates in solid-state structures, before concluding with some methods for scaling up a solid-state device to a full-scale quantum computer.

11.1 Basic concepts of solid-state systems

In order to understand the optical characteristics of semiconductors, we must first review some basic concepts from solid-state physics. In particular, we will need the form and properties of the electronic wave functions in a periodic crystal structure. Unfortunately, the calculation of electronic states in a solid is impossible to do exactly. This is due to the vast number of electrons in a solid, all of which are subject to Coulomb interactions with each other and with the nuclei in the crystal. In addition, the overall wave function of each many-electron state must obey Fermi–Dirac statistics, such that it is antisymmetric under the exchange of two single-electron parts. These factors lead to a highly correlated wavefunction that depends on a macroscopic number of variables and whose energy contains terms that arise from the wavefunction antisymmetry. It is impossible to represent such a wavefunction in an efficient way.

Fortunately, we do not need exact solutions: a great deal of insight into the behaviour of electrons in solids can be gained by making use of the 'one-electron approximation'. Here, solutions $\langle \mathbf{r}_i | \psi_i \rangle = \psi_i(\mathbf{r}_i)$ are calculated by first constructing the potential function for a single electron moving under the influence of all of the nuclei and other electrons, and then by solving the appropriate Schrödinger equation. The single-electron solutions can then be combined together in a total wavefunction $\Psi(\mathbf{r}_1, \dots, \mathbf{r}_N)$ that ensures antisymmetry. This form is given by the 'Slater determinant'

$$\Psi(\mathbf{r}_1, \ldots, \mathbf{r}_N) = \begin{vmatrix} \psi_1(\mathbf{r}_1) & \psi_2(\mathbf{r}_1) & \cdots & \psi_N(\mathbf{r}_1) \\ \psi_1(\mathbf{r}_2) & \ddots & & \psi_N(\mathbf{r}_2) \\ \vdots & & \psi_i(\mathbf{r}_j) & \vdots \\ \psi_1(\mathbf{r}_N) & \psi_2(\mathbf{r}_N) & \cdots & \psi_N(\mathbf{r}_N) \end{vmatrix}. \tag{11.1}$$

The procedure of using combinations of self-consistent single-electron solutions is known as 'Hartree–Fock theory'. We now explore the single-electron solutions in more detail.

11.1.1 Bloch's theorem

First, we introduce some basic definitions. Since crystal lattices are periodic, the single-electron potential must also be periodic. We can always define three primitive lattice-translation vectors, $\{\mathbf{a}_1, \mathbf{a}_2, \mathbf{a}_3\}$, that when combined in the form

$$\mathbf{T_i} = n_1\mathbf{a}_1 + n_2\mathbf{a}_2 + n_3\mathbf{a}_3, \quad \text{with} \quad \{n_1, n_2, n_3\} \in \mathbb{Z}, \tag{11.2}$$

can connect any lattice point to any other. The index $\mathbf{i}$ is a vector, consisting of the three numbers $\{n_1, n_2, n_3\}$, that characterizes one particular lattice-translation vector. As well as the real lattice, we can also define an orthogonal 'reciprocal lattice', whose three primitive vectors satisfy $\mathbf{b}_i \cdot \mathbf{a}_j = 2\pi \delta_{ij}$. The reciprocal lattice is a very useful tool when performing calculations of electronic structure in crystals.

To proceed, we write the periodic crystal potential $V(\mathbf{r}) = V(\mathbf{r} + \mathbf{T_i})$. The Schrödinger equation for the wavefunction $\psi(\mathbf{r})$ is then

$$\mathcal{H}(\mathbf{r})\psi(\mathbf{r}) = \left(-\frac{\hbar^2}{2m}\nabla^2 + V(\mathbf{r}) \right) \psi(\mathbf{r}) = E\psi(\mathbf{r}). \tag{11.3}$$

We can use the periodic symmetry of the crystal lattice to make an assertion about the form of the functions $\psi(\mathbf{r})$. To see this, let us first define the translation operator $\mathcal{T}_\mathbf{i}$

$$\mathcal{T}_\mathbf{i}\, g(\mathbf{r}) = g(\mathbf{r} + \mathbf{T_i}). \tag{11.4}$$

Applying this operator to our Schrödinger equation we obtain

$$\mathcal{T}_\mathbf{i}\,\mathcal{H}(\mathbf{r})\psi(\mathbf{r}) = \mathcal{H}(\mathbf{r} + \mathbf{T_i})\psi(\mathbf{r} + \mathbf{T_i}). \tag{11.5}$$

Using the periodicity of the potential function we find $\mathcal{H}(\mathbf{r} + \mathbf{T_i}) = \mathcal{H}(\mathbf{r})$ and therefore $\mathcal{T}_\mathbf{i}$ commutes with $\mathcal{H}$:

$$\mathcal{T}_\mathbf{i}\,\mathcal{H}(\mathbf{r})\psi(\mathbf{r}) = \mathcal{H}(\mathbf{r})\,\mathcal{T}_\mathbf{i}\,\psi(\mathbf{r}). \tag{11.6}$$

The $\psi(\mathbf{r})$ are therefore eigenstates of $\mathcal{T}_\mathbf{i}$ as well as $\mathcal{H}$. Let us define the eigenvalue of $\mathcal{T}_\mathbf{i}$ as $\alpha(\mathbf{T_i})$:

$$\mathcal{T}_\mathbf{i}\,\psi(\mathbf{r}) = \alpha(\mathbf{T_i})\psi(\mathbf{r}). \tag{11.7}$$

If we apply two translations $\mathcal{T}_\mathbf{i}$ and $\mathcal{T}_\mathbf{j}$ to our wavefunction, it makes no difference whether we apply $\mathcal{T}_\mathbf{i}$ first or $\mathcal{T}_\mathbf{j}$ first. Therefore

$$\mathcal{T}_\mathbf{i}\,\mathcal{T}_\mathbf{j} = \mathcal{T}_\mathbf{j}\,\mathcal{T}_\mathbf{i} = \mathcal{T}_{\mathbf{i+j}}\,, \tag{11.8}$$

which implies that the $\alpha(\mathbf{T_i})$ are exponential functions. We can decompose each translation operator into three primitive translations, $\mathcal{T}_1, \mathcal{T}_2, \mathcal{T}_3$, that correspond to each of the three vectors $\mathbf{a}_1, \mathbf{a}_2, \mathbf{a}_3$:

$$\mathcal{T}_\mathbf{i} = \mathcal{T}_1^{n_1}\,\mathcal{T}_2^{n_2}\,\mathcal{T}_3^{n_3}\,. \tag{11.9}$$

We can make the exponentiation explicit, so $\mathcal{T}_1 = e^{2i\pi x_1}$, etc. Then $\mathcal{T}_\mathbf{i} = e^{i\mathbf{k}\cdot\mathbf{T_i}}$, with $\mathbf{k} = x_1\mathbf{b}_1 + x_2\mathbf{b}_2 + x_3\mathbf{b}_3$. We have therefore proved that the electron eigenstates in a crystal must satisfy

$$\psi(\mathbf{r} + \mathbf{T_i}) = e^{i\mathbf{k}\cdot\mathbf{T_i}}\psi(\mathbf{r})\,, \tag{11.10}$$

and this is known as 'Bloch's theorem'. Another way to put this is that the wavefunction of an electron in a crystal must take the form

$$\psi(\mathbf{r}) = e^{i\mathbf{k}\cdot\mathbf{r}}U_{n\mathbf{k}}(\mathbf{r})\,, \tag{11.11}$$

where the $U_{n\mathbf{k}}(\mathbf{r})$ is a function, called the Bloch function, that takes the periodicity of the crystal lattice.

Exercise 11.1: Verify that the two forms for electron wavefunctions in a periodic crystal lattice in Eq. (11.10) and Eq. (11.11) are equivalent.

An important consequence of Bloch's theorem in Eq. (11.10) is that a reciprocal lattice vector defined by $\mathbf{G} = p_1\mathbf{b}_1 + p_2\mathbf{b}_2 + p_3\mathbf{b}_3$ (with p_1, p_2, and p_3 integers) can always be added to the wave vector $\mathbf{k}$ without affecting the form of the wavefunction. Therefore, $\mathbf{k}$ describes distinct states only within a restricted zone, called the 'Brillouin zone', around the origin of reciprocal space: points outside the zone can be translated into the zone by the addition of a particular choice of $\mathbf{G}$.

11.1.2 Wannier functions

To understand how a light field interacts with electrons in a solid, we will need to be more explicit about the nature of the periodic Bloch functions. In order to do this, let us start by considering a collection of well-separated single atoms, and then work out how their wavefunctions are modified as we bring them together and introduce interactions between them, eventually forming a fully periodic crystal. Let us start our discussion around the best-known semiconductor, silicon (Si). A free silicon atom has fourteen electrons in the electronic configuration $1s^2\,2s^2\,2p^6\,3s^2\,3p^2$. When silicon atoms are bought together into a crystal, it is the outer or 'valence' electrons $3s^2\,3p^2$ that begin to overlap most strongly. The other electrons are tightly bound to the nucleus and for the purposes of this discussion we can ignore them. All of the single-electron states are of course eigenstates of the atomic

Hamiltonian of silicon, $\mathcal{H}_{at}$. For example, an s electron with wavefunction χ_s satisfies $\mathcal{H}_{at}\chi_s = E_s\chi_s$.

When two (identical) silicon atoms are put together, the wavefunctions of the individual atoms start to overlap. The primitive unit cell of a silicon crystal, which is the smallest repeating unit in the lattice, contains two atoms called a 'basis' and we will consider this first. Since the atoms overlap, the two atomic s-state wavefunctions χ_{s1} and χ_{s2} are no longer eigenstates of the two-electron Hamiltonian $\mathcal{H}_2$. If the p states are sufficiently distant in energy, then the s states form two new eigenstates of the form $\chi_{\pm} = (\chi_{s1} \pm \chi_{s2})/\sqrt{2}$. These are called the 'bonding' and 'antibonding' orbitals, where the bonding orbital has lower energy. A similar effect occurs with the p states, although their three-fold degeneracy complicates matters slightly. Fig. 11.1 shows the new ordering of energy levels once the two atoms become closer to each other. Eight electrons must be accommodated in the new level structure: the two-fold spin degeneracy means that all of these can be accommodated in the lower-energy bonding levels. Importantly, the highest occupied states have p character, and the lowest unoccupied states have s character.

Using this knowledge, how can we take account of the larger crystal structure of silicon? We can extend our argument in a similar way to when we moved from one atom to two atoms, and use the knowledge that our extended crystal wavefunctions must satisfy Bloch's theorem. Assume first that the crystal potential has such a dilute spacing that each localized bonding orbital $\chi_+(\mathbf{r} + \mathbf{T_i})$ is still an eigenstate of the system (i.e., it does not overlap with

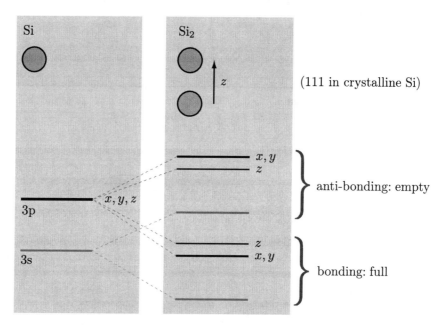

Fig. 11.1. The ordering of electronic energy levels for one and two silicon atoms. For a single atom, the valence states are $3s$ and the triply degenerate $3p$. When two atoms are brought together, bonding and antibonding orbitals form as the valence electrons are able to overlap. The highest occupied level in crystalline silicon has p character, while the lowest unoccupied level has s character.

neighbouring lattice sites). Any linear combination of these is of course also an eigenstate, so we can construct functions that satisfy Bloch's theorem

$$\psi(\mathbf{r}) = \sum_{\mathbf{T_i}} e^{i\mathbf{k} \cdot \mathbf{T_i}} \chi_+(\mathbf{r} + \mathbf{T_i}). \tag{11.12}$$

As might be expected, we find that the periodic part of the wavefunction is essentially a sum of the bonding orbitals.

Of course, we have made a gross approximation here: that the wavefunctions are non-overlapping. In this limit all of the eigenstates, Eq. (11.12), are degenerate regardless of $\mathbf{k}$. In reality, the wavefunctions between unit cells *do* overlap, but we can still describe the crystal eigenstates using a similar form to that in Eq. (11.12)

$$\psi(\mathbf{r}) = \sum_{\mathbf{T_i}} e^{i\mathbf{k} \cdot \mathbf{T_i}} \phi(\mathbf{r} + \mathbf{T_i}). \tag{11.13}$$

The functions $\phi(\mathbf{r} + \mathbf{T}_i)$ are now slightly modified from the original atomic eigenstates, and are called 'Wannier functions'. The Wannier functions can be found using perturbation theory: the atomic states are good approximations in the limit of a weak perturbation. Moreover, the states in Eq. (11.13) do now depend on $\mathbf{k}$ since neighbouring Wannier functions will generally overlap with each other.

The wave vector $\mathbf{k}$ is restricted to certain allowed values that are defined by the boundary conditions imposed by the crystal edges: essentially a whole number of wavelengths must 'fit' within the crystal. If one dimension of a crystal is L, then possible wave vectors are spaced by $2\pi/L$ in that dimension, and if the typical lattice spacing between atoms is a, then physically distinct wave vectors in that dimension only exist up to a maximum of $2\pi/a$.[1] Hence, there are $N = (L/a)^3$ distinct $\mathbf{k}$s, in a crystal of N unit cells. This means that the ground state of a crystal like silicon has $6N$ completely filled states with p-type Wannier functions, with the first empty states being of s type.

11.1.3 Conduction and valence bands

The onset of optical absorption in semiconductors is determined by the difference in energy between the highest occupied electron states and the lowest unoccupied electron states. Let us now look specifically at these, and develop our theory further. For silicon we found that the highest filled states are p-like, and there are six different Wannier functions (three orbital p states each with double spin degeneracy). The highest unoccupied states are s-like, and there are two Wannier functions in that case (a single orbital with spin degeneracy). This is true for a broad class of semiconductors. In particular, gallium arsenide (GaAs, an example of a III–V material) also has this type of structure, and for reasons we will come to later is used much more widely in optical semiconductor technology.

Let us expand our Hamiltonian in Eq. (11.3) in the three p orbitals (p_x, p_y, and p_z), and the one s-like orbital. All of these must have the Bloch form of Eq. (11.11). We will deal

---

[1] This is another statement of the restriction that all distinct $\mathbf{k}$s lie in the Brillouin zone.

with spin later; for now let us denote the four different orbital wavefunctions as:

$$\psi_j(\mathbf{r}) = U_j(\mathbf{r})\exp(i\mathbf{k}\cdot\mathbf{r}) \equiv \langle j | \mathbf{r} \rangle, \tag{11.14}$$

with $j \in \{s, p_x, p_y, p_z\}$. We have neglected the $\mathbf{k}$ dependence of the Bloch functions, which is a good approximation near the edges of the bands. The matrix elements are then

$$\mathcal{H}_{ij} = \langle \psi_i | \mathcal{H} | \psi_j \rangle$$

$$= \int d\mathbf{r}\, U_i^\dagger(\mathbf{r}) \left(-\frac{\hbar^2}{2m_0}\nabla^2 - i\frac{\hbar^2}{m_0}\mathbf{k}\cdot\nabla + \frac{\hbar^2 k^2}{2m_0} + V(\mathbf{r}) \right) U_j(\mathbf{r}). \tag{11.15}$$

By definition, the functions U_j are orthonormal, and have definite parity. They are the correct eigenstates for $\mathbf{k} = 0$, and therefore satisfy an eigenvalue equation

$$\left(-\frac{\hbar^2}{2m_0}\nabla^2 + V(\mathbf{r}) \right) U_i(\mathbf{r}) = E_{i0} U_i(\mathbf{r}). \tag{11.16}$$

This leads to an expression for the Hamiltonian

$$\mathcal{H}_{ij} = \delta_{ij}\left(\frac{\hbar^2 k^2}{2m_0} + E_{i0} \right) + \frac{\hbar}{m_0}\mathbf{k}\cdot\int d\mathbf{r}\, U_i^\dagger(\mathbf{r})\,\hat{\mathbf{p}}\, U_j(\mathbf{r}), \tag{11.17}$$

where the momentum operator $\hat{\mathbf{p}} \equiv -i\hbar\nabla$ has been used. Using the parity properties of the functions U_j, we find that the integral in this equation is zero, unless one of the Us is an s state and the other a p state. In the basis $\{s, p_x, p_y, p_z\}$ the Hamiltonian becomes

$$\mathcal{H} = \begin{pmatrix} E_s + \dfrac{\hbar^2 k^2}{2m_0} & Pk_x & Pk_y & Pk_z \\[2mm] P^* k_x & E_p + \dfrac{\hbar^2 k^2}{2m_0} & 0 & 0 \\[2mm] P^* k_y & 0 & E_p + \dfrac{\hbar^2 k^2}{2m_0} & 0 \\[2mm] P^* k_z & 0 & 0 & E_p + \dfrac{\hbar^2 k^2}{2m_0} \end{pmatrix}, \tag{11.18}$$

where we have defined $E_{s0} \equiv E_s$. We also assume that the bare p states are degenerate: $\{E_{p_x 0}, E_{p_y 0}, E_{p_z 0}\} \equiv E_p$. Furthermore, by rotational symmetry of the s state, and the directional symmetry of the three p states, we have

$$P \equiv \frac{\hbar}{m_0}\int d\mathbf{r}\, U_{p_x}^\dagger(\mathbf{r})\,\hat{\mathbf{p}}\, U_s(\mathbf{r})$$

$$= \frac{\hbar}{m_0}\int d\mathbf{r}\, U_{p_y}^\dagger(\mathbf{r})\,\hat{\mathbf{p}}\, U_s(\mathbf{r})$$

$$= \frac{\hbar}{m_0}\int d\mathbf{r}\, U_{p_z}^\dagger(\mathbf{r})\,\hat{\mathbf{p}}\, U_s(\mathbf{r}). \tag{11.19}$$

The energy difference between the s and p states, $\Delta \equiv E_s - E_p$, is typically much larger than the off-diagonal elements of the above matrix. The s and p states therefore do not mix strongly, and we can use perturbation theory to separate them. The s state is isolated and therefore a good eigenstate with an energy $E'_s = E_s + \hbar^2 k^2/2m_0 - |P|^2 |\mathbf{k}|^2/\Delta$, which is correct to second order. Importantly, the energy of the state depends on the magnitude of the wave vector k, which results in a set of very closely spaced energy levels in reciprocal space that is called a 'band'. For the electrons we have considered here, the 'band edge' is the state with lowest energy, i.e., it corresponds to $|\mathbf{k}| = 0$. As we discussed earlier, the values of $\mathbf{k}$ are restricted to the Brillouin zone, so the band has a finite width. The theory we have developed here is strictly valid only close to the $|\mathbf{k}| = 0$ band edge, where the energy depends quadratically on $|\mathbf{k}|$. Since a free electron would also have an energy proportional to $|\mathbf{k}|^2$, an electron close to the band minimum behaves like a free particle but with a modified mass, the 'effective mass'.

The p bands are, of course, more complicated. Nonetheless, using second-order perturbation theory, we can make some progress by eliminating the s state. In the $\{p_x, p_y, p_z\}$ basis, we have

$$
\mathcal{H}_p = \begin{pmatrix}
E_P + \dfrac{\hbar^2 k^2}{2m_0} + \dfrac{|P|^2 k_x^2}{\Delta} & \dfrac{|P|^2 k_x k_y}{\Delta} & \dfrac{|P|^2 k_x k_z}{\Delta} \\[2ex]
\dfrac{|P|^2 k_x k_y}{\Delta} & E_P + \dfrac{\hbar^2 k^2}{2m_0} + \dfrac{|P|^2 k_y^2}{\Delta} & \dfrac{|P|^2 k_y k_z}{\Delta} \\[2ex]
\dfrac{|P|^2 k_x k_z}{\Delta} & \dfrac{|P|^2 k_y k_z}{\Delta} & E_P + \dfrac{\hbar^2 k^2}{2m_0} + \dfrac{|P|^2 k_z^2}{\Delta}
\end{pmatrix}.
$$

(11.20)

The eigenvalues are $E_P + \hbar^2 k^2/2m_0$ (with multiplicity two) and $E_P + \hbar^2 k^2/2m_0 + |P|^2 k^2/\Delta$. We find again that parabolic bands are predicted, each with its own effective mass. There are three bands, and all three are degenerate at $|\mathbf{k}| = 0$.

As we mentioned above, in the ground state of the crystal the three p bands are occupied, whereas the s band is unoccupied. A band gap separates the highest occupied p band from the s band, and electrical conduction can occur only if electrons are promoted across this gap. It is common therefore to refer to the s band as the 'conduction band' and the p bands as the 'valence bands'.

One way of promoting electrons across the gap is for them to absorb a photon, and when this happens a state in the valence band is left unoccupied. It is straightforward to show that the behaviour of a band with one electron missing is equivalent to that of a band with a particle present, with energy, momentum, spin and charge opposite to the missing electron (see for example Kittel, 2005). This particle is called a 'hole'. In what follows we will work using an electron picture, except where explicitly stated. We next complete our description of single-particle states with the introduction of electron spin.

11.1.4 Spin

For the conduction band, the introduction of spin is trivial. Only s-like orbital states in the conduction band are populated, and these have zero angular momentum. The addition of the spin degree of freedom therefore leads to a doubly degenerate band corresponding to the two spin sublevels. These are usually labelled α and β, corresponding to $m_s = \pm\frac{1}{2}$, respectively:

$$|s_\alpha\rangle \equiv \left|\tfrac{1}{2}, \tfrac{1}{2}\right\rangle_c \quad \text{and} \quad |s_\beta\rangle \equiv \left|\tfrac{1}{2}, -\tfrac{1}{2}\right\rangle_c . \tag{11.21}$$

For the valence band, we have p-like orbital symmetry, and so have an angular momentum of $\hbar$. In this case the nuclei within the crystal cause the electrons to experience a magnetic field, which can interact with the spin of the electron. This is called 'spin–orbit coupling', and it gives rise to an interaction Hamiltonian

$$\mathcal{H}_{so} = \lambda \, \mathbf{L} \cdot \mathbf{S} . \tag{11.22}$$

Here, $\mathbf{L}$ and $\mathbf{S}$ are the orbital and spin angular-momentum operators, respectively, and to understand the spin-orbit effect we therefore switch from $\{p_x, p_y, p_z\}$ to eigenstates of the angular-momentum projection operator $L_z = \mathbf{L} \cdot \hat{\mathbf{z}}$

$$|\Uparrow\rangle = \frac{|p_x\rangle + i \, |p_y\rangle}{\sqrt{2}}, \quad |\Downarrow\rangle = \frac{|p_x\rangle - i \, |p_y\rangle}{\sqrt{2}}, \quad \text{and} \quad |\odot\rangle = |p_z\rangle . \tag{11.23}$$

These states correspond to the eigenvalues $L_z = +1, -1$, and 0, respectively. The inclusion of spin then leads to six different valence states, and $\mathcal{H}_{so}$ has two sets of degenerate eigenstates corresponding to total angular momentum of $J = L + S = \tfrac{3}{2}$ (four degenerate states) and $J = L + S = \tfrac{1}{2}$ (two degenerate states); the two sets are split by an energy $\tfrac{3}{2}\lambda$.

The four states in the degenerate $J = \tfrac{3}{2}$ manifold can be written in a basis corresponding to eigenstates of the J_z operator as follows:

$$\left|\tfrac{3}{2}, \tfrac{3}{2}\right\rangle_v = |\Uparrow\alpha\rangle \quad \text{and} \quad \left|\tfrac{3}{2}, -\tfrac{3}{2}\right\rangle_v = i \, |\Downarrow\beta\rangle , \tag{11.24}$$

and

$$\left|\tfrac{3}{2}, \tfrac{1}{2}\right\rangle_v = \frac{i}{\sqrt{3}}\left[|\Uparrow\beta\rangle - \sqrt{2} \, |\odot\alpha\rangle\right] \quad \text{and} \quad \left|\tfrac{3}{2}, -\tfrac{1}{2}\right\rangle_v = \frac{1}{\sqrt{3}}\left[|\Downarrow\alpha\rangle + \sqrt{2} \, |\odot\beta\rangle\right]. \tag{11.25}$$

In the degenerate $J = \tfrac{1}{2}$ manifold, the two J_z eigenstates are

$$\left|\tfrac{1}{2}, \tfrac{1}{2}\right\rangle_v = \frac{i}{\sqrt{3}}\left[|\odot\alpha\rangle + \sqrt{2} \, |\Uparrow\beta\rangle\right] \quad \text{and} \quad \left|\tfrac{1}{2}, -\tfrac{1}{2}\right\rangle_v = \frac{1}{\sqrt{3}}\left[|\odot\beta\rangle - \sqrt{2} \, |\Downarrow\alpha\rangle\right]. \tag{11.26}$$

The $J = \tfrac{1}{2}$ states are typically lower in energy than the $J = \tfrac{3}{2}$ states. They are sometimes called 'split-off' states, and we need not consider them further. In the basis $\{\left|\tfrac{3}{2}\right\rangle, \left|\tfrac{1}{2}\right\rangle, \left|-\tfrac{1}{2}\right\rangle, \left|-\tfrac{3}{2}\right\rangle\}$, the $J = \tfrac{3}{2}$ Hamiltonian reads

$$\mathcal{H} = \begin{pmatrix} h_{hh} & g_1 & g_2 & 0 \\ g_1^* & h_{lh} & 0 & g_2 \\ g_2^* & 0 & h_{lh} & -g_1 \\ 0 & g_2^* & -g_1^* & h_{hh} \end{pmatrix}, \tag{11.27}$$

with

$$h_{hh} = E_P + \frac{\lambda}{2} + \frac{\hbar^2 k^2}{2m_0} + \frac{|P|^2}{2\Delta}(k_x^2 + k_y^2), \tag{11.28}$$

$$h_{lh} = E_P + \frac{\lambda}{2} + \frac{\hbar^2 k^2}{2m_0} + \frac{|P|^2}{6\Delta}\left(k_x^2 + k_y^2 + 4k_z^2\right), \tag{11.29}$$

$$g_1 = -i\frac{|P|^2}{\sqrt{3}\Delta}(k_x - ik_y)k_z, \tag{11.30}$$

$$g_2 = \frac{|P|^2}{2\sqrt{3}\Delta}(k_x - ik_y)^2. \tag{11.31}$$

This is the well-known Luttinger–Kohn Hamiltonian. In fact, it is slightly simplified from its most general form, since we have here considered a highly symmetrical system. Systems with lower symmetry are easy to treat in the same way, and this results in slightly more complicated parameter expressions.

There are now two pairs of degenerate bands of states, with eigenvalues

$$E_1 = E_P + \frac{\lambda}{2} + \frac{\hbar^2 k^2}{2m_0}$$

$$E_2 = E_P + \frac{\lambda}{2} + \frac{\hbar^2 k^2}{2m_0} + \frac{2|P|^2 k^2}{3\Delta}. \tag{11.32}$$

Again, under the approximations we have used, parabolic bands are found and the effective-mass approximation can be used. There are two different effective masses for the two bands, one 'light' and one 'heavy'. In real systems, we have to take into account the different degeneracy splitting of the p states along different crystal axes (as shown in Fig. 11.1). This typically means that the heavy band is often dominated by the $J_z = \pm\frac{3}{2}$ angular-momentum states, whereas the light band has predominantly $J_z = \pm\frac{1}{2}$ character. The dependence of the energy on $|\mathbf{k}|$ is shown in Fig. 11.2.

We can now explain why silicon is not often used in optical quantum-computing experiments. The discussion here centred on the electronic properties near $\mathbf{k} = 0$ in both conduction and valence bands, and predicts a minimum band gap at $\mathbf{k} = 0$. In silicon, however, the behaviour of the bands far from $\mathbf{k} = 0$ is complex and leads to a conduction band minimum at the edge of the Brillouin zone, rather than in the middle of it. This complicates the optical properties of silicon, since the lowest possible energy transition across the band gap involves a change in $\mathbf{k}$. Silicon is therefore known as an 'indirect gap' semiconductor. GaAs, on the other hand, has no such problems and most of our following discussion of semiconductors will concentrate on GaAs.

11.1.5 Heterostructures and the envelope-function approximation

We have developed a picture of the electronic structure of semiconductors at or near the top of the valence band and bottom of the conduction band. Such materials in their bulk

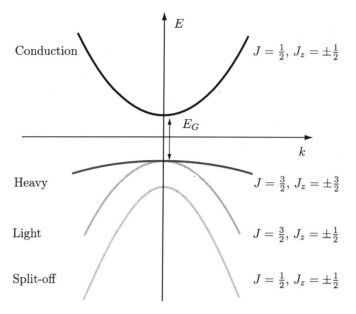

Fig. 11.2. In a bulk semiconductor, the optical properties are primarily determined by the electronic structure of the valence and conduction bands. The figure shows the wave vector dependence of the energy of these bands near the band edges. The conduction band has s-symmetry and this gives rise to one approximately parabolic band with spin $J = \frac{1}{2}$. The valence electrons have p-symmetry and this gives three bands, one of which is split off by spin-orbit coupling and characterized by $J = \frac{1}{2}$. The other two have $J = \frac{3}{2}$ character and are degenerate at the $|\mathbf{k}| = 0$ point. Their different curvatures give us the names 'light' and 'heavy'.

form are hard to use as qubits since they form a continuous spectrum of states. However, it is possible to create semiconductors with discrete energy states: this requires two or more types of semiconductor to be put together in a single 'heterostructure'.

A typical example material consists of GaAs, which has a band gap of 1.42 eV and InAs, whose band gap is 0.35 eV. If a layer of InAs is surrounded above and below by GaAs, then it is possible for electrons to be in certain energy states in the InAs that cannot exist in GaAs. The electrons then see a potential profile, as sketched in Fig. 11.3, and the states that can only exist in the InAs region are 'quantum confined'. These states can no longer be of the delocalized Bloch function form in Eq. (11.11). However, a simple procedure allows us to construct suitable eigenfunctions in this case.

We know that the bulk-dispersion relation for both electron states near the band edges is approximately parabolic – and therefore like a free particle – but with a modified mass, the effective mass m^*. Such states satisfy a Schrödinger equation:

$$\left(-\frac{\hbar^2}{2m^*}\nabla^2 + V_0 \right)\xi(\mathbf{r}) = E\xi(\mathbf{r}), \tag{11.33}$$

where V_0 is the energy corresponding to the band edge. The solutions are of the form $\xi(\mathbf{r}) \propto \exp(-i\mathbf{k} \cdot \mathbf{r})$, which coincide with the first factor on the right-hand side of the general solution for an electron in a periodic potential, given in Eq. (11.11).

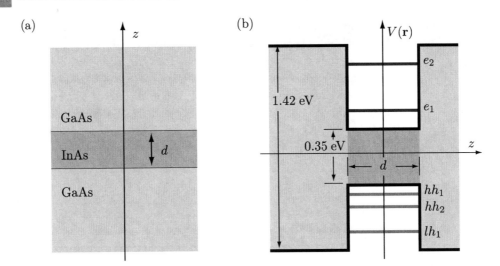

Fig. 11.3. (a) A two-dimensional quantum well structure; (b) the band gap, shown as the shaded regions, is modulated in one dimension when a narrow band gap material is sandwiched between two regions of wider band gap material. It gives rise to an effective confinement potential $V(\mathbf{r})$ in both conduction and valence bands, shown as the thick black line. This leads to confined states e_1, e_2, above the Fermi level, and hh_1, hh_2, and lh_1 below it.

In the case of a quantum heterostructure, we can find the form of $\xi(\mathbf{r})$ by keeping track of how the band edges vary in space by modifying the potential term in the Schrödinger equation:

$$\left(-\frac{\hbar^2}{2m}\nabla^2 + V(\mathbf{r})\right)\xi(\mathbf{r}) = E\xi(\mathbf{r}).\tag{11.34}$$

A *quantum well* is a sandwich structure where a uniform layer of InAs is the 'filling' and GaAs the 'bread'. The potential profile is modulated along, say, the z axis, shown in Fig. 11.3. The modulation creates a finite quantum well in both the conduction and valence bands. The solution to Eq. (11.34) for states in both bands is then a product of delocalized wavefunctions in the x and y directions, and a confined wavefunction in the z direction

$$\xi_{1D}(\mathbf{r}) \propto \xi_z(z)\exp(ik_x x + ik_y y).\tag{11.35}$$

The $\xi_z(z)$ are simply the solutions of the finite square well, which can be found in any undergraduate textbook on quantum mechanics. It can be shown that there always exists at least one confined state. The full solution must of course also include the Bloch function $U_n(\mathbf{r})$, and it is convenient to define the normalization such that the square of both functions on the right-hand side each separately integrates to unity over all space. We therefore have

$$\psi(\mathbf{r}) = \sqrt{V}\,\xi_{1D}(\mathbf{r})U_n(\mathbf{r}),\tag{11.36}$$

with V the volume of the crystal. The energy eigenvalues of the states defined in Eq. (11.35) are

$$E(k_x, k_y, nz) = V_0 + \frac{\hbar^2}{2m^*}(k_x^2 + k_y^2) + E_{nz}.\tag{11.37}$$

The E_{nz} represent the discrete energy eigenvalues of the finite square well problem and V_0 is the energy at the band extremum. Modulation in one direction therefore gives us a set of 'sub-bands', each corresponding to one value of nz, and each of which has a continuous set of energy eigenvalues corresponding to different values of k_x and k_y.

To define a qubit, we require discrete levels, and to achieve this we must have confinement in all three dimensions. We require a semiconductor with a small band gap to be completely surrounded by a material of larger band gap. Such structures are routinely produced and are called 'quantum dots'. The general solutions of the Schrödinger equation become:

$$\psi(\mathbf{r}) = \sqrt{V}\, \xi_{3D}(\mathbf{r}) U_n(\mathbf{r}), \tag{11.38}$$

where $\xi_{3D}(\mathbf{r})$ is now a function that describes completely localized states. It usually varies more slowly than $U_n(\mathbf{r})$, and is called the 'envelope function'.[2]

The finite square model is awkward to solve in three dimensions: a potential function describing such a box cannot be written as a sum of three potentials describing the variation in each spatial dimension, and so the Schrödinger equation cannot be solved using the separating of variables technique. A harmonic potential is often employed instead, since it can be written in separable form.

The discrete energy level structure of a quantum dot can be tuned in several ways, and such nanostructures are therefore often termed 'artificial atoms'. Different materials alter the sizes of the confinement potential, and shape and size can also be varied. Such tunability is useful for a variety of applications, for example in quantum-dot lasers. It also makes them extremely versatile candidate materials for quantum information processing technology, including for optical quantum computing. In the next section we explore the consequences of the electronic structure of quantum dots for their optical properties.

11.1.6 Optical selection rules

An electron in a valence state can be promoted across the band gap to a conduction state by absorption of an optical field. However, certain selection rules apply, which limit the number of possible transitions, and we will calculate them here.

In Chapter 7, we derived a Hamiltonian for an interaction of a two-level atom and a classical field. In a basis of the two-atom states $|g\rangle$ and $|e\rangle$ we found

$$\mathcal{H}(t) = \frac{\hbar}{2} \begin{pmatrix} \omega - \omega_0 & 2eE_0(\mathbf{r}_{eg} \cdot \boldsymbol{\epsilon})^*/\hbar \\ 2eE_0(\mathbf{r}_{eg} \cdot \boldsymbol{\epsilon})/\hbar & \omega_0 - \omega \end{pmatrix}, \tag{11.39}$$

where E_0, $\boldsymbol{\epsilon}$, and ω are the amplitude, polarization, and frequency of the field, respectively. The frequency of the atomic transition is denoted by ω_0, and $\mathbf{r}_{eg}$ is the transition dipole matrix element between the two-atom states. In order to calculate optical selection rules

---

[2] The argument we have described is quite simplistic, since we have completely decoupled the periodic part of the wavefunction from the envelope function. This is an approximation, called the 'envelope-function approximation'. For a detailed discussion see Burt (1999).

for our quantum-dot system, we must in principle evaluate the transition dipole between many-electron states of the form Eq. (11.1). However, it turns out that all the electrons that are not involved in the transition do not enter the calculation; they have the same states before and after the transition, so those parts of the many-body wavefunction do not affect the transition matrix elements. We assume that the electron involved in the transition has a wavefunction given by Eq. (11.36) and we will calculate interband transition matrix elements, since these typically correspond to optical frequencies. Denoting the interband transition dipole operators as $\mathbf{r}_{cv}$, we have

$$\mathbf{r}_{cv} = V \int_{\text{space}} d\mathbf{r}\, \xi_c^*(\mathbf{r}) U_c^*(\mathbf{r})\, \mathbf{r}\, \xi_v(\mathbf{r}) U_v(\mathbf{r})\,. \tag{11.40}$$

Using the Wannier functions defined earlier, we can write

$$U_p \equiv \frac{1}{\sqrt{N}} \sum_{\mathbf{i}} \phi_p(\mathbf{r} + \mathbf{T_i})\,, \tag{11.41}$$

where $p = c, v$, and N is the number of unit cells in the crystal. The $1/\sqrt{N}$ factor is included so that the Wannier functions are properly normalized.

Next, rewrite Eq. (11.40) as

$$\mathbf{r}_{cv} = V \sum_{\mathbf{i}} \int_{\text{space}} d\mathbf{r}\, \xi_c^*(\mathbf{r}) \phi_c^*(\mathbf{r} + \mathbf{T_i})\, \mathbf{r}\, \xi_v(\mathbf{r}) \phi_v(\mathbf{r} + \mathbf{T_i})\,, \tag{11.42}$$

where V is the unit cell volume. Hence

$$\mathbf{r}_{cv} = V \sum_{\mathbf{i}} \int_{\text{space}} d\mathbf{r}\, \xi_c^*(\mathbf{r} - \mathbf{T_i}) \phi_c^*(\mathbf{r}) (\mathbf{r} - \mathbf{T_i}) \xi_v(\mathbf{r} - \mathbf{T_i}) \phi_v(\mathbf{r})\,. \tag{11.43}$$

The functions ϕ_p are defined only within the unit cell centred on $\mathbf{r} = 0$, and each integral can therefore be restricted to this cell. Furthermore, we can assume that our envelope functions $\xi_p(\mathbf{r})$ vary slowly over each unit cell, and can be replaced by their average value. Therefore

$$\mathbf{r}_{cv} = V \sum_{\mathbf{i}} \xi_{c,\mathbf{i}}^* \xi_{v,\mathbf{i}} \int_{\text{cell}} \phi_c^*(\mathbf{r}) (\mathbf{r} - \mathbf{T_i}) \phi_v(\mathbf{r}) d\mathbf{r} \tag{11.44}$$

where the $\xi_{\mathbf{i}}$ now represent the average values of the envelope functions over cell i. Since ϕ_c and ϕ_v come from different bands, they are orthogonal to each other, which means, finally, that

$$\mathbf{r}_{cv} = \mathbf{t}_{cv} \int_{\text{space}} d\mathbf{r}\, \xi_c^*(\mathbf{r}) \xi_v(\mathbf{r})\,. \tag{11.45}$$

Here we have defined the function

$$\mathbf{t}_{cv} = \int_{\text{cell}} d\mathbf{r}\, \phi_c^*(\mathbf{r})\, \mathbf{r}\, \phi_v(\mathbf{r})\,, \tag{11.46}$$

which depends only on the periodic part of the wavefunctions.

Table 11.1. Selection rules for valence and conduction band transitions.

Valence J_z	Conduction J_z	ϵ_L	ϵ_R
3/2	1/2	0	r
3/2	−1/2	0	0
−3/2	1/2	0	0
−3/2	−1/2	ir	0
1/2	1/2	0	0
1/2	−1/2	0	$ir/\sqrt{3}$
−1/2	1/2	$r/\sqrt{3}$	0
−1/2	−1/2	0	0

11.1.7 Selection rules in quantum dots

The easiest way of showing how selection rules arise in quantum dots is by constructing an example. Let us suppose that we have circularly polarized light propagating in the z direction. Its polarization vector is $\epsilon_R = (\hat{\mathbf{x}} - i\hat{\mathbf{y}})/\sqrt{2}$. If the frequency of this optical field is tuned to the transition between the valence state $\left|\frac{3}{2}, \frac{3}{2}\right\rangle_v$ and the conduction state $\left|\frac{1}{2}, \frac{1}{2}\right\rangle_c$, the field creates an electron in the conduction band, and leaves a hole in the valence band. For this transition $\mathbf{t}_{cv}$ becomes

$$\mathbf{t}_{cv}\left(\left|\tfrac{3}{2}, \tfrac{3}{2}\right\rangle_v \rightarrow \left|\tfrac{1}{2}, \tfrac{1}{2}\right\rangle_c\right) = \frac{1}{\sqrt{2}}\left(r_{p_x}\hat{\mathbf{x}} + ir_{p_y}\hat{\mathbf{y}}\right) \tag{11.47}$$

with

$$r_i = \left|\int_{\text{cell}} \phi_i^*(\mathbf{r})\,\mathbf{r}\,\phi_s(\mathbf{r})\,d\mathbf{r}\right|, \tag{11.48}$$

for $i \in \{p_x, p_y, p_z\}$. Using the symmetry of the p states as before means that $r_{p_x} = r_{p_y} = r_{p_z} \equiv r$. We therefore find that $\epsilon_R \cdot \mathbf{t}_{cv}(\left|\tfrac{3}{2}, \tfrac{3}{2}\right\rangle_v \rightarrow \left|\tfrac{1}{2}, \tfrac{1}{2}\right\rangle_c) = r$.

If we had instead used left-handed circularly polarized light ϵ_L, the coupling matrix element of these two states would have been zero. We can think of this as a consequence of conservation of angular momentum: ϵ_L and ϵ_R polarized light have angular momentum values of $+1$ and -1, respectively. An electron in a $J_z = \frac{3}{2}$ state must *lose* one unit of angular momentum to be promoted into a $J_z = \frac{1}{2}$ state, and it can do this only by combining with ϵ_R light. We can also conclude that it is *impossible* to promote the $J_z = \frac{3}{2}$ electron to the $J_z = -\frac{1}{2}$ conduction band state, at least up to the dipole approximation. Using the same method allows us to calculate all of the matrix elements between valence and conduction band states. These are summarized in Table 11.1 and Fig. 11.4.

conduction states

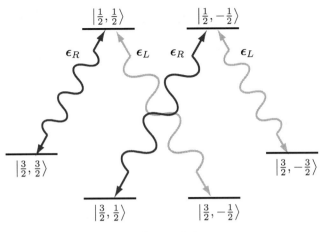

valence states

Fig. 11.4. Schematic drawing showing the allowed transitions between electronic states in a semiconductor quantum dot whose symmetry axis is along z. Transitions allowed in L and R circularly polarized light are shown.

11.2 Definition and optical manipulation of solid-state qubits

Having discussed in some detail the character of electronic states in solid-state nanostructures and their optical transition selection rules, we now discuss the possible representations of a qubit in such systems. In this section we consider excitonic qubits and spin qubits in quantum dots, and spin qubits in crystal defects.

11.2.1 Exciton qubits

An obvious way of representing one of the two-qubit basis states is to use the ground state of a quantum dot, where all the valence states are full and all the conduction states empty. A second, excited state must then be found to complete the qubit basis. The simplest optical implementation is to take the state that results when an electron is promoted from one of the highest-lying valence states (which usually correspond to $J_z = \pm\frac{3}{2}$) to one of the lowest-lying conduction states. A useful way of describing this excited state is as a composite particle, the 'exciton', consisting of an extra conduction electron and a missing valence electron (i.e., the hole). Remembering that the hole has opposite spin to the missing electron, we can select the spin of the exciton by choosing the polarization of the light we use, as shown in Fig. 11.4. Light propagating in the z direction with polarization ϵ_L or ϵ_R produces an exciton with net $J_z = 1$ or -1, respectively.

Exciton qubits can be manipulated using pulsed lasers. As an example, let us consider a qubit defined by $|0\rangle$, the ground state and $|1\rangle \equiv \left|\frac{1}{2}, \frac{1}{2}\right\rangle_{c,e} \otimes \left|\frac{3}{2}, -\frac{3}{2}\right\rangle_{v,h}$ where we have now added the labels e and h to denote the electron and the hole. The laser is resonantly tuned to the transition and is ϵ_R polarized. Let us also assume for generality that it has an extra fixed phase of θ and that the quantum dot lies in the $z = 0$ plane. The field acting on the quantum dot is then:

$$\mathbf{E} = E_0 \boldsymbol{\epsilon} \exp(i[\omega t + \theta]) + E_0 \boldsymbol{\epsilon}^* \exp(-i[\omega t + \theta]). \tag{11.49}$$

After moving to the rotating frame and making the rotating-wave approximation as before, we find our single-qubit Hamiltonian:

$$\mathcal{H} = eE_0 r \left(e^{-i\theta} |0\rangle \langle 1| + e^{i\theta} |1\rangle \langle 0| \right). \tag{11.50}$$

The eigenstates are $(|0\rangle \pm e^{i\theta} |1\rangle)/\sqrt{2}$, with energies $\pm eE_0 r$. The natural evolution of an arbitrary state is therefore a rotation about an axis in the xy plane of the Bloch sphere with azimuthal angle θ, at the Rabi frequency $\Omega = 2eE_0 r/\hbar$. If the laser is pulsed for a time τ the total angle of rotation is $\Theta_R = \Omega\tau$. This angle could be varied in the laboratory by changing τ, but in practice it is easier to change the intensity of the pulse, therefore altering Ω. The value of θ can also be varied, and by combining two independent rotations any single-qubit manipulation can be executed (see Section 5.1).

11.2.2 Electron spin

Exciton qubits are straightforward to define and easy to manipulate. However, they suffer from some serious drawbacks. In particular, they can decay through the same mechanism they were created by: the emission of a photon. The uncontrolled process can be devastating when it comes to preserving quantum information. For example, the coherence times of excitons in a InAs/GaAs system is only about 600 ps (see Borri *et al.* [2001]). Therefore, an alternative qubit representation is often called for.

The most successful of these to date is the electron spin. A single quantum dot can have an extra electron introduced into it, usually by growing a diode structure around the quantum dot. Under appropriate bias conditions, this results in a movement of the Fermi energy such that a single extra electron is introduced to the lowest-lying conduction state. This extra occupied state means that the quantum-dot system as a whole now has a net spin $J_z = \pm\frac{1}{2}$, and therefore constitutes a well-defined qubit. The most severe decoherence mechanism for the electron spin in the InAs/GaAs structure is usually the interaction with the nuclear spins of the atoms constituting the crystal, which is much weaker than the exciton–photon interaction. The electron spin then has a much longer coherence time than an exciton, typically over one microsecond (see for example Greilich *et al.*, 2006).

Single-qubit operations are a little less obvious to implement for this spin qubit. We could appeal to electron spin resonance methods to provide us with a direct way of performing the required manipulations. Applying a controlled microwave pulse that is resonant with the difference in the energies of the spin eigenstates in a magnetic field allows controlled rotations in exactly the same way as light does with exciton qubits. However, such manipulations tend to be much slower than optical methods, and any advantage gained in going

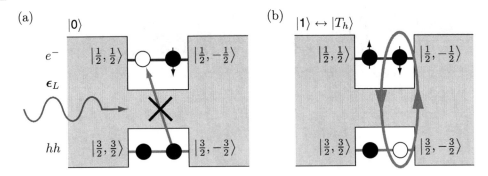

Fig. 11.5. The Pauli blocking effect. A qubit is defined by the spin of an electron in the conduction state of a quantum dot. When the structure is illuminated with resonant ϵ_L light, a trion can only be created from the qubit $|1\rangle$ state. In this figure, the valence-state angular momentum labels refer to the spin of an electron state, and not the hole created when the electron is removed.

to a long-lived species may be lost. Fortunately, we can couple an optical field to the spin degree of freedom, using the 'Pauli blocking effect', which we will now discuss.

Consider a spin qubit in state $|0\rangle \equiv \left|\frac{1}{2}, -\frac{1}{2}\right\rangle_{c,e}$ that is illuminated with ϵ_L polarized light with the correct energy to create a heavy-hole exciton in the quantum dot. Angular-momentum conservation dictates that the only exciton that could be created is $\left|\frac{1}{2}, -\frac{1}{2}\right\rangle_{c,e} \otimes \left|\frac{3}{2}, \frac{3}{2}\right\rangle_{v,h}$. However, our spin qubit is already occupying the state $\left|\frac{1}{2}, -\frac{1}{2}\right\rangle_{c,e}$, and Pauli's exclusion principle therefore forbids this possibility. On the other hand, if the qubit state is $|1\rangle \equiv \left|\frac{1}{2}, \frac{1}{2}\right\rangle_{c,e}$ the exciton can be created, and together with the qubit forms a 'trion state' $|T_h\rangle$, consisting of two electrons and one heavy hole. This is shown in Fig. 11.5.

Using the derivation of Eq. (11.50) for guidance, this behaviour can be captured through the three-state Hamiltonian (after transforming to the frame rotating with the frequency of the qubit–trion transition):

$$\mathcal{H} = \frac{\hbar\Omega}{2} \left(e^{-i\theta} |1\rangle \langle T_h| + e^{i\theta} |T_h\rangle \langle 1| \right). \tag{11.51}$$

The spin selective trion creation can then be exploited to give an arbitrary phase gate on the qubit. If our initial state is $|\psi\rangle = \alpha |1\rangle + \beta |0\rangle$, and we apply a laser pulse for a time $\tau = \pi/\Omega$, the result is the state $|\psi(\tau)\rangle = -ie^{i\theta}\alpha |T_h\rangle + \beta |0\rangle$. This kind of pulse, where all the population in one state is completely transferred onto another, is called a π pulse. Applying another such pulse with a different phase θ' completes the cycle back into the computational basis states, giving the state $|\psi(2\tau)\rangle = -e^{i(\theta-\theta')}\alpha |1\rangle + \beta |0\rangle$.

Exercise 11.2: Verify that the above sequence of two $\pi/2$ pulses with different phases θ and θ' allows any Z-rotation of the spin qubit.

The other required Bloch-sphere rotation presents more of a challenge. The trick of exciting the system from one of the qubit states into a higher level and then back again will always preserve the populations of the two-qubit states, and so can never do more than a Z-rotation. In order to move population between levels optically, the two-qubit states have to be coupled indirectly through the trion level, i.e., Raman-type transitions must be used.

Table 11.2. Selection rules for transition between valence band states and conduction band states using linearly polarized light with polarization axis along the symmetry axis of the crystal z

Valence J_z	Conduction J_z	Matrix Element
1/2	1/2	$-i\sqrt{2/3}r$
1/2	-1/2	0
-1/2	1/2	0
-1/2	-1/2	$\sqrt{2/3}r$

In the picture we have presented, it is impossible to couple either of the heavy-hole states to both possible electron-spin states optically, since one of these transitions will always involve an angular-momentum change of two units, which cannot be carried by photon polarization. However, light-hole trions, consisting of two electrons and a light-hole, present a viable alternative. The hole can have spin up or down and therefore we denote the two possible light-hole trion states as $|T_l^\uparrow\rangle$ and $|T_l^\downarrow\rangle$. We already know that using circularly polarized light creates an exciton of particular spin. If we combine equal components of left and right circularly polarized light (making horizontal polarization, for example), then $|0\rangle$ couples to $|T_l^\uparrow\rangle$ and $|1\rangle$ couples to $|T_l^\downarrow\rangle$.

We can complete a pair of Raman transitions by providing a laser that can couple the other spin state to each trion, and we can do this using z polarized light. Referring back to the decomposition of light holes into the angular-momentum eigenstates, Eq. (11.25), it is straightforward to find the matrix elements for z polarized light for the light-hole excitons, and these are summarized in Table 11.2.

Let us absorb the $\sqrt{2/3}r$ factors, together with the electric field strength, into a single parameter that describes the exciton–laser coupling Λ. We also include the horizontally polarized component of coupling strength Γ to our light field. Assuming that the laser is detuned by an amount Δ from resonance, and after moving to the rotating frame and making the rotating-wave approximation, we obtain:

$$\mathcal{H} = \Gamma \left(|0\rangle \langle T_l^\uparrow| + i|1\rangle \langle T_l^\downarrow| \right) + \Lambda \left(-i|0\rangle \langle T_l^\downarrow| + |1\rangle \langle T_l^\uparrow| \right)$$

$$+ \frac{\Delta}{2} \left(|T_l^\downarrow\rangle \langle T_l^\downarrow| + |T_l^\uparrow\rangle \langle T_l^\uparrow| \right) + \text{H.c.}. \tag{11.52}$$

Exercise 11.3: Assuming that $\Gamma \ll \Delta$ and $\Lambda \ll \Delta$ and using second-order degenerate perturbation theory show that the dynamics of the spin qubit can be represented by the effective Hamiltonian:

$$\mathcal{H}_{\text{eff}} = \frac{\Lambda^2 - \Gamma^2}{\Delta} \left(|0\rangle \langle 0| - |1\rangle \langle 1| \right) - \frac{2\Gamma\Lambda}{\Delta} \left(|0\rangle \langle 1| + |1\rangle \langle 0| \right). \tag{11.53}$$

If $\Gamma = \Lambda$ then the eigenstates are approximately $(|0\rangle \pm |1\rangle)/\sqrt{2}$, with eigenvalues $\pm 2\Lambda\Gamma/\Delta$. The evolution of the qubit is therefore an X-rotation in the Bloch sphere, with a frequency $4\Lambda\Gamma/\Delta$. By varying the polarization and intensity of the light fields it is then possible to achieve *any* single-qubit rotation, satisfying one of the requirements for universal quantum computing. In Section 11.3 we will show how to complete the universal gate set by exploiting interactions between qubits to perform two-qubit entangling gates.

Finally in this section we briefly mention that spin-selective optical transitions can be used for optical spin readout. The principle is illustrated as follows. If an ϵ_L optical π pulse is applied to a qubit in state $\alpha |1\rangle + \beta |0\rangle$, then we know that we achieve the superposition $\alpha |T_h\rangle + \beta |0\rangle$. If we now simply leave the system, there is an $|\alpha|^2$ probability that it will emit a photon. If we can detect this photon, we have performed a projective measurement of the qubit, with no photon detection corresponding to a projection into the $|0\rangle$ state. This works only with perfect detectors, but we can apply another π pulse after waiting for the first decay, and this gives us another shot at detection, improving the measurement fidelity. In fact, we can keep on doing this and supply a train of π pulses to improve the fidelity arbitrarily (assuming no other effects such as detector dark counts). Examples of experiments where spin readout has been achieved using optical techniques in quantum dots are Atatüre *et al.* (2007) and Berezovsky *et al.* (2008) .

11.2.3 Crystal defects

Our discussion of optical readout leads naturally to a discussion of another very important class of solid-state qubit implementation: the crystal defect. There are a myriad different types of defect in crystals. The simplest examples include single-atom substitutions within the lattice, extra atoms that position themselves at interstitial sites in the lattice, and missing atoms or 'vacancies'. However, one defect in particular has caught the attention of quantum information scientists in recent years, due to its exceptional spin and optical properties. It is the NV^- centre in diamond. It consists of two imperfections that lie side by side in the lattice. One of the carbon atoms of the lattice is replaced by a single nitrogen atom, and next to it is a vacant site. The structure is shown in Fig. 11.6.

The NV^- centre has six valence electrons: five from the nitrogen and a single extra electron that gives the centre its overall negative charge. Two of these electrons are immediately taken up by filling the two 2s orbitals in the nitrogen atom. This leaves four electrons, three of which associate themselves with the three carbon atoms nearest the vacancy. The remaining electron then goes into a singlet with one of the carbon electrons, leaving two unpaired spins which line up as a triplet, by Hund's rules. There are of course three possible singlet pairings and the lowest energy configuration becomes a superposition of all three. Since each component in the superposition is itself a triplet, the ground-state superposition remains a triplet.

Excited states are a singlet pairing of the two unpaired electrons, and other orthogonal superpositions of the unpaired triplet states. The level structure is sketched in Fig. 11.7. The ground level has a so-called A character in group notation,[3] and as mentioned above

---

[3] Crystallographers use group theory to characterize crystals and defect structures; for a discussion of the theory and terminology see Atkins and Friedman (2005), and Kuzmany (1998). The NV^- has C_{3v} point group symmetry. The symmetries of the six orbitals in the lowest energy configuration are usually denoted by $a_1^2 a_1'^2 e^2$.

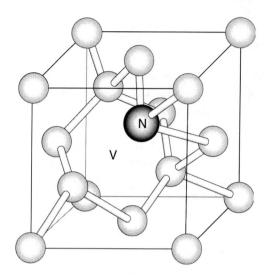

Fig. 11.6.　The crystal structure of the NV⁻ centre in diamond.

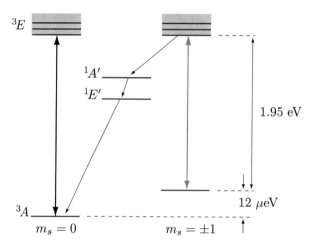

Fig. 11.7.　The energy-level structure of a single NV⁻ centre. The exact structure still causes controversy, but most of the features in this scheme are accepted. See Rogers *et al.* (2008) for more details.

is a triplet with $S = 1$. A corresponding A singlet lies somewhat higher in energy. This triplet is split by the crystal field into a degenerate doublet (corresponding to the $m_s = \pm 1$ Zeeman sublevels), and a single state (corresponding to the $m_s = 0$ sublevel). When linear polarized light is applied to the system with an energy greater than 1.95 eV (corresponding to a wavelength less than 637 nm), electrons can be promoted to the ^3E manifold. The

The collection of single electron levels in the NV⁻ centre then gives us combined states with overall symmetry $^3A_2$, $^1A_1$, and $^1E_1$. The next lowest energy configuration of $a_1^2 a_1' e^3$ provides combined states with symmetry representation ^3E and ^1E.

structure of this set of energy levels is quite complex and not all that well understood; in particular, there are vibronic levels associated with both the ^3E and the ^3A which complicate the structure. However, experimental evidence suggests that electrons promoted to the ^3E level decay quickly to the lowest level in that manifold and, importantly, after this series of transitions have the same spin as they had when they were in the ground level. The spin-preserving optical transition is the essential feature that makes the NV$^-$ centre so attractive as a solid-state qubit.

In order to define a qubit, we need to find two computational basis states $|0\rangle$ and $|1\rangle$ in the NV$^-$ centre. One of the qubit states can be chosen as $m_s = 0$. The second state can be chosen as either $m_s = 1$ or $m_s = -1$. However, these states are degenerate, which may cause addressability problems. To remove the degeneracy, a magnetic field is often applied to the centre in the z direction, which splits the $m_s = \pm 1$ doublet. Then one of these together with the $m_s = 0$ state forms the qubit. Initialization proceeds by optical pumping. Spin is preserved on excitation, but not always when the system decays back to the ground state. In particular, there is a small probability that the $m_s = \pm 1$ state will decay via the singlet levels to the $m_s = 0$ state in a so-called intersystem crossing. This is depicted as the dotted line in Fig. 11.7; the lowest singlet level is long lived and metastable. Pumping for long enough can lead to a spin polarization beyond 80%. See Jelezko and Wrachtrup (2006) for a more detailed discussion.

The asymmetric intersystem crossing also provides a mechanism for optical spin readout. Since there is a strong possibility that electrons in $m_s = \pm 1$ are 'shelved' in the singlet state soon after excitation, the fluorescence signal from the $m_s = 0$ state is around three orders of magnitude larger than for the $m_s = \pm 1$ state, which makes the NV$^-$ system perfect for spin readout, and for observing optically detected magnetic resonance. Perhaps the most promising method for scaling up the NV$^-$ centre qubit to many qubits is to use distributed quantum computing, which we will return to in Section 11.5.

11.3 Interactions in solid-state qubit systems

Controlled interactions between quantum bits lead to the possibility of two-qubit gate operations. In this section, we will describe two of the principal interactions that can exist between solid-state qubits: exciton–exciton and spin–spin coupling. Later, in Section 11.4 we will show that both of these can be exploited to generate quantum entanglement using only optical control.

11.3.1 Exciton–exciton interactions

Many solid-state optical implementations of quantum gates exploit the interactions between exciton-like states in some way. There are several different mechanisms that can be used, and here we will discuss two of the most important: the static dipole–dipole interaction and the transition dipole–dipole or 'Förster coupling'.

Static dipole–dipole interaction

An exciton consists of both a negative charge (the conduction-state electron) and a positive charge (the valence-state hole). If for some reason the electron and hole densities are not symmetrically distributed with respect to each other, a dipole moment can result. Since the spatial wavefunction of the hole must by definition be identical to the wavefunction of the missing electron, we can write this dipole moment as

$$\mathbf{p} = -e \int_{\text{space}} d\mathbf{r}\, |\xi_c(\mathbf{r}) U_c(\mathbf{r})|^2\, \mathbf{r} + e \int_{\text{space}} d\mathbf{r}\, |\xi_v(\mathbf{r}) U_v(\mathbf{r})|^2\, \mathbf{r}, \tag{11.54}$$

where we have made use of the envelope-function forms for the two wavefunctions. Since the parity of the Bloch functions is well defined, we can use a procedure similar to the one that led to Eq. (11.45) to simplify this expression

$$\mathbf{p} = -e \int_{\text{space}} d\mathbf{r}\, \left(|\xi_c(\mathbf{r})|^2 - |\xi_v(\mathbf{r})|^2 \right) \mathbf{r}. \tag{11.55}$$

There will be a finite dipole as long as the electron and hole are not centred at the same position within the nanostructure. This can occur naturally in some systems, or can be induced in others by applying an electric field to separate the electron and hole. If two neighbouring quantum dots A and B each have an exciton with a dipole moment, there is a resulting energy shift called the 'bi-exciton shift', due to the dipole–dipole interaction. It is given by

$$V_{XX} = \frac{1}{4\pi \epsilon_0 \epsilon_r R^3} \left(\mathbf{p}_A \cdot \mathbf{p}_B - \frac{3}{R^2} (\mathbf{p}_A \cdot \mathbf{r})(\mathbf{p}_B \cdot \mathbf{r}) \right), \tag{11.56}$$

where R is the distance between the two nanostructures, as shown in Fig. 11.8. Higher-order contributions, such as dipole–quadrupole, quadrupole–quadrupole, etc., are small as long as R is larger than the spatial dimensions of the nanostructures. For quantum dots spaced by only a few nanometres, V_{XX} can be on the order of a few meV.

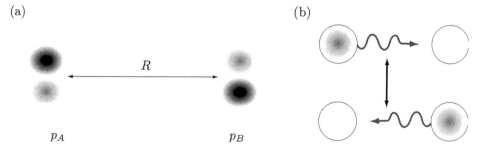

(a)   (b)

R

p_A p_B

Fig. 11.8. (a) The static exciton dipole–dipole interaction relies on the different spatial envelope functions of the electron (light shading) and hole (dark shading). (b) The transition dipole–dipole or Förster interaction relies on strong coupling to the electromagnetic field, which needs overlap between the electron and hole. The two particles are depicted here as perfectly overlapping (grey shading).

Transition dipole–dipole

This interaction occurs when an exciton couples strongly to an optical field. It is possible for an exciton on one quantum dot to emit a photon, thereby decaying to the ground state, and for the photon to be reabsorbed by a neighbouring dot. The dominant interaction of this type for closely spaced dots has an intermediate 'virtual photon'.

The matrix element can be calculated using the Coulomb operator, which couples the two quantum dots together via exchange of virtual photons:

$$V_F = \frac{e^2}{4\pi\epsilon_0\epsilon_r} \int\int d\mathbf{r}_A d\mathbf{r}_B \, \psi_c^*(\mathbf{r}_A)\psi_v^*(\mathbf{r}_B)\frac{1}{\mathbf{r}+\mathbf{r}_A-\mathbf{r}_B}\psi_v(\mathbf{r}_A)\psi_c(\mathbf{r}_B), \qquad (11.57)$$

where the labels $\mathbf{r}_A$ and $\mathbf{r}_B$ represent the position coordinates of the electrons on dots A and B with respect to the centres of those two dots, respectively, and $\mathbf{r}$ is the vector connecting the two centres. The Coulomb operator can be expanded in a Taylor series, and the lowest non-zero term is of dipole–dipole form:

$$V_F = \frac{e^2}{4\pi\epsilon_0\epsilon_r R^3}\left(\mathbf{r}_{cv,A}\cdot\mathbf{r}_{cv,B} - \frac{3}{R^2}(\mathbf{r}_{cv,A}\cdot\mathbf{r})(\mathbf{r}_{cv,B}\cdot\mathbf{r})\right). \qquad (11.58)$$

This time, however, the relevant dipole operators are the transition dipoles rather than the static dipoles. That is:

$$\mathbf{r}_{cv,A} = \int d\mathbf{r}_A \, \psi_c^*(\mathbf{r}_A)\,\mathbf{r}_A\,\psi_v(\mathbf{r}_A) \qquad (11.59)$$

with a similar expression for $\mathbf{r}_{cv,B}$. These transition dipoles are identical to the ones we found previously in Eq. (11.45) for the coupling of an optical field to a single exciton. We showed there that the dipole depends on two factors: first, the overlap of the envelope functions, and second the dipolar coupling of the Wannier functions for valence and conduction states. Under optimized conditions the Förster coupling can be as large as one meV.

The conditions that are needed for strong static dipole–dipole coupling are quite different from those needed for strong transition dipole–dipole coupling. For the former we need the electron and hole to be well separated, whereas the latter requires a large overlap between the envelope functions of the two. It is possible to move between these two different scenarios by choosing the right materials, or by applying an electric field to the sample. We will see in Section 11.4 how each type can be manipulated to create two-qubit gates with the entangling power needed for universal quantum computing.

11.3.2 Spin interactions

As we saw in Section 11.2, the electron spin is a very promising representation of a qubit in the solid state. We have so far discussed how spins can be manipulated optically, and we will see in the next section that interactions between optical excitations can give rise to indirect, controlled, spin–spin interactions. However, direct interactions can provide another way to create entangled states between spins, and can be controlled optically.

For the examples we will discuss later in the next section, we need to introduce only one type of spin coupling: exchange coupling. This coupling is the electrostatic interaction that arises because the combined wavefunction of two electrons must be antisymmetric under the exchange of the two particles. This means that the symmetries of the spatial and spin parts of the wavefunction are linked together. If the two electrons have single-electron spatial wavefunctions that overlap, the exchange symmetry changes the energy of the state – and therefore the symmetric and antisymmetric spin wavefunctions also have different energies. This is expressed mathematically by the Heisenberg Hamiltonian

$$\mathcal{H}_J = \hbar J_{12}\,\boldsymbol{\sigma}_1 \cdot \boldsymbol{\sigma}_2 \,. \tag{11.60}$$

The $\boldsymbol{\sigma}_j$ are the spin operators[4] and J_{12} denotes the strength of the exchange interaction.

Using the spin qubit notation defined in Chapter 2, i.e., $\sigma_z |0\rangle = Z |0\rangle = |0\rangle$ and $\sigma_z |1\rangle = Z |1\rangle = -|1\rangle$, we find that the eigenstates of the Heisenberg Hamiltonian are arranged into a non-degenerate singlet state $(|01\rangle - |10\rangle)/\sqrt{2}$ with eigenvalue $-3\hbar J_{12}$ and three degenerate triplet states, $\{(|10\rangle + |01\rangle)/\sqrt{2}, |00\rangle, |11\rangle\}$ with eigenvalue $\hbar J_{12}$. For more details on the exchange coupling and other possible interaction mechanisms we refer the reader to Ashcroft and Mermin (1976).

11.4 Entangling two-qubit operations

Having introduced some of the different types of interaction that can be found in solid-state systems, we will now discuss how to control these interactions such that they can provide a universal two-qubit gate. We will start with the system that is easiest to control optically, where the qubit is defined as the presence or absence of an exciton.

11.4.1 Optical gates with exciton qubits

As an example, let us consider our qubit $|0\rangle$ to be the ground state of a quantum dot. The lowest-lying exciton will be our qubit $|1\rangle$, and to remove the complications of spin let us assume that we always use ϵ_L polarized light to create a spin projection $J_z = +1$ heavy-hole exciton. Assume that we have two closely spaced quantum dots that are not necessarily the same size: each has its own exciton spectrum. Creating entanglement needs an interaction, and in this example we assume that there is a static dipole–dipole coupling between the two excitons, but negligible transition dipole–dipole interaction. The Hamiltonian for the two-dot system then becomes:

$$\mathcal{H} = \hbar\omega_1 |1\rangle \langle 1| \otimes \hat{\mathbb{I}} + \hbar\omega_2 \hat{\mathbb{I}} \otimes |1\rangle \langle 1| + V_{XX} |11\rangle \langle 11| \,. \tag{11.61}$$

---

[4] The spin operator is a vector of Pauli operators. In this chapter and the next, we will write σ_k for the Pauli matrices of general spin systems. Only when the spin system is also the qubit do we employ the notation X, Y, and Z for the Pauli operators.

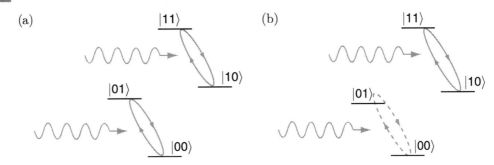

Fig. 11.9. The energy levels of two exciton qubits associated with two adjacent quantum dots. Each dot has a different exciton creation energy and so can be individually addressed. (a) There is no coupling here, so a laser tuned to the $|10\rangle \leftrightarrow |11\rangle$ transition will also be resonant with $|00\rangle \leftrightarrow |01\rangle$. (b) A dipole–dipole interaction is now included so that the $|10\rangle \leftrightarrow |11\rangle$ can be individually addressed. The laser now has little or no effect on the $|00\rangle$ state.

Since the interaction is diagonal in the computational basis, we can immediately see that the eigenstates are simply the computational basis states. Further, since the two dots have different single exciton creation energies, all of the eigenstate transitions are likely to be distinct, as is shown in Fig. 11.9. In particular, the two states $|10\rangle$ and $|11\rangle$ are separated by an energy of $\omega_2 - V_{XX}$. This transition is dipole-allowed and a laser on resonance will cause Rabi flopping between these two levels.

Exercise 11.4: Show that a π pulse of a laser tuned to the $|10\rangle \leftrightarrow |11\rangle$ transition effects the universal two-qubit gate:

$$U = \begin{pmatrix} 1 & 0 & 0 & 0 \\ 0 & 1 & 0 & 0 \\ 0 & 0 & 0 & -i \\ 0 & 0 & i & 0 \end{pmatrix}. \tag{11.62}$$

This is a CY gate, which is sometimes called a 'controlled rotation gate', and can be used to create maximally entangled states.

In order to pick out the $|10\rangle \leftrightarrow |11\rangle$ the laser bandwidth must be narrower than the coupling energy V_{XX}, and the CY gate can be executed in a time no shorter than $\sim \hbar/V_{XX}$. For a V_{XX} of a few meV, this results in a gating time of less than one picosecond.

Static dipole–dipole coupling is not the only interaction that allows the generation of an entangled state. In fact, almost all interactions will do it. However, generation of entanglement is not particularly useful unless it is controlled: it must be possible to modulate the coupling such that, once created, the entangling component of the evolution can be stopped and the entanglement preserved. This can be more difficult with non-diagonal interactions, but it is still possible as we will now see.

Instead of the diagonal V_{XX} coupling, we introduce the off-diagonal transition dipole–dipole, or Förster coupling V_F. Let us again consider two detuned quantum dots that are always excited by ϵ_L light. For two quantum dots stacked along the z direction, $\mathbf{r} = R\hat{\mathbf{z}}$, one can use the analysis of Section 11.3.1 to show that the Förster transfer preserves the spin of the exciton as it transfers. We can therefore again restrict our Hamitonian to excitons with spin $+1$, which are our qubit $|1\rangle$ states. Thus we obtain

$$\mathcal{H} = \hbar\omega_1 |1\rangle\langle1| \otimes \hat{\mathbb{I}} + \hbar\omega_2 \hat{\mathbb{I}} \otimes |1\rangle\langle1| + V_F(|10\rangle\langle01| + |10\rangle\langle01|). \tag{11.63}$$

Two computational basis states remain eigenstates: $|00\rangle$ and $|11\rangle$. The states $|10\rangle$ and $|01\rangle$ are mixed by the interaction; the degree of mixing depends on the value of V_F and the detuning $\nu = \omega_1 - \omega_2$. The eigenstates are

$$|\psi_+\rangle = \cos\frac{\theta}{2} |10\rangle + \sin\frac{\theta}{2} |01\rangle \qquad \text{and} \tag{11.64}$$

$$|\psi_-\rangle = -\sin\frac{\theta}{2} |10\rangle + \cos\frac{\theta}{2} |01\rangle\,, \tag{11.65}$$

which have energies

$$\lambda_\pm = \frac{\hbar\omega_1 + \hbar\omega_2}{2} \pm \sqrt{\frac{(\hbar\nu)^2}{4} + V_F^2}\,. \tag{11.66}$$

The mixing angle θ is given by

$$\theta = \arctan\left(\frac{2V_F}{\hbar\nu}\right). \tag{11.67}$$

We now look at some different parameter regimes. First, if $V_F \ll \hbar\nu$, the eigenstates are almost exactly the computational basis states $|10\rangle$ and $|01\rangle$. In this case an unentangled state will not evolve into an entangled one since the Förster coupling is effectively suppressed. However, in the opposite limit, $V_F \gg \hbar\nu$, the eigenstates become perfect Bell states: $(|10\rangle \pm |01\rangle)/\sqrt{2}$. In this case, entanglement generation can occur, and we can understand this as follows.

Since the $|00\rangle$ and $|11\rangle$ states are effectively uncoupled from all other states, we know that any wavefunction amplitude associated with them will simply evolve in time by accumulating a phase. The interesting dynamics happen in the subspace spanned by $|10\rangle$ and $|01\rangle$, which is two-dimensional and so can be represented on a Bloch sphere, as shown in Fig. 11.10. Let the poles of the sphere be the two unentangled states $|10\rangle$ and $|01\rangle$. Then all points on the equator are equal superpositions of $|10\rangle$ and $|01\rangle$: maximally entangled states.

Imagine we initialize in state $|10\rangle$ and prepare the system in the $V_F \ll \hbar\nu$ regime; nothing happens apart from our state accumulating a phase. Now imagine that we can suddenly switch to the $V_F \gg \hbar\nu$ regime. The eigenstates are then the maximally entangled states, represented by diametrically opposite points on the equator. The time evolution is now a precession around the axis connecting the two eigenstates at a frequency given by their energy difference $2V_F\hbar$. Therefore, our initial state starts to rotate towards the equator. After a time $\tau = \hbar\pi/(4V_F)$ the state has reached the equator and is maximally entangled. We can prevent it going any further by switching back to the $V_F \ll \hbar\nu$ regime, whence the state simply evolves in the equatorial plane. This sequence of controlled entanglement generation is equivalent to a universal two-qubit gate.

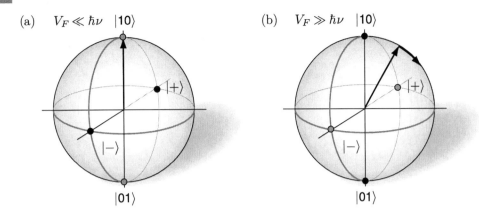

Fig. 11.10. The Bloch-sphere representation of the two-dimensional subspace spanned by $|01\rangle$ and $|10\rangle$. The poles represent the two product states and the equator the set of maximally entangled states. (a) In the limit $V_F \ll \hbar\nu$ the eigenstates (marked as grey circles) are at the poles, so the state $|10\rangle$ (shown as an arrow) only accumulates an overall phase as time passes. (b) In the limit $V_F \gg \hbar\nu$ the eigenstates are on the equator, so $|10\rangle$ rotates towards $|01\rangle$.

How can we move between the two regimes so quickly? The most obvious strategy might be to modulate V_F in some way. For example, one could apply an electric field to the quantum-dot structure that would have the effect of moving the electron and hole part, thus reducing their transition dipole. However, only a moderate amount of modulation can be achieved with reasonable fields, and this process is quite slow. An alternative is to change ν. This can be done quickly and the modulation can be quite large if we exploit the so-called AC Stark effect: if a detuned laser is applied to an excitonic transition with a coupling strength Ω and a detuning δ, then second-order perturbation theory shows us that this shifts the transition energy by an amount $\hbar\Omega^2/4\delta$. Excitons in adjacent quantum dots typically have different oscillator strength and energy, and so different values of Ω and δ; their energy shifts are therefore distinct and it is possible to tune the two excitons in and out of resonance in this way (see Fig. 11.11). Exploiting the AC Stark shift is a particularly promising idea since the control is provided by a laser that can be pulsed straightforwardly on the timescales needed to create entanglement (typically picoseconds). In Fig. 11.12 we show a numerical simulation in this two-dot system for an entangling laser pulse of $\hbar\pi/4V_F \approx 5$ ns. The figure of merit we use is the entanglement of formation (see Chapter 3).

11.4.2 Optical gates with spin qubits

Let us now turn to the longer-lived spin qubit, and see how we can use excitonic interactions to create two-qubit entanglement in this system. It is possible to use both static and transition dipole–dipole coupling to do this, but we will not go through the details of both; rather we will focus on the latter. We refer to Pazy *et al.* (2003) for a discussion of static dipole–dipole gates.

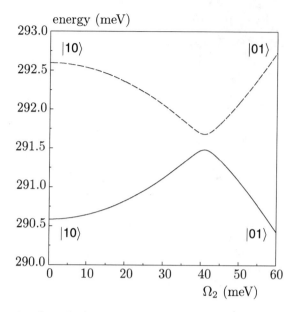

The eigenstate energies of two Förster coupled quantum dots as a function of the coupling strength of an applied laser field to the second dot (Ω_2). A constant ratio of couplings to the two dots $\Omega_1/\Omega_2 = 0.55$ is assumed. The two detunings are $\hbar\delta_1 = 292.59$ meV and $\hbar\delta_2 = 290.59$ meV. $V_F = 0.1$ meV. (Figure adapted from Nazir *et al.* 2004b.)

Using the results of previous sections we can write down the Hamiltonian for two Förster-coupled quantum dots under illumination with ϵ_L light. Remember that only heavy-hole trions with spin +1 can be produced from the $|1\rangle$ state, and assume that the dots are arranged such that the angular momentum of the trion is preserved when it is transferred. We obtain:

$$
\mathcal{H}(t) = \hbar\omega_T \left(|T_h\rangle \langle T_h| \otimes \hat{\mathbb{I}} + \hat{\mathbb{I}} \otimes |T_h\rangle \langle T_h| \right)
$$

$$
+ \hbar\Omega \cos\omega_l t \left(|1\rangle \langle T_h| \otimes \hat{\mathbb{I}} + |T_h\rangle \langle 1| \otimes \hat{\mathbb{I}} + \hat{\mathbb{I}} \otimes |1\rangle \langle T_h| + \hat{\mathbb{I}} \otimes |T_h\rangle \langle 1| \right)
$$

$$
+ V_F \left(|1, T_h\rangle \langle T_h, 1| + |T_h, 1\rangle \langle 1, T_h| \right) , \tag{11.68}
$$

where ω_T is the heavy-hole trion creation energy, ω_l is the laser frequency.

Let us examine the eigenstate structure of this Hamiltonian, which now spans a nine-dimensional Hilbert space. This space splits into four non-interacting subspaces: $\{|00\rangle\}$, $\{|11\rangle, |1, T_h\rangle, |T_h, 1\rangle, |T_h T_h\rangle\}$, $\{|01\rangle, |0, T_h\rangle\}$, and $\{|10\rangle, |T_h, 0\rangle\}$, which considerably simplifies the analysis. Fig. 11.13 shows the spectrum of eigenstates in each of the four subspaces *with the laser turned off.*

The figure shows that the coupling V_F has no effect on the subspaces that contain $|10\rangle$, $|01\rangle$, and $|00\rangle$. On the other hand, the two single-trion states contained in the $|11\rangle$ subspace have their degeneracy lifted by V_F such that the eigenstates become $(|1, T_h\rangle \pm |T_h, 1\rangle)/\sqrt{2}$.

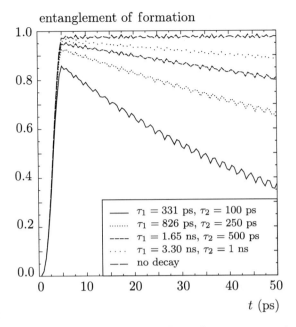

entanglement of formation

Fig. 11.12. Entanglement of formation of a system prepared in the product state $|01\rangle$ and subjected to a laser pulse of duration 5 ns. The pulse is designed to induce AC Stark shifts in the two dots that tune them into resonance, allowing the Förster coupling to induce resonant exciton transfer. The evolution is stopped when the state is maximally entangled. E_F is plotted for a range of exciton decay times τ. (Figure adapted from Nazir *et al.* 2004b.)

Exercise 11.5: Assuming that the transition dipoles for the two quantum dots are equal, show that the transition $|11\rangle \leftrightarrow (|1, T_h\rangle - |T_h, 1\rangle)/\sqrt{2}$ is dipole-forbidden, but that $|11\rangle \leftrightarrow (|1, T_h\rangle + |T_h, 1\rangle)/\sqrt{2}$ is dipole-allowed.

The energy level shifts of the single-exciton states in the $|11\rangle$ space mean that we can uniquely address the transition $|11\rangle \leftrightarrow (|1, T_h\rangle - |T_h, 1\rangle)/\sqrt{2}$ by choosing a laser frequency $\omega_l = \omega_T - V_F$, and this is the basis of our two-qubit gate. Starting out in a computational basis state, then applying a 2π pulse with the tuned laser, we find the following transformation

$$|00\rangle \rightarrow |00\rangle \,,$$
$$|01\rangle \rightarrow |01\rangle \,,$$
$$|10\rangle \rightarrow |10\rangle \,,$$
$$|11\rangle \rightarrow -|11\rangle \,. \tag{11.69}$$

This is exactly a cz gate, performed on the spin qubits, using an interaction between excitons. The time in which the gate can be performed is limited only by the size of V_F: the laser coupling strength must be weaker than V_F to avoid unwanted excitations in the other subspaces. For a typical system this would give a minimum gate time of a few picoseconds.

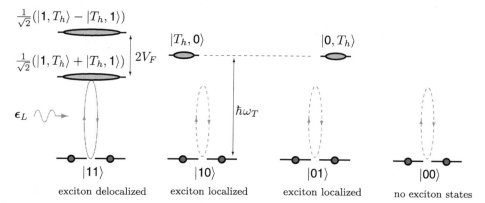

Fig. 11.13. Eigenstate spectrum for the four uncoupled subspaces that describe two Förster-coupled quantum dots, each of which house a spin qubit. See Nazir *et al.* (2004b) for details.

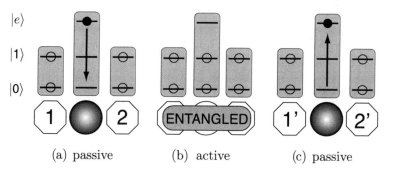

(a) passive   (b) active   (c) passive

Fig. 11.14. Diagram showing the mechanism for entangling two logical qubits 1 and 2. (a) The system evolution starts when the central control bit is brought down optically from the excited level $|e\rangle$. (b) The natural spin–spin dynamics in the system causes the three spins to become entangled with one another. (c) After a characteristic 'revival time' τ the control returns to $|0\rangle$ and is immediately shelved into $|e\rangle$ to prevent further evolution. Figure adapted from Benjamin *et al.* (2004b).

11.4.3 Passivating spin–spin interactions optically

Another possible route to performing optical entangling gates with spin qubits is to take an interacting chain of spins and to modulate the interaction using optics. This idea followed that of Benjamin and Bose (2004). Imagine, for example, the three-spin chain depicted in Fig. 11.14. The interactions between the neighbouring spins in the chain might be described by the Heisenberg Hamiltonian in Eq. (11.60). If the spins are placed in a magnetic field, they will have a Zeeman splitting E_j,[5] such that the overall Hamiltonian for the spin chain is:

$$\mathcal{H} = \sum_{j=1}^{3} E_j \sigma_j^Z + \sum_{j=1}^{2} J \boldsymbol{\sigma}_j \cdot \boldsymbol{\sigma}_{j+1}$$

$$\equiv \sum_{j=1}^{3} E_j Z_j + J \sum_{j=1}^{2} \left(X_j X_{j+1} + Y_j Y_{j+1} + Z_j Z_{j+1} \right). \qquad (11.70)$$

An important property of a system described in this way is that it *conserves total spin projection*. Therefore, if we define spin down as $|0\rangle$ and spin up as $|1\rangle$ for all three spins, then if we initialize the system in $|100\rangle$, say, it can only evolve into states with amplitudes for $|010\rangle$ and $|001\rangle$. This conservation law makes the analysis of the dynamics much simpler, since we can treat the four subspaces corresponding to total spin $-\frac{3}{2}, -\frac{1}{2}, \frac{1}{2}$ and $\frac{3}{2}$ separately.

We proceed by defining two logical qubits at the ends of the short chain (i.e., spins 1 and 3). The central qubit can now act as a 'control' species, which through suitable manipulation can be used to induce an entangling gate on the two logical qubits. In addition, we can arrange the interaction such that the control qubit does not become entangled with the logical qubits. Each logical qubit can of course start in any state, but we can work out the unitary evolution that must occur by taking each computational basis state in turn. For simplicity, let us assume that the control qubit starts out in the state $|0\rangle$, and that the magnetic field, and therefore the Zeeman splitting of all the spins, is zero.

The initial logical state $|00\rangle_L$ lives in the one-dimensional subspace that consists only of $|000\rangle$. This eigenstate has eigenvalue zero, so $|00\rangle_L$ stays the same for all time. The initial logical states $|10\rangle_L$ and $|01\rangle_L$ live in the same subspace, which has three component basis states $\{|100\rangle, |010\rangle, |001\rangle\}$. We calculate the evolution in this subspace in the next exercises.

Exercise 11.6: Write out the Hamiltonian for this three-dimensional subspace in a matrix form and show that one of the eigenstates must have the form $|\psi^-\rangle \equiv (|100\rangle - |001\rangle)/\sqrt{2}$. Next, write out the Hamiltonian in the basis $\{|\psi^+\rangle |\psi^-\rangle, |\phi\rangle\} \equiv \{(|100\rangle + |001\rangle)/\sqrt{2}, (|100\rangle - |001\rangle)/\sqrt{2}, |010\rangle\}$. Third, show therefore that the state $|\psi_+\rangle$ returns to itself after a time $\tau = \pi\hbar/3J$, but with an accrued phase. The time τ is a 'revival' time, at which point the control bit returns to state $|0\rangle$ and is unentangled with its neighbours.

Exercise 11.7: Demonstrate that the effective unitary gate that occurs between $|10\rangle_L$ and $|01\rangle_L$ at time τ is

$$U_2 = -\frac{1}{2} \begin{pmatrix} \exp(i\pi/3) - 1 & \exp(i\pi/3) + 1 \\ \exp(i\pi/3) + 1 & \exp(i\pi/3) - 1 \end{pmatrix}. \qquad (11.71)$$

Finally, the initial state $|11\rangle_L$ lives in another three-dimensional subspace, but again revives to itself after the time τ, undergoing the transformation $|11\rangle_L \rightarrow$

---

[5] Zeeman splitting is the name given to the separation of spin energy levels that results from the application of a magnetic field.

$-\exp(-i\pi/3)\,|11\rangle_L$. The combined unitary evolution of all four logical qubits after the time τ is a universal entangling gate.

We have so far ignored the most difficult challenge here: how do we stop the evolution once the gate has been executed? In fact, a number of methods have been proposed for this, including a direct modulation of the Zeeman splitting of the control spin, which can prevent energy transfer between spins under certain conditions. A possibly simpler method is optical control. If the central spin has an addressable optical transition between $|0\rangle$ and some higher state $|e\rangle$, then we might resonantly excite that transition once the central qubit has revived; a resonant π pulse promotes all population to $|e\rangle$, which in an ideal system would switch off all spin–spin interactions. A potential problem with this kind of scheme is that spontaneous emission could return the control back to its active spin states. However, there are more sophisticated techniques which never require population of $|e\rangle$: optical control is used to shift levels that are not populated, thus bringing the three spins out of resonance with one another. For more details, see Benjamin *et al.* (2004).

11.4.4 Optical control of spin–spin exchange interactions

We have seen how exchange interactions can be passivated using an excitation to a higher-lying energy level. Another approach, which has much in common with this, is to *generate* an exchange interaction using excitation to a higher-lying level.

Certain defects in solids, for example bismuth or phosphorus in silicon, have well-localized ground states $|g\rangle$, but somewhat more delocalized optically excited states $|e\rangle$. This is a similar effect to that observed in the hydrogen atom, for example, whose more energetic electronic states have a much larger mean distance from the nucleus. Imagine now that one of these defects acts as a control species C, and is placed near two defects that act as spin qubits Q and Q', and which do not have such properties. If the control is in state $|g\rangle$ there is no overlap with the nearby qubits, and from the argument of Section 11.3, there is no exchange interaction. On the other hand, if the control is excited to $|e\rangle$ an exchange coupling would be expected. We would describe their Hamiltonian as

$$\mathcal{H} = E_Q \sigma_z^Q + E_C \sigma_z^C + E_{Q'} \sigma_z^{Q'} \tag{11.72}$$
$$+ \left(\hbar J_{QC} \boldsymbol{\sigma}^Q \cdot \boldsymbol{\sigma}^C + \hbar J_{Q'C} \boldsymbol{\sigma}^{Q'} \cdot \boldsymbol{\sigma}^C + \hbar \omega_0 \right) |e\rangle \langle e|.$$

E_Q represents the Zeeman splitting of spin Q (with similar notation for the other spins); J_{QC} ($J_{Q'C}$) is the exchange interaction, assumed isotropic, between Q (Q') and C' which only exists when C is in its excited state $|e\rangle$. ω_0 is the optical excitation energy of C.

A gate operation can be performed in a similar way to that discussed in the previous section. The control is initialized in some well-defined state and it is then optically excited, turning the exchange coupling on. The system is now allowed to evolve naturally and after a certain period of time the control returns to its initial state – at which point a second optical pulse returns it to state $|g\rangle$. It can be shown that, for $J_{QC} = J_{Q'C}$ there are revival times at which a non-trivial two-qubit gate operation is applied to Q and Q'.

11.5 Scalability of solid-state devices

So far, we have focused our attention on small quantum devices consisting of at most two qubits. This is of course far from what is needed for a useful quantum computer, even for small-scale applications. It is therefore necessary to scale up to many qubits. This presents quite a difficulty: it is not obvious how any of the ideas presented so far could be scaled. All of our single-qubit operations require the individual addressing of a single nanostructure with a laser, but the beam waist of a laser is at least on the order of an optical wavelength, i.e., a few hundred nanometres. However, if there is to be a sizable qubit–qubit interaction, the spacing between nanostructures must be only a few nanometres. Therefore individual addressing is impossible using only spatial addressing methods. Some progress can be made by employing optical fields with different frequencies to address individual nanostructures, although this is inherently limited by the bandwidth of the optical pulses, especially if they are to be delivered quickly enough to make viable quantum gates. A combination of both spatial and spectral selectivity may provide a possible route to proper scalability. However, there are some interesting alternatives to this strategy, which require a shift of design perspective.

11.5.1 Path erasure

One alternative route to scalability is to use the entangling operations based on path-erasure techniques developed in Chapter 7. A spin qubit in a quantum dot maps perfectly onto the double-heralding scheme. The basic requirement for such a scheme to work is for an 'L'-type energy-level structure: two low-lying levels that define the qubit and a single level lying at an optical energy above the first two. This higher level must couple only optically to one of the two-qubit basis states. The higher-lying level for the spin qubit is a heavy-hole trion, and the optical selectivity is provided by using a single resonant laser with circular polarization. The NV$^-$ centre can also be used in this way. The qubit is the $m_s = 0$ and one of the $m_s = \pm 1$ ground electron-spin states, split in a magnetic field. The combination of magnetic and zero field splitting means that a spin-dependent optical transition can be found using frequency tuning.

11.5.2 Global control

An alternative to measurement-based techniques is so-called 'global control'. It is a natural extension of the system described by the Hamiltonian in Eq. (11.70), but with a slightly modified interaction term, and with a much longer chain of interacting spins. The global-control technique was designed to circumvent the need for individual addressing, and it is possible to perform universal quantum computing in a repeating chain of physical qubits simply by applying control commands globally, i.e., to the entire chain simultaneously.

A variety of schemes have been proposed in this vein, and following Benjamin (2000) we will focus here on one of the simpler ones that is most applicable to a chain of optically active solid-state nanostructures. The repeating unit of the chain is a pair of distinct qubits, and we assume that these are excitons in two adjacent quantum dots, which have different transition energies. The two will be labelled 'A' and 'B', and have energies ω_A and ω_B, respectively. This chain then repeats, forming an '$ABABAB\ldots$' structure. Suppose that two adjacent spin systems are coupled via the Ising interaction, which is of the form $\sigma_{z1}\sigma_{z2}$. A small ABA section of chain then has the Hamiltonian

$$\mathcal{H} = \hbar\omega_A(Z_1 + Z_3) + \hbar\omega_B Z_2 + \hbar J(Z_1 Z_2 + Z_2 Z_3)\,. \tag{11.73}$$

If we now want to apply, say, a π pulse on the 'B' qubit, the frequency that our laser needs to have is determined by the state of the neighbouring 'A' qubits. If both are in state $|0\rangle$, the required frequency is $\omega_B + 2J$; if both are $|1\rangle$ it is $\omega_B - 2J$, and otherwise it is ω_B. These neighbour-dependent frequencies allow us to apply global update rules to all of the B qubits. For example, we could get all of the 'B' qubits to flip (i.e., have an X operation applied to them) as long as their neighbours had opposite spin orientations simply by applying a pulse of frequency ω_B.

The two spins at the ends of the chain are unique since they have only one neighbour, and this can be used to feed an initial state onto the chain using a 'shift register' set of global update instructions, as shown in Fig 11.15. The chain is initially cooled to the ground state with all qubits in $|1\rangle$. The far left end qubit can then be addressed using a laser of frequency $\omega_A - J$, and any end-qubit superposition can be made with a suitable laser pulse. Let the resulting state be

$$|\psi_1\rangle = (\alpha\,|0\rangle + \beta\,|1\rangle) \otimes |1111\ldots\rangle = \alpha\,|01111\ldots\rangle + \beta\,|11111\ldots\rangle\,. \tag{11.74}$$

The state of the chain end spin can now be shifted along by applying two global pulses. The first applies the rule 'flip if your neighbours have opposite states to each other' to *all* the B spins, giving

$$|\psi\rangle = \alpha\,|00111\ldots\rangle + \beta\,|11111\ldots\rangle\,. \tag{11.75}$$

The second addresses the chain end spin once more, applying the rule 'flip if your neighbour is in state $|0\rangle$', giving

$$|\psi\rangle = |1\rangle \otimes (\alpha\,|0\rangle + \beta\,|1\rangle) \otimes |1111\ldots\rangle\,. \tag{11.76}$$

This state is simply the initial state shifted onto the second qubit. Similarly, the information can be shifted to the third qubit using two further flip pulses, one with frequency ω_A and the second with frequency ω_B. Transfer to the fourth qubit then needs pulses at ω_A, then ω_B. These last four pulses can then be repeated to move the information to the desired position along the chain.

The clever trick we have just discussed gives the appearance of individual addressability, even though in reality we can only uniquely address the chain ends. This is a very powerful tool, which goes beyond simple state initialization. Using an encoding of four physical qubits (P) per logical qubit (L), with $|0\rangle_L \equiv |0011\rangle_P$ and $|1\rangle_L \equiv |1100\rangle_P$, Benjamin showed that a universal set of quantum gates can be performed efficiently using just global pulses of the type we have just discussed.

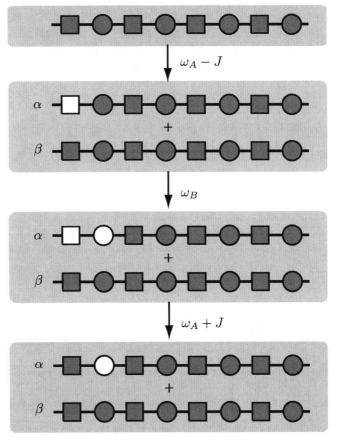

Fig. 11.15. A chain of alternating qubit species 'A' (squares) and 'B' (circles) can serve as a universal quantum computer. The topmost figure shows the initial state, with all qubits in the state $|1\rangle$ (dark grey). Applying appropriate pulses with frequency $\omega_A - J$ allows one to prepare an arbitrary state of the chain end species: the $|0\rangle$ component is shown in white. A series of two further global pulses allows this state to be transferred to the adjacent 'B' qubit, and this process can be repeated to prepare any initial state using only global pulses.

11.6 References and further reading

There are many good books on solid-state physics. For an introductory level book, we recommend Myers (1997) and for a more advanced treatment we refer to Ashcroft and Mermin (1976). The optical properties of semiconductors are covered in Basu (1997) and quantum heterostructures are discussed in Harrison (2005); the envelope-function approximation is discussed in Burt (1999). For a comprehensive review of crystal defects, we refer to the book by Stoneham (1975).

There are many schemes for quantum computing using excitons in quantum dots. Selected papers include Biolatti *et al.* (2002), Troiani *et al.* (2000), and Lovett *et al.*

(2003). Creating two-qubit entanglement using optical control of spin qubits has been discussed by Pazy *et al.* (2003), Calarco *et al.* (2003), and Nazir *et al.* (2004b). Control of spin–spin interactions using optics has, for example, been proposed by Stoneham *et al.* (2003). Finally, global-control protocols were envisaged by Lloyd (1993) and developed by Benjamin (2000).

Solid-state systems, by their very nature, have a vast number of different possible quantum degrees of freedom. In Chapter 11, we saw that some of these degrees of freedom make good qubits. However, there are plenty more which are less suitable, since they cannot easily be localized and externally controlled. Once the qubit has been chosen, it is important to think about how it interacts with the other, uncontrolled quantum excitations in its environment. Such an interaction leads to unpredictable behaviour and can cause decoherence – the irretrievable loss of quantum information from the qubit – and this will be the topic of this chapter. The most obvious decoherence mechanism for any optical manipulation scheme is the spontaneous emission of photons. The theory behind this follows analogously from the theory we discussed in Chapter 7, with a suitable definition of a transition dipole for the relevant transitions. However, solid-state systems bring with them lattice vibrations, or phonons, which have no direct atomic analogue. We will therefore focus on phonons in this chapter, first discussing how we model them, and second how they interact with the electron-based qubit that we discussed in the last chapter. Later, we will see how this leads to a loss of coherence, and how optical methods can be used to slow the rate of coherence loss. Phonon interactions are complex and not easy to model exactly, but we will show that with certain approximations very successful theories can be developed. In the last part of the chapter we will discuss a particular phonon–electron system that can be solved exactly.

12.1 Phonons

Let us consider first the simplest system which captures the essential physics of lattice vibrations: a uniform linear chain of N atoms. The atoms have a mass m and have equilibrium positions that are spaced by a distance a. We label the displacement from equilibrium of atom l by u_l, and assume that nearest-neighbour atoms interact electrostatically (Fig. 12.1). If we expand this interaction energy in a Taylor series in the displacement and keep terms up to second order, we obtain the Hamiltonian:

$$\mathcal{H} = \sum_l \left(\frac{1}{2m} p_l^2 + \frac{g}{2} (u_l - u_{l+1})^2 \right), \tag{12.1}$$

where p_l is the momentum operator conjugate to u_l. Cutting off at second order is called the 'harmonic approximation', and the 'force constant' g parameterizes the second-order term in the expansion. The first-order term vanishes, since we assume that the net force on the atoms at equilibrium is zero.

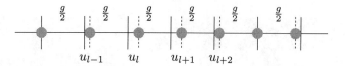

Fig. 12.1. Schematic diagram illustrating the simple model of a one-dimensional chain of atoms. The solid lines represent the equilibrium position of each atom, and the dashed lines the instantaneous positions which are displaced from equilibrium.

It is straightforward to solve this Hamiltonian by transforming to the 'normal modes' of the chain; this removes the coupling between oscillators and allows one to recast the problem in terms of independent modes. We define

$$u_k = \frac{1}{\sqrt{N}} \sum_l e^{-ikal} u_l \quad \text{and} \tag{12.2}$$

$$p_k = \frac{1}{\sqrt{N}} \sum_l e^{ikal} p_l \,. \tag{12.3}$$

The allowed wave vectors are found by applying 'periodic boundary conditions', where the N sites are imagined to be joined up in a circle.[1] This leads to $u_l = u_{l+N}$, yielding $k = 2\pi n/Na$. We then rewrite Eq. (12.1) as

$$\mathcal{H} = \frac{1}{2m} \sum_k \left(p_k p_{-k} + m^2 \omega_k^2 u_k u_{-k} \right), \tag{12.4}$$

with $\omega_k^2 = (4g/m) \sin^2(ka/2)$.

Exercise 12.1: Show that the Hamiltonian in Eq. (12.4) is the normal mode expansion of the Hamiltonian in Eq. (12.1).

Next, we quantize the normal modes by promoting the position and momentum variables to operators $\hat{u}_k$ and $\hat{p}_k$. We can use the usual commutation relations of conjugate variables to show that the operators

$$\hat{a}_k = \sqrt{\frac{m\omega_k}{2\hbar}} \left(\hat{u}_k + \frac{i}{m\omega_k} \hat{p}_{-k} \right) \quad \text{and} \tag{12.5}$$

$$\hat{a}_k^\dagger = \sqrt{\frac{m\omega_k}{2\hbar}} \left(\hat{u}_k - \frac{i}{m\omega_k} \hat{p}_{-k} \right) \tag{12.6}$$

have the standard bosonic commutators

$$\left[\hat{a}_k, \hat{a}_{k'}^\dagger \right] = \delta_{k',k} \quad \text{and} \quad [\hat{a}_k, \hat{a}_{k'}] = \left[\hat{a}_k^\dagger, \hat{a}_{k'}^\dagger \right] = 0 \,. \tag{12.7}$$

[1] See Ashcroft and Mermin (1976) for a detailed discussion of periodic boundary conditions.

They allow us to write the Hamiltonian in the form

$$\mathcal{H} = \hbar \sum_k \omega_k \left(\hat{a}_k^\dagger \hat{a}_k + \frac{1}{2} \right) . \tag{12.8}$$

This is entirely analogous to the Hamiltonian we derived in Chapter 1 for photons in the quantum theory of light. The main difference here is that the Hamiltonian describes a discrete set of modes. We can therefore think of $\hat{a}_k$ and $\hat{a}_k^\dagger$ as the annihilation and creation operators for lattice vibrations. Each quantum of excitation in mode k has an energy ω_k and is called a phonon. More specifically it is an 'acoustic phonon'.[2] The eigenstates are products of the equivalent of Fock states for phonons:

$$|\psi\rangle = |n_1, n_2, ..., n_N\rangle = \prod_k \frac{(\hat{a}_k^\dagger)^n}{\sqrt{n_k!}} |0\rangle . \tag{12.9}$$

In three dimensions the theory is slightly more complicated, since neighbours in all directions must be taken into account. Our momenta and displacement operators therefore become vectors and the Hamiltonian becomes:

$$\mathcal{H} = \frac{1}{2m} \sum_l \mathbf{p}_l^* \cdot \mathbf{p}_l + \frac{1}{2} \sum_{l,l'} \mathbf{u}_l \cdot \mathbf{g}_{ll'} \cdot \mathbf{u}_{l'}. \tag{12.10}$$

The 'force constant' is now a tensor $\mathbf{g}$, which accounts for the influence of each displacement on all of the surrounding displacements. This equation looks pretty complicated, but we can more or less follow the same prescription to solve it as before. First, we define some normal modes that are characterized by a three-dimensional wave vector $\mathbf{k}$:

$$\mathbf{u_k} = \frac{1}{\sqrt{N}} \sum_l e^{-i\mathbf{k}\cdot\mathbf{R}_{l0}} \mathbf{u}_l , \tag{12.11}$$

$$\mathbf{p_k} = \frac{1}{\sqrt{N}} \sum_l e^{i\mathbf{k}\cdot\mathbf{R}_{l0}} \mathbf{p}_l , \tag{12.12}$$

where $\mathbf{R}_{l0}$ is the equilibrium position of atom l. Applying periodic boundary conditions in three dimensions leads to allowed $\mathbf{k}$ vectors only within the first Brillouin zone of the crystal (see Chapter 11). After some manipulation, we obtain

$$\mathcal{H} = \frac{1}{2} \sum_\mathbf{k} \mathbf{p_k}^* \cdot \mathbf{p_k} + \mathbf{u_k}^* \cdot \mathbf{g}(\mathbf{k}) \cdot \mathbf{u_k} , \tag{12.13}$$

where

$$\mathbf{g}(\mathbf{k}) = \sum_n \mathbf{g}_{rn} e^{i\mathbf{k}\cdot(\mathbf{R}_{n0}-\mathbf{R}_{r0})} . \tag{12.14}$$

[2] The other kind of phonon is called an optical phonon, and can occur only when the repeating unit in a lattice consists of more than one atom.

This expression is independent of the label r, since we assume that each atom is in the same environment as every other atom, and we ignore edge effects. Also, $\mathbf{g}(\mathbf{k})$ does not necessarily point in the same direction as the $\mathbf{u_k}$ and so our Hamiltonian is not yet in the diagonal form we desire. However, there are three directions in which the displacements are in the same direction as $\mathbf{g}(\mathbf{k})$: they are the principal axes of the original $\mathbf{g}$ tensor. By *defining* three axes of vibration λ along these directions, we do obtain the diagonal form we need. Then the three-dimensional creation and annihilation operators are

$$\hat{a}_{\mathbf{k}\lambda} = \sqrt{\frac{m\omega_{\mathbf{k}\lambda}}{2\hbar}} \left(\hat{u}_{\mathbf{k}\lambda} + \frac{i}{m\omega_{\mathbf{k}\lambda}} \hat{p}_{-\mathbf{k}\lambda} \right), \tag{12.15}$$

$$\hat{a}_{\mathbf{k}\lambda}^{\dagger} = \sqrt{\frac{m\omega_{\mathbf{k}\lambda}}{2\hbar}} \left(\hat{u}_{\mathbf{k}\lambda} - \frac{i}{m\omega_{\mathbf{k}\lambda}} \hat{p}_{-\mathbf{k}\lambda} \right), \tag{12.16}$$

where $\omega_{\mathbf{k}\lambda}$ is the eigenenergy of the mode characterized by $\mathbf{k}$ and λ. The Hamiltonian then takes the form:

$$\mathcal{H} = \hbar \sum_{\mathbf{k},\lambda} \omega_{\mathbf{k}\lambda} \left(\hat{a}_{\mathbf{k}\lambda}^{\dagger} \hat{a}_{\mathbf{k}\lambda} + \frac{1}{2} \right). \tag{12.17}$$

12.2 Electron–phonon coupling

Having established the Hamiltonian that describes the vibrations of the lattice, we must now look at how phonons affect the coherence of qubits in solid-state environments. Let us therefore think about how an electron in a solid responds when the lattice vibrates.

In the last chapter we used Bloch's theorem to find the form of electron eigenstates in the periodic potential of a regular array of atoms in a crystal. However, our model ignored any movements in the atoms, which we must now consider. It is reasonable to expect that the interaction between an electron at position $\mathbf{r}$ and an atom at position $\mathbf{R}$ depends only on the distance between them. We therefore write for all the atoms and electrons in a crystal

$$\mathcal{H}_{ea} = \sum_{jl} V(\mathbf{r}_j - \mathbf{R}_l). \tag{12.18}$$

Let us explicitly put in the atom equilibrium positions $\mathbf{R}_{l0}$, since we already know the electron solution to the resulting periodic potential. We will also assume that the displacements from this equilibrium are small, so that we can expand in a Taylor series:

$$\mathcal{H}_{ea} = \sum_{jl} V(\mathbf{r}_j - \mathbf{R}_{l0}) + \mathbf{u}_l \cdot \nabla V(\mathbf{r}_j - \mathbf{R}_{l0}) + O(u^2). \tag{12.19}$$

The first term here is the periodic potential that appeared in the electron Hamiltonian we studied in the last chapter, and whose eigenstates are Bloch wavefunctions. The phonons couple to the electrons via the first derivative and the higher-order terms. Let us assume displacements small enough that we may take only the second term. The electron–phonon interaction then becomes

$$\mathcal{H}_{ep} = \sum_{jl} \mathbf{u}_l \cdot \nabla V(\mathbf{r}_j - \mathbf{R}_{l0}) \,. \tag{12.20}$$

The phonon states are given in terms of wave vectors, and it is therefore natural to take the Fourier transform of the interaction function V

$$V(\mathbf{r}) = \frac{1}{N} \sum_{\mathbf{q}} v(\mathbf{q}) e^{i\mathbf{q} \cdot \mathbf{r}} \,. \tag{12.21}$$

The electron–phonon interaction then becomes

$$\begin{aligned}
\mathcal{H}_{ep} &= \frac{i}{N} \sum_{\mathbf{q}j} v(\mathbf{q}) e^{i\mathbf{q} \cdot \mathbf{r}_j} \mathbf{q} \cdot \sum_l \mathbf{u}_l e^{-i\mathbf{q} \cdot \mathbf{R}_{l0}} \\
&= \frac{i}{\sqrt{N}} \sum_{\mathbf{q}j} v(\mathbf{q}) e^{i\mathbf{q} \cdot \mathbf{r}_j} \mathbf{q} \cdot \mathbf{u}_{\mathbf{q}} \,.
\end{aligned} \tag{12.22}$$

Since the electron–atom potential can have Fourier amplitudes anywhere, and not just inside the first Brillouin zone, the vector $\mathbf{q}$ extends over all reciprocal space. However, we know that $\mathbf{u}_{\mathbf{q}}$ is uniquely defined only within the first zone. Therefore, we translate those $\mathbf{q}$ lying outside the first zone by the reciprocal lattice vector $\mathbf{G}$, which brings it back inside the zone to obtain a new wave vector $\mathbf{k} = \mathbf{q} + \mathbf{G}$. We then find

$$\mathcal{H}_{ep} = \frac{i}{\sqrt{N}} \sum_{\mathbf{k},\mathbf{G},j} v(\mathbf{k} + \mathbf{G}) e^{i(\mathbf{k}+\mathbf{G}) \cdot \mathbf{r}_j} (\mathbf{k} + \mathbf{G}) \cdot \mathbf{u}_{\mathbf{k}} \,. \tag{12.23}$$

Referring again to the three principal polarization axes λ and writing the displacement operator in terms of the creation and annihilation operators, Eq. (12.6), we find

$$\mathcal{H}_{ep} = i \sum_{\mathbf{k},\mathbf{G},j,\lambda} \sqrt{\frac{\hbar}{2mN\omega_{\mathbf{k}\lambda}}} v(\mathbf{k} + \mathbf{G}) e^{i(\mathbf{k}+\mathbf{G}) \cdot \mathbf{r}_j} (\mathbf{k} + \mathbf{G}) \cdot \boldsymbol{\xi}_{\mathbf{k}+\mathbf{G},\lambda} (\hat{a}_{\mathbf{k}\lambda} + \hat{a}_{\mathbf{k}\lambda}^\dagger) \,, \tag{12.24}$$

where $\boldsymbol{\xi}_{\mathbf{k}+\mathbf{G},\lambda}$ is the polarization of the mode characterized by $\mathbf{k} + \mathbf{G}$ and λ.

At the moment things look pretty complicated, but we will now make some simplifications. First, we assume that the electrons in the solid have an associated charge density operator $\varrho(\mathbf{r})$.[3] Using this, we can transform the sum over j above by an integral over $\mathbf{r}$ and after integration obtain

$$\mathcal{H}_{ep} = i \sum_{\mathbf{k},\mathbf{G},\lambda} \sqrt{\frac{\hbar}{2mN\omega_{\mathbf{k}\lambda}}} v(\mathbf{k} + \mathbf{G})(\mathbf{k} + \mathbf{G}) \cdot \boldsymbol{\xi}_{\mathbf{k}+\mathbf{G},\lambda} \mathcal{R}(\mathbf{k} + \mathbf{G})(\hat{a}_{\mathbf{k}\lambda} + \hat{a}_{\mathbf{k}\lambda}^\dagger) \,, \tag{12.25}$$

---

[3] $\varrho(\mathbf{r})$ represents the number of electrons per unit volume at each position $\mathbf{r}$. If an electron with wavefunction $\psi_j(\mathbf{r})$ has creation operator $b_j^\dagger$, then $\varrho(\mathbf{r}) = \sum_{jj'} \psi_j^*(\mathbf{r})\psi_{j'}(\mathbf{r})b_j^\dagger b_{j'}$.

where $\mathcal{R}$ is the Fourier transform of ϱ. It is customary to split up this expression by defining a matrix element

$$M_{\mathbf{q}+\mathbf{G},\lambda} = i\sqrt{\frac{\hbar}{2mN\omega_{\mathbf{k}\lambda}}}\,v(\mathbf{k}+\mathbf{G})(\mathbf{k}+\mathbf{G})\cdot\boldsymbol{\xi}_{\mathbf{k}+\mathbf{G}\lambda}\,. \tag{12.26}$$

such that

$$\mathcal{H}_{ep} = \sum_{\mathbf{k},\mathbf{G},\lambda} M_{\mathbf{q}+\mathbf{G},\lambda}\mathcal{R}(\mathbf{k}+\mathbf{G})(\hat{a}_{\mathbf{k}\lambda} + \hat{a}_{\mathbf{k}\lambda}^{\dagger})\,. \tag{12.27}$$

This is the starting point for nearly all calculations of electron–phonon coupling. Different forms for M are found in different regimes, but the general structure of the coupling interaction is unchanged. Next, we consider a particular form of electron–phonon coupling.

12.2.1 Deformation potential coupling

Qubits in semiconductor quantum dots tend to be associated with low energy states close to the band edge, that have relatively long wavelength when compared to the lattice spacing. Such states will interact most strongly with small wave vector (i.e., long-wavelength) lattice distortions, and we are therefore primarily concerned with calculating the matrix element M in this limit. We therefore ignore all contributions from outside the first Brillouin zone, setting $\mathbf{G} = 0$.

There are several ways in which a strain on the crystal lattice can produce a change in an electron energy. The strongest mechanism in the examples we will discuss in this chapter (using GaAs as a model semiconductor material) is due to the inhomogeneous displacement of lattice positions. This locally alters the band gap for electrons and gives a position-dependent energy and is called 'deformation potential coupling'.

For these long-wavelength interactions, it is possible to come up with a phenomenological model rather than calculating the interaction strength from first principles. For deformation potential coupling, we take the lowest-order approximation for the coupling potential v: we assume it is a constant, $\mathscr{D}$. Further, only longitudinal (L) excitations are important. Long-wavelength transverse (T) modes do not change the energy of short wave vector electrons through inhomogeneous strain, since their effect averages. These approximations give an interaction of the form

$$\mathcal{H}_{ep,\mathscr{D}} = i\mathscr{D}\sum_{\mathbf{k}}\left(\frac{\hbar}{2mN\omega_{\mathbf{k}L}}\right)^{\frac{1}{2}}\mathcal{R}(\mathbf{k})|\mathbf{k}|(\hat{a}_{\mathbf{k}L} + \hat{a}_{\mathbf{k}L}^{\dagger})\,, \tag{12.28}$$

where $\mathscr{D}$ can be measured in experiments, and is called the deformation potential coupling constant.

There are other types of electron–phonon coupling. For example, in some crystals a homogeneous strain can change the energy of an electron since it produces an electric field through the 'piezoelectric effect'.[4] However, these other mechanisms are generally much

---

[4] This can only happen in crystals that are non-centrosymmetric.

weaker than deformation potential coupling and we will not consider them further here. We refer to Ziman (2001) for more details.

12.2.2 Qubit–phonon coupling

What effect will the electron–phonon interaction have on the solid-state qubits we discussed in Chapter 11? Since the phonons couple to the electron charge density, there will be no direct interaction between an electron spin qubit and the lattice vibrations, since each spin state has the same charge distribution.[5] An exciton qubit does, however, have a different charge configuration for the $|0\rangle$ and $|1\rangle$ states. Call the Fourier transforms of these $\mathcal{R}_{0\mathbf{q}}$ and $\mathcal{R}_{1\mathbf{q}}$. The charge-density operator can then be written, in the qubit Hilbert space

$$
\begin{aligned}
\mathcal{R}(\mathbf{q}) &= \mathcal{R}_{0\mathbf{q}} |0\rangle \langle 0| + \mathcal{R}_{1\mathbf{q}} |1\rangle \langle 1| , \\
&= \frac{\mathcal{R}_{0\mathbf{q}} + \mathcal{R}_{1\mathbf{q}}}{2} (|0\rangle \langle 0| + |1\rangle \langle 1|) + \frac{\mathcal{R}_{0\mathbf{q}} - \mathcal{R}_{1\mathbf{q}}}{2} (|0\rangle \langle 0| - |1\rangle \langle 1|).
\end{aligned}
\tag{12.29}
$$

Since the first term is proportional to the identity, its only effect is a slight perturbation of the phonon energies, and it has no effect on the qubit states. The exciton–phonon coupling Hamiltonian may therefore be written as

$$
\mathcal{H}_{xp} = Z \sum_{\mathbf{k}} g_{\mathbf{k}} (\hat{a}_{\mathbf{k}} + \hat{a}_{\mathbf{k}}^{\dagger}) ,
\tag{12.30}
$$

where we have dropped the label L for simplicity, and the exciton–phonon coupling constant is

$$
g_{\mathbf{k}} \equiv i\mathscr{D} \sqrt{\frac{\hbar}{2mN\omega_{\mathbf{k}L}}} |\mathbf{k}| \, (\mathcal{R}_{0\mathbf{k}} - \mathcal{R}_{1\mathbf{k}}) .
\tag{12.31}
$$

Eq. (12.30) is an interaction Hamiltonian in the form we first discussed in Chapter 7 in the context of open quantum systems where our knowledge of the environment is limited. If the electron–phonon coupling is weak, we can use the general methodology that we developed in Section 7.3 and derive a master equation for our exciton qubit. This will allow us to calculate qubit dynamics even when we have no knowledge of the phonon bath, and to determine what sort of decoherence is caused by phonons.

12.3 The master equation for electrons and phonons

We would like to know what the effect of phonons is on an exciton qubit, and in particular what effect phonons might have on a single-qubit gate operation. We therefore allow an optical field to be applied to the qubit close to resonance. We know that the way to analyze

---

[5] In fact, this is not always the case since spin–orbit coupling can change the charge distribution for different spin states and therefore give rise to an electron–phonon coupling. However, this effect will be weak compared with the phonon effects on exciton states, so we do not discuss it here.

such a situation is to move into a frame rotating with the optical field frequency and to make the rotating-wave approximation. Allowing for the possibility of the field being detuned from the qubit transition energy and using a Pauli matrix notation to describe the exciton qubit, we obtain

$$\mathcal{H}_Q = \frac{\hbar}{2}(\Delta Z + \Omega X). \tag{12.32}$$

2Δ is the detuning and 2Ω is the transition dipole coupling between the field and the qubit. We have studied this kind of Hamiltonian before in Chapter 7. Its energy eigenvalues are $\pm\frac{\hbar}{2}W = \pm\frac{\hbar}{2}\sqrt{\Delta^2 + \Omega^2}$, with corresponding eigenvectors

$$|+\rangle = \cos\frac{\theta}{2}|0\rangle + \sin\frac{\theta}{2}|1\rangle \,,$$

$$|-\rangle = -\sin\frac{\theta}{2}|0\rangle + \cos\frac{\theta}{2}|1\rangle \,, \tag{12.33}$$

where

$$\theta = \arctan\left(\frac{\Omega}{\Delta}\right) , \tag{12.34}$$

and

$$W = \sqrt{\Delta^2 + \Omega^2} \tag{12.35}$$

is the effective Rabi frequency. The exciton–phonon interaction Hamiltonian is unaffected by the transformation into the rotating frame, so it still takes the form of Eq. (12.30).[6] The bare phonon Hamiltonian was found in Eq. (12.1) and its form tells us that the transformation of the creation and annihilation operators of the phonons in the interaction picture is $\hat{a}_\mathbf{k} \rightarrow \hat{a}_\mathbf{k}e^{-i\omega_\mathbf{k}t}$. In Chapter 7 we developed the general theory of open quantum systems. In particular, we showed how to derive a master equation to describe the system dynamics in the Born–Markov approximation. With reference to the result we obtained in Eq. (7.97), we decompose the system operator Z in the interaction Hamiltonian into three parts

$$\Upsilon_0(0) = (\cos^2\frac{\theta}{2} - \sin^2\frac{\theta}{2})(|+\rangle\langle+| - |-\rangle\langle-|) \,, \tag{12.36}$$

$$\Upsilon_{-W}(-W) = -2\cos\frac{\theta}{2}\sin\frac{\theta}{2}|+\rangle\langle-| \,, \tag{12.37}$$

$$\Upsilon_W(W) = -2\cos\frac{\theta}{2}\sin\frac{\theta}{2}|-\rangle\langle+| \,. \tag{12.38}$$

In all three cases the corresponding bath operator is the same, namely

$$\Lambda(t) = \sum_\mathbf{k} g_\mathbf{k}(\hat{a}_\mathbf{k}e^{-i\omega_\mathbf{k}t} + \hat{a}_\mathbf{k}^\dagger e^{i\omega_\mathbf{k}t}) \,, \tag{12.39}$$

which has a correlation function given by

$$\gamma(\omega) = \frac{1}{\hbar^2}\sum_{\mathbf{k},\mathbf{k}'}\int_0^\infty dt\, e^{i\omega t}\left\langle g_\mathbf{k}^*\left(\hat{a}_\mathbf{k}e^{-i\omega_\mathbf{k}t} + \hat{a}_\mathbf{k}^\dagger e^{i\omega_\mathbf{k}t}\right)g_{\mathbf{k}'}\left(\hat{a}_{\mathbf{k}'} + \hat{a}_{\mathbf{k}'}^\dagger\right)\right\rangle. \tag{12.40}$$

---

[6] This can be verified by applying the general time-dependent unitary transformation, Eq. (7.19), to the combined Hamiltonian $\mathcal{H}_Q + \mathcal{H}_{xp}$.

Assuming a thermal occupation of modes for the phonons and making use of the standard boson commutation relations, we find[7]

$$\gamma(\omega) = \pi \sum_{\mathbf{k}} N(\omega_{\mathbf{k}}) \frac{|g_{\mathbf{k}}|^2}{\hbar^2} \delta(\omega + \omega_{\mathbf{k}}) + \pi \sum_{\mathbf{k}} [N(\omega_{\mathbf{k}}) + 1] \frac{|g_{\mathbf{k}}|^2}{\hbar^2} \delta(\omega - \omega_{\mathbf{k}}), \quad (12.41)$$

where

$$N(\omega) = \frac{1}{\exp(\hbar\omega/k_B T) - 1} \quad (12.42)$$

is the usual Bose–Einstein occupation number at temperature T.

Exercise 12.2: Using Eq. (12.40) calculate the correlation function $\gamma(\omega)$ in Eq. (12.41).

We can proceed further by defining the spectral density $J(\omega)$

$$J(\omega) = 2\pi \sum_{\mathbf{k}} \frac{|g_{\mathbf{k}}|^2}{\hbar^2} \delta(\omega - \omega_{\mathbf{k}}). \quad (12.43)$$

Using the fact that $\Upsilon_W(W) = \Upsilon_{-W}^\dagger(-W)$, we obtain the following equation for the density operator of the exciton qubit ρ:

$$\dot{\rho} = \sum_{\beta \in \{0, W\}} J(\beta) \left\{ [N(\beta) + 1] \mathcal{D}[\Upsilon_\beta]\rho + N(\beta)\mathcal{D}[\Upsilon_\beta^\dagger]\rho) \right\}. \quad (12.44)$$

where we have defined the dissipator

$$\mathcal{D}[\Upsilon_\beta]\rho \equiv \Upsilon_\beta \rho \Upsilon_\beta^\dagger - \frac{1}{2} \left(\Upsilon_\beta^\dagger \Upsilon_\beta \rho + \rho \Upsilon_\beta^\dagger \Upsilon_\beta \right). \quad (12.45)$$

Eq. (12.44) is the master equation we have been looking for. There are two terms in the sum. The first describes absorption of energy β from the bath, and the second describes emission of energy β into the bath. The rate of both processes is proportional to the spectral density at the energy β, i.e., to both the number of bath modes with an energy β and to the ability of the system to couple to modes of that energy. Of course, the $\beta = 0$ term does not correspond to any energy exchange but rather constitutes a 'pure dephasing' process.

The spectral-density function gives us all the information we need about the bath in order to calculate system dynamics in the Born–Markov approximation. This is a rather powerful result, especially when one considers the complexity of the problem at hand. In order to give a little more insight into how it works, we calculate the spectral density for the specific problem of deformation potential coupling of phonons to an exciton qubit.

First, we generalize Eq. (12.31) slightly, and allow the electron that is promoted when an exciton is created to have different deformation potential coupling constants $\mathcal{D}_v$ and $\mathcal{D}_c$, depending on whether it is in its ground (valence) state $|0\rangle$ or excited (conduction) state $|1\rangle$. In order to find the electron density Fourier transforms we must use a particular form

---

[7] There are also Lamb shift terms (see Section 7.3), which are assumed to be small relative to the bare qubit energy, and have been dropped.

for the conduction and valence wavefunctions, which reflect the confinement induced by the quantum dot. The simplest choice is the ground solution of a harmonic potential

$$\psi_j(\mathbf{r}) = (d_j\sqrt{\pi})^{-\frac{3}{2}} \exp\left(-\frac{r^2}{2d_j^2}\right),$$ (12.46)

where $j = v, c$ denotes the valence or conduction states and the d_j are constants that reflect the typical extent of the valence and conduction wavefunctions. They can be different since the two states are characterized by distinct effective masses. We therefore find that

$$\mathcal{D}_v R_{0\mathbf{q}} - \mathcal{D}_c R_{1\mathbf{q}} = \mathcal{D}_c \exp\left(-\frac{d_c^2 q^2}{4}\right) - \mathcal{D}_v \exp\left(-\frac{d_v^2 q^2}{4}\right).$$ (12.47)

By converting the sum over $\mathbf{k}$ to an integral and assuming a three-dimensional density of phonon states,[8] we find that

$$J(\omega) = \frac{\omega^3}{2\pi\mu c_s^5}\left(\mathcal{D}_v^2 e^{-\omega^2/\omega_v^2} + \mathcal{D}_c^2 e^{-\omega^2/\omega_c^2} - 2\mathcal{D}_c\mathcal{D}_v e^{-\omega^2/\omega_{cv}^2}\right).$$ (12.48)

The low-energy phonons have a linear dispersion relation characterized by the sound velocity $c_s = \omega_{\mathbf{k}}/|\mathbf{k}|$ and we have defined three characteristic cut-off frequencies $\omega_j = c_s\sqrt{2}/d_j$ for $j = c, v$, and $\omega_{cv} = 2c_s/\sqrt{d_v^2 + d_c^2}$. Here, μ is the mass density of the semiconductor material.

The spectral density at $\omega = 0$ is zero, and we can therefore drop this term in our master equation, Eq. (12.44). We can conclude therefore that $J(W)$, a single value, completely characterizes the rate of decoherence due to phonons in this system. Importantly, W can be controlled and manipulated: it depends on the Rabi frequency and detuning, which can be altered by changing the laser frequency and intensity. In the next section, we will use the results derived here to show that we can control the decoherence rate for an exciton qubit by altering some basic properties of the laser with which we drive the system.

12.4 Overcoming decoherence

12.4.1 Exploiting the spectral-density function

We can now ask the question: what will be the optimal strategy for performing an optical quantum gate on this exciton qubit? To answer this question, consider Fig. 12.2, which is a plot of the spectral density for a typical quantum dot. At low frequencies, the leading term in $J(\omega)$ is proportional to ω^3, and if W is kept low the phonon decoherence mechanism is ineffective. This parameter regime corresponds to a weak laser that is closely resonant

---

[8] Since allowed wave vectors are evenly distributed in reciprocal space and each take up a reciprocal volume of $(2\pi)^3/V$ with V the sample volume, this amounts to $\sum_{\mathbf{k}} \to (2\pi)^{-3}V \int d\mathbf{k}$.

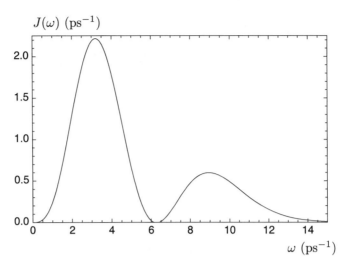

Fig. 12.2. The spectral-density function for an exciton qubit coupled to acoustic phonons through the deformation potential. The parameters used in this calculation are typical for a GaAs quantum dot: $c_s = 4.8 \times 10^3$ ms^{-1}, $\mu = 5300$ kgm^{-3}, $\mathscr{D}_c = 14.6$ eV, $\mathscr{D}_v = 4.8$ eV, $\omega_c = 3.135$ ps^{-1}, $\omega_v = 4.708$ ps^{-1}. Figure adapted from Gauger *et al.* (2008).

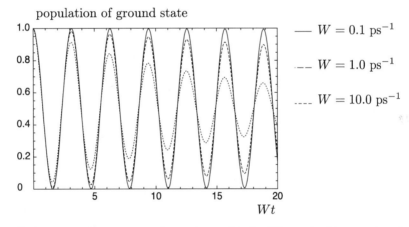

Fig. 12.3. Rabi oscillations at zero temperature of an exciton qubit, for different Rabi frequencies. Phonon decoherence gets worse at larger values of W, since the spectral density function increases approximately as W^3. Parameters typical of GaAs have been used in the simulation.

with the exciton transition. This is shown in Fig. 12.3, where Rabi oscillations of an exciton qubit are shown for different values of W. The time axis has been suitably scaled, such that the oscillations lie on top of each other. A Rabi oscillation is an example of a single-qubit manipulation for the exciton, and even though an oscillation takes longer for a smaller W, the loss of population per oscillation is reduced since the effect of phonons is greatly suppressed.

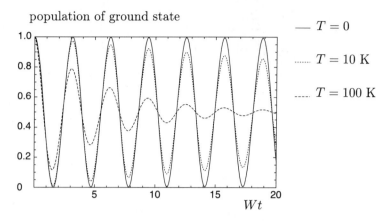

population of ground state

Fig. 12.4. Rabi oscillations of an exciton qubit, for different temperatures, at $W = 0.1$ ps^{-1}. A higher temperature means a larger number of phonons are present and so decoherence gets worse. Parameters typical of GaAs have been used in the simulation.

We have ignored the effect of spontaneous emission of photons in our analysis so far. A slower gate-operation time means an increased probability of photon emission, so phonon and photon decoherence processes are minimized in quite different regimes. They play off against each other, but optimal values can be found in which their combined effect is minimized.

A further consideration is temperature, and in Fig. 12.4 we display Rabi oscillations at a constant driving for different temperatures. A higher temperature excites more phonons, which leads to a more rapid loss of exciton coherence. In laboratory experiments the temperature is often as low as 4 K and this effect is suppressed, even at lower driving rates.

One may wonder whether an alternative strategy might be employed: to drive our system very strongly indeed, beyond the cut-off frequencies for the phonon spectral function. We would then again find very low decoherence rates in this model. However, our Markovian master equation is no longer valid in this regime: higher-order effects become important when the laser driving is strong. We will discuss such effects later in the chapter.

12.4.2 Phonon decoherence for optically controlled spin qubits

As we discussed in the last chapter, spontaneous emission of photons means that exciton qubits will never be more than a test system. For scalable quantum computing, an optically controlled spin qubit is a better candidate, and by making judicious choices of optical control parameters we can mitigate the effect of phonons in this case, too.

Let us focus on the simplest problem of a single-qubit under optical control. As we discussed in Section 11.2, a spin qubit with a higher optical trion level $|T_h\rangle$ can be described by the Hamiltonian

$$\mathcal{H} = \hbar\Omega \left(|1\rangle \langle T_h| + e^{i\theta} |T_h\rangle \langle 1| \right), \tag{12.49}$$

where we have set the relative phase factor to zero, and Ω is the usual coupling of the laser to the (in this case resonant) optical transition. The qubit state $|0\rangle$ does not appear here since we make the simplifying assumption that it does not couple at all to $|T_h\rangle$, and is degenerate with $|1\rangle$.

There are mechanisms by which the spin can decohere that do not involve the optical level at all. For example, if there is a spin–orbit interaction, the two spin levels have slightly different charge distributions, and would therefore have a different deformation potential coupling to phonons, leading to a spin–phonon coupling. There may also be other (electron or nuclear) spins in the vicinity of the qubit spin that can couple through the exchange, hyperfine, or dipole–dipole interactions. All of these mechanisms play an important role in quantum gates that involve direct manipulation of the spin with, for example, microwave resonance. However, we are concerned with optical techniques for spin manipulation, and it is then the optical transition to the higher level that gives the strongest decoherence channel. Again, we are faced with two principal problems: photon emission and phonon coupling.

It is straightforward to transfer the analysis of the previous section to the trion–phonon coupling problem. When a trion is created from an electron spin qubit state, the change in the electron charge-density function is the same as it was in the exciton case. One can therefore carry over the analysis exactly, and so associate the Pauli operators for the exciton qubit with the $\{|1\rangle, |T_h\rangle\}$ subspace here. Nonetheless, the effect on the qubit decoherence characteristics here is going to be quite different since we are no longer concerned with the relative coherence of the $|1\rangle$ and $|T_h\rangle$, but rather on that of our spin, here labelled $|0\rangle$ and $|1\rangle$.

Consider the following problem: how might we best perform a π rotation around the z axis of the Bloch sphere for our spin qubit? Without any environmental interactions, the answer was given in the last chapter, and is quite simple. Starting in an arbitrary spin state (with no population in $|T_h\rangle$), we induce a complete Rabi oscillation between states $|1\rangle$ and $|T_h\rangle$ such that all the population returns to $|1\rangle$. Quantum mechanics tells us that this gives rise to a relative phase of π between $|0\rangle$ (which is unaffected by the pulse) and $|1\rangle$. This is the Z gate we set out to construct.

In the presence of phonons and photon emission we have seen that Rabi oscillations become imperfect as coherence is lost. At low enough temperature, the system will always end up back in the $\{|0\rangle, |1\rangle\}$ subspace following the control pulse, since photon decay will always eventually remove any $|T_h\rangle$ population. Indeed, the relative populations of the two-qubit states will be unchanged by the operation. However, the coherence of $|0\rangle$ and $|1\rangle$ is affected by a pure dephasing process. We can quantify this effect by considering an initial state $|\psi_{in}\rangle = (|0\rangle + |1\rangle)/\sqrt{2}$ and then simulating the effect of the Z-gate operation including decoherence. The result is a qubit density operator ρ_{out}, which is compared to the ideal final state $|\psi_{out}\rangle = (|0\rangle - |1\rangle)/\sqrt{2}$ using the fidelity

$$F = \langle \psi_{out}| \rho_{out} |\psi_{out}\rangle . \qquad (12.50)$$

The result, plotted as a function of Rabi strength for different temperatures, is displayed in Fig. 12.5.

As expected, F becomes lower as the temperature is increased for all displayed driving strengths. This is due to the temperature dependence of the Bose–Einstein factor for

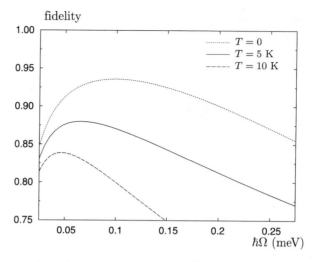

Fig. 12.5. Fidelity of a Z gate performed on a quantum-dot spin qubit by coherently driving an optical transition on resonance, as a function of the resonant Rabi energy $\hbar\Omega$. Parameters typical of GaAs have been used in the simulation. Figure adapted from Gauger *et al.* (2008).

phonons. The more interesting feature of the displayed plots is that they have a peaked structure: there is an optimal value of Ω that maximizes the fidelity of operation. This is the result of the competing dependencies of phonon and photon decoherence channels. As the driving strength increases, the time the system spends in the state $|T_h\rangle$ decreases, and fewer spontaneous photon-emission events occur on average. By contrast, the phonon spectral-density function increases for higher driving and so the phonon-induced decoherence rate goes up.

Even when the parameters are optimized at $T = 0$ there is still a significant loss of fidelity during a gate operation, since both photon and phonon effects cannot be eliminated simultaneously. In the next section we will introduce an alternative control strategy that can result in significant fidelity improvements for some optically active systems.

12.4.3 Adiabatic gating

If it were possible to avoid populating the state $|T_h\rangle$, or to run the gate operation such that the phonon spectral density was sampled beyond the cut-off frequencies, improved gate fidelities could be achieved. With an alternative adiabatic approach to system driving, both of these are possible. Consider a more general version of the three-level Hamiltonian in Eq. (12.49)

$$\mathcal{H} = \hbar\Omega(|1\rangle\langle T_h| + |T_h\rangle\langle 1|) + 2\hbar\Delta\,|T_h\rangle\langle T_h| \tag{12.51}$$

where 2Δ is now the detuning of the laser from the optical transition. The eigenenergies are displayed in Fig. 12.6.

In the eigenstate picture, the dynamic gate above is implemented with a rapid switch from a condition $|\Delta/\Omega| \gg 0$ to $\Delta = 0$. Under the first condition, $|0\rangle$ and $|1\rangle$ are degenerate

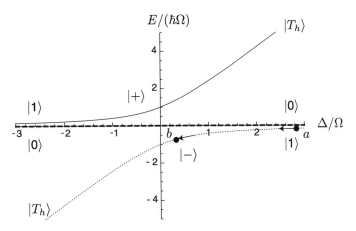

Fig. 12.6. The energies of the three eigenstates of the spin–trion system described by Hamiltonian, Eq. (12.51), as a function of the ratio Δ/Ω. The uncoupled $|0\rangle$ is always an eigenstate. At the far left and far right of the plot, $|1\rangle$ and $|T_h\rangle$ are approximate eigenstates; towards the centre, two of the eigenstates become mixtures of $|1\rangle$ and $|T_h\rangle$.

eigenstates and the time evolution is a simple build-up of global phase. Under the second condition, $|0\rangle$ remains an eigenstate, but $|1\rangle$ becomes a superposition of eigenstates $|+\rangle$ and $|-\rangle$. The subsequent time evolution is then a Rabi oscillation between $|1\rangle$ and $|T_h\rangle$, and switching back to the first condition after one oscillation gives the relative phase on $|1\rangle$ with respect to $|0\rangle$. An alternative strategy is to move *slowly* from the far detuned condition towards resonance, and then slowly away again. If the change is slow enough, it is adiabatic, and so-called eigenstate following occurs. Imagine an initial state $|1\rangle$ when $|\Delta/\Omega| \gg 0$. This would be represented by point 'a' in Fig. 12.6. A gradual reduction in the ratio $|\Delta/\Omega|$ results in a following of the $|-\rangle$ eigenstate line, until the state eventually reaches the point 'b'. The ratio is then increased again, taking us slowly back to point 'a'. The net result is a phase accumulation on $|1\rangle$. On the other hand, if the system were prepared as $|0\rangle$ and the same procedure is executed, the phase accumulated is different from that for $|1\rangle$, since the eigenstate trajectory for $|0\rangle$ simply follows the horizontal axis in Fig. 12.6. A qubit-superposition state therefore undergoes a nontrivial Z gate under this operation:

$$|0\rangle \to |0\rangle \quad \text{and} \quad |1\rangle \to e^{-i\phi} |1\rangle \,, \tag{12.52}$$

with

$$\phi = \int_0^\tau dt\, \omega_{0+}(t) \,. \tag{12.53}$$

Here, $\omega_{0+} = \sqrt{\Delta^2 + \Omega^2} - \Delta$ is the energy difference between the two eigenstates involved in the procedure and τ is the time taken to return to the initial parameter configuration.

This kind of control can be achieved experimentally by using laser pulses whose intensity profile or detuning varies slowly (the latter type is known as a 'chirped' pulse). Fig. 12.7 shows the fidelity of a phase gate for a pulse with constant detuning Δ and intensity profile $\Omega(t) = \Omega_0 \exp[-(t/\tau)^2]$. The simulations are performed from a time $t = -3\tau$ to $t = +3\tau$,

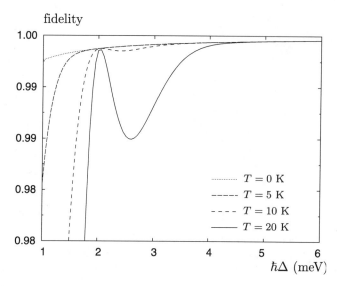

Fig. 12.7. The fidelity of a phase gate performed using the adiabatic-control technique, as a function of detuning Δ and for various temperatures. The driving strength $\hbar\Omega_0$ is always 1 meV. The local minimum is due to the shape of the spectral density. Parameters typical of GaAs have been used in the simulation. Figure adapted from Gauger *et al.* (2008).

where τ is set to create the correct π phase shift and so is longer for a more detuned pulse. Both phonon interaction through deformation potential coupling and photon spontaneous emission are taken into account using the Markovian master-equation approach, using values typical of GaAs.

The fidelities obtained for this adiabatic approach are much improved with respect to those for the dynamic gating strategy presented in Fig. 12.5. Since the laser is somewhat detuned for all displayed curves, the trion level is never fully excited, which restricts the photon-emission probability. In this approach, going to large values of detuning allows one to probe the small spectral-density values beyond the cut-off frequencies, without creating fast dynamics that violate the assumptions of the Markov approximation. The local minimum in the fidelity (most clearly seen in the 20 K plot) occurs when the detuning value corresponds to the first minimum in the spectral-density function. The typical timescale for performing the adiabatic approach is perhaps a few tens of picoseconds, whereas in the dynamic approach the time taken can be smaller, perhaps only a few picoseconds. However, the adiabatic approach has the significant advantage of an in-built tolerance to the phonon decoherence mechanism for optically controlled spins.

12.5 Strong coupling effects

So far we have exclusively looked at open quantum systems by making the Born–Markov approximation, and indeed this does work very well in situations with a weak system–bath coupling. However, the weak coupling limit is not always appropriate, and in this section we will introduce an exact solution to a particular Hamiltonian in the strong coupling regime. This gives a nice illustration of some effects that cannot be captured in the Born–Markov approach. Moreover, this exactly solvable model can be used as a basis for solving more difficult problems.

12.5.1 An exact model: independent bosons

Our starting point will be the Hamiltonian of an exciton qubit coupled to a single-phonon mode. Using the coupling derived for Eq. (12.30) for a single mode with bosonic operators $\hat{a}$ and $\hat{a}^\dagger$, we obtain

$$\mathcal{H} = \epsilon Z + \hbar\omega\hat{a}^\dagger a + \hbar g Z(\hat{a} + \hat{a}^\dagger), \tag{12.54}$$

where ϵ is the exciton creation energy and we have not included any laser driving term. First, we change the form of this Hamiltonian slightly (the reason for this will become apparent later). We introduce the unitary displacement operator for phonons

$$D(\delta) = \exp[\delta\hat{a}^\dagger - \delta^*\hat{a}]. \tag{12.55}$$

This operator is formally identical to the displacement operator for optical quantum fields, which we encountered in Chapter 1, and used extensively in Chapters 8 and 9.

Exercise 12.3: Show that the effect of the displacement operator on our usual phonon annihilation and creation operators is given by

$$D^\dagger(\delta)\hat{a}D(\delta) = \hat{a} + \delta \quad \text{and} \quad D^\dagger(\delta)\hat{a}^\dagger D(\delta) = \hat{a}^\dagger + \delta^*. \tag{12.56}$$

Making this transformation to our Hamiltonian in Eq. (12.54) and choosing δ real, we obtain

$$\mathcal{H}_1 = \epsilon Z + \hbar\omega(\hat{a}^\dagger\hat{a} + \delta^2) + \hbar(\hat{a} + \hat{a}^\dagger)(gZ + \omega\delta). \tag{12.57}$$

Setting $\delta = -g/\omega$ and removing the irrelevant constant factor from the Hamiltonian gives

$$\mathcal{H}_1 = \epsilon Z + \hbar\omega\hat{a}^\dagger\hat{a} + \hbar G|1\rangle\langle 1|(\hat{a} + \hat{a}^\dagger), \tag{12.58}$$

where $G \equiv -2g$. This has not made much difference to our original Hamiltonian, except that the newly shifted phonon operators interact only with the $|1\rangle$ state of the exciton qubit. We could have used Eq. (12.58) as our starting point, but we have included the extra step to show the simple connection to Eq. (12.54). We can now use our displacement operator in a more powerful way to completely eliminate the coupling term. We make its application conditional on the state of the qubit, i.e., we apply the unitary transformation

$$U = |0\rangle\langle 0| + |1\rangle\langle 1|D(\alpha) = e^S, \tag{12.59}$$

where the definition in terms of the operator $S = \alpha(\hat{a}^\dagger - \hat{a})|1\rangle\langle 1|$ will be useful in some future manipulations. The transformation, Eq. (12.59), represents a controlled displacement of the phonon field. Now

$$
\begin{aligned}
\mathcal{H}_0 &= U\mathcal{H}_1 U^\dagger \\
&= \hbar\omega\,\hat{a}^\dagger\hat{a} + \epsilon\,|0\rangle\langle 0| + \hbar\left(\hat{a}^\dagger + \hat{a}\right)(G - \omega\alpha)\,|1\rangle\langle 1| \\
&\quad + \left(\hbar\alpha^2\omega - \epsilon - 2\hbar G\alpha\right)|1\rangle\langle 1| \,.
\end{aligned}
\tag{12.60}
$$

We can choose our parameter α to give us the simplest possible form for the Hamiltonian. Setting $\alpha = G/\omega$ then allows us to completely remove the coupling term and finally

$$
\mathcal{H}_0 = \epsilon' Z + \omega\hat{a}^\dagger\hat{a}\,,
\tag{12.61}
$$

where the renormalized qubit energy is $\epsilon' = \epsilon + \hbar G^2/(2\omega)$. This Hamiltonian is diagonal: its eigenstates are products of the transformed qubit- and phonon-number states. Of course, one must reverse the transformation, Eq. (12.59), in order to interpret these eigenstates in terms of the particles we defined in our original Hamiltonian. With these original definitions the eigenstates are entangled qubit–phonon states which are called 'polarons'.

Let us now try to obtain some physical insight into what we have done in applying this polaron transformation. The displacement operator shifts both the creation and annihilation operator by the same amount α. In terms of an observable quantity we can think of it as a global shift in the position operator of a simple harmonic oscillator, since this is proportional to $(\hat{a} + \hat{a}^\dagger)$. The shift is needed only if the qubit is in state $|1\rangle$, and it is therefore common to picture the system as two harmonic potentials, one that is appropriate for $|0\rangle$ and the other for $|1\rangle$, which are shifted relative to each other. The eigenstate wavefunctions for phonons are always solutions of the harmonic potential, but the solutions for $|0\rangle$ are displaced from those for $|1\rangle$; see Fig. 12.8. It therefore becomes impossible to factor the phonon and qubit dynamics when an initial superposition of $|0\rangle$ and $|1\rangle$ is prepared, leading to a pure dephasing of the qubit. We will return to derive the form of this dephasing shortly, but let us digress for a moment and calculate the optical properties of this coupled system.

12.5.2 The optical spectrum

Assuming that our system is in its ground state (i.e. that we are in the zero temperature limit), then all the lines in the optical absorption spectrum result from transitions between the lowest energy level in the $|0\rangle$ potential and the set of levels in the shifted $|1\rangle$ potential. Let us label the ground vibrational level in the lower well $|V_l\rangle$ and the upper well $|V_u\rangle$, and the nth excited phonon level $|n_l\rangle$ and $|n_u\rangle$ for the lower and upper well, respectively. To avoid confusion we will use the $\hat{a}$ and $\hat{a}^\dagger$ operators for the lower well, with $\hat{b}$ and $\hat{b}^\dagger$ in the upper well. The intensity of an absorption line is proportional to the square of the dipole matrix element connecting the two levels involved in the transition, M_p. For the ground level in the lower well and the nth excited level in the upper well, we obtain

$$
M_p = \langle\psi_i|\,\mathbf{p}\,|\psi_f\rangle = \langle 0, V_l|\,\mathbf{p}\,|1, n_u\rangle \,.
\tag{12.62}
$$

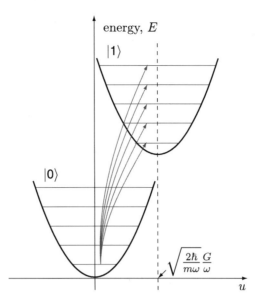

Fig. 12.8. The exact qubit–phonon model discussed in this section can be represented by two harmonic oscillators, one associated with qubit $|0\rangle$ and centred at displacement $u = 0$, and one is associated with $|1\rangle$ and is shifted away from the original by an amount proportional to the coupling strength G. Since the oscillator positions are different, there are many dipole-allowed transitions and the zero-temperature absorption lines correspond to the indicated transitions.

With the usual definition of a Fock state given in Eq. (1.62), we can rewrite the matrix element

$$M_p = \frac{\langle 0, V_l | \, \mathbf{p}(\hat{b}^\dagger)^n \, | 1, V_u \rangle}{\sqrt{n!}} = \frac{p_{01}}{\sqrt{n!}} \langle V_l | (\hat{b}^\dagger)^n | V_u \rangle \,, \tag{12.63}$$

where we have defined p_{01} as the optical dipole element between qubit states $|0\rangle$ and $|1\rangle$.[9] To proceed further, we can write the b creation operators in terms of $\hat{a}$, since the two are related by the displacement discussed above:

$$\hat{b} = \hat{a} - \frac{G}{\omega} \quad \text{and} \quad \hat{b}^\dagger = \hat{a}^\dagger - \frac{G}{\omega}. \tag{12.64}$$

Using the properties of boson operators it is then straightforward to show that

$$M_p = \frac{p_{01}}{\sqrt{n!}} \left(-\frac{G}{\omega} \right)^n \langle V_l | V_u \rangle \,. \tag{12.65}$$

The transition between the lowest levels in the lower and upper wells is called the 'zero phonon line' (ZPL). We label its intensity I_{ZPL}. We can easily relate the intensities of the more energetic absorption lines to this one:

$$I_n = I_{\text{ZPL}} \left(\frac{G^2}{\omega^2} \right)^n \frac{1}{n!}. \tag{12.66}$$

---

[9] For simplicity we assume the dipole aligns with the optical field polarization, and we drop its vector character.

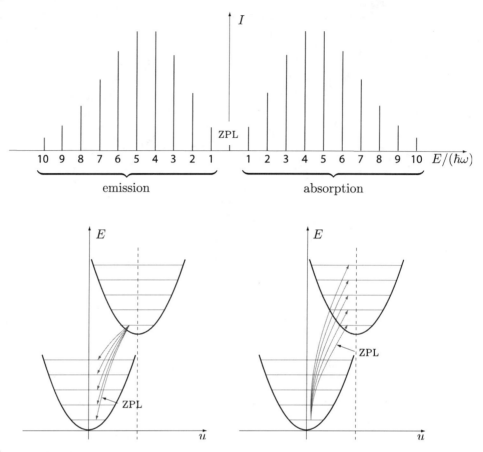

Fig. 12.9. For a single phonon interacting with our exciton qubit the optical spectrum consists of a series of discrete lines. The absorption spectrum at zero temperature has lines corresponding to the transitions indicated in Fig. 12.8: a ZPL and a series of equally spaced, blue-shifted, phonon-assisted transitions. The emission spectrum is the mirror image of the absorption spectrum, as described in the text.

This is a Poisson distribution of intensity, which is displayed in Fig. 12.9. The emission spectrum is also shown in this figure. At zero temperature, we would expect any excitation in the system to quickly decay, non-radiatively, into the lowest state in the upper well. The emission spectrum must then be the mirror image of the absorption spectrum. The highest energy transition in emission is the ZPL, with other lines corresponding to the emission of various numbers of phonon energies.

In certain systems with rather localized, discrete-phonon modes, some qualitative spectral features are described reasonably well by this model. For example, the NV$^-$ defect in diamond has local vibronic modes, and the lowest energy optical transition is characterized by a zero phonon line, with higher sidebands corresponding to phonon-assisted transitions. These sidebands are quite a headache for the kinds of distributed entanglement scheme discussed in the previous chapter, since they give an uncertainty in the frequency of the emitted photons. This can be overcome by filtering out all photons apart from the ones with the frequency of the ZPL, but this greatly reduces the success probability of these entanglement-generation protocols. An alternative to filtering is using an optical cavity in the weak coupling regime, which is tuned to the ZPL transition. Since the cavity imposes a structure on the photon bath modes, it enhances emission into the ZPL.

12.5.3 Dephasing

We now need to calculate the nature and form of the dephasing that is predicted by this exact model. We will follow the method of Hohenester (2007). The time-evolution operator for the Hamiltonian in Eq. (12.54) is easiest to write in an interaction picture, where the non-interacting part of the Hamiltonian is

$$\mathcal{H}_0 = \epsilon' Z + \hbar\omega\,\hat{a}^\dagger\hat{a} = e^S \mathcal{H}_1 e^{-S}. \tag{12.67}$$

We find that

$$
\begin{aligned}
U(t, t') &= e^{i\mathcal{H}_0 t} e^{i\mathcal{H}_1(t-t')} e^{-i\mathcal{H}_0 t'} \\
&= e^{i\mathcal{H}_0 t} e^{-S} e^{-i\mathcal{H}_0 t} e^{i\mathcal{H}_0 t'} e^S e^{-i\mathcal{H}_0 t'}.
\end{aligned}
\tag{12.68}
$$

The first and last sets of three terms are similar and straightforward to calculate

$$e^{i\mathcal{H}_0 t} e^{-S} e^{-i\mathcal{H}_0 t} = |0\rangle\langle 0| + D(-\alpha e^{i\omega t})|1\rangle\langle 1|. \tag{12.69}$$

Finally, we can set $t' = 0$ for simplicity and use the fact that $D(\alpha)D(\alpha') = D(\alpha + \alpha')\exp[i\mathrm{Im}(\alpha\alpha')]$ to obtain

$$U(t, 0) = |0\rangle\langle 0| + \exp\left(-i\alpha^2 \sin\omega t\right) D\left[\alpha(1 - e^{i\omega t})\right]|1\rangle\langle 1|. \tag{12.70}$$

As we have already discussed, the qubit–phonon Hamiltonian cannot cause qubit relaxation. However, dephasing is possible. Let us therefore consider an initial state that is an equal superposition of $|0\rangle$ and $|1\rangle$, and look at how its coherence, expressed as a density matrix element ρ_{10} behaves as a function of time. If the phonon mode is initially in thermal equilibrium, with a density matrix ρ_{th}, we have

$$\rho_{10}(t) = \frac{1}{2}\exp[-i\alpha^2 \sin\omega t]\mathrm{Tr}_{ph}[\rho_{th}D[\alpha(1 - e^{i\omega t})]]. \tag{12.71}$$

The thermal average of the displacement function can be found with some effort, and we will not reproduce the derivation here. The interested reader should refer to Barnett and Radmore (1997) for more details. The result is:

$$\mathrm{Tr}_{ph}[\rho_{th}D(\delta)] = \exp\left[\left(-|\delta|^2\left(\bar{n} + \frac{1}{2}\right)\right)\right], \tag{12.72}$$

where $\bar{n} = (\exp[\hbar\omega\beta] - 1)^{-1}$ and $\beta = (k_B T)^{-1}$. Substituting Eq. (12.72) into Eq. (12.71) yields

$$\rho_{10}(t) = \frac{1}{2} \exp\left(-\frac{G^2}{\omega^2}\left[i\sin\omega t + (1 - \cos\omega t)\coth\left(\frac{\hbar\omega\beta}{2}\right)\right]\right). \qquad (12.73)$$

The first term in the exponent is purely imaginary, so it affects the phase accumulation of our qubit, but does not lead to loss of coherence. The other term is real, and corresponds to a dephasing process. The modulus of ρ_{10} oscillates between its maximum value and a minimum value with an amplitude that depends on temperature, coupling strength, and phonon frequency. The coherence returns to its maximum value periodically, and in this sense it is never truly lost from the qubit. Rather, it moves between the qubit- and single-phonon mode. Such an effect could not have been predicted in the Born–Markov approach, where the bath (in this case the single mode) is assumed to have no memory.

In most solids, however, it is of course very unrealistic to assume that the environment consists of a single mode. We must therefore extend our model to cover all possible modes $\mathbf{q}$. In fact, this turns out to be relatively straightforward since each mode acts independently of the others; hence the name 'independent boson model'.

Exercise 12.4: Show that for a bath of phonon modes $\mathbf{q}$, with internal Hamiltonian $\sum_{\mathbf{q}} \omega_{\mathbf{q}} a_{\mathbf{q}}^\dagger a_{\mathbf{q}}$, and coupling $|1\rangle\langle 1| \sum_{\mathbf{q}} G_{\mathbf{q}}(a_{\mathbf{q}}^\dagger + a_{\mathbf{q}})$, we can write the time development of the coherence ρ_{10} as

$$\rho_{10}(t) = \frac{1}{2} \exp\left(-\sum_{\mathbf{q}} \frac{G_{\mathbf{q}}^2}{\omega_{\mathbf{q}}^2}\left[i\sin\omega_{\mathbf{q}}t + (1 - \cos\omega_{\mathbf{q}}t)\coth\left(\frac{\hbar\omega_{\mathbf{q}}\beta}{2}\right)\right]\right). \qquad (12.74)$$

In order to study the coherence of our superposition state, we need only look at the real argument of the exponential in Eq. (12.74). Let us call this function $\Gamma(t)$. We go back to the definition of the spectral-density function that we derived earlier in Eq. (12.43) to recast this in an integral form:

$$\Gamma(t) = -\int_0^\infty \frac{J(\omega)}{2\pi\omega^2}(1 - \cos\omega t)\coth\left(\frac{\hbar\omega\beta}{2}\right)d\omega. \qquad (12.75)$$

Various exciton–phonon interactions can now be modelled, each giving rise to a different form for the spectral-density function. As an example, let us look again at the deformation potential to acoustic phonons given by Eq. (12.48). For simplicity, we assume that all of the cut-off frequencies are equal ($\omega_c = \omega_v = \omega_{cv} = \omega_0$). Then

$$J(\omega) = C\omega^3 e^{-\omega^2/\omega_0^2}, \qquad (12.76)$$

with $C = (\mathcal{D}_e - \mathcal{D}_v)^2/(2\pi\mu c_s^5)$.

A straightforward analytical expression can be found for $\Gamma(t)$ in the limit of zero temperature:

$$\Gamma(t) = -i\frac{C\sqrt{\pi}}{4}\omega_0^3 \exp\left(-\frac{\omega_0^2 t^2}{4}\right)\text{Erf}\left(\frac{i\omega_0 t}{2}\right). \qquad (12.77)$$

The resulting time dependence of the absolute value of the coherence function $|\rho_{10}|$ is displayed in Fig. 12.10. It displays several interesting features. First, there is dephasing in

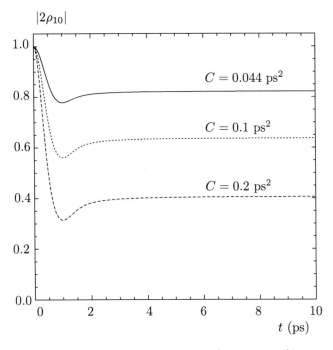

$|2\rho_{10}|$

Fig. 12.10. The time evolution of the modulus of the coherence $|\rho_{10}|$ of an exciton qubit, interacting with phonons in the independent boson model. The interaction is assumed to be a three-dimensional phonon bath with a deformation potential coupling. All curves assume equal cut-off frequencies for conduction and valence states and use $\omega_0 = 3$ ps^{-1}. $C = 0.044$ ps^{-1} is appropriate for coupling to phonons in GaAs, with larger values being plotted for comparison.

this case, even at zero temperature. Second, unlike in the single-mode model the coherence does not revive this time, which is a consequence of the infinite Poincaré revival time expected for the continuum of bath energies. Third, the loss of coherence is not, in general, complete (it approaches a finite plateau) and fourth, the initial decay is Gaussian. The first, third, and fourth features could never have been captured in the Markovian model, whose history of independent dynamics gives rise to an exponential decay to zero.

The theory of optical control of decoherence in this sort of model is in its infancy. The main difficulty is that when a laser driving term is included in the Hamiltonian, its structure changes fundamentally and an exact solution is no longer available. However, some results have been obtained using numerical methods in the driven case, and show that appropriately tailored laser pulses can improve the coherence characteristics, even beyond the Markovian limit.

12.6 References and further reading

There are many excellent texts that discuss the general theory of open quantum systems, for example Breuer and Petruccione (2002) and Weiss (2008). Both of these texts introduce many more techniques than we have been able to cover here. The interested reader should also consult the classic review article of Leggett *et al.* (1987). 'Many-particle physics' by Mahan (2000) contains a wealth of information about electron–phonon coupling, as does the book by Ziman (2001) on electrons and phonons. For a more basic introduction to the quantization of lattice vibrations, see Ziman (1969). For a comprehensive treatment of several techniques, including master equations and polaron tranforms, and their application to quantum-dot systems, see Brandes (2005) and for control of polarons in the independent boson model we refer to Hohenester (2007).

Quantum metrology

In this chapter we consider the physical limits to information extraction. This is an important aspect of optical quantum information processing in that many high-precision experiments (such as gravitational wave detection) are implemented in optical systems, i.e., interferometers. It is not surprising that just as in computation and communication, the use of quintessentially quantum mechanical properties allows us to improve the sensitivity in interferometry. We start this chapter with a derivation of the Fisher information and the Cramér–Rao bound, which tell us how much information we can extract about a parameter in a set of measurements. In Section 13.2 we introduce the statistical distance between two probability distributions. This can in turn be used to determine how many times the system needs to be queried before we can determine which probability distribution governs the system. In addition, we make a connection between the statistical distance and the angle between states in Hilbert space. In Sections 13.3 and 13.4 we derive bounds on how fast quantum states evolve to orthogonal states, and how entangled states can be used to improve parameter estimation. Finally, in Section 13.5 we present a number of approaches for implementing quantum metrology in optical systems, most importantly in optical interferometers.

13.1 Parameter estimation and Fisher information

In the theory of computation, discrete variables have the benefit that a practically perfect readout is often possible. Readout of continuous variables, on the other hand, always has an intrinsic precision (as we saw in Chapter 9). If the continuous variable is related to a physical parameter we wish to measure, we need to maximize the precision of the measurement. There are generally two aspects to parameter estimation, namely the encoding of the (unknown) parameter in the state of a probe system, and the measurement of the system that reveals information about the parameter. The entire process must be optimized in order to prescribe the best procedure for parameter estimation. In this section, we construct a quantity called the Fisher information, which tells us how much information about the parameter we can extract from the state of the probe system given the chosen measurement procedure (see Fig. 13.1). The Fisher information is related to the statistical distance between two probability distributions, and also to the angles between rays (states) in Hilbert space. This allows us to talk about the dynamical speed of evolution, and leads to two bounds on the time needed to distinguish between processes that give rise to different quantum states. We will see that entangled states can have an increased speed of dynamical evolution, and have

Fig. 13.1. Schematic of the parameter-estimation procedure. A procedure S prepares a system in state $\rho(0)$, the process under study evolves the state to $\rho(\theta)$, and a measurement M extracts information about θ.

the potential to give a better precision in parameter estimation than separable probe states, given comparable resources.

13.1.1 Bounds on precision measurements

On a practical level, experimental science is primarily concerned with making precision measurements. Often, such experiments are properly understood as the estimation of a set of parameters $\{\theta_1, \ldots, \theta_n\}$. Here, we consider the estimation of a single parameter θ. The experiment involves the measurement of an observable, with each measurement event yielding an outcome x that depends on the value of θ. In general, there will be errors and uncertainties in the measurement, which means that the relation between θ and the measurement outcome x is a probability distribution $p(\theta|x)$. We do not have direct access to this probability distribution. Rather, we have the measurement outcome x. We can model the experiment using a description of the state of the system $\rho(\theta)$ and the POVM $\hat{E}_x$ associated with measurement outcome x. The Born rule then stipulates that

$$p(x|\theta) = \mathrm{Tr}\left[\hat{E}_x \rho(\theta)\right] , \tag{13.1}$$

with $\int dx\, \hat{E}_x = \hat{\mathbb{1}}$. The uncertainties of the measurement procedure are encapsulated in the POVM $\hat{E}_x$, while any uncertainties in the state-preparation procedure are captured by the mixed state $\rho(\theta)$. We can relate $p(\theta|x)$ and $p(x|\theta)$ via Bayes' rule

$$p(\theta|x) = \frac{p(x|\theta)\, p(\theta)}{p(x)} \equiv L(\theta|x) , \tag{13.2}$$

where L is the *likelihood* of θ given the experimental data x. We assume that $p(\theta)$ and $p(x)$ are known or can be inferred. A large part of this chapter is devoted to the construction of such measurements with quantum optics.

In this section, we relate $p(x|\theta)$ to the procedure for extracting a value of θ from the data x using an estimator $T(x)$, and determine the error in θ. An estimator is a function that tells us how to translate the measurement data x into a value of θ. This will lead to the introduction of the Fisher information and the Cramér–Rao bound. We follow the derivation of Braunstein and Caves (1994). Throughout we will assume that x is a continuous variable, but the argument also holds for discrete variables, in which case the integrals over x must be replaced by sums over a discrete set of measurement outcomes x_j.

Let $T(x)$ be an estimator for θ based on the measurement outcome x, and let $\Delta T \equiv T(x) - \langle T \rangle_\theta$ with

$$\langle T \rangle_\theta \equiv \int dx\, p(x|\theta) T(x) . \tag{13.3}$$

The estimator is called 'unbiased' when $\langle T \rangle_\theta = \theta$. For *any* estimator T (biased or unbiased), and for N independent samples yielding measurement outcomes $x_1, \ldots, x_N$ we have

$$\int dx_1 \cdots dx_N\, p(x_1|\theta) \cdots p(x_N|\theta)\, \Delta T = 0 . \tag{13.4}$$

This follows from the definition of ΔT. Taking the derivative to θ and using the chain rule yields

$$\sum_{i=1}^{N} \int dx_1 \cdots dx_N\, p(x_1|\theta) \cdots p(x_N|\theta) \frac{1}{p(x_i|\theta)} \frac{\partial p(x_i|\theta)}{\partial \theta} \Delta T - \left\langle \frac{d \langle T \rangle_\theta}{d\theta} \right\rangle = 0 , \tag{13.5}$$

where we have used the fact that T does not depend on θ, but $\langle T \rangle_\theta$ does. This can be rewritten as

$$\int dx_1 \cdots dx_N\, p(x_1|\theta) \cdots p(x_N|\theta) \left(\sum_{i=1}^{N} \frac{\partial \ln p(x_i|\theta)}{\partial \theta} \right) \Delta T = \left\langle \frac{d \langle T \rangle_\theta}{d\theta} \right\rangle . \tag{13.6}$$

To this equation we can now apply the Cauchy–Schwarz inequality:

$$|\langle f, g \rangle|^2 \leq \|f\|_1^2 \|g\|_2^2 , \tag{13.7}$$

where f and g are duals with norms $\|f\|_1^2$ and $\|g\|_2^2$, respectively. We substitute

$$f = \sum_{i=1}^{N} \frac{\partial \ln p(x_i|\theta)}{\partial \theta} \quad \text{and} \quad g = \Delta T , \tag{13.8}$$

yielding

$$\int dx_1 \cdots dx_N\, p(x_1|\theta) \cdots p(x_N|\theta) \left(\sum_{i=1}^{N} \frac{\partial \ln p(x_i|\theta)}{\partial \theta} \right)^2$$
$$\times \int dx_1 \cdots dx_N\, p(x_1|\theta) \cdots p(x_N|\theta)\, (\Delta T)^2 \geq \left| \left\langle \frac{d \langle T \rangle_\theta}{d\theta} \right\rangle \right|^2 . \tag{13.9}$$

We can simplify this inequality by introducing the 'Fisher information' $F(\theta)$:

$$F(\theta) \equiv \int dx\, p(x|\theta) \left(\frac{\partial \ln p(x|\theta)}{\partial \theta} \right)^2 = \int dx \frac{1}{p(x|\theta)} \left(\frac{\partial p(x|\theta)}{\partial \theta} \right)^2 . \tag{13.10}$$

This is a measure of the information that the experimental set-up (consisting of the state $\rho(\theta)$ and the POVM $\{\hat{E}_x\}$) reveals about θ. Writing the integral over $(\Delta T)^2$ as $\langle (\Delta T)^2 \rangle_\theta$ and removing the superfluous brackets in the derivative then yields the following inequality:

$$N F(\theta) \langle (\Delta T)^2 \rangle_\theta \geq \left| \frac{d \langle T \rangle_\theta}{d\theta} \right|^2 . \tag{13.11}$$

This is almost a useful inequality. However, rather than the average error in the estimator ΔT, we want to relate the Fisher information to the average error in the *actual* value of θ. We therefore need to find a relationship between $\langle (\Delta T)^2 \rangle_\theta$ and $\langle (\Delta \theta)^2 \rangle_\theta$.

The error $\Delta \theta$ in the parameter θ for a single data point x is related to the estimator $T(x)$ in the following way:

$$\Delta \theta \equiv \frac{T(x)}{|d \langle T \rangle_\theta / d\theta|} - \theta \, . \tag{13.12}$$

If the estimator T is biased, then $\langle \Delta \theta \rangle_\theta$ is non-zero. Note that $\Delta \theta$ is not the same as ΔT, since $\Delta \theta$ is related to the *actual value* of θ, while ΔT is completely determined by the data point x, the estimator function $T(x)$, and the calculated probability distribution $p(x|\theta)$. The derivative corrects for the way $\langle T \rangle_\theta$ changes with θ. For example, if we record a millisecond process such as the popping of a balloon with a high-speed camera, we can measure the time it takes for the balloon to deflate by timing the slowed-down version on the film. This is the estimator. To find the real time it takes to deflate, we have to divide the measured time by the slow-down factor of the high-speed camera. For the error in the deflating time we need to make the same adjustment. It is a translation from the measured timescale of seconds, to the real timescale of milliseconds, or a change in units.

In order to derive the Cramér–Rao bound, we need to relate the error $\Delta \theta$ to the variance of the estimator ΔT. Using $\Delta T = T(x) - \langle T \rangle_\theta$, we find

$$\langle (\Delta T)^2 \rangle_\theta = \left| \frac{d \langle T \rangle_\theta}{d\theta} \right|^2 \left(\left\langle (\Delta \theta)^2 \right\rangle_\theta - \langle \Delta \theta \rangle_\theta^2 \right) \, . \tag{13.13}$$

We can eliminate the factor $\langle (\Delta T)^2 \rangle_\theta$ using Eq. (13.11), and we find

$$\langle (\Delta \theta)^2 \rangle_\theta \geq \frac{1}{N \, F(\theta)} + \langle \Delta \theta \rangle_\theta^2 \geq \frac{1}{N \, F(\theta)} \, . \tag{13.14}$$

This is the celebrated Cramér–Rao bound, which puts a minimum value on the error in θ. In any conceivable experiment (classical or quantum) the variance must always be greater than the inverse of (N times) the Fisher information. It is clear that the precision improves when the Fisher information increases. We therefore have to design experiments such that the Fisher information is as large as possible. Since $F(\theta)$ is a function of $p(x|\theta)$, which in turn is determined by the state of the system $\rho(\theta)$ and a POVM $\hat{E}_x$, this means we need to construct suitable states and observables to measure. However, for a given experiment, there is no guarantee that this bound can actually be attained. Generally, the Cramér–Rao bound is the theoretical minimum for the size of an error, even though in practice the error may be larger.

One important case of the Cramér–Rao bound occurs when the Fisher information is approximately constant (and of order one). That is, the amount of information about θ in a single measurement is constant. In this case, we can write the uncertainty $\delta \theta$ in the parameter θ as

$$\delta \theta \equiv \sqrt{\langle (\Delta \theta)^2 \rangle_\theta} \gtrsim \frac{1}{\sqrt{N}} \, . \tag{13.15}$$

In other words, the error in a parameter θ after N independent samples scales as $N^{-\frac{1}{2}}$. This is called the 'standard quantum limit' (SQL) or the 'shot-noise limit'.

Exercise 13.1: Given an input state $|\psi\rangle = (|0\rangle + \exp(i\theta)\,|1\rangle)/\sqrt{2}$ and a measurement of the Pauli X operator, calculate the Fisher information. Repeat this for the Pauli Z operator and the Hadamard. What is the minimum error in θ after N such measurements?

13.2 The statistical distance

A problem related to parameter estimation is hypothesis testing. Suppose we have two theories, predicting different values for our parameter, θ or θ'. We again set up an experiment consisting of the parameter-dependent state and the measurement procedure. Rather than trying to measure the parameter precisely, we define two probability distributions $p(x|\theta)$ and $p(x|\theta')$, and ask how well we can distinguish between the two probability distributions in the experiment. This means that we have to introduce the concept of a distance between two probability distributions, called the 'statistical distance'. In this section we will review some general properties of metric spaces and use these to derive a distance function on the space of probability distributions. We relate the statistical distance to the Fisher information, and give a Cramér–Rao bound for distinguishing two probability distributions. We give a classical derivation of the statistical distance first, and then extend the discussion to the quantum mechanical case.

13.2.1 Distance between classical probability distributions

We wish to construct a *space* of probability distributions with a *distance* defined on it. A function $s(a, b)$ is a distance between two points a and b if

(i) $s(a, b) \geq 0$
(ii) $s(a, b) = 0 \quad \Leftrightarrow \quad a = b$
(iii) $s(a, b) = s(b, a)$
(iv) $s(a, c) \leq s(a, b) + s(b, c) \qquad$ (the triangle inequality).

Furthermore, the distance in a metric space is determined by a metric g_{kl}. In infinitesimal form, this becomes

$$ds^2 = \sum_{kl} g_{kl}\, da^k da^l \,, \tag{13.16}$$

where the da^k are the components of the tangent vector to a. For a Euclidean space, the metric is the identity tensor, and the distance between points a and $a + da$ becomes Pythagoras' theorem:

$$ds^2 = \sum_k da_k da^k \,. \tag{13.17}$$

Note that there is a difference between the upper and the lower indices. We call the mathematical object with upper indices a 'vector', and the object with lower indices a '1-form' (in older literature the vector is often called 'contravariant' and the 1-form 'covariant', which refers to the way they behave under linear transformations). We can represent this pictorially as shown in Fig. 13.2. For more information about this geometrical representation of vectors, see Weinreich (1998), and for a more detailed introduction to vectors and 1-forms see Carroll (2004). The scalar product is a product (a 'contraction') between a vector and a 1-form (upper and lower indices). It corresponds to the number of sheets in the stack that are pierced by the arrow. It is now easy to see that transformations on the space leave the scalar product invariant: distorting the arrow (vector) and the stack of sheets (1-form) in equal measure will not change the number of sheets that are pierced by the arrow. The 1-form is the 'dual' of the vector (and vice versa), and they are related via the metric:

$$a_j = \sum_k g_{jk}\, a^k \quad \text{and} \quad a^j = \sum_k g^{jk}\, a_k \,. \tag{13.18}$$

In other words, the metric comes with upper or lower indices (or a mixture, if necessary), and acts as 'raising' or 'lowering' operators on the indices. Pictorially, the vectors transform as shown in Fig. 13.3. Moreover, the vectors that lie *in* the surfaces of the 1-form (at point a) are all orthogonal to the dual vector to the 1-form (at point a).

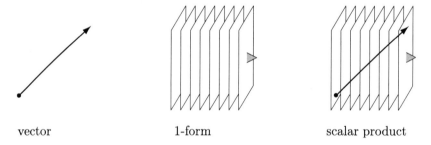

vector   1-form   scalar product

Fig. 13.2. Vectors and 1-forms. The inner product is invariant under space transformations.

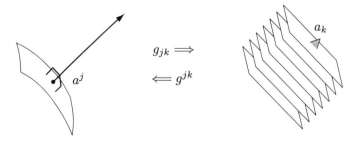

Fig. 13.3. Using the metric to transform between vectors and 1-forms.

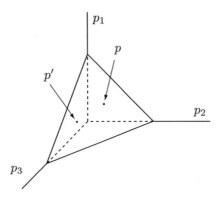

Fig. 13.4. **The probability simplex for probability distributions over three possible events** E_1, E_2, and E_3, **occurring with probabilities** p_1, p_2, and p_3, **respectively. A point** p **in the simplex uniquely determines a probability distribution.**

From the raising and lowering of the indices, we can derive a key property of the metric:

$$a_j = \sum_k g_{jk} \left(\sum_l g^{kl} a_l \right) = \sum_{kl} g_{jk}\, g^{kl}\, a_l \quad \Rightarrow \quad \sum_l g_{jk}\, g^{kl} = \delta_j{}^l \equiv \delta_{jl}, \qquad (13.19)$$

where the second identity follows from the linear independence of the a_j, and δ_{jl} is the Kronecker delta. The metric with upper indices g^{jk} is therefore the inverse of the metric g_{jk} with lower indices (and vice versa).

Next, the 'probability simplex' is the convex space of probability distributions (see Fig. 13.4). This is a metric space, but the (Euclidean) distance between the two points p and p' is not necessarily the most useful distance function for probability distributions. We can write down a general distance function between two probability distributions p and $p' = p + dp$ in terms of the metric:

$$ds^2 = \sum_{jk} g_{jk}\, dp^j\, dp^k \ . \qquad (13.20)$$

In general, the sum may be an integral.

Our next task is to find a 'natural metric' g_{jk} of the simplex. To this end we construct the natural dual to the tangent vectors dp of a probability distribution p, namely a random variable. In particular, an observable A can be considered a (classical) random variable with expectation value

$$\langle A \rangle \equiv \sum_j A_j p^j \ , \qquad (13.21)$$

where A_j are the possible values of A. The points of constant expectation value form surfaces in the dual space to dp. Consequently, surfaces with incrementally increasing expectation values form a stack, which makes for a natural 1-form.

An explicit expression for the metric can be determined by a quadratic form of (tangent) vectors, or alternatively, a quadratic form of 1-forms. We choose to derive the metric from the latter. For a pair of observables A and B, the natural quadratic form is the so-called 'correlation'

$$\sum_{jk} A_j B_k \, g^{jk} = \langle AB \rangle = \sum_j A_j B_j p^j \, . \tag{13.22}$$

From this we see immediately that $g^{jk} = \delta_j{}^k p^j$. Using $\sum_k g^{jk} g_{kl} = \delta^j{}_l$ we have

$$g_{jk} = \frac{\delta^j{}_k}{p^j} \, , \tag{13.23}$$

and therefore

$$ds^2 = \sum_{jk} g_{jk} \, dp^j \, dp^k = \sum_j \frac{(dp^j)^2}{p^j} \tag{13.24}$$

is a 'statistical distance' between two infinitesimally close probability distributions p and $p + dp$. We derived this expression using the correlation in Eq. (13.22), which is the only natural quadratic form for classical observables. We therefore claim that ds^2 is the *natural* statistical distance between two infinitesimally close probability distributions, up to an overall constant factor. The mathematical form of ds^2 is slightly peculiar, which hints at a fundamental property of probability distributions. We can substitute $p^j = (r^j)^2$ and use $d(r^j)^2 = 2r^j \, dr^j$, which yields

$$ds^2 = 4 \sum_j (dr^j)^2 \, . \tag{13.25}$$

We have now expressed the statistical distance as the natural Euclidean distance for 'probability amplitudes'! Note that we have not talked about quantum mechanics at all at this point: probability amplitudes arise naturally in classical probability theory. We should also point out the factor four in Eq. (13.25). This can be interpreted as a scale factor due to a difference in units. It will appear again in the next section, when we discuss the dynamical evolution of states.

We can relate the statistical distance to the Fisher information in the following way. Divide both sides of Eq. (13.24) by $d\theta^2$

$$\left(\frac{ds}{d\theta} \right)^2 = \sum_j \frac{1}{p^j} \left(\frac{dp^j}{d\theta} \right)^2 = F(\theta) \, . \tag{13.26}$$

This is the Fisher information for a probability distribution over a discrete variable, given in Eq. (13.10). When p is defined over a continuum x, we can write this as

$$\left(\frac{ds}{d\theta} \right)^2 = \int dx \, \frac{1}{p(x|\theta)} \left(\frac{\partial p(x|\theta)}{\partial \theta} \right)^2 = F(\theta) \, . \tag{13.27}$$

A Taylor expansion of s in θ can be written as

$$s(\theta) = s(0) + \frac{ds}{d\theta} \delta\theta + O(\delta\theta^2) \, . \tag{13.28}$$

If we set $ds = s(\theta) - s(0)$ and take the square of the expansion to lowest order, we find

$$ds^2 = \left(\frac{ds}{d\theta} \right)^2 \delta\theta^2 \,. \tag{13.29}$$

From the Cramér–Rao bound in Eq. (13.14) and the expression of the Fisher information in Eq. (13.27), we derive

$$ds^2 = \left(\frac{ds}{d\theta} \right)^2 \delta\theta^2 \geq \left(\frac{ds}{d\theta} \right)^2 \frac{1}{N F(\theta)} = \frac{1}{N} \,. \tag{13.30}$$

The criterion for distinguishing nearby probability distributions $p(x|\theta)$ and $p(x|\theta')$ is therefore given by

$$N ds^2 \gtrsim 1 \,. \tag{13.31}$$

Here, N is the number of times we have to query the system. In other words, it is the size of our measurement data. We state this relation in infinitesimal form as an approximate inequality, since the distinguishability criterion is tight only for infinitesimal distances. However, this formula does provide useful practical estimates for distinguishing probability distributions separated by a small finite distance.

Consider now the special case where one of the probabilities p^j in the probability distribution is zero, while $p^{j'}$ is non-zero. The statistical distance diverges, since $(dp^j)^2$ is non-zero and the denominator p^j is zero. What does this mean? Assume that our system can be described by either one of two probability distributions p_1 or p_2. When p_1 assigns a zero probability to a certain event, while p_2 assigns a finite probability to the same event, a single observation of the event is sufficient to determine that our system obeys p_2, rather than p_1. It does not matter that the chance of seeing the event $(p^{j'})$ may be very small. The *possibility* of distinguishing p_1 and p_2 *with certainty* in a single shot is enough to make the statistical distance infinitely large. In practice, infinities will cause problems, and we tend to exclude the boundaries of the simplex in practical situations, such as when performing numerical calculations.

13.2.2 Distance between density operators

The derivation of the statistical distance between two probability distributions has so far been completely classical, and in particular the correlation we used in Eq. (13.22) is a quadratic form based on taking classical averages. In quantum theory, averages are calculated by taking the expectation value of observables with respect to the state ρ:

$$\langle \hat{A} \rangle = \text{Tr}(\rho \hat{A}) \,. \tag{13.32}$$

We are especially interested in the case where $\hat{A}$ is a Hermitian operator, or an observable. This introduces a subtlety for the quantum mechanical correlation between two observables $\hat{A}$ and $\hat{B}$, since the correlation is itself an observable and should therefore be Hermitian.

However, the product of two Hermitian operators is not Hermitian when $\hat{A}$ and $\hat{B}$ do not commute:

$$(\hat{A}\hat{B})^{\dagger} = \hat{B}^{\dagger}\hat{A}^{\dagger} = \hat{B}\hat{A} \neq \hat{A}\hat{B}. \tag{13.33}$$

One way to solve this is to define the natural quadratic form for quantum mechanical observables as the expectation value of the anti-commutator $\frac{1}{2}(\hat{A}\hat{B} + \hat{B}\hat{A}) = \frac{1}{2}\{\hat{A}, \hat{B}\}$:

$$\frac{1}{2}\left\langle\{\hat{A}, \hat{B}\}\right\rangle = \frac{1}{2}\mathrm{Tr}\left[\{\hat{A}, \hat{B}\}\rho\right], \tag{13.34}$$

where the factor $1/2$ is chosen such that the autocorrelation $\hat{A}^2 = \frac{1}{2}\{\hat{A}, \hat{A}\}$ is included in the quadratic form. Using the cyclic property of the trace, the correlation can be written as

$$\frac{1}{2}\left\langle\{\hat{A}, \hat{B}\}\right\rangle = \frac{1}{2}\mathrm{Tr}\left[\hat{A}\{\rho, \hat{B}\}\right] \equiv \mathrm{Tr}\left[\hat{A}\,\mathscr{R}_{\rho}(\hat{B})\right], \tag{13.35}$$

where we have defined the superoperator $\mathscr{R}_{\rho}(\hat{B})$ as

$$\mathscr{R}_{\rho}(\hat{B}) = \frac{1}{2}\{\rho, \hat{B}\} = \frac{1}{2}\sum_{jk}(p_j + p_k)B_{jk}\,|j\rangle\langle k|, \tag{13.36}$$

and the density operator is written in the diagonal basis as $\rho = \sum_j p_j\,|j\rangle\langle j|$. The number B_{jk} is a matrix element of $\hat{B}$ in the basis $\{|j\rangle\}$. The operator $\mathscr{R}_{\rho}$ is analogous to the metric in Eq. (13.22). It can therefore be interpreted as a raising operator, just as g^{jk} raised the index from a 1-form to a vector. The inverse of $\mathscr{R}_{\rho}$ must then be analogous to the lowering operator $\mathscr{L}_{\rho} = \mathscr{R}_{\rho}^{-1}$. Using Eq. (13.36) it is not too difficult to show that (see Exercise 13.2):

$$\mathscr{L}_{\rho}(\hat{B}) = \sum_{jk}\frac{2B_{jk}}{p_j + p_k}\,|j\rangle\langle k|. \tag{13.37}$$

This is the direct analog of Eq. (13.23). Just as in the classical case, we exclude the boundaries of the probability simplex in order to ensure that the inverse $\mathscr{L}_{\rho}$ exists.

Exercise 13.2: Show that $\mathscr{L}_{\rho}\left(\mathscr{R}_{\rho}(\hat{B})\right) = \mathscr{R}_{\rho}\left(\mathscr{L}_{\rho}(\hat{B})\right) = \hat{B}$ for any $\hat{B}$. This proves that $\mathscr{L}_{\rho}$ is the inverse of $\mathscr{R}_{\rho}$, and vice versa.

The observables $\hat{A}$ and $\hat{B}$ are the natural 1-forms (where the surfaces of constant expectation value form the stack in Fig. 13.3). Similarly, the density operator ρ plays the role of the probability distribution, which is naturally a vector with components p^j. The raising and lowering operators $\mathscr{R}_{\rho}$ and $\mathscr{L}_{\rho}$ can then be used to turn the 1-forms into vectors, and vice versa. This is represented pictorially in Fig. 13.5, and is a direct analogue to Fig. 13.3. Note that we use $d\rho$ as a vector, rather than ρ. It is the vector from one density matrix ρ to another $\rho + d\rho$.

We generalize the statistical distance for classical probability distributions to a statistical distance ds_{ρ}^2 between the two density operators ρ and $\rho + d\rho$:

$$ds^2 = \sum_{jk}g_{jk}dp^j dp^k \quad \Rightarrow \quad ds_{\rho}^2 = \mathrm{Tr}\left[d\rho\,\mathscr{L}_{\rho}(d\rho)\right]. \tag{13.38}$$

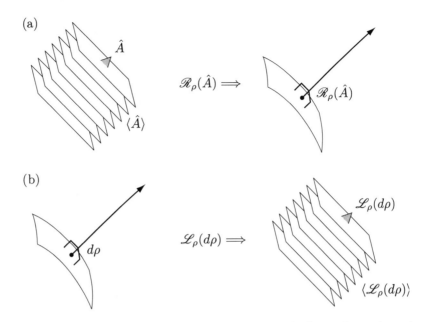

(a)

$\mathscr{R}_\rho(\hat{A}) \Longrightarrow$

$\hat{A}$

$\langle \hat{A} \rangle$

$\mathscr{R}_\rho(\hat{A})$

(b)

$d\rho$

$\mathscr{L}_\rho(d\rho) \Longrightarrow$

$\mathscr{L}_\rho(d\rho)$

$\langle \mathscr{L}_\rho(d\rho) \rangle$

Fig. 13.5. The raising and lowering operators $\mathscr{R}_\rho$ and $\mathscr{L}_\rho$ relate vectors and 1-forms. This applies to both the operators $\hat{A}$ and the density matrices ρ. The stacks (1-forms) are surfaces of constant expectation value.

Using Eq. (13.27) we can relate the statistical distance along a path parameterized by θ to the Fisher information and the derivative of ρ with respect to θ:

$$F(\theta) = \left(\frac{ds_\rho}{d\theta}\right)^2 = \mathrm{Tr}\left[\rho'\,\mathscr{L}_\rho(\rho')\right] \quad \text{with} \quad \rho' = \frac{d\rho}{d\theta}. \tag{13.39}$$

This relation allows us to derive a bound on the Fisher information. This is important, because so far the Cramér–Rao bound on the error $\delta\theta$ can be arbitrarily small given large enough Fisher information, as is evident from Eq. (13.14). Physically, a bound on the Fisher information means that there is a limit to the amount of information about θ that one can extract in any given measurement.

In order to derive the bound on $F(\theta)$, we recall that in quantum mechanics the Heisenberg equations of motion can be written as

$$\frac{d\rho}{d\theta} = -\frac{i}{\hbar}\left[\hat{K}, \rho\right], \tag{13.40}$$

where the operator $\hat{K}$ is the generator of translations in θ. For example, if θ is a time translation, $\hat{K}$ is the Hamiltonian; if θ is a position, $\hat{K}$ is the momentum operator, etc. In the equations of motion we have assumed that the generator does not depend on θ, which is a natural assumption in parameter estimation (any unitary evolution $\rho(0) \to \rho(\theta)$ satisfies this). Furthermore, we can replace $\hat{K}$ by $\Delta\hat{K} = \hat{K} - \langle\hat{K}\rangle$ without changing the equations of motion, since the expectation value $\langle\hat{K}\rangle$ is a number, and therefore always commutes with ρ:

$$\frac{d\rho}{d\theta} = -\frac{i}{\hbar}\left[\Delta\hat{K}, \rho\right]. \tag{13.41}$$

We can now use the explicit form of $\mathscr{L}_\rho$ in Eq. (13.37) to calculate

$$F(\theta) = \text{Tr}\left[\rho' \mathscr{L}_\rho(\rho')\right] = \frac{2}{\hbar^2}\sum_{jk}(p_j + p_k)\left(\frac{p_j - p_k}{p_j + p_k}\right)^2 |\Delta\hat{K}_{jk}|^2$$

$$\leq \frac{4}{\hbar^2}\langle(\Delta\hat{K})^2\rangle_\theta \equiv \frac{4}{\hbar^2}(\delta\hat{K})^2, \tag{13.42}$$

in our usual notation. This means that the Fisher information is bounded by the variance of the generator $\hat{K}$ of translations in the parameter θ. We will use this important result later in this chapter.

Compare Eq. (13.42) to a single-shot parameter-estimation procedure with $N = 1$. The Cramér–Rao bound in Eq. (13.14) then dictates that

$$(\delta\theta)^2 \geq \frac{1}{F(\theta)}, \tag{13.43}$$

and by eliminating $F(\theta)$ from both equations, we find

$$\delta\theta\,\delta\hat{K} \geq \frac{\hbar}{2}. \tag{13.44}$$

This is the 'Heisenberg uncertainty relation' for a parameter θ and its generator $\hat{K}$. The most important examples are the energy–time uncertainty relation, where θ is time and $\hat{K}$ is the Hamiltonian, and the number–phase uncertainty relation, where θ is the phase of a quantum mechanical oscillator, and $\hat{K} = \hat{n}$ is its number operator. The interpretation of the uncertainty principle, which often leads to debates, can now be inferred: the error in the measurement of a parameter times the uncertainty in its generator for the measured system can never be smaller than a finite value. For this relation to be meaningful, we need a well-defined measurement procedure leading to $\delta\theta$, and a well-defined Hermitian generator $\hat{K}$. Note that the derivation given here is different from the usual textbooks on quantum mechanics, which derive the Heisenberg uncertainty relations for non-commuting observables. The two derivations complement each other.

Exercise 13.3: Verify Eqs. (13.42) and (13.44).

We conclude this section with a discussion on the interpretation of the statistical distance when we confine ourselves to pure quantum states. Consider two (possibly non-orthogonal) states $|\psi_1\rangle$ and $|\psi_2\rangle$. There will be some preparation device that produces either $|\psi_1\rangle$ or $|\psi_2\rangle$, and we have to find out which one it is. The preparation device may be some physical process that changes the state of a probe system. There are many possible observables that may give us some information about these states, and we are of course interested in the optimal measurement.

The statistical distance is defined for probability distributions, and in quantum mechanics we need both a quantum state and an observable to construct a probability distribution via the

Born rule. We here therefore define the statistical distance for the two states relative to the measurement. If the (non-degenerate) measurement basis is denoted by $\{|\phi_1\rangle, \ldots, |\phi_m\rangle\}$, the two probability distributions become

$$p_1(j) = |\langle \psi_1 | \phi_j \rangle|^2 \quad \text{and} \quad p_2(j) = |\langle \psi_2 | \phi_j \rangle|^2, \tag{13.45}$$

and the corresponding amplitudes are

$$r_1(j) = |\langle \psi_1 | \phi_j \rangle| \quad \text{and} \quad r_2(j) = |\langle \psi_2 | \phi_j \rangle|. \tag{13.46}$$

The probability amplitudes form the components of a unit vector in a hypersphere, and the Euclidean distance between two such vectors is given by the angle between the vectors:

$$s(\psi_1, \psi_2) = \max\left[\arccos\left(\sum_j |\langle \psi_1 | \phi_j \rangle| |\langle \phi_j | \psi_2 \rangle|\right)\right]. \tag{13.47}$$

The optimal measurement occurs when one of the basis vectors of the measurement observable is equal to one of the states, say $|\phi_1\rangle = |\psi_1\rangle$, such that

$$s(\psi_1, \psi_2) = \arccos\left(|\langle \psi_1 | \psi_2 \rangle|\right). \tag{13.48}$$

This is the angle between the two rays $|\psi_1\rangle$ and $|\psi_2\rangle$ in Hilbert space. Notice that there is a difference of a factor 2 between the statistical distance $s(\psi_1, \psi_2)$, which is a Euclidean distance, and the statistical distance derived in Eq. (13.25). This is not a problem, since both definitions satisfy the criteria for a distance function given at the beginning of this section. However, it becomes important when we use the two definitions side-by-side in the next section.

13.3 The dynamical evolution of states

Since we have a well-defined concept of distance between quantum states, we can ask how quickly a statistical distance may be travelled in a quantum evolution. In other words, we wish to derive the maximum dynamical speed of evolution, and relate this to the distinguishability of probability distributions. The speed of evolution v can be formally defined as

$$v(t) \equiv \frac{ds}{dt}. \tag{13.49}$$

However, we can be more general by considering θ, rather than t as the dynamical parameter. We have already defined the Hermitian operator $\hat{K}$ as the generator of translations in θ in the previous section. We therefore write

$$v(\theta)^2 = \left(\frac{ds}{d\theta}\right)^2 = F(\theta) \leq \frac{4}{\hbar^2}(\delta\hat{K})^2, \tag{13.50}$$

where we used Eq. (13.42) for the inequality. This leads to

$$\frac{ds}{d\theta} \leq \frac{2\delta\hat{K}}{\hbar} , \tag{13.51}$$

which can be integrated to give

$$\int_0^\theta d\theta' \geq \frac{\hbar}{2\delta\hat{K}} \int_0^\pi ds \quad \Rightarrow \quad \theta \geq \frac{\pi}{2} \frac{\hbar}{\delta\hat{K}} . \tag{13.52}$$

Note that we have integrated the distance s from 0 to π. From the Euclidean distance between quantum states we know that orthogonality is reached when $s = \pi/2$, since s is the angle between two unit vectors in Hilbert space. However, in the definition of $v(\theta)$ we used the statistical distance of Eq. (13.25), which means that we need to multiply $\pi/2$ by a factor 2. In the special case of $\theta = t$, the generator $\hat{K}$ is the Hamiltonian, and we write $\delta\hat{K}$ as the variance of the energy ΔE:

$$t \geq \frac{\pi}{2} \frac{\hbar}{\Delta E} . \tag{13.53}$$

This bound is known as the Mandelstam–Tamm inequality, and was first derived in 1945 by Mandelstam and Tamm. It is another example of a Cramér–Rao bound. There exists an extensive literature on exactly what this inequality means. From the preceding discussions, we see that it gives the minimum time in which a system can evolve from one state to an orthogonal state. See Bhattacharyya (1983) for a discussion of other interpretations.

The inequality is not applicable when ΔE tends to zero or infinity. An important example where this bound fails is for processes with a Lorentzian spectrum

$$|\langle\omega|\psi(\omega_0)\rangle| = \frac{\gamma}{\pi} \frac{1}{(\omega - \omega_0)^2 + \gamma^2} . \tag{13.54}$$

The variance in energy is infinite, but the linewidth γ does define a sensible lifetime $\tau = \gamma^{-1}$. States such as these occur naturally in spontaneous-emission processes.

Exercise 13.4: Calculate $\langle E\rangle$ and ΔE for a single-photon state with a Lorentzian spectrum.

In the derivation of the Mandelstam–Tamm inequality above we have used the relation between $ds/d\theta$ and the Fisher information $F(\theta)$. However, we also defined the statistical distance as the angle between two quantum states $s = \arccos(|\langle\psi_0|\psi_\theta\rangle|)$. Assume that $|\psi_0\rangle$ is some initial state with $\theta = 0$, and $|\psi_\theta\rangle$ is the state after an evolution generated by $\hat{K}$, parameterized by θ,

$$|\psi_\theta\rangle = \exp\left(-\frac{i}{\hbar}\hat{K}\theta\right)|\psi_0\rangle . \tag{13.55}$$

We can now explicitly calculate the derivative of the statistical distance with respect to the parameter θ

$$\frac{ds}{d\theta} = \frac{d}{d\theta}\arccos\left(|\langle\psi_0|\psi_\theta\rangle|\right) = -\frac{1}{\sqrt{1 - |\langle\psi_0|\psi_\theta\rangle|^2}}\frac{d}{d\theta}|\langle\psi_0|\psi_\theta\rangle| . \tag{13.56}$$

If we use

$$\frac{1}{\sqrt{1 - |x|^2}} \geq 1 \quad \text{for all } x, \tag{13.57}$$

we obtain the inequality

$$\frac{ds}{d\theta} \leq -\frac{d}{d\theta} |\langle \psi_0 | \psi_\theta \rangle| . \tag{13.58}$$

Next, we show that

$$-\frac{d}{d\theta} |\langle \psi_0 | \psi_\theta \rangle| \leq \left| \frac{d}{d\theta} \langle \psi_0 | \psi_\theta \rangle \right| . \tag{13.59}$$

Given the generalized Schrödinger equation $i\hbar d/d\theta |\psi\rangle = \hat{K} |\psi\rangle$, we can express the left-hand side as

$$
\begin{aligned}
\frac{d}{d\theta} |\langle \psi_0 | \psi_\theta \rangle| &= \frac{d}{d\theta} \sqrt{\langle \psi_0 | \psi_\theta \rangle \langle \psi_\theta | \psi_0 \rangle} \\
&= \frac{-i \langle \psi_0 | \hat{K} | \psi_\theta \rangle \langle \psi_\theta | \psi_0 \rangle + i \langle \psi_0 | \psi_\theta \rangle \langle \psi_\theta | \hat{K} | \psi_0 \rangle}{2\hbar |\langle \psi_0 | \psi_\theta \rangle|} \\
&= \frac{\text{Im}(\langle \psi_0 | \hat{K} | \psi_\theta \rangle \langle \psi_\theta | \psi_0 \rangle)}{\hbar |\langle \psi_0 | \psi_\theta \rangle|} \\
&\leq \frac{| \langle \psi_0 | \hat{K} | \psi_\theta \rangle \langle \psi_\theta | \psi_0 \rangle |}{\hbar |\langle \psi_0 | \psi_\theta \rangle|} .
\end{aligned} \tag{13.60}
$$

The right-hand side of Eq. (13.59) can be calculated as

$$
\begin{aligned}
\left| \frac{d}{d\theta} \langle \psi_0 | \psi_\theta \rangle \right| &= \frac{1}{\hbar} | \langle \psi_0 | \hat{K} | \psi_\theta \rangle | = \frac{| \langle \psi_0 | \hat{K} | \psi_\theta \rangle | \cdot | \langle \psi_0 | \psi_\theta \rangle |}{\hbar |\langle \psi_0 | \psi_\theta \rangle|} \\
&\geq \frac{| \langle \psi_0 | \hat{K} | \psi_\theta \rangle \langle \psi_\theta | \psi_0 \rangle |}{\hbar |\langle \psi_0 | \psi_\theta \rangle|} ,
\end{aligned} \tag{13.61}
$$

where in the last line we have used the Cauchy–Schwarz inequality. Comparing the left-hand side and the right-hand side we just calculated, we obtain

$$\frac{d}{d\theta} |\langle \psi_0 | \psi_\theta \rangle| \leq \frac{| \langle \psi_0 | \hat{K} | \psi_\theta \rangle \langle \psi_\theta | \psi_0 \rangle |}{\hbar |\langle \psi_0 | \psi_\theta \rangle|} \leq \left| \frac{d}{d\theta} \langle \psi_0 | \psi_\theta \rangle \right| . \tag{13.62}$$

Since $\arccos x$ is a monotonically decreasing function, its derivative is (strictly) negative, and the derivative of $|\langle \psi_0 | \psi_\theta \rangle|$ to θ is strictly positive. Consequently, we have proved that Eq. (13.59) is true. We continue our derivation by substituting this into Eq. (13.58)

$$\frac{ds}{d\theta} \leq \left| \frac{d}{d\theta} |\langle \psi_0 | \psi_\theta \rangle| \right| \leq \frac{| \langle \psi_0 | \hat{K} | \psi_\theta \rangle |}{\hbar} \leq \frac{| \langle \psi_0 | \hat{K} | \psi_0 \rangle |}{\hbar} \equiv \frac{|\langle \hat{K} \rangle|}{\hbar} . \tag{13.63}$$

Separating again the variables s and θ, we find

$$\int_0^\theta d\theta' \geq \frac{\hbar}{|\langle \hat{K} \rangle|} \int_0^{\pi/2} ds . \tag{13.64}$$

After integration, the inequality becomes

$$\theta \geq \frac{\pi}{2} \frac{\hbar}{|\langle \hat{K} \rangle|}. \tag{13.65}$$

When $\hat{K}$ is the Hamiltonian of the system, and $\theta = t$. This inequality is known as the Margolus–Levitin inequality

$$t \geq \frac{\pi}{2} \frac{\hbar}{\langle E \rangle}, \tag{13.66}$$

where $\langle E \rangle$ is the average energy in the system at $t = 0$. This inequality was discovered in 1998, more than fifty years after the formulation of the inequality of Mandelstam and Tamm, and it has since played an important role in quantum parameter estimation and general quantum information theory. The proof of the Margolus–Levitin inequality for the unitary evolution of mixed states can be found via state purification. Alternative proofs are given in Giovannetti *et al.* (2003) and in Levitin and Toffoli (2009). Levitin and Toffoli also proved that the only states that can achieve the maximum dynamical speed limit are pure states of the form

$$|\psi\rangle = \frac{1}{\sqrt{2}} \left(|\psi_0\rangle + e^{i\varphi} |\psi_1\rangle \right), \tag{13.67}$$

where $|\psi_0\rangle$ and $|\psi_1\rangle$ are (possibly degenerate) energy eigenstates, and φ is an arbitrary phase.

In order to find the dynamical speed of evolution, we need to know both $\langle E \rangle$ and ΔE to find out whether the Mandelstam–Tamm or the Margolus–Levitin bound applies. For metrology, we need to choose our states such that both $\langle E \rangle$ and ΔE are favourable. If we look at a single optical mode with a sharply peaked central frequency ω, the phase change can be written as $\Delta \theta = \omega t$, and the two inequalities become

$$\Delta \theta \geq \frac{\pi}{2} \frac{\hbar \omega}{\Delta E} \quad \text{and} \quad \Delta \theta \geq \frac{\pi}{2} \frac{\hbar \omega}{\langle E \rangle}, \tag{13.68}$$

so we have two expressions for the minimal resolvable phase $\Delta \theta$, given an interaction time t and Hamiltonian $\mathcal{H}$.

We now consider two examples, namely the coherent state and a NOON state. The coherent state $|\alpha\rangle$ of a single mode with frequency ω has mean energy $\langle E \rangle = \hbar \omega \langle n \rangle$, and energy uncertainty $\Delta E = \hbar \omega \sqrt{\langle n \rangle}$. The two inequalities then yield

$$t_{\text{MT}} \geq \frac{\pi}{2} \frac{1}{\omega \sqrt{\langle n \rangle}} \quad \text{and} \quad t_{\text{ML}} \geq \frac{\pi}{2} \frac{1}{\omega \langle n \rangle}. \tag{13.69}$$

Therefore $t_{\text{MT}} \geq t_{\text{ML}}$, and the Mandelstam–Tamm inequality is the equality that restricts the dynamical speed of evolution for coherent states. Similarly, the minimum phase resolution given a fixed interaction time t is determined by the Mandelstam–Tamm inequality, and reads

$$\Delta \theta \geq \frac{\pi}{2} \frac{1}{\sqrt{\langle n \rangle}}. \tag{13.70}$$

Therefore, coherent states can in principle be used to measure phases at the shot-noise limit.

NOON states, on the other hand, are written as $(|N,0\rangle + |0,N\rangle)/\sqrt{2}$, and have average energy $\langle E \rangle = N\hbar\omega$ and uncertainty $\Delta E = 0$. Therefore, the Mandelstam–Tamm inequality is ill defined, while the Margolus–Levitin inequality yields

$$t_{\mathrm{ML}} \geq \frac{\pi}{2} \frac{1}{N\omega} .$$

(13.71)

If $\langle n \rangle$ and N are measures for the resources of the states, then the dynamical speed of evolution for NOON states is much faster than that for coherent states. Similarly, given a fixed interaction time t, the phase change $\Delta\theta$ that can be resolved with a NOON state is given by

$$\Delta\theta \geq \frac{\pi}{2} \frac{1}{N} .$$

(13.72)

In other words, NOON states can be used in principle to reach the Heisenberg limit. However, we will see that both the state preparation of these states and the optimal measurements that give the Heisenberg limit are extremely challenging to implement in practice.

13.4 Entanglement-assisted parameter estimation

We have seen above that NOON states, in which N photons are entangled with a spatial degree of freedom, can in principle attain the Heisenberg limit. In this section, we will consider how the Heisenberg limit can be reached with a large number of entangled qubits. We first calculate the precision in the estimation of a phase θ given N independent qubits, and we compare this with the situation where the N qubits are entangled in a GHZ state. In addition, we calculate the Fisher information in both cases.

First we consider the 'classical' case without any entanglement. We prepare a qubit in the state $|+\rangle = (|0\rangle + |1\rangle)/\sqrt{2}$. The evolution of the qubit is given by

$$|0\rangle \rightarrow |0\rangle \quad \text{and} \quad |1\rangle \rightarrow \exp(i\theta)|1\rangle .$$

(13.73)

This produces the state

$$|+\rangle \rightarrow |\theta\rangle = \frac{|0\rangle + e^{i\theta}|1\rangle}{\sqrt{2}} .$$

(13.74)

We need to find a suitable observable to estimate θ. A promising candidate is the Pauli X operator:

$$\langle X \rangle = \langle \theta | X | \theta \rangle = \cos\theta .$$

(13.75)

A measurement of X has two possible outcomes, ± 1. The probability $p(\pm|\theta)$ is given by the Born rule

$$p(\pm|\theta) = |\langle \pm | \theta \rangle|^2 = \frac{1}{2}(1 \pm \cos\theta) .$$

(13.76)

The Fisher information is readily calculated to be

$$F(\theta) = \sum_{\pm} \frac{1}{p(\pm|\theta)} \left(\frac{\partial p(\pm|\theta)}{\partial\theta} \right)^2 = 1 .$$

(13.77)

If we repeat this procedure with N qubits, we find $\langle X \rangle_N^2 = \sum_{j=1}^{N} \langle X_j \rangle^2 = N \cos^2 \theta$. Using $X^2 = \hat{\mathbb{1}}$, we get $\sum_{j=1}^{N} \langle (X_j^2) \rangle = N$ and the variance becomes

$$(\Delta X)_N^2 = N - N \cos^2 \theta = N \sin^2 \theta \,. \tag{13.78}$$

Using Eq. (13.13) we can relate the variance $(\Delta X^{\otimes N})^2$ to the average error in the phase $\delta\theta$:

$$\delta\theta = \left| \frac{d \langle X \rangle_N}{d\theta} \right|^{-1} (\Delta X)_N = \frac{1}{\sqrt{N}} \,. \tag{13.79}$$

This is the standard quantum limit. If we calculate the Cramér–Rao bound using the Fisher information, we find

$$\delta\theta \geq \frac{1}{\sqrt{N\,F(\theta)}} = \frac{1}{\sqrt{N}} \,, \tag{13.80}$$

which is consistent with Eq. (13.79).

Next, we consider an N-qubit GHZ state as the input state, and assume the same evolution $|0\rangle \to |0\rangle$ and $|1\rangle \to e^{i\theta} |1\rangle$. The GHZ state can be written as

$$|+_N\rangle \equiv \frac{1}{\sqrt{2}} (|0, 0, \ldots, 0\rangle + |1, 1, \ldots, 1\rangle) \tag{13.81}$$

with the same evolution as before. Afterwards, the state can be written as

$$|\theta_N\rangle \equiv \frac{1}{\sqrt{2}} \left(|0, 0, \ldots, 0\rangle + e^{iN\theta} |1, 1, \ldots, 1\rangle \right) \,. \tag{13.82}$$

For convenience, we redefine $|\psi_0\rangle = |0, 0, \ldots, 0\rangle$ and $|\psi_1\rangle = |1, 1, \ldots, 1\rangle$. Now our N-qubit system is mathematically equivalent to a single (non-local!) system with a relative phase shift $N\theta$. If we choose the observable $X_N = |\psi_0\rangle \langle \psi_1| + |\psi_1\rangle \langle \psi_0|$, then we calculate

$$\langle \theta_N | X_N | \theta_N \rangle = \cos N\theta \quad \text{and} \quad \langle \theta_N | X_N^2 | \theta_N \rangle = 1 \,. \tag{13.83}$$

Therefore, $\Delta X_N = \sqrt{1 - \cos^2 N\theta} = \sin N\theta$, and we find for the phase resolution

$$\delta\theta = \left| \frac{d \langle X_N \rangle}{d\theta} \right|^{-1} \Delta X_N = \frac{1}{N} \,. \tag{13.84}$$

This is the Heisenberg limit. The Fisher information is calculated using

$$p(\pm_N | \theta_N) = \frac{1}{2} (1 \pm \cos N\theta) \,, \tag{13.85}$$

yielding

$$F(\theta) = \sum_{\pm} \frac{1}{p(\pm | \theta)} \left(\frac{\partial p(\pm | \theta)}{\partial \theta} \right)^2 = N^2 \,. \tag{13.86}$$

The Cramér–Rao bound now reads

$$\delta\theta \geq \frac{1}{\sqrt{F(\theta)}} = \frac{1}{N} \,, \tag{13.87}$$

which is consistent with Eq. (13.84). Since all N qubits are used in a single measurement, the Fisher information depends on N, while the number of independent samples is one. Since the error $\delta\theta$ in Eq. (13.84) coincides with the minimum error derived via the Fisher information, this is an optimal parameter-estimation procedure given this type of measurement.

Exercise 13.5: Verify the Fisher information in Eq. (13.86).

Even though the measurement of a highly entangled observable X_N is in principle possible in the quantum mechanical formalism, it is not likely that this can be implemented in practice with optical systems. Instead, we want to rewrite this problem in terms of two-qubit entangling operations and single-qubit measurements. The universality of single- and two-qubit gates for quantum computing means that this rewriting can be done (at least in principle). It is convenient to use the stabilizer formalism (see Chapter 2). We construct a star-shaped cluster of N qubits, where each qubit is prepared in the $|+\rangle$ state, and $N-1$ CZ gates are applied between qubit 1 and qubits $j \neq 1$. The stabilizer generators are then given by

$$S_1 = X_1 \prod_{j \neq 1} Z_j \quad \text{and} \quad S_j = X_j Z_1 \,. \tag{13.88}$$

Next, the star cluster is converted to a GHZ state by means of (single-qubit) Hadamard operators on qubits $j \neq 1$. This leads to the stabilizer

$$S_1' = \prod_{j=1}^{N} X_j \quad \text{and} \quad S_j' = Z_j Z_1 \,. \tag{13.89}$$

The state is given by Eq. (13.81), and the evolution $U_{\theta,j}$ of each qubit j again induces a relative phase shift θ, leading to the state $|\theta_N\rangle$. In terms of the stabilizer, this evolution can be written as

$$S_j'' = U_{\theta,j} U_{\theta,1} \, Z_j Z_1 \, U_{\theta,1}^\dagger U_{\theta,j}^\dagger = Z_j Z_1 = S_j' \tag{13.90}$$

and

$$S_1'' = \prod_{j=1}^{N} U_{\theta,j} X_j U_{\theta,j}^\dagger = \prod_{j=1}^{N} (\cos\theta \, X_j + \sin\theta \, Y_j) \,. \tag{13.91}$$

Applying the Hadamard operators on qubits $j \neq 1$ and the $N-1$ disentangling CZ operators, the stabilizer becomes

$$\bar{S}_j = X_j \quad \text{and} \quad \bar{S}_1 = \cos(N\theta) \, X_1 + \sin(N\theta) \, Y_1 \,. \tag{13.92}$$

In other words, the qubits $j \neq 1$ are all back in the state $|+\rangle$, and qubit 1 is in the state stabilized by $\bar{S}_1$:

$$\bar{S}_1 |\psi\rangle = |\psi\rangle \quad \Longleftrightarrow \quad |\psi\rangle = \frac{1}{\sqrt{2}} \left(|0\rangle + e^{iN\theta} |1\rangle \right) \,. \tag{13.93}$$

Following the standard procedure above, a single Pauli X measurement on qubit 1 then estimates the value of θ with a precision $\delta\theta = N^{-1}$. When the measurement yields $+1$, we estimate $\theta = 0$, and when the measurement yields the outcome -1, we estimate $\theta = \pi/N$.

Exercise 13.6: Show that the Mandelstam–Tamm and Margolus–Levitin bounds can be violated in general CP maps, by considering the isolated, non-unitary evolution of qubit 1 in Eq. (13.93).

So far, we have used entangled states to achieve the Heisenberg limit. However, one may want to use probe states that are separable to begin with, and perform a non-local measurement on all the qubits after the evolution U_θ has been applied to the separable state. The benefit of this approach is that the probe state is less sensitive to decoherence as it undergoes the evolution U_θ. The question is whether this also allows us to achieve the Heisenberg limit. Surprisingly, the answer is negative. To see this, assume that we wish to measure a time parameter t. Suppose we have N systems labelled j, and each system has d energy levels. The difference between the smallest and the largest energy eigenvalue is ϵ. Each system is prepared in a state $|\psi\rangle_j$ and undergoes a unitary evolution $U_j(t) = \exp[-i\mathcal{H}_j t/\hbar]$. The total Hamiltonian for the N systems can be written as $\mathcal{H} = \sum_j \mathcal{H}_j$. If we use the N qubits in a single-shot parameter-estimation procedure, the Heisenberg uncertainty relation dictates that

$$\delta t \, \delta\mathcal{H} \geq \frac{\hbar}{2}. \tag{13.94}$$

In order to minimize δt, we need to maximize $\Delta\mathcal{H}$:

$$\delta\mathcal{H} = \sqrt{\sum_j (\Delta\mathcal{H}_j)^2} = \sqrt{N(\Delta\mathcal{H}_{\max})^2} = \sqrt{N}\,\frac{\epsilon}{2}, \tag{13.95}$$

where $(\Delta\mathcal{H}_{\max})^2$ is the maximum variance for a single system. The average error in the parameter θ is therefore

$$\delta t \geq \frac{\hbar}{\epsilon\sqrt{N}}. \tag{13.96}$$

This is the standard quantum limit. Therefore, when we start with a separable input of N quantum systems, regardless of the measurement procedure, the average error in θ is bounded by the SQL, which can be attained in a strictly classical setting. This derivation applies not only to energy and time, but to any parameter θ and its generator $\hat{K}$. Consequently, in order to reach the Heisenberg limit, we need entangled states as our resources. This places a powerful restriction on implementations of quantum metrology.

13.5 Optical quantum metrology

So far, we have presented a theoretical approach for estimating parameters that take small values, and we have seen that entanglement can be used to improve the precision from the standard quantum limit to the Heisenberg limit. However, it is thus far not clear how these protocols can be implemented in practice. In particular, we are interested how to implement quantum metrology with optical interferometry. This is the subject of the present section.

13.5.1 The Jordan–Schwinger representation

To translate the quantum-metrology protocols of the previous section to optical interferometers, we use the Stokes operators:

$$\hat{J}_x = \frac{\hbar}{2}\left(\hat{a}_1^\dagger \hat{a}_2 + \hat{a}_1 \hat{a}_2^\dagger\right)$$

$$\hat{J}_y = -\frac{i\hbar}{2}\left(\hat{a}_1^\dagger \hat{a}_2 - \hat{a}_1 \hat{a}_2^\dagger\right)$$

$$\hat{J}_z = \frac{\hbar}{2}\left(\hat{a}_1^\dagger \hat{a}_1 - \hat{a}_2^\dagger \hat{a}_2\right). \tag{13.97}$$

In addition, we can define the total-number operator

$$\hat{N} = \hat{a}_1^\dagger \hat{a}_1 + \hat{a}_2^\dagger \hat{a}_2 . \tag{13.98}$$

We make two crucial observations about these observables. First, the Stokes operators obey the commutation relations for the Pauli operators $[\hat{J}_i, \hat{J}_j] = i\hbar\epsilon_{ijk}\hat{J}_k$, and second, they generate the unitary transformations of beam splitters and phase shifters. Combining these two properties, we can translate the qubit-based metrology of the previous section into optical interferometers. However, we know from Chapter 6 that we cannot perform deterministic two-qubit gates, and so we have to proceed with caution.

The operators $\hat{J}_x$ and $\hat{J}_y$ generate the unitary transformation of (generalized) beam splitters. In particular, we will use

$$U_{\text{BS}} = \exp\left(\frac{i}{\hbar}\phi\hat{J}_x\right) = \begin{pmatrix} \cos\frac{\phi}{2} & i\sin\frac{\phi}{2} \\ i\sin\frac{\phi}{2} & \cos\frac{\phi}{2} \end{pmatrix}, \tag{13.99}$$

acting on modes a_1 and a_2. Similarly, we can generate a relative phase shift between two modes a_1 and a_2 using the operator $\hat{J}_z$

$$U_\theta = \exp\left(\frac{i}{\hbar}\theta\hat{J}_z\right) = \begin{pmatrix} e^{-i\theta/2} & 0 \\ 0 & e^{i\theta/2} \end{pmatrix}. \tag{13.100}$$

This means that interferometers (constructed on modes a_1 and a_2) can be generated by the Stokes operators. In particular, if the input state of a Mach–Zehnder interferometer (see Fig. 13.6) is given by $|\psi_{\text{in}}\rangle$, then the output state can be written as

$$|\psi_{\text{out}}\rangle = \exp\left(\frac{i\pi\hat{J}_x}{2\hbar}\right)\exp\left(-\frac{i}{\hbar}\theta\hat{J}_z\right)\exp\left(-\frac{i\pi\hat{J}_x}{2\hbar}\right)|\psi_{\text{in}}\rangle$$

$$= \exp\left(-\frac{i}{\hbar}\theta\hat{J}_y\right)|\psi_{\text{in}}\rangle . \tag{13.101}$$

A typical measurement of the output modes of the Mach–Zehnder interferometer is the intensity difference between the two output modes, which is given by $\hat{J}_z$. This operator

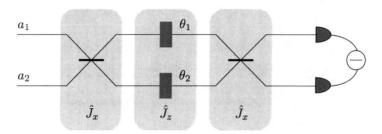

Fig. 13.6. The Mach–Zehnder interferometer. The beam-splitter transformations are generated by the operator J_x, and the relative phase shift $\theta = \theta_1 - \theta_2$ is generated by $\hat{J}_z$.

therefore both generates the relative phase shift, and measures the intensity difference of the output. Using Eq. (13.101) we can write

$$\langle \psi_{\text{out}} | \hat{J}_z | \psi_{\text{out}} \rangle = \langle \psi_{\text{in}} | \hat{J}_z' | \psi_{\text{in}} \rangle \,, \tag{13.102}$$

with

$$\hat{J}_z' = e^{i\pi \hat{J}_x/2\hbar} e^{i\theta \hat{J}_z/\hbar} e^{-i\pi \hat{J}_x/2\hbar} \, \hat{J}_z \, e^{i\pi \hat{J}_x/2\hbar} e^{-i\theta \hat{J}_z/\hbar} e^{-i\pi \hat{J}_x/2\hbar}$$

$$= \cos\theta \, \hat{J}_z - \sin\theta \, \hat{J}_x \,. \tag{13.103}$$

From this equation, we can relate the variance in $\hat{J}_z'$ to the variance in $\hat{J}_z$ and $\hat{J}_x$:

$$(\Delta \hat{J}_z')^2 = \sin^2\theta (\Delta \hat{J}_x)^2 + \cos^2\theta (\Delta \hat{J}_z)^2 - \sin 2\theta \left(\langle \hat{J}_x \hat{J}_z \rangle - \langle \hat{J}_x \rangle \langle \hat{J}_z \rangle \right) \,. \tag{13.104}$$

The error in the phase $\delta\theta$ can then be written as

$$\delta\theta = \left| \frac{d \langle \hat{J}_z' \rangle}{d\theta} \right|^{-1} \Delta \hat{J}_z' \,. \tag{13.105}$$

This is as far as we can go at this juncture without specifying the input state of the interferometer.

Exercise 13.7: For a coherent state in one input mode of a Mach–Zehnder interferometer, and vacuum in the other, $|\alpha, 0\rangle$, show that the phase sensitivity scales as

$$\delta\theta = \frac{1}{|\alpha|} = \frac{1}{\sqrt{\langle n \rangle}} \,. \tag{13.106}$$

The second observation about the Stokes operators was that they have the same structure as the angular-momentum operators in quantum mechanics. We therefore can use our knowledge of the angular-momentum eigenstates in the description of interferometers, namely the relations

$$\hat{J}^2 |j, m\rangle = j(j+1)\hbar^2 |j, m\rangle \quad \text{and} \quad \hat{J}_z |j, m\rangle = m\hbar |j, m\rangle \,. \tag{13.107}$$

This is the Jordan–Schwinger representation of interferometers.

Suppose that the input state to the interferometer has a well-defined number of photons n. We can then write

$$\hat{N} |\psi_{\text{in}}\rangle = n |\psi_{\text{in}}\rangle \ . \tag{13.108}$$

Furthermore, using the definition of the Stokes operators in Eq. (13.97), we can express the operator $\hat{J}^2$ in terms of $\hat{N}$:

$$\hat{J}^2 = \hat{J}_x^2 + \hat{J}_y^2 + \hat{J}_z^2 = \frac{\hat{N}}{2} \left(\frac{\hat{N}}{2} + 1 \right) \hbar^2 \ , \tag{13.109}$$

which in turn means that

$$\hat{J}^2 |\psi_{\text{in}}\rangle = j(j+1)\hbar^2 |\psi_{\text{in}}\rangle = \frac{n}{2} \left(\frac{n}{2} + 1 \right) \hbar^2 |\psi_{\text{in}}\rangle \ . \tag{13.110}$$

Therefore, the input state on modes a_1 and a_2 with total photon number n corresponds to an 'angular-momentum' state $j = n/2$. Similarly, the second quantum number m takes integer values between $-j$ and $+j$:

$$\hat{J}_z |\psi_{\text{in}}\rangle = m\hbar |\psi_{\text{in}}\rangle \quad \text{with} \quad -\frac{n}{2} \leq m \leq \frac{n}{2} \ . \tag{13.111}$$

The benefit of this approach is that we can calculate the expectation values of functions of the Stokes operators for large photon-number states very easily. In particular, we can readily calculate the variance $\Delta \hat{J}_z$ without having to transform large powers of the creation and annihilation operators.

Exercise 13.8: Verify Eqs. (13.103) and (13.109).

Next, we consider the input state $|j, m\rangle$. We evaluate

$$\langle \hat{J}_z' \rangle = \cos\theta \langle \hat{J}_z \rangle - \sin\theta \langle \hat{J}_x \rangle = m\hbar \cos\theta \ . \tag{13.112}$$

The variance is

$$(\Delta \hat{J}_z')^2 = \sin^2\theta \, (\Delta \hat{J}_x)^2 \tag{13.113}$$

where we have used the fact that the remaining terms in Eq. (13.104) are zero for eigenstates of the $\hat{J}_z$ operator. Exploiting the cylindrical symmetry along the z axis, we can write $\hat{J}_x^2 = (\hat{J}^2 - \hat{J}_z^2)/2$, which yields the variance

$$(\Delta \hat{J}_z')^2 = \frac{\hbar^2}{2} \sin^2\theta \left[j(j+1) - m^2 \right] \ . \tag{13.114}$$

Substituting this into Eq. (13.105) we obtain the error in the phase θ:

$$\delta\theta = \left| \frac{d\langle \hat{J}_z' \rangle}{d\theta} \right|^{-1} \Delta \hat{J}_z' = \sqrt{\frac{j(j+1)}{2m^2} - \frac{1}{2}} \ . \tag{13.115}$$

If we want to minimize the error in θ, we should choose m as large as possible, i.e., $m = \pm j$. Using $j = m = n/2$, this leads to the standard quantum limit

$$\delta\theta = \frac{1}{\sqrt{n}}. \tag{13.116}$$

When $m = n/2$, all photons are in mode a_1, and no photons are in mode a_2. In other words, n input photons in the same input mode (and no photons in the second input mode) behave as n independent particles. When $m = 0$, the mean $\langle \hat{J}'_z \rangle$ is zero and the error $\delta\theta$ diverges, so we cannot make definite statements about this input state at this stage. We will return to this later.

Exercise 13.9: Calculate the Fisher information given the input state $j = n/2$, $m = n/2$, and a measurement of the intensity difference. How is the sensitivity affected by the detector efficiency?

Next, we consider input states that approach the Heisenberg limit. The eigenstates $|j, m\rangle$ are separable two-mode states: both the photon number and the number difference of the modes is sharp, which means that we can write $|j, m\rangle$ as a tensor product of two Fock states $|n_1\rangle$ and $|n_2\rangle$ on modes a_1 and a_2, respectively

$$|j, m\rangle = |n_1, n_2\rangle \quad \text{with} \quad n_1 = j + m, \quad n_2 = j - m. \tag{13.117}$$

We can consider (entangled) superpositions of these states that may achieve the Heisenberg limit. We need to keep the total photon number $2j$ fixed, otherwise the eigenstates will not interfere, and we can write

$$\left| \text{YMCK}_{j,mm'} \right\rangle = \frac{1}{\sqrt{2}} \left(|j, m\rangle + |j, m'\rangle \right). \tag{13.118}$$

We call these 'YMCK' states, after Yurke, McCall, and Klauder, who first described them. We now have to determine the expectation value $\langle \hat{J}'_z \rangle$ and the variance $\Delta \hat{J}'_z$. The derivation is straightforward but rather lengthy, and the results for arbitrary m and m' do not come in a nice compact form. We will therefore consider two special cases, one where $m = 0$ and $m' = 1$, and secondly the NOON state ($m = -m' = j$).

When $m = 0$ and $m' = 1$, the state takes the form

$$\left| \text{YMCK}_{j,01} \right\rangle = \frac{1}{\sqrt{2}} \left(\left| \frac{n}{2}, \frac{n}{2} \right\rangle + \left| \frac{n}{2} + 1, \frac{n}{2} - 1 \right\rangle \right), \tag{13.119}$$

and the expectation value $\langle \hat{J}'_z \rangle$ is evaluated as

$$\langle \hat{J}'_z \rangle = \frac{\hbar}{2} \cos\theta - \frac{\hbar}{2} \sin\theta \sqrt{j(j+1)} \tag{13.120}$$

and the variance is

$$\left(\Delta \hat{J}'_z \right)^2 = \frac{\hbar^2}{4} \left[\cos^2\theta + \sin^2\theta \, (j^2 + j - 1) \right]. \tag{13.121}$$

The error in the phase can then be calculated as

$$(\delta\theta)^2 = \left|\frac{d\langle\hat{J}_z'\rangle}{d\theta}\right|^{-2} \left(\Delta\hat{J}_z'\right)^2 = \frac{\cos^2\theta + \sin^2\theta(j^2 + j - 1)}{(\sin\theta + \cos\theta\sqrt{j(j+1)})^2}. \tag{13.122}$$

This error depends on the value of θ, and we therefore need to estimate θ around the value that yields the smallest error. This is found to be $\theta = 0$, and the effective error becomes

$$(\delta\theta)^2 = \frac{1}{j(j+1)} \quad\rightarrow\quad \delta\theta \approx \frac{2}{n}, \tag{13.123}$$

where we have used $j = n/2$ and set n large. Up to a constant factor, this input state for the Mach–Zehnder interferometer reaches the Heisenberg limit.

Another important state for interferometry is the NOON state, and more generally the YMCK state with $m = -m'$

$$\left|\text{YMCK}_{j,m,-m}\right\rangle = \frac{1}{\sqrt{2}}\left(\left|\frac{n}{2} + m, \frac{n}{2} - m\right\rangle + \left|\frac{n}{2} - m, \frac{n}{2} + m\right\rangle\right). \tag{13.124}$$

The NOON state is recovered when $m = j = n/2$. The expectation value of the intensity difference at the output modes for large m becomes

$$\langle\hat{J}_z'\rangle = \hbar\cos\theta\left(\frac{m + m'}{2}\right) = \hbar\cos\theta\left(\frac{m - m}{2}\right) = 0. \tag{13.125}$$

Since the measured expectation value does not depend on θ, no information about θ can be learned in such an experiment. Also, the variance around $\theta = 0$ is $\Delta\hat{J}_z' = |m|\hbar$, which means that the larger the separation m, the bigger the spread.

One may wonder why the NOON state (and all other states with $m = -m'$) behaves this badly in a Mach–Zehnder interferometer. After all, it was designed for its maximal sensitivity to phase shifts. The answer is that the NOON state yields the Heisenberg limit when the phase shift is applied *directly* to it. Here, we use the NOON state as the input to a 50:50 beam splitter, which changes the state to a form that is insensitive to phase differences in the interferometer. When only two photons are involved, it is clear why this is the case. A two-photon NOON state $|2, 0\rangle + |0, 2\rangle$ transforms under a 50:50 beam splitter into the state $|1, 1\rangle$, which is of course completely insensitive to any relative phase shift in the two modes.

This observation leads us back to the question of what can be achieved with input states of the form $|j, 0\rangle$, in other words, when an equal number of photons enter the Mach–Zehnder interferometer in each input mode. Suppose that the input state to the Mach–Zehnder interferometer is given by

$$|\psi_{\text{in}}\rangle = \left|\frac{n}{2}, \frac{n}{2}\right\rangle, \tag{13.126}$$

such that the total photon number is n. Assuming no losses in the photodetectors, and perfect photon-number resolution, we can calculate the probability distribution that the

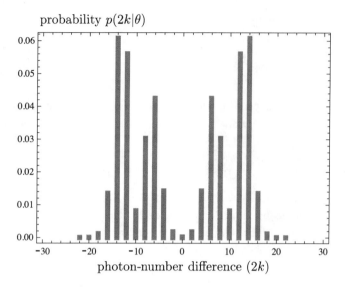

probability $p(2k|\theta)$

photon-number difference $(2k)$

Fig. 13.7. **Probability for finding a photon-number difference of $2k$, given an input state $|n/2, n/2\rangle$ in a Mach–Zehnder interferometer.**

photon-number difference between the two output modes is $2k$:

$$p(2k|\theta) = \left| \left\langle \frac{n}{2} + k, \frac{n}{2} - k \right| \exp\left(-\frac{i}{\hbar} \theta \hat{J}_j \right) \left| \frac{n}{2}, \frac{n}{2} \right\rangle \right|^2 = J_k^2\left(\frac{n\theta}{2} \right), \qquad (13.127)$$

where $J_n(x)$ is a Bessel function, and θ is again the relative phase shift in the arms of the interferometer. This probability distribution is shown in Fig. 13.7. The width of the distribution can be estimated at $|k| < \theta n/2$, which approaches the Heisenberg limit. For more details, see Holland and Burnett (1993).

13.5.2 The creation of YMCK and NOON states

How can we create the states that yield the Heisenberg limit? In particular, we wish to create the state in Eq. (13.119) with $n/2 = N$:

$$|\psi_{\text{in}}\rangle = \frac{1}{\sqrt{2}} \left(|N, N\rangle + |N + 1, N - 1\rangle \right). \qquad (13.128)$$

If we consider only linear-optical elements, there is no way to create this state with unit probability. However, there is a procedure that creates a close approximation to this state with less than unit probability (see Fig. 13.8). Assuming that we can create Fock states (which is still a very tall order), we can start with an input state $|N + 1, N\rangle$. Both modes are sent through a beam splitter with very small reflection coefficient. The two reflected modes are sent into a 50:50 beam splitter, the output modes of which are detected. When the two detectors register a single photon, we have removed it from the input, but we do

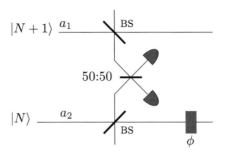

Fig. 13.8. The probabilistic creation of the state $(|N, N\rangle + |N + 1, N - 1\rangle)/\sqrt{2}$. Two beam splitters with small reflection coefficients are placed into the input modes. The reflected beams are mixed in a 50:50 beam splitter. When a single photon is detected, the required state is created, up to a relative phase shift $e^{i\pi}$, which is corrected by ϕ.

not know from which mode it came. Adjusting the beam-splitter reflectivities for the fact that the input photon number is different, we can obtain the required state with exceedingly high fidelity. When the photodetectors are bucket detectors, in other words they cannot distinguish between one or more photons, then we can achieve arbitrary high fidelity at the cost of reduced success probability by lowering the reflectivity of the two beam splitters.

Exercise 13.10: Calculate the optimal beam-splitter reflectivities for Fig. 13.8, assuming perfect photodetectors.

Similarly, we can construct a probabilistic method for creating NOON states. To this end we write the NOON state as

$$\frac{1}{\sqrt{2}} (|N, 0\rangle + |0, N\rangle) = \frac{\hat{a}_1^{\dagger N} + \hat{a}_2^{\dagger N}}{\sqrt{2N!}} |0, 0\rangle = \frac{\hat{a}_1^N + \hat{a}_2^N}{\sqrt{2N!}} |N, N\rangle . \qquad (13.129)$$

In the last equation we have used the fact that we can create a NOON state by removing photons from a separable state $|N, N\rangle$. The removal of photons can then be translated into detection, just as for the YMCK states. However, now we need to repeat this procedure N times:

$$\hat{a}_1^N + \hat{a}_2^N = \prod_{k=1}^{N} \left(\hat{a}_1 + e^{i\phi_k} \hat{a}_2 \right) , \qquad (13.130)$$

where $\exp(i\phi_k) = \exp(2\pi i k/N)$ are the Nth roots of unity. Given a *single* photon in the two detectors in Fig. 13.8, the detection event can be interpreted as an operator $\hat{a}_1 \pm e^{i\phi} \hat{a}_2$ (depending on which detector registers the photon). Repeating this structure N times, and assuming a successful detection at each stage, the output state is a NOON state (see Fig. 13.9). The overall success probability is exponentially small, but feed-forward techniques can be used to reduce the overhead to polynomial size in N. One may argue that such a high overhead should be counted towards the necessary resources, in which case a square-root improvement in the precision is insignificant compared to the overhead, be it exponential or polynomial. However, there may still be a role for these techniques in situations where

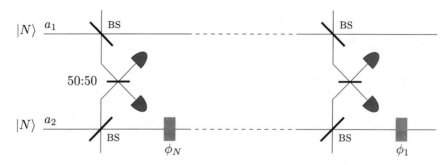

Fig. 13.9.   The probabilistic creation of the NOON state $(|N, 0\rangle + |0, N\rangle)/\sqrt{2}$.

high-intensity light causes damage to a particular sample. A light source with the maximum allowed intensity operating at the shot-noise limit may simply not be accurate enough for phase estimation.

The creation of the YMCK state $(|j, 0\rangle + |j, 1\rangle)/\sqrt{2}$ and the NOON state are special cases of a general procedure for creating multi-mode optical states with well-defined photon number. Any such state can be described by a polynomial of creation operators acting on a multi-mode vacuum state. We can translate this into a polynomial of annihilation operators acting on a multi-mode number state $|n\rangle^{\otimes m}$. This polynomial can be decomposed into its roots, which are themselves linear polynomials. In turn, these can be implemented successively with linear-optical elements according to the technique described above.

Exercise 13.11: Assume that N is even. Instead of post-selecting on a single photon in each detector pair in Fig. 13.9, we post-select on detector coincidences (one photon in each detector). What is the success probability of creating a size N NOON state?

Exercise 13.12: Starting from N identical single-photon states in N different modes, prove using Eq. (6.25) that the theoretical upper limit for creating a NOON state with linear-optical elements and photodetection is $p = 2/N$.

13.5.3 Squeezed-state metrology

So far, we have considered Mach–Zehnder interferometers with input states that have a definite number of photons. Depending on the distribution of the photons over the two input modes, we have found a phase sensitivity that scales either as the standard quantum limit, or as the Heisenberg limit. However, it is very difficult to create states with a definite number of photons N, especially if N is large. We have also seen (in Exercise 13.7) that coherent states give rise to the standard quantum limit. This should not be surprising, since the coherent states are in some sense the classical states of light. The question is therefore whether squeezed states can help achieve the Heisenberg limit in a practical way.

In order to investigate this question, we will first modify the Jordan–Schwinger representation presented in the previous section. The natural operators in the context of squeezed states are

$$\hat{K}_x = \frac{\hbar}{2}\left(\hat{a}_1^\dagger \hat{a}_2^\dagger + \hat{a}_1 \hat{a}_2\right)$$

$$\hat{K}_y = -\frac{i\hbar}{2}\left(\hat{a}_1^\dagger \hat{a}_2^\dagger - \hat{a}_1 \hat{a}_2\right)$$

$$\hat{K}_z = \frac{\hbar}{2}\left(\hat{a}_1^\dagger \hat{a}_1 + \hat{a}_2^\dagger \hat{a}_2 + 1\right), \tag{13.131}$$

the first two of which ($\hat{K}_x$ and $\hat{K}_y$) generate the two-mode squeezing operator

$$S(\xi) = \exp\left(-\xi \hat{a}_1^\dagger \hat{a}_2^\dagger + \xi^* \hat{a}_1 \hat{a}_2\right) = \exp\left[-\frac{i}{\hbar}\left(2\xi_2 \hat{K}_x + 2\xi_1 \hat{K}_y\right)\right], \tag{13.132}$$

where $\xi = \xi_1 + i\xi_2$. Note that the operator $\hat{K}_z$ is linearly related to the total-number operator, rather than the number difference $\hat{J}_z$. This is because we want the commutation relations of $\hat{K}_x$, $\hat{K}_y$, and $\hat{K}_z$ to form a closed algebra

$$\left[\hat{K}_x, \hat{K}_y\right] = -i\hbar \hat{K}_z, \quad \left[\hat{K}_y, \hat{K}_z\right] = i\hbar \hat{K}_x, \quad \text{and} \quad \left[\hat{K}_z, \hat{K}_x\right] = i\hbar \hat{K}_y \tag{13.133}$$

and this is not possible with just the operators $\hat{K}_x$, $\hat{K}_y$, and $\hat{J}_z$. This closed algebra in turn allows us to calculate the evolution of states and operators generated by the $\hat{K}_i$. We find that

$$e^{i\zeta \hat{K}_x/\hbar}\, \hat{K}_z \, e^{-i\zeta \hat{K}_x/\hbar} = \cosh\zeta \, \hat{K}_z + \sinh\zeta \, \hat{K}_y,$$

$$e^{i\theta \hat{K}_z/\hbar}\, \hat{K}_y \, e^{-i\theta \hat{K}_z/\hbar} = \cos\theta \, \hat{K}_y + \sin\theta \, \hat{K}_x,$$

$$e^{i\zeta \hat{K}_x/\hbar}\, \hat{K}_y \, e^{-i\zeta \hat{K}_x/\hbar} = \cosh\zeta \, \hat{K}_y + \sinh\zeta \, \hat{K}_z. \tag{13.134}$$

These are the Baker–Campbell–Hausdorff relations, and a general method for calculating the more complicated ones is presented in Appendix 1.

Next, we translate the Mach–Zehnder interferometer, as described by the angular-momentum operators $\hat{J}_i$, into a squeezing interferometer by replacing $\hat{J}_i$ with $\hat{K}_i$ throughout the protocol. In other words, as shown in Fig. 13.10, the beam splitters generated by $\hat{J}_x$ are replaced by two-mode squeezing operators generated by $\hat{K}_x$, the phase is encoded not with

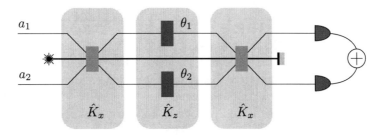

Fig. 13.10. Translating the Mach–Zehnder interferometer into a squeezing interferometer by replacing $\hat{J}_i$ with $\hat{K}_i$. We define $\theta = -(\theta_1 + \theta_2)$.

a regular phase shift but with the $\hat{K}_z$ operator, and the measured observable in the output modes is the total photon number $\hat{N} = \hat{a}_1^\dagger \hat{a}_1 + \hat{a}_2^\dagger \hat{a}_2$, which can be written as $2\hat{K}_z - 1$. Such an interferometer produces an output state

$$|\psi_{\text{out}}\rangle = \exp\left(\frac{i}{\hbar}\zeta\hat{K}_x\right) \exp\left(-\frac{i}{\hbar}\theta\hat{K}_z\right) \exp\left(-\frac{i}{\hbar}\zeta\hat{K}_x\right) |\psi_{\text{in}}\rangle . \tag{13.135}$$

Using the same method as above, we calculate the expectation values of the operators with respect to the input state, rather than the output state. Using the relations in Eq. (13.134), we find that

$$\langle\hat{N}\rangle = \langle\psi_{\text{out}}| 2\hbar\hat{K}_z - 1 |\psi_{\text{out}}\rangle = (1 - \cos\theta)\sinh^2\zeta , \tag{13.136}$$

and

$$(\Delta\hat{N})^2 = \left[\sin^2\theta + (1 - \cos\theta)^2 \cosh^2\zeta\right]\sinh^2\zeta . \tag{13.137}$$

The precision in the phase is then given by the standard procedure

$$(\delta\theta)^2 = \frac{\sin^2\theta + (1 - \cos\theta)^2 \cosh^2\zeta}{\sin^2\theta \sinh^2\zeta} = \frac{1}{\sinh^2\zeta}\bigg|_{\theta=0} . \tag{13.138}$$

We have again evaluated the precision around $\theta = 0$, where the error $\delta\theta$ reaches a minimum. To compare this with the interferometers described so far, we need to relate $\sinh\zeta$ to the average number of photons generated by the two-mode squeezers. There is, however, a subtlety in determining this number. From Eq. (13.136) we deduce that the average number of detected photons at $\theta = 0$ is zero. This is expected, since in the absence of the phase shift θ the second two-mode squeezer is the inverse of the first two-mode squeezer. We therefore have to compare the precision with the average number of photons *inside* the interferometer, i.e., the average number of photons created by the first two-mode squeezer. For a vacuum input state, this is readily calculated as

$$\langle\hat{N}\rangle = 2\hbar\left\langle e^{i\zeta\hat{K}_x}\hat{K}_z\, e^{-i\zeta\hat{K}_x}\right\rangle - 1 = \cosh\zeta - 1 . \tag{13.139}$$

From this, we find $\sinh^2\zeta = \langle\hat{N}\rangle(\langle\hat{N}\rangle + 2)$, and the precision in the phase becomes

$$\delta\theta = \frac{1}{\sqrt{\langle\hat{N}\rangle(\langle\hat{N}\rangle + 2)}} \simeq \frac{1}{\langle\hat{N}\rangle} , \tag{13.140}$$

which is the Heisenberg limit. The best current levels of squeezing have a strength of about 10 dB, and the average photon number is approximately $\langle N \rangle = 2$. This is not enough to make a meaningful improvement in parameter estimation in the above configuration. Other difficulties in implementing this protocol are mode matching in the second two-mode squeezer, the reduced collection and detection efficiency of the detectors, and photon loss inside the interferometer (if the probe medium is dispersive).

Exercise 13.13: Verify Eqs. (13.136) and (13.137).

13.5.4 The Mach–Zehnder interferometer revisited

The final route to Heisenberg-limited parameter estimation we consider here is a Mach–Zehnder interferometer that receives as input states a coherent state and squeezed vacuum.[1] We again wish to measure the relative phase shift θ between the two interior modes of the interferometer, and as before, we measure the photon-number difference in the output modes. We can write the input and output states as

$$|\psi_{\text{in}}\rangle = |\alpha, \xi\rangle \quad \text{and} \quad |\psi_{\text{out}}\rangle = \exp\left(-\frac{i}{\hbar}\theta \hat{J}_y\right) |\psi_{\text{in}}\rangle, \qquad (13.141)$$

where $|\alpha\rangle$ is a coherent state with average photon number $\langle n \rangle = |\alpha|^2$ and $|\xi\rangle$ is a squeezed vacuum state with average photon number $\sinh^2 r$, and $\xi = re^{i\phi}$. Given the measurement of $\hat{J}_z$ in the output modes and the state of the input modes, we can calculate the Fisher information

$$F(\theta) = |\alpha|^2 e^{2r} + \sinh^2 r. \qquad (13.142)$$

This leads to the Cramér–Rao bound for the average error in θ

$$\delta\theta \geq \frac{1}{\sqrt{M F(\theta)}} = \frac{1}{\sqrt{M}} \frac{1}{\sqrt{|\alpha|^2 e^{2r} + \sinh^2 r}}, \qquad (13.143)$$

where M is the number of times the experiment is repeated (independently).

Exercise 13.14: Verify Eq. (13.142).

In the discussion above, using the YMCK states as input states to the Mach–Zehnder interferometer, we saw that the sensitivity of the procedure depended on the actual value of θ. In particular, θ had to be close to zero or π in order to achieve the Heisenberg limit. When the phase shift we expect to measure is small, this may not be such an important restriction. However, when the phase shift is completely unknown, we would like to have a parameter-estimation procedure that attains the Heisenberg limit independent of the value of θ. From Eq. (13.142) we see that the Cramér–Rao bound allows for this possibility. Maximum-likelihood techniques are known to saturate the Cramér–Rao bound asymptotically for large M, and we therefore have found a technique that attains the Heisenberg limit for arbitrary values of θ.

There are several important regimes in this set-up, and we will briefly review them here. In each case, the error $\delta\theta$ is expressed in terms of the average photon number $\langle n \rangle$ in the experiment.

(i) $r = 0$: There is no squeezing, and the bound in Eq. (13.143) reduces to

$$\delta\theta \geq \frac{1}{\sqrt{M|\alpha|^2}} = \frac{1}{\sqrt{M\langle n \rangle}}. \qquad (13.144)$$

This is the standard quantum limit, as expected for coherent states.

---

[1] Historically, this set-up was first considered by Caves in 1981, and is considered the start of Heisenberg-limited parameter estimation.

(ii) $\alpha = 0$: In this case, there is only squeezing, which leads to the bound

$$\delta\theta \geq \frac{1}{\sqrt{M \sinh^2 r}} = \frac{1}{\sqrt{M}} \frac{1}{\sqrt{\langle n \rangle}}, \qquad (13.145)$$

where $\langle n \rangle = \sinh^2 r$.

(iii) $\sinh^2 r \ll |\alpha|^2$: When there is only a small amount of squeezing, the bound becomes

$$\delta\theta \geq \frac{e^{-r}}{\sqrt{M \langle n \rangle}}, \qquad (13.146)$$

where $\langle n \rangle \simeq |\alpha|^2$. This is sub-shot-noise precision, but does not reach the Heisenberg limit. It is also the regime that can be probed with current technology.

(iv) $\sinh^2 r \gg |\alpha|^2$: In this regime, the bound on the error in θ becomes

$$\delta\theta \geq \frac{1}{\sqrt{M \langle n \rangle (4|\alpha|^2 + 1)}}, \qquad (13.147)$$

where $\langle n \rangle \simeq \sinh^2 r$.

(v) $\sinh^2 r \simeq |\alpha|^2$: Finally, the optimal Cramér–Rao bound is obtained when the intensities of the coherent mode and the squeezed vacuum are approximately the same:

$$\delta\theta \geq \frac{1}{\sqrt{M}} \frac{1}{\langle n \rangle}, \qquad (13.148)$$

where $\langle n \rangle \simeq 2|\alpha|^2 \simeq 2\sinh^2 r$. However, given current levels of experimentally achievable squeezing, it will be very difficult to achieve this level of precision in the measurement of θ for large $\langle n \rangle$.

For a given number of detected photons N, the optimal regime (5) provides two-mode states in the interior that exhibit the largest contribution of NOON components $|N, 0\rangle + |0, N\rangle$ in the total superposition. It is therefore the amount of entanglement between the two interior modes that is responsible for the Heisenberg-limited precision, rather than the reduced quadrature noise of the squeezed input state.

13.6 References and further reading

As has become clear in this chapter, constructing a practical implementation for Heisenberg-limited quantum metrology is extremely challenging. First, we require entangled quantum states of some sort, which can then be used to encode the parameter we wish to estimate. Second, we need to maintain coherence in the probing stage, when the parameter is imprinted onto the entangled state. And finally, we need the ability to measure the quantum state with high fidelity. In the case of photon counting, low detector efficiencies can be disastrous. There have been many theoretical proposals for achieving the Heisenberg limit, but currently only proof-of-principle experiments have been carried out (that is, experiments that show a square-root improvement, but for rather small N). It seems that large-scale quantum metrology is as hard to implement as large-scale quantum computing.

For an introduction to classical probability theory, see Jaynes (2003). The theory of parameter estimation in quantum mechanics was pioneered by Helstrom (1976) and Holevo (1982, 2001). The statistical distance for pure quantum states was introduced by Wootters (1981), and expanded for mixed states by Braunstein and Caves (1994). For the geometrical description of quantum parameter estimation, see Braunstein *et al.* (1996), who also discuss generalized uncertainty relations. The energy bound on the dynamical evolution of states was found by Margolus and Levitin (1998), and this complemented the earlier bound by Mandelstam and Tamm (1945).

Optical metrology was pioneered by Caves (1981), and Yurke *et al.* (1986). It was shown by Giovannetti *et al.* (2006) that entanglement is essential in the input state when the Heisenberg limit is to be attained. Noon states were introduced by Bollinger *et al.* (1996), and named by Kok and Dowling. For the state preparation of noon states, see Kok *et al.* (2002a), and Cable and Dowling (2007). The performance of various schemes to attain the Heisenberg limit were analyzed in the presence of photon loss by Huelga *et al.* (1997) and recently by Pezzé and Smerzi (2008) and Demkowicz-Dobrzanski *et al.* (2009).

In this appendix, we prove the operator ordering formula in Eq. (1.203):

$$e^{\kappa K_+ - \kappa^* K_-} = e^{\tau K_+}\, e^{-2\nu K_0}\, e^{-\tau^* K_-}. \tag{A1.1}$$

We follow the derivation by Truax (1985). The ordering formula is completely determined by the commutation relations of the operators K_+, K_-, and K_0, which are given by

$$[K_-, K_+] = 2K_0 \quad \text{and} \quad [K_0, K_\pm] = \pm K_\pm, \tag{A1.2}$$

together with

$$K_+^\dagger = K_- \quad \text{and} \quad K_0^\dagger = K_0. \tag{A1.3}$$

Operators that obey this algebra generate a Lie group that is known as SU(1,1).

First, we define two unitary operators, $U_1(\lambda)$ and $U_2(\lambda)$ according to

$$U_1(\lambda) = e^{\lambda(\kappa K_+ - \kappa^* K_-)},$$
$$U_2(\lambda) = e^{p_+(\lambda)K_+}\, e^{p_0(\lambda)K_0}\, e^{p_-(\lambda)K_-}. \tag{A1.4}$$

The operators must satisfy $U_1(0) = U_2(0) = \hat{\mathbb{1}}$. This leads to the boundary conditions for the functions $p_k(\lambda)$ with $k = +, -, 0$:

$$p_+(0) = p_0(0) = p_-(0) = 0. \tag{A1.5}$$

We will now derive three coupled differential equations for the $p_k(\lambda)$, and find the solutions for these boundary conditions. We then set $\lambda = 1$ to arrive at the operator ordering formula of Eq. (1.203). In principle, this technique can be used to find numerous other operator ordering formulas, and we will give another example towards the end of this appendix.

In order to find the coupled differential equations, we require that $U_1(\lambda) = U_2(\lambda)$, and we differentiate both sides with respect to λ:

$$\left(\kappa K_+ - \kappa^* K_-\right) U_1(\lambda) = \left(\kappa K_+ - \kappa^* K_-\right) U_2(\lambda)$$
$$= \frac{dp_+}{d\lambda} K_+\, e^{p_+ K_+} e^{p_0 K_0} e^{p_- K_-}$$
$$+ \frac{dp_0}{d\lambda} e^{p_+ K_+}\, K_0\, e^{p_0 K_0} e^{p_- K_-}$$
$$+ \frac{dp_-}{d\lambda}\, e^{p_+ K_+} e^{p_0 K_0}\, K_-\, e^{p_- K_-}. \tag{A1.6}$$

Multiplying this from the right with $U_2^{-1}(\lambda)$ then gives the equation

$$\kappa K_+ - \kappa^* K_- = \frac{dp_+}{d\lambda} K_+ + \frac{dp_0}{d\lambda} e^{p_+ K_+} K_0 \, e^{-p_+ K_+}$$
$$+ \frac{dp_-}{d\lambda} e^{p_+ K_+} e^{p_0 K_0} K_- \, e^{-p_0 K_0} e^{-p_+ K_+}. \qquad (A1.7)$$

Next, we use the standard Baker–Campbell–Hausdorff relation

$$e^A B e^{-A} = B + [A, B] + \frac{1}{2!} [A, [A, B]] + \dots \qquad (A1.8)$$

to evaluate the transformation of the operators K_0 and K_-:

$$e^{p_+ K_+} K_0 \, e^{-p_+ K_+} = K_0 - p_+ K_+, \qquad (A1.9)$$

and

$$e^{p_+ K_+} e^{p_0 K_0} K_- e^{-p_0 K_0} e^{-p_+ K_+} = \left(K_- + p_+^2 K_+ - 2 p_+ K_0 \right) e^{-p_0}. \qquad (A1.10)$$

This leads to the equation

$$\kappa K_+ - \kappa^* K_- = \left(\frac{dp_+}{d\lambda} - p_+ \frac{dp_0}{d\lambda} + p_+^2 e^{-p_0} \frac{dp_-}{d\lambda} \right) K_+$$
$$+ \left(\frac{dp_0}{d\lambda} - 2 p_+ e^{-p_0} \frac{dp_-}{d\lambda} \right) K_0 + e^{-p_0} \frac{dp_-}{d\lambda} K_-. \qquad (A1.11)$$

Collecting the coefficients of the three different operators gives the three coupled differential equations

$$\kappa^* = -\frac{dp_-}{d\lambda} e^{-p_0}, \qquad (A1.12)$$

$$0 = \frac{dp_0}{d\lambda} - 2 p_+ e^{-p_0} \frac{dp_-}{d\lambda}, \qquad (A1.13)$$

$$\kappa = \frac{dp_+}{d\lambda} - p_+ \frac{dp_0}{d\lambda} + p_+^2 e^{-p_0} \frac{dp_-}{d\lambda}. \qquad (A1.14)$$

These equations can be rearranged to give

$$\frac{dp_0}{d\lambda} + 2\kappa^* p_+ = 0 \quad \text{and} \quad \frac{dp_+}{d\lambda} + \kappa^* p_+^2 = \kappa. \qquad (A1.15)$$

We can solve the latter equation by making two substitutions: first we choose $p_+ = y/\kappa^*$ with $y(0) = 0$, and then we choose $y = u'/u$ with $u'(0) = 0$. The resulting second-order differential equation is

$$u'' - |\kappa|^2 u = 0. \qquad (A1.16)$$

This equation can be solved using standard techniques. We find that

$$p_+(\lambda) = \frac{\kappa}{|\kappa|} \tanh(\lambda |\kappa|), \qquad (A1.17)$$

which in turn yields

$$p_0(\lambda) = -2\ln\left[\cosh(\lambda|\kappa|)\right] \quad \text{and} \quad p_-(\lambda) = -\frac{\kappa^*}{|\kappa|}\tanh(\lambda|\kappa|). \tag{A1.18}$$

When we substitute these functions into Eq. (A1.4) we obtain the operator ordering formula in Eq. (A1.1) and Eq. (1.203).

The technique described above is very general, and can be used to derive various other operator ordering formulas. For example, we can change the order of the exponentials in $U_2(\lambda)$ and rederive the functions p_+, p_-, and p_0. Alternatively, we can choose a different set of commutation relations and derive a new ordering formula. One such alternative set is

$$[K_-, K_+] = -2K_0 \quad \text{and} \quad [K_0, K_\pm] = \pm K_\pm, \tag{A1.19}$$

together with

$$K_+^\dagger = K_- \quad \text{and} \quad K_0^\dagger = K_0. \tag{A1.20}$$

Operators that obey this algebra generate a Lie group that is known as SU(2). The algebra is almost identical to that of SU(1,1), except for the minus sign in $[K_-, K_+]$. Following the procedure given above, we find that the operator ordering formula for these operators is

$$e^{\kappa K_+ - \kappa^* K_-} = e^{\tau K_+}\, e^{-2\nu K_0}\, e^{-\tau^* K_-}, \tag{A1.21}$$

with

$$\tau = \frac{\kappa}{|\kappa|}\tan|\kappa| \quad \text{and} \quad \nu = \ln(\cos|\kappa|). \tag{A1.22}$$

In this appendix, we present the original argument by Knill, Laflamme, and Milburn (KLM). It consists of three parts: (1) a probabilistic CZ gate based on the NS gate discussed in Chapter 6; (2) a method of teleporting the probabilistic gate into a circuit to achieve arbitrary high success probability; and (3) an error-correction protocol that facilitates efficient encoding with respect to the level of accuracy of the gates. Here, we discuss the gate-teleportation aspect of the KLM protocol. A longer introduction is given by Myers and Laflamme (2005).

We first present the gate-teleportation protocol due to Gottesman and Chuang (1999), and then show how it was adapted for linear-optical implementations by Knill, Laflamme, and Milburn (2001). We show how to teleport single- and two-qubit gates, which requires the capability of perfect Bell measurements. In quantum optics, perfect Bell measurements are not possible, and the remainder of this appendix is devoted to the adaptation of the gate-teleportation protocol to single- and multi-photon teleportation.

We can write the teleportation of a single qubit schematically as

$$|\psi\rangle_1 |\Psi_{00}\rangle_{23} \to X_3^n Z_3^m |\Psi_{nm}\rangle_{12} |\psi\rangle_3 , \tag{B2.1}$$

where $|\Psi_{00}\rangle_{23}$ is the (unnormalized) Bell state $|00\rangle + |11\rangle$ on modes 2 and 3, and $|\Psi_{nm}\rangle_{12}$ is the Bell state found in Alice's Bell measurement. The indices $n, m = \{0, 1\}$ denote the measurement outcomes (sent to Bob), that determine how the teleported state differs from the input state. Now consider what happens if the entangled pair is in a slightly different state, such that $|\Psi_{00}\rangle_{23} \to \hat{\mathbb{I}}_2 \otimes U_3 |\Psi_{00}\rangle_{23}$. The teleportation protocol then yields

$$\hat{\mathbb{I}}_1 \otimes \hat{\mathbb{I}}_2 \otimes U_3 |\psi\rangle_1 |\Psi_{00}\rangle_{23} \to U_3 X_3^n Z_3^m |\Psi_{nm}\rangle_{12} |\psi\rangle_3 , \tag{B2.2}$$

and the output state is given by

$$|\psi_{\text{out}}\rangle = U X^n Z^m |\psi\rangle , \tag{B2.3}$$

where we suppressed the subscript 3 for the single-qubit operators. If Bob applies his corrective Pauli operations, X^{-n} first, followed by Z^{-m}, the output state becomes

$$|\psi_{\text{out}}\rangle = Z^{-m} X^{-n} U X^n Z^m |\psi\rangle = \tilde{U} |\psi\rangle . \tag{B2.4}$$

For certain choices of known U we can use this outcome for the so-called gate teleportation of U. In other words, U is not implemented directly: it is induced by employing a different entangled state in the teleportation procedure. Hence, the gate operation is reduced to state preparation, which may be easier in some cases. The teleportation circuit is given by

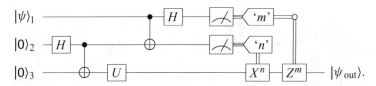

In order for the gate teleportation to work, we must be able to commute U through the Pauli operators X and Z. In particular, the commutator should not be proportional to an arbitrary unitary operator, because Bob has to 'undo' the commutator $[U, X^n Z^m]$. If this is again an arbitrary single-qubit operator, Bob may as well directly apply U. We assume that U is part of the Clifford group, which means that

$$UX^n Z^m = e^{i\phi(n,m)} X^{n'} Z^{m'} U, \tag{B2.5}$$

where $\phi(n, m) \in [0, 2\pi)$ is a global phase factor. We can then write the gate teleportation (up to a global phase) as

$$|\psi_{\text{out}}\rangle = X^k Z^l U |\psi\rangle \rightarrow U |\psi\rangle, \tag{B2.6}$$

for some k, l that can be worked out by Bob if he knows U, n, and m. This protocol is useful if Bob can perform deterministic Pauli operators, but not arbitrary single-qubit gates. In addition, the protocol assumes that the entangled input state can be created, and that Alice can perform a deterministic Bell measurement.

It is straightforward to extend this technique to two-qubit gates, where the advantage becomes very important. We teleport both qubits, and apply the two-qubit operation to the two entangled resource states for teleportation. The double teleportation protocol can be written as

$$|\psi\rangle_1 |\Psi_{00}\rangle_{23} |\Psi_{00}\rangle_{45} |\phi\rangle_6 \rightarrow |\Psi_{km}\rangle_{12} \left(X_3^k Z_3^m |\psi\rangle_3 \right) \left(X_4^n Z_4^l |\phi\rangle_4 \right) |\Psi_{nl}\rangle_{56}. \tag{B2.7}$$

Applying a two-qubit gate U_{34} to the entangled state $|\Psi_{00}\rangle_{23} |\Psi_{00}\rangle_{45}$ prior to teleportation produces the output state

$$|\Psi_{\text{out}}\rangle = U_{34} X_3^k Z_3^m X_4^n Z_4^l |\psi, \phi\rangle_{34}. \tag{B2.8}$$

In the circuit model, this can be written as

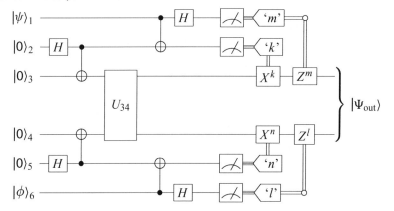

When the two-qubit unitary U_{34} is part of the Clifford group, Bob can apply corrective Pauli gates such that the output state is $U_{34}|\psi,\phi\rangle_{34}$. In particular, when the two-qubit unitary is a cz gate, the circuit becomes

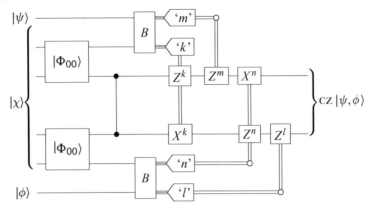

where B denotes the Bell measurement, and the two-qubit entangled resource states are denoted by the boxed $|\Phi_{00}\rangle$. The advantage of gate teleportation now becomes clear: it is possible to deterministically implement interesting two-qubit gates, such as the cx and the cz gates, via the state preparation of a suitable four-qubit entangled state. Since the four-qubit entangled state must be prepared in advance, this may be a probabilistic process, as long as the preparation stage tells us that we succeeded. Bob does not need the capability of deterministic two-qubit gates, only single-qubit gates. The protocol does, however, require the capability to perform complete deterministic Bell measurements.

A perfect deterministic Bell measurement of single-photon qubits is impossible, as shown by Lütkenhaus *et al.* (1999). Perfect distinguishability of the four two-qubit Bell states has a success probability of at most one-half. Gate teleportation would then also be probabilistic, and the advantage of teleporting probabilistic gates into a quantum computation is squandered. However, if we allow for an arbitrarily small imperfection in the teleportation protocol, we can circumvent the assumptions of the no-go theorem, and we can perform gate teleportation with arbitrary large success probability.

Teleportation of a single photonic qubit with high success probability requires a large number of auxiliary modes (see Fig. 2.1). The single-photon qubit exists as an excitation in two optical modes, namely

$$|\psi\rangle = c_0|0\rangle + c_1|1\rangle = c_0|1,0\rangle + c_1|0,1\rangle. \tag{B2.9}$$

We will be teleporting only *one* of these modes (called 'mode teleportation'), which we will denote by $\alpha|0\rangle_0 + \beta|1\rangle_0$ in mode 0. This is *not* the reduced state of Eq. (B2.9), but the linearity of quantum mechanics ensures that mixed states can be teleported with the same protocol. Both modes, constituting the entire qubit state, can be teleported using mode teleportation of each mode. The entangled resource for mode teleportation is a $2N$-mode (unnormalized) state $|t_N\rangle$

$$|t_N\rangle = \sum_{j=1}^{N} |1\rangle^j |0\rangle^{N-j} \, |0\rangle^j |1\rangle^{N-j} \equiv \sum_{j=1}^{N} |0\rangle^j |1\rangle^{N-j}, \tag{B2.10}$$

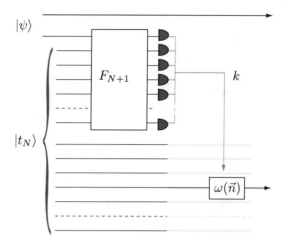

Fig. 2.1.   Mode teleportation.

where we used that $|0\rangle = |1, 0\rangle$ and $|1\rangle = |0, 1\rangle$, and $|0\rangle^j \equiv |0\rangle_1 \otimes \ldots \otimes |0\rangle_j$, etc. The teleportation protocol proceeds by sending the input mode 0 and the first N modes of the entangled resource into a balanced $N + 1$-port, where any which-path information of the input photons is erased. All $N + 1$ output modes are measured in the number basis, recording a string of numbers $\vec{n} = (n_0, \ldots, n_N)$, where n_i indicates the number of detected photons in mode i. Let the total-photon number be given by $k = \sum_i n_i$. The input mode is then teleported to mode $N + k$ with relative phase shift $\omega(\vec{n})$ between $|0\rangle$ and $|1\rangle$ given by

$$\omega(\vec{n}) = \exp\left(\frac{2\pi i}{N + 1} \sum_{j=0}^{N} j n_j\right). \tag{B2.11}$$

To show that the input mode is teleported to mode $N + k$, consider the following argument. Given that k photons are detected, only two terms in the superposition in $|\psi\rangle_0 |t_N\rangle$ contribute to the detection event, namely

$$\alpha |0\rangle_0 |1\rangle^k \quad |0\rangle^{N-k} \quad |0\rangle^{k-1} |0\rangle_{N+k} |1\rangle^{N-k}$$
$$+ \beta |1\rangle_0 |1\rangle^{k-1} |0\rangle^{N-k+1} |0\rangle^{k-1} |1\rangle_{N+k} |1\rangle^{N-k}, \tag{B2.12}$$
$$\uparrow$$

where we have aligned the two terms vertically to show most clearly how the teleportation arises. The first $N + 1$ modes are the detected modes, and the mode $N + k$ is the teleported mode, as indicated by the arrow. Moreover, the remaining undetected modes are unentangled with the qubit. The trailing $N - k$ modes can be recycled as single-photon states. It is essential that the correct number of photons is detected. This places very strong requirements on the photodetectors that are used in this protocol. In Chapter 6 we have seen how we can circumvent this requirement. Here, we ignore these requirements and treat the KLM protocol as a proof of principle.

The protocol for mode teleportation relies on the fact that the total number of detected photons k can arise in two distinct ways: either mode 0 had no photons and all k photons were due to the term $|1\rangle^k |0\rangle^{N-k} |0\rangle^k |1\rangle^{N-k}$ in the state $|t_N\rangle$, or mode 0 had a photon and the remaining $k-1$ photons came from the term $|1\rangle^{k-1} |0\rangle^{N-k+1} |0\rangle^{k-1} |1\rangle^{N-k+1}$ in the state $|t_N\rangle$. What happens when $k=0$ or $k=N+1$? In that case we can infer that mode 0 had zero or one photons, respectively. In turn, this means that we accidentally made a measurement of the input qubit in the computational basis. The probability of failure for mode teleportation is therefore

$$p_{\text{fail}} = \frac{|\alpha|^2}{N+1} + \frac{|\beta|^2}{N+1} = \frac{1}{N+1}, \tag{B2.13}$$

where we used the fact that $|\alpha|^2 + |\beta|^2 = 1$.

Putting together the two-qubit gate-teleportation protocol with the linear-optical implementation, we can perform gate teleportation of the cz gate by using the $4N$-mode entangled state

$$|cz_N\rangle = \sum_{j,k=0}^{N} (-1)^{(N-j)(N-k)} |1\rangle^j |0\rangle^{N-j} |0\rangle^j |1\rangle^{N-j} |1\rangle^k |0\rangle^{N-k} |0\rangle^k |1\rangle^{N-k}$$

$$= \sum_{j,k=0}^{N} (-1)^{(N-j)(N-k)} |0\rangle^j |1\rangle^{N-j} |0\rangle^k |1\rangle^{N-k}, \tag{B2.14}$$

where again $|0\rangle = |1,0\rangle$ and $|1\rangle = |0,1\rangle$. This state can be created from the two-mode state $|t_N, t_N\rangle$ by applying N successful cz operations between modes $N+i$ and $2N+i$, for all $i=1,\ldots,N$. The success probability for creating such a state with the probabilistic cz gate introduced by Knill, Laflamme, and Milburn is 16^{-N}. Since two modes must be teleported, the success probability of the gate teleportation given $|cz_N\rangle$ is

$$p_{cz_N} = \frac{N^2}{(N+1)^2}. \tag{B2.15}$$

If this probability is larger than the fault-tolerant threshold, scalable quantum computing with single photons, linear optics, and photodetection are possible, albeit with a large overhead per cz gate.

Creating a cz gate with an error rate that is within the fault-tolerant threshold of at most a few per cent will require a very large overhead in auxiliary modes N. For example, for a failure probability of about 2% we require $N=100$. These overheads are not practical, and the question is whether this can be improved. Due to the nature of the error, a measurement of the qubit in the computational basis, this can indeed be done via an added level of error correction.

A code that protects against accidental measurements in the computational basis is the parity code introduced in Section 6.3.3. There, we showed explicitly how a universal set of quantum gates can be performed with the probabilistic fusion gates. In the original KLM protocol, the role of the fusion gates were played by probabilistic cz gates implemented according to the method described above.

Appendix C
Cross-Kerr nonlinearities for single photons

In this appendix, we give a more realistic model for the interaction of optical modes with a cross-Kerr nonlinearity, and show that this has important ramifications for any application that uses single photons to induce large phase shifts. In several chapters throughout the book we have used a simple single-mode model for cross-phase modulation between modes a_1 and a_2, described by the Hamiltonian

$$\mathcal{H}_K = \hbar\kappa\, \hat{a}_1^\dagger\hat{a}_1\, \hat{a}_2^\dagger\hat{a}_2, \tag{C3.1}$$

which leads to the mode transformations

$$\hat{a}_1 \to \hat{b}_1 = e^{i\kappa\hat{a}_2^\dagger\hat{a}_2}\,\hat{a}_1 \quad \text{and} \quad \hat{a}_2 \to \hat{b}_2 = e^{i\kappa\hat{a}_1^\dagger\hat{a}_1}\,\hat{a}_2. \tag{C3.2}$$

In this model we assume there is no photon loss, no dispersion, and no self-phase modulation. The input and output mode operators all obey the bosonic commutation relations

$$[\hat{a}_i,\hat{a}_j^\dagger] = [\hat{b}_i,\hat{b}_j^\dagger] = \delta_{ij}, \tag{C3.3}$$

and

$$[\hat{a}_i,\hat{a}_j] = [\hat{a}_i^\dagger,\hat{a}_j^\dagger] = 0 \quad \text{and} \quad [\hat{b}_i,\hat{b}_j] = [\hat{b}_i^\dagger,\hat{b}_j^\dagger] = 0. \tag{C3.4}$$

This is the so-called instantaneous model of the cross-Kerr nonlinearity, since it does not take into account the finite time it takes to build up the phase. We will now consider what happens when both the single-photon wave packet and the response of the optical fields to the medium have a finite time duration.

When we wish to consider the temporal behaviour of the field modes, we need to consider the time-dependent mode operators $\hat{a}_j(t)$ and $\hat{b}_j(t)$ and their Hermitian conjugates. Again, we assume no photon loss, dispersion, or self-phase modulation. Following Shapiro (2006), the response of the medium is given by a convolution of the intensity operator $\hat{a}_j^\dagger\hat{a}_j$ with a function $h(t)$, such that the accumulated phase in the instantaneous model is replaced by the operator $\exp[i\hat{\mu}_j(t)]$:

$$e^{i\kappa\hat{a}_j^\dagger\hat{a}_j} \to \exp\left[i\kappa\int_{-\infty}^{\infty} d\tau\, h(t-\tau)\,\hat{a}_j^\dagger(\tau)\hat{a}_j(\tau)\right] \equiv \exp\left[i\hat{\mu}_j(t)\right]. \tag{C3.5}$$

Here, the convolution $h(t)$ will be non-zero only for a finite duration Δ, and is normalized such that

$$\int_{-\infty}^{\infty} dt\, h(t) = 1. \tag{C3.6}$$

We must apply the substitution rule in Eq. (C3.5) to Eq. (C3.2) in order to find the non-instantaneous, or 'causal' mode transformations. Since we want to treat the outgoing modes as proper quantum-field modes, we demand that the commutation relations in Eqs. (C3.3) and (C3.4) still hold for modes b_1 and b_2. This is not satisfied trivially, and we must include operators $e^{i\hat{\xi}_1(t)}$ and $e^{i\hat{\xi}_2(t)}$ in the mode operators $\hat{b}_1$ and $\hat{b}_2$, respectively. The mode transformations then become

$$\hat{a}_1(t) \rightarrow \hat{b}_1(t) = e^{i\hat{\xi}_1(t)}\, e^{i\hat{\mu}_2(t)}\, \hat{a}_1(t)$$

$$\hat{a}_2(t) \rightarrow \hat{b}_2(t) = e^{i\hat{\xi}_2(t)}\, e^{i\hat{\mu}_1(t)}\, \hat{a}_2(t). \tag{C3.7}$$

Demanding that the modes b_1 and b_2 obey the bosonic commutation relations implies that the operators $\hat{\xi}_1(t)$ and $\hat{\xi}_2(t)$ obey the commutation relation

$$\left[\hat{\xi}_1(t), \hat{\xi}_2(t')\right] = i\kappa \left[h(t'-t) - h(t-t')\right]. \tag{C3.8}$$

The operators $\hat{\xi}_1$ and $\hat{\xi}_2$ are Langevin noise operators, and they will in general affect the fidelity of the quantum gates based on single-photon cross-Kerr nonlinearities.

Next, consider a single photon in mode a_1 with temporal mode shape defined by the wave packet

$$|1\rangle_1 = \int_{-\infty}^{\infty} dt\, \psi(t)\, \hat{a}_1^{\dagger}(t)\, |0\rangle_1 , \tag{C3.9}$$

where $|0\rangle_1$ is the multi-mode vacuum at different times for the spatial mode a_1. The second mode a_2 may carry another single-photon state or a multi-photon state, depending on the particular application. Typically, this mode must have the same temporal mode shape as a_1, since the photons may be used in subsequent interferometry. The (average) phase shift that the single photon in mode a_1 produces in mode a_2 can be calculated using Eqs. (C3.7) and (C3.5), and is given by

$$\langle e^{i\hat{\mu}_1(t)} \rangle = \int_{-\infty}^{\infty} d\tau\, |\psi(\tau)|^2\, e^{i\kappa h(t-\tau)}. \tag{C3.10}$$

We can now consider two types of situation, namely the regime in which the Kerr medium has a fast response compared to the photon wave packet, and the regime of slow response.

In the fast-response regime, the temporal width of the photon wave packet τ_0 is much larger than the duration Δ of the nonlinear interaction of the Kerr medium. Consequently, the temporal region where $h(t) \neq 0$ is only a fraction of the temporal wave packet, and the phase shift occurs only in this small region of the wave packet. If we think of the photon as a superposition of the form in Eq. (C3.9), only those terms with t such that $h(t) \neq 0$ will experience the phase shift $\kappa h(t)$. All other terms do not experience any phase shift. Therefore, the phase shift cannot be taken out of the integral over t, and the nonlinear medium does not produce an overall phase shift that can be set to π. The fast-response regime is therefore ruled out, and we must consider the slow-response regime.

In the slow-response regime the photon wave packet is much shorter than the interaction time, $\tau_0 \ll \Delta$. For a reasonably well-behaved $\psi(\tau)$ we can then approximate $|\psi(\tau)|^2$ with a delta function $\delta(\tau - t_0)$, for some t_0 chosen in the region where h is relatively large. The induced phase in mode a_2 then becomes

$$\langle e^{i\hat{\mu}_1(t)} \rangle = e^{i\kappa h(t-t_0)} \quad \text{for } \tau_0 \ll \Delta. \tag{C3.11}$$

In principle, this can be made arbitrary large. However, large κ also implies a strong noise term due to $\hat{\xi}_j(t)$, as indicated by Eq. (C3.8). As a result, the noise will degrade the fidelity of the operation considerably.

The nature of the noise operator $\hat{\xi}_j(t)$ depends on the physical implementation of the cross-Kerr interaction. In Shapiro (2006) and Shapiro and Razavi (2007) it is shown that single-photon gates based on the cross-Kerr nonlinearity experience a fidelity degradation that is too severe to be useful in quantum information processing.

References

Achilles, D., Silberhorn, C., Sliwa, C., Banaszek, K., and Walmsley, I. A. (2003). Fiber-assisted detection with photon-number resolution. *Opt. Lett.*, 28:2387.

Adami, C. and Cerf, N. J. (1999). Quantum computation with linear optics. *Lect. Notes Comput. Sc.*, 1509:391.

Akopian, N., Lindner, N. H., Poem, E., Berlatzky, Y., Avron, J., Gershoni, D., Gerardot, B. D., and Petroff, P. M. (2006). Entangled photon pairs from semiconductor quantum dots. *Phys. Rev. Lett.*, 96:130501.

Allen, L., Barnett, S. M., and Padgett, M. J. (2003). *Optical Angular Momentum*. Institute of Physics.

Aoki, T., Takahashi, G., Kajiya, T., Yoshikawa, J., Braunstein, S. L., van Loock, P., and Furusawa, A. (2008). Quantum error correction beyond qubits. *arXiv:0811.3734*.

Armani, D. K., Kippenberg, T. J., Spillane, S. M., and Vahala, K. J. (2003). Ultra-high-Q toroid microcavity on a chip. *Nature*, 421:925.

Ashcroft, N. W. and Mermin, N. D. (1976). *Solid State Physics*. Thompson Learning.

Atatüre, M., Dreiser, J., Badolato, A., and Imamoglu, A. (2007). Observation of Faraday rotation from a single confined spin. *Nature Physics*, 3:101.

Atkins, P. W. and Friedman, R. S. (2005). *Molecular Quantum Mechanics, Fourth Edition*. Oxford University Press.

Banaszek, K., Dragan, A., Wasilewski, W., and Radzewicz, C. (2004). Experimental demonstration of entanglement-enhanced classical communication over a quantum channel with correlated noise. *Phys. Rev. Lett.*, 97:257901.

Barnett, S. M. and Radmore, P. M. (1997). *Methods in Theoretical Quantum Optics*, volume 15 of *Oxford Series in Optical and Imaging Science*. Clarendon Press.

Barrett, S. D. and Kok, P. (2005). Efficient high-fidelity quantum computation using matter qubits and linear optics. *Phys. Rev. A*, 71:060310(R).

Barrett, M. D., Chiaverini, J., Schaetz, T., Britton, J., Itano, W. M., Jost, J. D., Knill, E., Langer, C., Leibfried, D., Ozeri, R., and Wineland, D. J. (2004). Deterministic quantum teleportation of atomic qubits. *Nature*, 429:737.

Barrett, S. D., Rohde, P. P., and Stace, T. M. (2008). *arXiv:0804.0q62*.

Bartlett, S. D., Rudolph, T., and Spekkens, R. W. (2007). Reference frames, superselection rules, and quantum information. *Rev. Mod. Phys.*, 79:555.

Basché, T., Moerner, W. E., Orrit, M., and Talon, H. (1992). Photon antibunching in the fluorescence of a single dye molecule trapped in a solid. *Phys. Rev. Lett.*, 69:1516.

Bason, M. G., Mohapatra, A. K., Weatherill, K. J., and Adams, C. S. (2009). Narrow absorptive resonances in a four-level atomic system. *J. Phys. B*, 42:075503.

Basu, P. K. (1997). *Theory of Optical Processes in Semiconductors*. Oxford University Press.

Beausoleil, R. G., Munro, W. J., and Spiller, T. P. (2004). Applications of coherent population transfer to quantum information processing. *J. Mod. Opt.*, 51:1559.

Beijersbergen, M. W., Allen, L., van der Veen, H. E. L. O., and Woerdman, J. P. (1993). Astigmatic laser mode converters and transfer of orbital angular momentum. *Opt. Commun.*, 96:123.

Benjamin, S. C. (2000). Schemes for parallel quantum computation without local control of qubits. *Phys. Rev. A*, 61:020301(R).

Benjamin, S. C. and Bose, S. (2004). Quantum computing in arrays coupled by 'always-on' interactions. *Phys. Rev. A*, 70:032314.

Benjamin, S. C., Browne, D. E., Fitzsimons, J., and Morton, J. J. L. (2006). Brokered graph state quantum computing. *New J. Phys.*, 8:141.

Benjamin, S. C., Lovett, B. W., and Reina, J. H. (2004). Optical quantum computation with perpetually coupled spins. *Phys. Rev. A*, 70:060305(R).

Bennett, C. H. and Brassard, G. (1984). Quantum cryptography: public key distribution and coin tossing. *Proc. IEEE Int. Conference on Computers, Systems and Signal Processing*, page 175.

Bennett, C. H., Brassard, G., Crépeau, C., Jozsa, R., Peres, A., and Wootters, W. K. (1993). Teleporting an unknown quantum state via dual classical and Einstein–Podolsky–Rosen channels. *Phys. Rev. Lett.*, 70:1895.

Bennett, C. H., DiVincenzo, D. P., Smolin, J. A., and Wootters, W. K. (1996). Mixed-state entanglement and quantum error correction. *Phys. Rev. A*, 54:3824.

Berezovsky, J., Mikkelsen, M. H., Stoltz, N. G., Coldren, L. A., and Awschalom, D. D. (2008). Picosecond coherent optical manipulation of a single electron spin in a quantum dot. *Science*, 320:349.

Bhattacharyya, K. (1983). Quantum decay and the Mandelstam–Tamm time-energy inequality. *J. Phys. A*, 16:2993.

Bigelow, M. S., Lepeshkin, N. N., and Boyd, R. W. (2003). Superluminal and slow light propagation in a room temperature solid. *Science*, 301:200.

Biolatti, E., D'Amico, I., Zanardi, P., and Rossi, F. (2002). Electro-optical properties of semiconductor quantum dots: applications to quantum information processing. *Phys. Rev. B*, 65:075306.

Bjorken, J. D. and Drell, S. D. (1965). *Relativistic Quantum Fields*. International series in pure and applied physics. McGraw-Hill.

Bollinger, J. J., Itano, W. M., Wineland, D. J., and Heinzen, D. J. (1996). Optimal frequency measurements with maximally correlated states. *Phys. Rev. A*, 54:R4649.

Borri, P., Langbein, W., Schneider, S., Woggon, U., Sellin, R. L., Ouyang, D., and Bimberg, D. (2001). Long-lived coherence in self-assembled quantum dots. *Phys. Rev. Lett.*, 87:157401.

Boschi, D., Branca, S., DeMartini, F., Hardy, L., and Popescu, S. (1998). Experimental realization of teleporting an unknown pure quantum state via dual classical and Einstein–Podolsky–Rosen channels. *Phys. Rev. Lett.*, 80:1121.

Bose, S., Knight, P. L., Plenio, M. B., and Vedral, V. (1999a). Proposal for teleportation of an atomic state via cavity decay. *Phys. Rev. Lett.*, 83:5158.

Bose, S., Vedral, V., and Knight, P. L. (1999b). Purification via entanglement swapping and conserved entanglement. *Phys. Rev. A*, 60:194.

Bouwmeester, D., Pan, J. W., Mattle, K., Eibl, M., Weinfurter, H., and Zeilinger, A. (1997). Experimental quantum teleportation. *Nature*, 390:575.

Boyd, R. W. (2003). *Nonlinear Optics. Second edition*. Academic Press.

Brandes, T. (2005). Coherent and collective quantum optical effects in mesoscopic coherent and collective quantum optical effects in mesoscopic systems. *Phys. Rep.*, 408:315.

Braunstein, S. L. (1996). Geometry of quantum inference. *Phys. Lett. A*, 219:169.

Braunstein, S. L. (1998a). Error correction for continuous quantum variables. *Phys. Rev. Lett.*, 80:4084.

Braunstein, S. L. (1998b). Quantum error correction for communication with linear optics. *Nature*, 394:47.

Braunstein, S. L. (2005). Squeezing as an irreducible resource. *Phys. Rev. A*, 71:055801.

Braunstein, S. L. and Caves, C. M. (1994). Statistical distance and the geometry of quantum states. *Phys. Rev. Lett.*, 72:3439.

Braunstein, S. L. and Kimble, H. J. (1998). Teleportation of continuous quantum variables. *Phys. Rev. Lett.*, 80:869.

Braunstein, S. L. and Pati, A. K., editors (2003). *Quantum Information with Continuous Variables*. Kluwer Academic Publishers.

Braunstein, S. L. and van Loock, P. (2005). Quantum information with continuous variables. *Rev. Mod. Phys.*, 77:513.

Braunstein, S. L., Caves, C. M., and Milburn, G. J. (1996). Generalized uncertainty relations: theory, examples, and Lorentz invariance. *Ann. Phys.*, 247:135.

Brennen, G. K., Caves, C. M., Jessen, P. S., and Deutsch, I. H. (1999). Quantum logic gates in optical lattices. *Phys. Rev. Lett.*, 82:1060.

Breuer, H.-P. and Petruccione, F. (2002). *The Theory of Open Quantum Systems*. Oxford University Press.

Briegel, H. J., Dür, W., Cirac, J. I., and Zoller, P. (1998). Quantum repeaters: the role of imperfect local operations in quantum communication. *Phys. Rev. Lett.*, 81:5932.

Brillouin, L. (1960). *Wave Propagation and Group Velocity*. Academic Press.

Brion, E., Molmer, K., and Saffman, M. (2007). Quantum computing with collective ensembles of multilevel systems. *Phys. Rev. Lett.*, 99:260501.

Brokmann, X., Coolen, L., Dahan, M., and Hermier, J. P. (2004). Measurement of the radiative and nonradiative decay rates of Single CdSe nanocrystals through a controlled modification of their spontaneous emission. *Phys. Rev. Lett.*, 93:107403.

Browne, D. E. and Rudolph, T. (2005). Resource-efficient linear optical quantum computation. *Phys. Rev. Lett.*, 95:010501.

Browne, D. E., Eisert, J., Scheel, S., and Plenio, M. B. (2003). Driving non-Gaussian to Gaussian states with linear optics. *Phys. Rev. A*, 67:062320.

Browne, D. E., Plenio, M. B., and Huelga, S. F. (2003). Robust creation of entanglement between ions in spatially separate cavities. *Phys. Rev. Lett.*, 91:067901.

Brunel, C., Lounis, B., Tamarat, P., and Orrit, M. (1999). Triggered source of single photons based on controlled single molecule fluorescence. *Phys. Rev. Lett.*, 83:2722.

Bruß, D. and Leuchs, G., editors (2007). *Lectures on Quantum Information*. Wiley-VCH.

Burt, M. G. (1999). Fundamentals of envelope function theory for electronic states and photonic modes in nanostructures. *J. Phys. Condens. Matter*, 11:R53.

Cable, H. and Dowling, J. P. (2007). Efficient generation of large number-path entanglement using only linear optics and feed-forward. *Phys. Rev. Lett.*, 99:163604.

Cabrillo, C., Cirac, J. I., García-Fernández, P., and Zoller, P. (1999). Creation of entangled states of distant atoms by interference. *Phys. Rev. A*, 59:1025.

Calarco, T., Datta, A., Fedichev, P., Pazy, E., and Zoller, P. (2003). Spin-based all-optical quantum computing with quantum dots: understanding and suppressing decoherence. *Phys. Rev. A*, 68:012310.

Carmichael, H. J. (1999). *Statistical Methods in Quantum Optics: Master Equations and Fokker–Planck Equations*. Springer-Verlag.

Carroll, S. M. (2004). *Spacetime and Geometry*. Addison-Wesley.

Caves, C. M. (1981). Quantum-mechanical noise in an interferometer. *Phys. Rev. D*, 23:1693.

Cerf, N. J., Adami, C., and Kwiat, P. G. (1998). Optical simulation of quantum logic. *Phys. Rev. A*, 57:R1477.

Cerf, N. J., Leuchs, G., and Polzik, E. S. (2007). *Quantum Information with Continuous Variables of Atoms and Light*. Imperial College Press.

Cerf, N. J., Lévy, M., and Assche, G. V. (2002). Quantum distribution of Gaussian keys using squeezed states. *Phys. Rev. A*, 63:052311.

Dawson, C. M., Haselgrove, H. L., and Nielsen, M. A. (2006). Noise thresholds for optical cluster-state quantum computation. *Phys. Rev. A*, 73:052306.

Demkowicz-Dobrzanski, R., Dorner, U., Smith, B. J., Lundeen, J. S., Wasilewski, W., Banaszek, K., and Walmsley, I. A. (2009). Quantum phase estimation with lossy interferometers. *arXiv:0904.0456*.

Dieks, D. (1982). Communication by EPR devices. *Phys. Lett. A*, 92:271.

Divochiy, A., Marsili, F., Bitauld, D., Gaggero, A., Leoni, R., Mattioli, F., Korneev, A., Seleznev, V., Kaurova, N., Minaeva, O., Gol'tsman, G., Lagoudakis, K. G., Benkhaoul, M., Lévy, F., and Fiore, A. (2008). Superconducting nanowire photon-number-resolving detector at telecommunication wavelengths. *Nature Photonics*, 2:302.

Duan, L.-M., Giedke, G., Cirac, J. I., and Zoller, P. (2000a). Entanglement purification of Gaussian continuous variable quantum states. *Phys. Rev. Lett.*, 84:4002.

Duan, L.-M., Giedke, G., Cirac, J. I., and Zoller, P. (2000b). Physical implementation for entanglement purification of Gaussian continuous-variable quantum states. *Phys. Rev. A*, 62:032304.

Duan, L.-M., Lukin, M. D., Cirac, J. I., and Zoller, P. (2001). Long-distance quantum communication with atomic ensembles and linear optics. *Nature*, 414:413.

Eisert, J. (2005). Optimizing linear optics quantum gates. *Phys. Rev. Lett.*, 95:040502.

Eisert, J., Browne, D. E., Scheel, S., and Plenio, M. B. (2004). Distillation of continuous-variable entanglement with optical means. *Ann. Phys.*, 311:431.

Eisert, J., Scheel, S., and Plenio, M. B. (2002). Distilling Gaussian states with Gaussian operations is impossible. *Phys. Rev. Lett.*, 89:137903.

Ekert, A. K. (1991). Quantum cryptography based on Bell's theorem. *Phys. Rev. Lett.*, 67:661.

Enderlein, J. and Pampaloni, F. (2004). Unified operator approach for deriving Hermite–Gaussian and Laguerre–Gaussian laser modes. *J. Opt. Soc. Am. A*, 21:1553.

Fitch, M. J., Jacobs, B. C., Pittman, T. B., and Franson, J. D. (2003). Photon-number resolution using time-multiplexed single-photon detectors. *Phys. Rev. A*, 68:043814.

Fiurášek, J. (2002). Gaussian transformations and distillation of entangled Gaussian states. *Phys. Rev. Lett.*, 89:137904.

Fiurášek, J., Marek, P., Filip, R., and Schnabel, R. (2007). Experimentally feasible purification of continuous-variable entanglement. *Phys. Rev. A*, 75:050302.

Fiurášek, J. L., Mišta, J., and Filip, R. (2003). Entanglement concentration of continuous-variable states. *Phys. Lett. A*, 67:022304.

Flammia, S. T., Menicucci, N. C., and Pfister, O. (2009). The optical frequency comb as a one-way quantum computer. *J. Phys. B*, 42:114009.

Fleischhauer, M. and Lukin, M. D. (2000). Dark-state polaritons in electromagnetically induced transparency. *Phys. Rev. Lett.*, 84:5094.

Fleischhauer, M. and Lukin, M. D. (2002). Quantum memory for photons: dark-state polaritons. *Phys. Rev. A*, 65:022314.

Fleischhauer, M., Imamoglu, A., and Marangos, J. P. (2005). Electromagnetically induced transparency: optics in coherent media. *Rev. Mod. Phys.*, 77:633.

Fox, A. M. (2006). *Quantum Optics: An Introduction*. Oxford University Press.

Franson, J. D., Jacobs, B. C., and Pittman, T. B. (2004). Quantum computing using single photons and the Zeno effect. *Phys. Rev. A*, 70:062302.

Furusawa, A., Sørensen, J. L., Braunstein, S. L., Fuchs, C. A., Kimble, H. J., and Polzik, E. S. (1998). Unconditional quantum teleportation. *Science*, 282:706.

Garrison, J. C. and Chiao, R. Y. (2008). *Quantum Optics*. Oxford University Press.

Gauger, E. M., Benjamin, S. C., Nazir, A., and Lovett, B. W. (2008). High-fidelity all-optical control of quantum dot spins: detailed study of the adiabatic approach. *Phys. Rev. B*, 77:115322.

Gerry, C. C. and Knight, P. L. (2005). *Introductory Quantum Optics*. Cambridge University Press.

Giedke, G. and Cirac, J. I. (2002). Characterization of Gaussian operations and distillation of Gaussian states. *Phys. Rev. A*, 66:032316.

Gilchrist, A., Hayes, A. J. F., and Ralph, T. C. (2007). Efficient parity-encoded optical quantum computing. *Phys. Rev. A*, 75:052328.

Giovannetti, V., Lloyd, S., and Maccone, L. (2003). The role of entanglement in dynamical evolution. *Europhys. Lett.*, 62:615.

Giovannetti, V., Lloyd, S., and Maccone, L. (2006). Quantum metrology. *Phys. Rev. Lett.*, 96:010401.

Goldstein, H. (1980). *Classical Mechanics*. Addison-Wesley.

Gottesman, D. (1996). Class of quantum error-correcting codes saturating the quantum hamming bound. *Phys. Rev. A*, 54:1862.

Gottesman, D. (1997). *Stabilizer codes and quantum error correction*. PhD thesis, California Institute of Technology.

Gottesman, D. and Chuang, I. L. (1999). Demonstrating the viability of universal quantum computation using teleportation and single-qubit operations. *Nature*, 402:390.

Gottesman, D. and Preskill, J. (2001). Secure quantum key distribution using squeezed states. *Phys. Rev. A*, 63:022309.

Gottesman, D., Kitaev, A., and Preskill, J. (2001). Encoding a qubit in an oscillator. *Phys. Rev. A*, 64:012310.

Grangier, P., Levenson, J. A., and Poizat, J. P. (1998). Quantum non-demolition measurements in optics. *Nature*, 396:537.

Greilich, A., Yakovlev, D. R., Shabaev, A., Efros, A. L., Yugova, I. A., Oulton, R., Stavarache, V., Reuter, D., Wieck, A., and Bayer, M. (2006). Mode locking of electron spin coherences in singly charged quantum dots. *Science*, 313:341.

Gross, D., Kieling, K., and Eisert, J. (2006). Potential and limits to cluster-state quantum computing using probabilistic gates. *Phys. Rev. A*, 74:042343.

Gruber, A., Dräbenstedt, A., Tietz, C., Fleury, L., Wrachtrup, J., and von Borczyskowski, C. (1997). Scanning confocal optical microscopy and magnetic resonance on single defect centers. *Science*, 276:2012.

Gu, M., Weedbrook, C., Menicucci, N. C., Ralph, T. C., and van Loock, P. (2009). Quantum computing with continuous-variable clusters. *arXiv:0903.3233*.

Hafenbrak, R., Ulrich, S. M., Michler, P., Wang, L., Rastelli, A., and Schmidt, O. G. (2007). Triggered polarization-entangled photon pairs from a single quantum dot up to 30 K. *New J. Phys.*, 9:315.

Harrison, P. (2005). *Quantum Wells, Wires and Dots*. Wiley-Interscience.

Hayes, A. J. F., Gilchrist, A., and Ralph, T. C. (2008). Loss-tolerant operations in parity-code linear optics quantum computing. *Phys. Rev. A*, 77:012310.

Heckenberg, N. R., McDuff, R., Smith, C. P., and White, A. G. (1992). Generation of optical phase singularities by computer-generated holograms. *Opt. Lett.*, 17:221.

Heersink, J., Marquardt, C., Dong, R., Filip, R., Lorenz, S., Leuchs, G., and Andersen, U. L. (2006). Distillation of squeezing from non-Gaussian quantum states. *Phys. Rev. Lett.*, 96:253601.

Hein, M., Eisert, J., and Briegel, H. J. (2004). Multiparty entanglement in graph states. *Phys. Rev. A*, 69:062311.

Helstrom, C. W. (1976). *Quantum Detection and Estimation Theory*. Academic Press.

Hillery, M. (2000). Quantum cryptography with squeezed states. *Phys. Rev. A*, 61:022309.

Hochstadt, H. (1971). *The Functions of Mathematical Physics*, volume XXIII of *Pure and Applied Mathematics*. Wiley-Interscience.

Hohenester, U. (2007). Quantum control of polaron states in semiconductor quantum dots. *J. Phys. B*, 40:S315.

Holevo, A. S. (1982). *Probabilistic and Statistical Aspects of Quantum Theory*. North-Holland Publishing Company.

Holevo, A. S. (2001). *Statistical Structure of Quantum Theory*. Lecture notes in physics. Springer-Verlag.

Holland, M. J. and Burnett, K. (1993). Interferometric detection of optical phase shifts at the Heisenberg limit. *Phys. Rev. Lett.*, 71:1355.

Huelga, S. F., Macchiavello, C., Pellizzari, T., Ekert, A. K., Plenio, M. B., and Cirac, J. I. (1997). Improvement of frequency standards with quantum entanglement. *Phys. Rev. Lett.*, 79:3865.

Hwang, W.-Y. (2003). Quantum key distribution with high loss: toward global secure communication. *Phys. Rev. Lett.*, 91:057901.

Imoto, N., Haus, H. A., and Yamamoto, Y. (1985). Quantum nondemolition measurement of the photon number via the optical Kerr effect. *Phys. Rev. A*, 32:2287.

Jaksch, D., Briegel, H.-J., Cirac, J. I., Gardiner, C. W., and Zoller, P. (1999). Entanglement of atoms via cold controlled collisions. *Phys. Rev. Lett.*, 82:1975.

Jaynes, E. T. (2003). *Probability Theory, the Logic of Science*. Cambridge University Press.

Jelezko, F. and Wrachtrup, J. (2006). Single defect centres in diamond: a review. *Phys. Stat. Sol. A*, 203:3207.

Jiang, L. A., Dauler, E. A., and Chang, J. T. (2007). Photon-number-resolving detector with 10 bits of resolution. *Phys. Rev. A*, 75:062325.

Jozsa, R. (1994). Fidelity for mixed quantum states. *J. Mod. Opt.*, 41:2315.

Kasapi, A., Jain, M., Yin, G. Y., and Harris, S. E. (1995). Electromagnetically induced transparency: propagation dynamics. *Phys. Rev. Lett.*, 74:2447.

Kittel, C. (2005). *Introduction to Solid State Physics, 8th Edition*. Wiley.

Knill, E. (2003). Bounds on the probability of success of postselected nonlinear sign shifts implemented with linear optics. *Phys. Rev. A*, 68:064303.

Knill, E., Laflamme, R., and Milburn, G. J. (2001). A scheme for efficient quantum computation with linear optics. *Nature*, 409:46.

Kogelnik, H. and Li, T. (1966). Laser beams and resonators. *Appl. Opt.*, 5:1550.

Kok, P. and Braunstein, S. L. (2000). Post-selected versus non-post-selected quantum teleportation using parametric down-conversion. *Phys. Rev. A*, 61:042304.

Kok, P. and Braunstein, S. L. (2001). Detection devices in entanglement-based optical state preparation. *Phys. Rev. A*, 63:033812.

Kok, P., Lee, H., and Dowling, J. P. (2002a). The creation of large photon-number path entanglement conditioned on photodetection. *Phys. Rev. A*, 65:052104.

Kok, P., Lee, H., and Dowling, J. P. (2002b). Single-photon quantum nondemolition detectors constructed with linear optics and projective measurements. *Phys. Rev. A*, 66:063814.

Kok, P., Munro, W. J., Nemoto, K., Ralph, T. C., Dowling, J. P., and Milburn, G. J. (2007). Linear optical quantum computing with photonic qubits. *Rev. Mod. Phys.*, 79:135.

Kraus, K. (1983). *States, Effects and Operations: Fundamental Notions of Quantum Theory*. Springer-Verlag.

Kraus, B., Hammerer, K., Giedke, G., and Cirac, J. I. (2003). Entanglement generation and Hamiltonian simulation in continuous-variable systems. *Phys. Rev. A*, 67:042314.

Kuzmany, H. (1998). *Solid State Spectroscopy*. Springer-Verlag.

Kuzmich, A., Bowen, W. P., Boozer, A. D., Boca, A., Chou, C. W., Duan, L.-M., and Kimble, H. J. (2003). Generation of nonclassical photon pairs for scalable quantum communication with atomic ensembles. *Nature*, 423:731.

Kwiat, P. G., Mattle, K., Weinfurter, H., Zeilinger, A., Sergienko, A. V., and Shih, Y. (1995). New high-intensity source of polarization-entangled photon pairs. *Phys. Rev. Lett.*, 75:4337.

Langford, N. K., Dalton, R. B., Harvey, M. D., O'Brien, J. L., Pryde, G. J., Gilchrist, A., Bartlett, S. D., and White, A. G. (2004). Measuring entangled qutrits and their use for quantum bit commitment. *Phys. Rev. Lett.*, 93:053601.

Lapaire, G. G., Kok, P., Dowling, J. P., and Sipe, J. E. (2003). Conditional linear-optical measurement schemes generate effective photon nonlinearities. *Phys. Rev. A*, 68:042314.

Leach, J., Courtial, J., Skeldon, K., Barnett, S. M., Franke-Arnold, S., and Padgett, M. J. (2004). Interferometric methods to measure orbital and spin, or the total angular momentum of a single photon. *Phys. Rev. Lett.*, 92:013601.

Leach, J., Padgett, M. J., Barnett, S. M., Franke-Arnold, S., and Courtial, J. (2002). Measuring the orbital angular momentum of a single photon. *Phys. Rev. Lett.*, 88:257901.

Leggett, A. J., Chakravarty, S., Dorsey, A. T., Fisher, M. P., Garg, A., and Zwerger, W. (1987). Dynamics of the dissipative two-state system. *Rev. Mod. Phys.*, 59:1.

Leonhardt, U. (1997). *Measuring the Quantum State of Light*. Cambridge Studies in Modern Optics. Cambridge University Press.

Leung, P. M. and Ralph, T. C. (2007). Optical Zeno gate: bounds for fault-tolerant operation. *New J. Phys.*, 9:224.

Levitin, L. B. and Toffoli, T. (2009). The fundamental limit on the rate of quantum dynamics: the unified bound is tight. *arXiv:0905.3417*.

Lim, Y. L., Barrett, S. D., Beige, A., Kok, P., and Kwek, L. C. (2006). Repeat-until-success quantum computing using stationary and flying qubits. *Phys. Rev. A*, 73:012304.

Lim, Y. L., Beige, A., and Kwek, L. C. (2005). Repeat-until-success linear optics distributed quantum computing. *Phys. Rev. Lett.*, 95:030505.

Lindblad, G. (1976). On the generators of quantum dynamical semigroups. *Communications in Mathematical Physics*, 48(2):119–130.

Lloyd, S. (1993). A potentially realizable quantum computer. *Science*, 261:1569.

Lloyd, S. and Braunstein, S. L. (1999). Quantum computation over continuous variables. *Phys. Rev. Lett.*, 82:1784.

Lloyd, S. and Slotine, J. J. E. (1998). Analog quantum error correction. *Phys. Rev. Lett.*, 80:4088.

Lodewyck, J., Debuisschert, T., Garcá-Patrón, R., Tualle-Brouri, R., Cerf, N. J., and Grangier, P. (2007). Experimental implementation of non-Gaussian attacks on a continuous-variable quantum-key-distribution system. *Phys. Rev. Lett.*, 98:030503.

Lo, H. and Chau, H. F. (1999). Unconditional security of quantum key distribution over arbitrarily long distances. *Science*, 283:2050.

Lo, H., Ma, X., and Chen, K. (2005). Decoy state quantum key distribution. *Phys. Rev. Lett.*, 94:230504.

Lounis, B. and Moerner, W. E. (2000). Single photons on demand from a single molecule at room temperature. *Nature*, 407:491.

Lounis, B. and Orrit, M. (2005). Single photon sources. *Rep. Prog. Phys.*, 68:1129.

Lounis, B., Bechtel, H. A., Gerion, D., Alivisatos, P., and Moerner, W. E. (2000). Photon antibunching in single CdSe/ZnS quantum dot fluorescence. *Chem. Phys. Lett.*, 329:399.

Lovett, B. W., Reina, J. H., Nazir, A., and Briggs, G. A. D. (2003). Optical schemes for quantum computation in quantum dot molecules. *Phys. Rev. B*, 68:205319.

Lukin, M. D. (2003). Colloquium: trapping and manipulating photon states in atomic ensembles. *Rev. Mod. Phys.*, 75:457.

Lukin, M. D., Fleischhauer, M., Cote, R., Duan, L. M., Jaksch, D., Cirac, J. I., and Zoller, P. (2001). Dipole blockade and quantum information processing in mesoscopic atomic ensembles. *Phys. Rev. Lett.*, 87:037901.

Lütkenhaus, N., Calsamiglia, J., and Suominen, K. A. (1999). Bell measurements for teleportation. *Phys. Rev. A*, 59:3295.

Mahan, G. D. (2000). *Many Particle Physics, Third Edition*. Kluwer.

Mandelstam, L. and Tamm, I. (1945). The uncertainty relation between energy and time in non-relativistic quantum mechanics. *J. Phys. (USSR)*, 9:249.

Margolus, N. and Levitin, L. B. (1998). The maximum speed of dynamical evolution. *Phys. D*, 120:188.

McKeever, J., Boca, A., Boozer, A. D., Miller, R., Buck, J. R., Kuzmich, A., and Kimble, H. J. (2004). Deterministic generation of single photons from one atom trapped in a cavity. *Science*, 303:1992.

Menicucci, N. C., van Loock, P., Gu, M., Weedbrook, C., Ralph, T. C., and Nielsen, M. A. (2006). Universal quantum computation with continuous-variable cluster states. *Phys. Rev. Lett.*, 97:110501.

Mermin, N. D. (2007). *Quantum Computer Science*. Cambridge University Press.

Metz, J., Trupke, M., and Beige, A. (2006). Robust entanglement through macroscopic quantum jumps. *Phys. Rev. Lett.*, 97:040503.

Moehring, D. L., Maunz, P., Olmschenk, S., Younge, K. C., Matsukevich, D. N., Duan, L. M., and Monroe, C. (2007). Entanglement of single-atom quantum bits at a distance. *Nature*, 449:68.

Munro, W. J., Nemoto, K., and Spiller, T. P. (2005). Weak nonlinearities: a new route to optical quantum computation. *New J. Phys.*, 7:137.

Myers, C. R. and Laflamme, R. (2005). Linear optics quantum computation: an overview. In *Quantum Computers, Algorithms and Chaos*. International School of Physics 'Enrico Fermi'.

Myers, H. P. (1997). *Introductory Solid State Physics*. CRC Press.

Nazir, A., Lovett, B. W., Barrett, S. D., Spiller, T. P., and Briggs, G. A. D. (2004a). Selective spin coupling through a single exciton. *Phys. Rev. A*, 93: 150502.

Nazir, A., Lovett, B. W., and Briggs, G. A. D. (2004b). Creating excitonic entanglement in quantum dots through the optical Stark effect. *Phys. Rev. A*, 70:052301.

Nielsen, M. A. (2004). Optical quantum computation using cluster states. *Phys. Rev. Lett.*, 93:040503.

Nielsen, M. A. (2006). Cluster-state quantum computation. *Rep. Math. Phys.*, 57:147.

Nielsen, M. A. and Chuang, I. L. (2000). *Quantum Computation and Quantum Information*. Cambridge University Press.

Niset, J., Fiurášek, J., and Cerf, N. J. (2009). No-go theorem for Gaussian quantum error correction. *Phys. Rev. Lett.*, 102:120501.

Nogues, G., Rauschenbeutel, A., Osnaghi, S., Brune, M., Raimond, J. M., and Haroche, S. (1999). Seeing a single photon without destroying it. *Nature*, 400:239.

O'Brien, J. L., Pryde, G. J., White, A. G., Ralph, T. C., and Branning, D. (2003). Demonstration of an all-optical quantum controlled-NOT gate. *Nature*, 426:264.

Pan, J. W., Bouwmeester, D., Weinfurter, H., and Zeilinger, A. (1998). Experimental entanglement swapping: entangling photons that never interacted. *Phys. Rev. Lett.*, 80:3891.

Parker, S., Bose, S., and Plenio, M. B. (2000). Entanglement quantification and purification in continuous-variable systems. *Phys. Rev. A*, 61:032305.

Pazy, E., Biolatti, E., Calarco, T., D'Amico, I., Zanardi, P., Rossi, F., and Zoller, P. (2003). Spin-based optical quantum computation via Pauli blocking in semiconductor quantum dots. *Europhys. Lett.*, 62:175.

Perelomov, A. (1986). *Generalized Coherent States and Their Applications*. Springer-Verlag.

Pezzé, L. and Smerzi, A. (2008). Mach–Zehnder interferometry at the Heisenberg limit with coherent and squeezed vacuum light. *Phys. Rev. Lett.*, 100:073601.

Pirandola, S., Mancini, S., Vitali, D., and Tombesi, P. (2004). Constructing finite-dimensional codes with optical continuous variables. *Europhys. Lett.*, 68:323.

Pittman, T. B., Jacobs, B. C., and Franson, J. D. (2001). Probabilistic quantum logic operations using polarizing beam splitters. *Phys. Rev. A*, 64:062311.

Pittman, T. B., Jacobs, B. C., and Franson, J. D. (2002). Single photons on pseudodemand from stored parametric down-conversion. *Phys. Rev. A*, 66:042303.

Plenio, M. B. and Virmani, S. (2007). An introduction to entanglement measures. *Quant. Inf. Comp.*, 7:1.

Ralph, T. C. (2000). Continuous variable quantum cryptography. *Phys. Rev. A*, 61:010303(R).

Ralph, T. C., White, A. G., Munro, W. J., and Milburn, G. J. (2002a). Simple scheme for efficient linear optics quantum gates. *Phys. Rev. A*, 65:012314.

Ralph, T. C., Langford, N. K., Bell, T. B., and White, A. G. (2002b). Linear optical controlled-NOT gate in the coincidence basis. *Phys. Rev. A*, 65:062324.

Raussendorf, R. and Briegel, H. J. (2001). A one-way quantum computer. *Phys. Rev. Lett.*, 86:5188.

Raussendorf, R. and Harrington, J. (2007). Fault-tolerant quantum computation with high threshold in two dimensions. *Phys. Rev. Lett.*, 98:190504.

Raussendorf, R., Browne, D. E., and Briegel, H. J. (2003). Measurement-based quantum computation on cluster states. *Phys. Rev. A*, 68:022312.

Raussendorf, R., Harrington, J., and Goyal, K. (2006). A fault-tolerant one-way quantum computer. *Ann. Phys.*, 321:2242.

Raussendorf, R., Harrington, J., and Goyal, K. (2007). Topological fault-tolerance in cluster state quantum computation. *New J. Phys.*, 9:199.

Reck, M., Zeilinger, A., Bernstein, H. J., and Bertani, P. (1994). Experimental realization of any discrete unitary operator. *Phys. Rev. Lett.*, 73:58.

Riebe, M., Häffner, H., Roos, C. F., Hänsel, W., Benhelm, J., Lancaster, G. P. T., Körber, T. W., Becher, C., Schmidt-Kaler, F., James, D. F. V., and Blatt, R. (2004). Deterministic quantum teleportation with atoms. *Nature*, 429:734.

Rogers, L. J., Armstrong, S., Sellars, M. J., and Manson, N. B. (2008). New infrared emission of the NV centre in diamond: Zeeman and uniaxial stress studies. *arXiv:0806.0895*.

Rosenberg, D., Lita, A. E., Miller, A. J., and Nam, S. W. (2005). Noise-free high-efficiency photon-number-resolving detectors. *Phys. Rev. A*, 71:061803.

Rudolph, T. and Pan, J. W. (2001). A simple gate for linear optics quantum computing. *ArXiv:quant-ph/0108056*.

Santori, C., Vatal, D., Vuckovic, J., Salomon, G. S., and Yamamoto, Y. (2002). Indistinguishable photons from a single-photon device. *Nature*, 419:594.

Scheel, S. and Lütkenhaus, N. (2004). Upper bounds on success probabilities in linear optics. *New J. Phys.*, 6:51.

Scheel, S., Munro, W. J., Eisert, J., Nemoto, K., and Kok, P. (2006). Feed-forward and its role in conditional linear optical quantum dynamics. *Phys. Rev. A*, 73:034301.

Scheel, S., Nemoto, K., Munro, W. J., and Knight, P. L. (2003). Measurement-induced nonlinearity in linear optics. *Phys. Rev. A*, 68:032310.

Shapiro, J. H. (2006). Single-photon Kerr nonlinearities do not help quantum computation. *Phys. Rev. A*, 73:062305.

Shapiro, J. H. and Razavi, M. (2007). Continuous-time cross-phase modulation and quantum computation. *New J. Phys.*, 9:16.

Shor, P. W. and Preskill, J. (2000). Simple proof of security of the BB84 quantum key distribution protocol. *Phys. Rev. Lett.*, 85:441.

Steane, A. M. (2003). Overhead and noise threshold of fault-tolerant quantum error correction. *Phys. Rev. A*, 68:042322.

Stevenson, R. M., Young, R. J., Atkinson, P., Cooper, K., Ritchie, D. A., and Shields, A. J. (2006). A semiconductor source of triggered entangled photon pairs. *Nature*, 439:179.

Stoneham, A. M. (1975). *Theory of Defects in Solids*. Oxford University Press.

Stoneham, A. M., Fisher, A. J., and Greenland, P. T. (2003). Optically driven silicon-based quantum gates with potential for high-temperature operation. *J. Phys. Condens. Matter*, 15:L447.

Thé, G. A. P. and Ramos, R. V. (2007). Multiple-photon number-resolving detector using fiber ring and single-photon detector. *J. Mod. Opt.*, 54:1187.

Treussart, F., Alléaume, R., Le Floc'h, V., Xiao, L. T., Courty, J. M., and Roch, J. F. (2002). Direct measurement of the photon statistics of a triggered single photon source. *Phys. Rev. Lett.*, 89:093601.

Troiani, F., Hohenester, U., and Molinari, E. (2000) Exploiting exciton–exciton interactions in semiconductor quantum dots for quantum-information processing. *Phys. Rev. B*, 62:R2263.

Truax, D. R. (1985). Baker–Campbell–Hausdorff relations and unitarity of SU(2) and SU(1,1) squeeze operators. *Phys. Rev. D*, 31:1988.

Ulu, G., Sergienko, A. V., and Ünlü, M. S. (2000). Influence of hot-carrier luminescence from avalanche photodiodes on time-correlated photon detection. *Optics Letters*, 25:758.

Vaidman, L. and Yoran, N. (1999). Methods for reliable teleportation. *Phys. Rev. A*, 59:116.

van Enk, S. J. and Nienhuis, G. (1992). Eigenfunction description of laser beams and orbital angular momentum of light. *Opt. Commun.*, 94:147.

van Loock, P. (2007). Examples of Gaussian cluster computation. *J. Opt. Soc. Am. B*, 24:340.

van Loock, P. and Braunstein, S. L. (2000). Unconditional teleportation of continuous-variable entanglement. *Phys. Rev. A*, 61:010302(R).

van Loock, P. and Lütkenhaus, N. (2004). Simple criteria for the implementation of projective measurements with linear optics. *Phys. Rev. A*, 69:012302.

van Loock, P., Weedbrook, C., and Gu, M. (2007). Building Gaussian cluster states by linear optics. *Phys. Rev. A*, 76:032321.

Varnava, M., Browne, D. E., and Rudolph, T. (2006). Loss tolerance in one-way quantum computation via counterfactual error correction. *Phys. Rev. Lett.*, 97:120501.

Waks, E., Diamanti, E., Sanders, B. C., Bartlett, S. D., and Yamamoto, Y. (2004). Direct observation of nonclassical photon statistics in parametric down-conversion. *Phys. Rev. Lett.*, 92:113602.

Waks, E., Inoue, K., Oliver, W. D., Diamanti, E., and Yamamoto, Y. (2003). High-efficiency photon-number detection for quantum information processing. *IEEE J. Sel. Top. Quant.*, 9:1502.

Walls, D. and Milburn, G. J. (2008). *Quantum Optics. Second edition*. Springer.

Walther, P., Resch, K. J., Rudolph, T., Weinfurter, H., Vedral, V., Aspelmeyer, M., and Zeilinger, A. (2005). Experimental one-way quantum computing. *Nature*, 434:169.

Weinreich, G. (1998). *Geometrical Vectors*. Chicago Lecture in Physics. University of Chicago Press.

Weiss, U. (2008). *Quantum Dissipative Systems, Third Edition*. World Scientific.

Wootters, W. K. (1981). Statistical distance and Hilbert space. *Phys. Rev. D*, 23:357.

Wootters, W. K. and Zurek, W. H. (1982). A single quantum cannot be cloned. *Nature*, 299:802.

Yoran, N. and Reznik, B. (2003). Deterministic linear optics quantum computation with single photon qubits. *Phys. Rev. Lett.*, 91:037903.

Yurke, B., McCall, S. L., and Klauder, J. R. (1986). SU(2) and SU(1,1) interferometers. *Phys. Rev. A*, 33:4033.

Zambra, G., Bondani, M., Spinelli, A. S., Paleari, F., and Andreoni, A. (2004). Counting photoelectrons in the response of a photomultiplier tube to single picosecond light pulses. *Rev. Sci. Instrum.*, 75:2762.

Zanardi, P. and Rasetti, M. (1997). Noiseless quantum codes. *Phys. Rev. Lett.*, 79:3306.

Zhang, J. (2008a). Graphical description of local Gaussian operations for continuous-variable weighted graph state. *Phys. Rev. A*, 78:052307.

Zhang, J. (2008b). Local complementation rule for continuous variable four-mode unweighted graph states. *Phys. Rev. A*, 78:034301.

Zhang, J. and Braunstein, S. L. (2006). Continuous-variable Gaussian analog of cluster states. *Phys. Rev. A*, 73:032318.

Zhang, L., Silberhorn, C., and Walmsley, I. A. (2008). Secure quantum key distribution using continuous variables of single photons. *Phys. Rev. Lett.*, 100:110504.

Ziman, J. M. (1969). *Elements of Advanced Quantum Theory*. Cambridge University Press.

Ziman, J. M. (2001). *Electrons and Phonons, Oxford Classic*. Oxford University Press.

Zou, X. and Mathis, W. (2005). Scheme for optical implementation of orbital angular momentum beam splitter of a light beam and its application in quantum information processing. *Phys. Rev. A*, 71:042324.

Zwierz, M. and Kok, P. (2009). High-efficiency cluster-state generation with atomic ensembles via the dipole-blockade mechanism. *Phys. Rev. A*, 79:022304.

Index

Operators in quantum optics

Creation and annihilation operators:

$$\left[\hat{a}_j, \hat{a}_k^\dagger\right] = \delta_{jk} \quad \text{and} \quad \left[\hat{a}_j, \hat{a}_k\right] = \left[\hat{a}_j^\dagger, \hat{a}_k^\dagger\right] = 0$$

Position and momentum quadrature operators:

$$\hat{q} = \sqrt{\frac{\hbar}{2\omega}}\left(\hat{a} + \hat{a}^\dagger\right) \qquad \hat{p} = -i\sqrt{\frac{\hbar\omega}{2}}\left(\hat{a} - \hat{a}^\dagger\right)$$

$$\hat{a} = \sqrt{\frac{\omega}{2\hbar}}\,\hat{q} + \frac{i}{\sqrt{2\hbar\omega}}\,\hat{p} \qquad \hat{a}^\dagger = \sqrt{\frac{\omega}{2\hbar}}\,\hat{q} - \frac{i}{\sqrt{2\hbar\omega}}\,\hat{p}$$

Position–momentum commutation relation:

$$[\hat{q}, \hat{p}] = i\hbar$$

The displacement operator:

$$D(\alpha) = \exp\left(\alpha\hat{a}^\dagger - \alpha^*\hat{a}\right)$$

The single-mode squeezer:

$$S_1(\xi) = \exp\left(-\frac{\xi}{2}\hat{a}^{\dagger 2} + \frac{\xi^*}{2}\hat{a}^2\right)$$

The two-mode squeezer:

$$S_2(\xi) = \exp\left(-\xi\hat{a}_1^\dagger\hat{a}_2^\dagger + \xi^*\hat{a}_1\hat{a}_2\right)$$

The Baker–Campbell–Hausdorff relations:

$$e^{\mu B}Ae^{-\mu B} = A + \mu[A,B] + \frac{\mu^2}{2!}[B,[B,A]] + \ldots$$

$$\exp(A)\exp(B) = \exp\left(A + B + \frac{1}{2}[A,B]\right) \quad \text{if } [A,[A,B]] = [B,[A,B]] = 0$$

The Wigner function:

$$W_\rho(q,p) = \frac{1}{2\pi\hbar}\int_{-\infty}^{\infty} dx\, \exp\left(\frac{i}{\hbar}xp\right)\left\langle q - \frac{x}{2}\right|\rho\left|q + \frac{x}{2}\right\rangle$$

The overlap formula for Wigner functions:

$$\text{Tr}(AB) = 2\pi\hbar\int_{-\infty}^{\infty}\int_{-\infty}^{\infty} dq\, dp\, W_A(q,p)W_B(q,p)$$

Continuous variable operators

Heisenberg–Weyl operators for qunats:

$$X(q) = \exp\left(-\frac{i}{\hbar}q\hat{p}\right)$$

$$Z(p) = \exp\left(\frac{i}{\hbar}p\hat{q}\right)$$

Fourier transform $\mathcal{F}$:

$$\mathcal{F} = \exp\left[\frac{i}{\hbar}\frac{\pi}{4}(\hat{q}^2 + \hat{p}^2)\right]$$

Phase gate $\Phi(\theta)$:

$$\Phi(\theta) = \exp\left(\frac{i}{\hbar}\frac{\theta\hat{q}^2}{2}\right)$$

Conjugation relations:

$$\mathcal{F}X(q)\mathcal{F}^\dagger = Z(q) \qquad \mathcal{F}Z(p)\mathcal{F}^\dagger = X(-p) = X^{-1}(p)$$

$$\Phi(\theta)X(q)\Phi^\dagger(\theta) = X(q)Z\left(\frac{q\theta}{\hbar}\right)\exp\left(\frac{i\theta q^2}{2\hbar^2}\right)$$

$$\Phi(\theta)Z(p)\Phi^\dagger(\theta) = Z(p)$$

Two-qunat operations:

$$\mathrm{cx}_{ij} = \exp\left(-\frac{i}{\hbar}\hat{q}_i \otimes \hat{p}_j\right)$$

$$\mathrm{cz}_{ij} = \exp\left(\frac{i}{\hbar}\hat{q}_i \otimes \hat{q}_j\right)$$

Qubit operators

The Pauli operators:

$$X = \begin{pmatrix} 0 & 1 \\ 1 & 0 \end{pmatrix} \qquad Y = \begin{pmatrix} 0 & -i \\ i & 0 \end{pmatrix} \qquad Z = \begin{pmatrix} 1 & 0 \\ 0 & -1 \end{pmatrix}$$

Single-qubit rotations:

$$U_X(\theta) = \exp(-i\theta\,X) = \begin{pmatrix} \cos\theta & -i\sin\theta \\ -i\sin\theta & \cos\theta \end{pmatrix}$$

$$U_Y(\phi) = \exp(-i\phi\,Y) = \begin{pmatrix} \cos\phi & -\sin\phi \\ \sin\phi & \cos\phi \end{pmatrix}$$

$$U_Z(\varphi) = \exp(-i\varphi\,Z) = \begin{pmatrix} e^{-i\varphi} & 0 \\ 0 & e^{i\varphi} \end{pmatrix}$$

The Hadamard H, the phase operator Φ, and the $\pi/8$ gate T:

$$H = \frac{1}{\sqrt{2}} \begin{pmatrix} 1 & 1 \\ 1 & -1 \end{pmatrix} \qquad \Phi = \begin{pmatrix} 1 & 0 \\ 0 & i \end{pmatrix} \qquad T = \begin{pmatrix} 1 & 0 \\ 0 & e^{i\pi/4} \end{pmatrix}$$

Important commutation relations:

$$XZ = -ZX \qquad YX = -XY \qquad ZY = -YZ$$

$$HX = ZH \qquad HZ = XH \qquad HY = -YH$$

$$\Phi X \Phi^\dagger = Y \qquad \Phi Y \Phi^\dagger = -X \qquad \Phi Z \Phi^\dagger = Z$$

Two-qubit gates:

$$\text{CX} = \begin{pmatrix} 1 & 0 & 0 & 0 \\ 0 & 1 & 0 & 0 \\ 0 & 0 & 0 & 1 \\ 0 & 0 & 1 & 0 \end{pmatrix} \;(=\text{CNOT}) \qquad \text{CZ} = \begin{pmatrix} 1 & 0 & 0 & 0 \\ 0 & 1 & 0 & 0 \\ 0 & 0 & 1 & 0 \\ 0 & 0 & 0 & -1 \end{pmatrix}$$

Advanced quantum mechanical relations

Von Neumann entropy:

$$S(\rho) = -\mathrm{Tr}(\rho \log_2 \rho) = -\sum_j \lambda_j \log_2 \lambda_j$$

Positive operator-valued measures of projectors P_μ:

$$\hat{E}_\nu = \sum_\mu \lambda_\mu^\nu P_\mu \quad \text{with} \quad \sum_\nu \hat{E}_\nu = \hat{\mathbb{I}}$$

In terms of Kraus operators:

$$\hat{E}_\nu = \sum_\mu \mathcal{A}_{\mu\nu} \mathcal{A}_{\mu\nu}^\dagger \quad \text{with} \quad \mathcal{A} = \sum_{\mu\nu} \alpha_\mu^\nu |\mu\rangle \langle \nu| \quad \text{and} \quad |\alpha_\mu^\nu|^2 = \lambda_\mu^\nu$$

The fidelity between two density operators:

$$F = \left[\mathrm{Tr} \left(\sqrt{\rho_2^{1/2} \rho_1 \rho_2^{1/2}} \right) \right]^2$$

The Lindblad master equation:

$$\frac{d\rho}{dt} = -\frac{i}{\hbar}[\rho, \mathcal{H}] - \sum_{n,m} \gamma_{nm} \left(\rho L + m L_n + L_m L_n \rho - 2 L_n \rho L_m \right) + \text{H.c.}$$

The Fisher information:

$$F(\theta) = \int_{-\infty}^{\infty} dx \, \frac{1}{p(x|\theta)} \left(\frac{\partial p(x|\theta)}{\partial \theta} \right)^2 = \int_{-\infty}^{\infty} dx \, p(x|\theta) \left(\frac{\partial \ln p(x|\theta)}{\partial \theta} \right)^2$$

Printed in the United States
by Baker & Taylor Publisher Services